Disputed Inheritance

Disputed Inheritance

THE BATTLE OVER MENDEL AND THE FUTURE OF BIOLOGY

Gregory Radick

THE UNIVERSITY OF CHICAGO PRESS
CHICAGO AND LONDON

The University of Chicago Press, Chicago 60637
The University of Chicago Press, Ltd., London

Published 2022
Printed in the United States of America

32 31 30 29 28 27 26 25 24 23   1 2 3 4 5

ISBN-13: 978-0-226-82270-9 (cloth)
ISBN-13: 978-0-226-82272-3 (paper)
ISBN-13: 978-0-226-82271-6 (e-book)
DOI: https://doi.org/10.7208/chicago/9780226822716.001.0001

Published with support of the Susan E. Abrams Fund.

Library of Congress Cataloging-in-Publication Data

Names: Radick, Gregory, author.
Title: Disputed inheritance : the battle over Mendel and the future of biology / Gregory Radick.
Description: Chicago : The University of Chicago Press, 2023. | Includes bibliographical references and index.
Identifiers: LCCN 2022031515 | ISBN 9780226822709 (cloth) | ISBN 9780226822723 (paperback) | ISBN 9780226822716 (ebook)
Subjects: LCSH: Bateson, William, 1861–1926. | Weldon, Walter Frank Raphael. | Mendel, Gregor, 1822–1884. | Genetics—History. | BISAC: SCIENCE / History | SCIENCE / Life Sciences / Genetics & Genomics
Classification: LCC QH428 .R335 2023 | DDC 576.5—dc23/eng/20220803
LC record available at https://lccn.loc.gov/2022031515

♾ This paper meets the requirements of ANSI/NISO Z39.48-1992 (Permanence of Paper).

FOR ESTHER ERMAN AND STUART RADICK

Contents

Illustrations

Introduction

It seems to me, quite apart from my own share in the matter, that the present is a rather interesting and important moment. There is a "boom" in a quite unstatistical theory of inheritance, which is so simple that everyone can understand it, and is stated so confidently that all sorts of people are getting interested in it. We can make it ridiculous, and I think we must. It is really the first time the unstatistical folk have fairly recognised that there is a fundamental antithesis, and have accepted battle on that issue. The side which can now get a vulgarly dramatic "score" will have a better hearing presently.

W. F. R. WELDON TO KARL PEARSON, 23 JUNE 1902

The Mendelian Revelation

TITLE OF A 1909 REVIEW IN THE *PALL MALL GAZETTE* OF WILLIAM BATESON'S *MENDEL'S PRINCIPLES OF HEREDITY*

A Scotch soldier, when I was lecturing in Y.M.C.A. huts, said: "Sir, what ye're telling us is nothing but Scientific Calvinism.*" Sometimes I think that would serve.*

WILLIAM BATESON, 1920, ON WHAT TO CALL A VOLUME COLLECTING HIS "LAY" PAPERS ON MENDELISM

"Scientific Calvinism" never took hold as a name for the science of inheritance that boomed its way into biology in the early years of the twentieth century. "Mendelism" fared much better, and "genetics"—William Bateson's coinage in 1905—better still, along with an associated word that arrived in 1909, "gene." Nowadays, talk of genes is everywhere, in and out of biology: genes for aggression, alcoholism, and autism; for baldness, blue eyes, and breast cancer; for caffeine consumption, curly hair, and cystic fibrosis. You name it, and there is, we are told, a gene for it, invisibly pulling the strings, determining how our bodies grow and our lives go. How and why did such talk become so pervasive?[1]

A familiar answer runs like this. We came to talk of inheritance as genic for the same reason that we came to talk of species as evolved and matter as atomic: because genes and evolution and atoms are real, and whatever is real is bound to get discovered eventually. Gregor Mendel

(1822–84) discovered the gene through experiments that he conducted with varieties of the garden pea in the garden of the monastery where he lived and worked in Brünn, in the Austrian Empire (now Brno, in the Czech Republic). He reported his discovery in 1865, publishing it the next year. But alas, Mendel was ahead of his time, and his achievement went unrecognized. Only in 1900, well after his death, were biologists ready to appreciate the significance of what he had done. They made up for lost time, however, swiftly establishing a new science based on the gene. The rest is history: of science, but also of medicine, agriculture, and domains almost undreamt of in Mendel's day, such as forensics. What we now know about inheritance, and what we know how to do with it, we owe to the discovery of the gene. Mendel got there first, but the discovery was inevitable, along with the central role that biologists assign to genes in understanding and controlling life.

This book aims to replace this answer with a rather different one. We came to talk of inheritance as genic, I will suggest, not because the invisible string-pullers are real but because, in the early years of the twentieth century, a debate over Mendel's experiments and their interpretation went as it did. The debate was centered in England, indeed, in its two ancient universities, Cambridge and Oxford. On the one side was the Cambridge-based William Bateson (1861–1926). From 1900, he made it his mission to reshape biology in the image of Mendel's experiments. For Bateson, what Mendel had done in that monastery garden was to sweep away all of the hitherto distorting complexity surrounding inheritance, exposing its underlying simplicity: a vision kept alive in our textbooks. On the other side was the Oxford-based Walter Frank Raphael Weldon (1860–1906). Admiring Mendel's experiments in a limited way, Weldon nevertheless regarded them as profoundly misleading if taken as a guide to inheritance as such. In Weldon's view, Mendel had shown not how heredity at its most basic works, but what it looks like in lineages from which almost all ordinary sources of variability, internal and external, have been eliminated. To generalize from Mendelian experiments, in which context had been made to look ignorable and hereditary characters made to look well described by *x*-or-*y* (yellow-or-green, round-or-wrinkled, purple-flowered-or-white-flowered) categories, was thus to mistake the exception for the rule.

The debate between Bateson and Weldon drew in others, but no one else cared as deeply, fought as bitterly, or—in consequence—thought as creatively about what was at stake in reorganizing the science of heredity around Mendel's peas. Nor was the Mendelian victory a foregone conclusion. Indeed, by winter 1905–6, Weldon seemed, in Bateson's eyes, to be worryingly close to winning the argument. But then, in spring 1906,

Weldon died, with a book manuscript setting out his alternative vision unfinished. I shall argue that had Weldon lived, and had Bateson's fears been realized, our present understanding of how heredity works, and our ability to make use of that knowledge in ways we value, would have been none the poorer. Indeed, in some crucial respects, we might now be better off. We might still talk of "genes," but without even a hint of that string-puller notion, as though the presence of a particular DNA variant in itself, and by itself, can determine whether or not someone is born to be aggressive, alcoholic, or autistic; to be bald, blue-eyed, or doomed to breast cancer; to crave caffeine, have curly hair, or suffer from cystic fibrosis. Instead, our gene talk would be routinely hedged with talk of contexts, internal and external, because our concept of the gene—"our" meaning *everybody's*, no matter how glancing their contact with biological science—would be of an entity whose effects on bodies and minds can be variable depending on the mix of other causes in play. Change those other causes, and you potentially change a gene's effect, or even extinguish it: something you would hardly guess if you think of inheritance as Mendel's peas writ large. No wonder that, on hearing Bateson lecture, a Scottish soldier recalled fatalist theology at its most grim.

Ever since Bateson's day, introductory lectures in genetics typically deal early on with Mendel's experiments. Mendel, the student will learn, was a scientifically inclined monk who, in the mid-1850s, having become interested in the mystery of the biological ties binding offspring to their parents—the mystery of inheritance—turned the garden of his monastery into a laboratory devoted to experimental hybridizing. With uncommon shrewdness, he judged the garden pea to be an especially suitable choice for these experiments because it has a suite of easily tracked, sharply differentiated, either/or hereditary characters. The color of a garden-pea seed is either yellow or green. Similarly, the surface of the seed is either smoothly round or unsmoothly wrinkled; the color of the flower on the plant that grows from the seed is either purple or white; the plant itself is either tall or "dwarf"; and so on, with seven such characters tracked in all. Next, and again shrewdly, before beginning to make hybrid plants, Mendel spent a very long time purifying his originating stocks, making sure that—to stick with flower color for now—his purple-flowered pea plants only ever gave purple-flowered progeny, and his white-flowered pea plants only ever gave white-flowered progeny. The stocks became, in other words, "true-breeding." Only at that point did he begin the experiments per se: first cross-fertilizing; then collecting the hybrid seeds produced; and then planting them. He did all this not with a handful of plants but (showing yet another sign of shrewdness) with lots of them—

indeed, ultimately, over the eight years of his work in the garden, over ten thousand of them. And what he found is that the plants grown from the hybrid seeds, when they produced their own flowers, produced flowers that were not lilac, nor mottled purple and white, nor a mixture of purple and white, but uniformly purple.

Here was a remarkable empirical discovery: a regularity new to science, uncovered thanks to the care Mendel took to purify his starting stocks, and thus—as the student is encouraged to see it—to have removed the baffling clutter, the signal-muffling noise that defeated previous investigators. And once Mendel had made this discovery, he went on to make others, in the course of explaining that regularity simply and powerfully. Suppose, he reasoned, that underlying the purpleness of the purple-flowered pea plants there is a purple-making factor, "*P*." Since his purple-flowered stocks only ever gave purple-flowered offspring (thanks to his purification efforts), there seemed to be nothing in those stocks, when it came to flower color, except *P*. Similarly, for his white-flowered stocks, there seemed to be nothing in them colorwise but a factor for whiteness, "*p*." Now, on those suppositions, what happened in the course of cross-fertilization was that *P* and *p* were brought together, making *Pp* hybrid plants. And yet, as noted, those plants all had purple flowers only. According to Mendel, what that showed was that, in the pairing of *P* and *p*, the effects of the former are visible while the effects of the latter are not. In Mendel's enduring terms, purpleness is thus "dominant" and whiteness "recessive." As he saw, an important corollary immediately followed. The dominant version of a character can arise in two ways: if the dominant factor alone is present (as in the purple-flowered parents), or if there is a mixture of dominant and recessive factors (as in the purple-flowered offspring). The recessive version of a character, however, can arise only if the recessive factor alone is present (as in the white-flowered parents).

From that impressive opener, Mendel proceeded, as described in his remarkable 1866 paper on his inquiry, to uncover and explain further regularities within his experimental lineages by further extending this same set of basic methods and concepts. After he had bred into being all those purple-flowered hybrid plants, he let them self-fertilize, and got a generation in which, along with purple flowers, white flowers came back, in the ratio of 3 purple to 1 white. (Mendel's methodological innovations, the student learns, were not merely to purify and experiment and scale up, but to count.) That ratio, in turn, Mendel explained by way of two additional suppositions. The first concerned what he called "segregation": that is, when the purple-flowered hybrid plants generated pollen and egg cells, each pollen grain and each egg cell contained either a *P* factor or a *p* factor, but not both factors together. The second concerned chanciness: it was a matter of chance whether a particular sex cell (or "gamete") got a *P* factor or a *p* factor, and also a matter of chance which factor came together with which in the union of a pollen grain and an egg cell in the making of an offspring plant. Summing the possible outcomes, one thus expects, Mendel reasoned, to find equal numbers of four factor combinations in the next generation: male *P* with female *P*, male *P* with female *p*, male *p* with female *P*, and male *p* with female *p*. Since *P* is dominant to *p*, that gives flowers that are, respectively, purple, purple, purple, and white, or 3 purple-flowered plants to 1 white-flowered plant (see fig. I.1).

Anyone even remotely susceptible to the charms of explanatory science will, at this point, feel a little pop of pleasure. So that is what is going on! How elegantly simple on nature's part, as on Mendel's. But even those not so susceptible will, sooner or later, encounter Mendel's peas. They are a staple of scientific education all the way through, in formal schooling as well as outside of it. My children have a *Horrible Science Annual* that conveys the Mendelian essentials, purple flowers and all, in a good-humored way, including the inevitable pea jokes. (Brother Mendel ladles out pea soup but withholds it from his naughtier brethren, since, he admonishes, there can be "no peas for the wicked.")[2] In the wider culture too, Mendel's position as begetter of genetics is reinforced in all kinds of ways, subtle and not-so-subtle. There is Mendelian kitsch: ties, mouse mats, and baseball caps bearing his face, above and below the commandment "Obey Mendelian Principles: It's the Laws of Inheritance." In newspapers and their online successors, a standard accompaniment to the latest biotech wonder-story is a history lesson on Mendel and those who have built on the foundation he laid. Such linkings of the Mendelian past to the biotech present can be found well beyond the sphere of popular science. "Since Gregor Mendel first suggested the existence of a gene in the 1860s, genomics research has progressed at an incredible speed. Recent technological advances have moved genomics out of research labs and into the real world." So reads a posting on an investment blog, advising readers to consider the commercial potential of genetic tests, and noting their recent applications ranging from use in identifying future champion athletes to the much-publicized decision by Angelina Jolie to have her breasts removed after a test revealed a mutation associated with an increased risk of breast cancer.[3]

Thus does a traditional, Mendel-venerating, Mendelism-based education in genetics, informal as well as formal, tend to strengthen the notion that traits are "all in the genes," even when teachers and textbook writers would disclaim any such agenda. And the lessons stick. A friend of mine told me about a married couple she knows. Both husband and wife have light-colored eyes, green and blue respectively. They had a daughter with

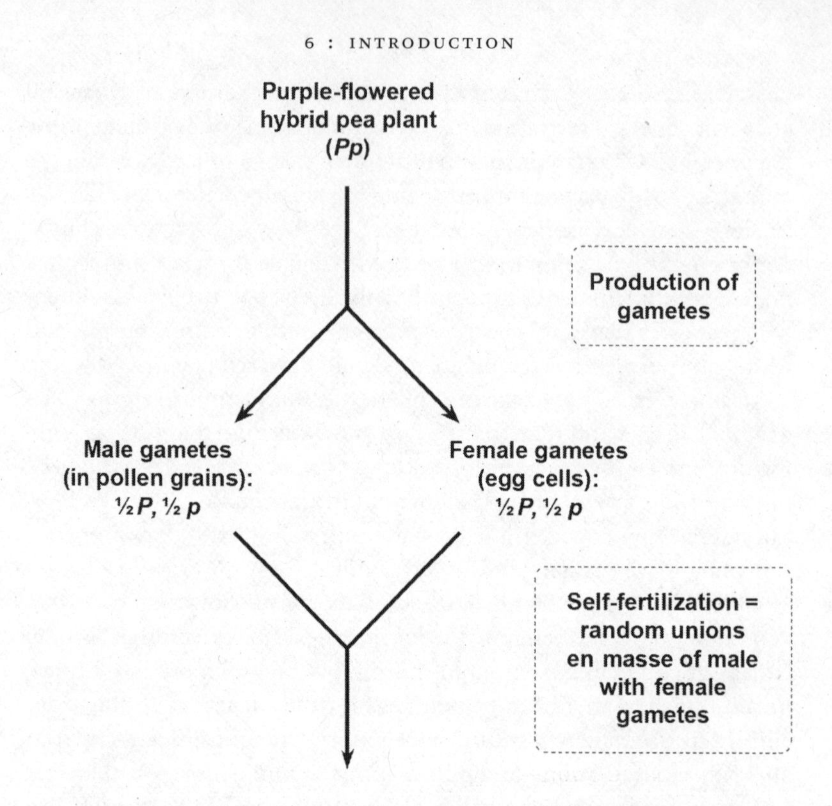

1) Male *P* can combine with female *P* = *PP* → purple flower

2) Male *P* can combine with female *p* = *Pp* → purple flower

3) Male *p* can combine with female *P* = *pP* → purple flower

4) Male *p* can combine with female *p* = *pp* → white flower

with these four combinations produced with equal probability

FIG. 1.1. A schematic overview of how Mendel derived and explained the famous 3:1 ratio

brown eyes. On learning that fact, the husband's mother blew the whistle. According to Mendelian principles, she announced, no child of that couple could possibly have brown eyes, since, according to the version of those principles she (like everyone else) learned in school, dark-colored eyes in humans are dominant and light-colored eyes are recessive, so that light-eyed people, including her son and his wife, could only ever pass on the gene variants for light-colored eyes. Any other outcome implied a violation, and probably not of Mendel's laws . . . Her son and daughter-in-law, of course, reacted furiously to the implied insinuation, and did what anyone else nowadays would do in their shoes: they turned to the internet. After some searching, they found the reassurance they sought, to the effect that eye color did not, in fact, always follow the Mendelian rules. "Blue-eyed parents can have brown-eyed kids and other eye-oddities" is the title of a currently available online newspaper column, which goes on to explain that, contrary to the simple Mendelian explanation of eye color so widely believed, eye color in humans is not determined by just one gene in one of two states, light-making or dark-making. Like more or less every other hereditary character, eye color is multifactorial in its causation and variable in its expression. To look back on the Bateson-Weldon debate over Mendel is to return to the moment when such complexity, rather than becoming unquestioned common knowledge, became permanently surprising.[4]

I have grouped the chapters that follow into three parts, titled "Before," "Battle," and "Beyond." The "Before" chapters furnish background in the form of historical essays centered on the 1860s, '70s, '80s, and '90s successively. Here I introduce some of the people, proposals, and places that—along with heredity itself, as a new object of scientific knowledge and public fascination—went on to figure in the debate over Mendel. The "Battle" chapters slow the pace in order to project the reader into the thick of the debate as it unfolded, with each chapter covering at most a year or two between 1900 and 1906. Although Bateson and Weldon are central, the cast of characters, as well as the countries involved, gradually expands as Mendelism extends its reach. Throughout, I try to do justice not only to the intellectual to-and-fro but to the role that the politics of heredity, notably in connection with the Anglo-Boer War (1899–1902) and the new prominence of eugenics, played in it. The "Beyond" chapters are, like the "Before" chapters, a set of interlinked historical essays, but more adventuresome in their means and ends. Touching on everything from the philosophy of explanation to the molecular biology of pea-seed shape, and from the history of eugenics to an experiment in teaching genetics as if Weldon's vision for the future of biology had been realized, they represent my best shot at probing the significance of the debate for what followed, from 1906 to the present. At the end, I provide a brief overview of my conclusions.

We start with two Olympians of biology, Gregor Mendel and Charles Darwin. In 1865, both put forward ideas that became important for later students of heredity. But where Mendel is lauded for solving the puzzle of heredity correctly, Darwin is lambasted for his incorrect solution, "pangenesis." We shall see that neither man thought of himself as addressing heredity first and foremost, let alone as setting out to establish new foundations for its study. For Mendel, the puzzle was not heredity but hybrids—indeed, a particular class of hybrid plants whose characters, like flower color in the garden-pea plant, do not stay fixedly, uniformly the same down the generations, but exhibit returns to the characters of the ancestral plants. For Darwin, that puzzle was just one of an astounding range of what he judged to be connected problems, from how wounds heal to how bodies develop to how changes acquired during an adult's lifetime can get transmitted to offspring (so-called Lamarckian inheritance), for which he sought a single, unifying account. Published in 1868, his hypothesis of pangenesis posited the existence of tiny particles constantly being shed by every part of the body. Pangenesis convinced no one. Yet it inspired a great deal of creative research, by people who increasingly thought of themselves as students of heredity. When, at century's end, they encountered Mendel's paper, they read it as if he were one of their own.

Stripping away the layers of anachronism that now obscure our vision of Mendel's and Darwin's projects takes some doing. The same is true of Francis Galton's projects, dealt with in chapter 2. Unlike Mendel and Darwin, Galton really did see himself as aiming to illuminate heredity, and from 1865, he threw himself publicly into inventing the new statistical methods that he reckoned were needed for the task. Furthermore, he regarded that task as urgent, since he believed that the only way to stave off imminent national ruin was to breed into being better leaders, intellectually and morally superior to the current ones; and that the only way to make that plan for eugenics (Galton's coinage) convincing was to demonstrate its scientific bona fides. Not for nothing has Galton come to be remembered as a doctrinaire, ultimately sinister hereditarian. Accordingly, it has become hard to imagine how anyone could want what Galton wanted and yet stress the huge variability of inherited characters, the multifactorial nature of their causation, and the large role that context and chance play in how organisms—very much including people—develop. Those emphases, however, were exactly Galton's, most conspicuously in his work in the early to mid-1870s, in writings now largely forgotten, but also in a device now used the world over in teaching statistics. Called the "quincunx," it went public in the same 1874 lecture on inheritance in which Galton introduced the phrase "nature and nurture." For Galton, "nature" included not only the germinal basis of a hereditary character but the range of other causes, germinal and otherwise, impinging on the developing embryo before birth; while "nurture" took in the range of new causes impinging after birth. Without the germinal basis, there could be no character. But the character's expression—

indeed, whether it was expressed at all—depended on the interactions with all the impinging causes.

Mendel, Darwin, and Galton belonged to a scientific world run by cleric naturalists and gentleman amateurs. By contrast, Bateson and Weldon belonged to the first generation of biology professionals. Chapter 3 tracks their converging and diverging paths through the 1880s. They met as students near the start of the decade at Cambridge, where Bateson's father was the master of their college, and where they learned about evolutionary embryology from its then-greatest practitioner in Britain, the young, aristocratic Francis Balfour. The working out of the genealogy of life—the Darwinian family tree—on the basis of commonalities in the embryos of different species, especially ones easily studied in the new marine biological laboratories sprouting up around Europe and elsewhere, was the era's defining research problem in biology. Distinctively, Balfour was as interested in the adaptive processes behind the genealogical patterns as he was in the patterns themselves, and encouraged a new thoughtfulness about embryos as products of natural selection and, therefore, of heredity and environments, ancestral and present-day. But whereas Weldon took up those preoccupations and made them his own, Bateson, by decade's end, was moving in the opposite direction, toward a vision of organisms as having the forms they do not because of evolutionary history (which he came to regard as unknowable) or adaptation to environments (which he had begun to suspect was exaggerated) but because of internal dynamics, operating independently of history and environments. Weldon, wanting to Darwinize form, looked outward from the organism, and Bateson, wanting to geometrize form, looked inward.

Throughout the 1890s, when Weldon was based at University College in London, Weldon and Bateson encountered each other most often at the Royal Society of London. Then as now, being elected a fellow of the Royal Society was a sign that, scientifically, one had arrived. Chapter 4 uses the elections of Weldon in 1890, Bateson in 1894, and Karl Pearson (a University College colleague and then friend of Weldon's) in 1896 as invitations into the research that made their professional reputations. In each case, that research was indebted to Galton's oeuvre, above all to his concerns with, respectively, the effects of natural selection on variation in populations of wild organisms, the possibility that evolution proceeds by discontinuous leaps, and the mathematical theory behind the curves that emerge when variation is plotted on a graph. But the Royal Society, as we will see, bears on their story in other ways. Anxieties about whether the society's meetings were too dull led to the deliberate pitting of Weldon and Bateson against each other in public at an 1895 meeting to discuss Weldon's latest work. The fallout brought further deterioration to a once friendly but increasingly frosty relationship. Then, over Weldon's objections, Galton invited Bateson to join the Royal Society committee on the statistical study of evolution in whose name Weldon had presented his work. Early in 1900, Weldon led a mass exodus from the committee. It carried on under Bateson's leadership, in effect becoming the Mendelism committee.

The "Battle" chapters (5–9) take full advantage of the truly astonishing wealth of documents, published and unpublished, that survive from both sides of the Bateson-Weldon debate. No previous study has drawn as comprehensively on letters, manuscripts, and related materials from archives around Britain and elsewhere. The resulting reconstruction of the weave of public and private actions, interactions, and reflections—at points chronicling events in day-by-day detail—departs from the received version of what historians of science have labeled the "biometrician-Mendelian controversy" in several ways.[5] One often gets the impression, for example, that hardly anyone outside the most specialized circles had even heard of chromosomes, let alone taken them seriously as the cellular bearers of hereditary material, until well into the 1910s, when T. H. Morgan and his students at Columbia University showed, via brilliant Mendelian experiments with fruit flies, that Mendelian genes could literally be mapped onto chromosomes. But a major role for chromosomes in heredity was already the stuff of popular-science lectures in London in spring 1900, and would continue to be taken for granted through the 1900s by many people, Weldon included.

"Treasure your exceptions!" is a famous one-liner from Bateson.[6] As we shall see, however, a large part of what made Mendelism under his leadership so powerful was precisely his genius for neutralizing the intellectual force of exceptions to Mendelian rules. Bringing into focus that element of Bateson's achievement—his multilayered shielding of Mendelism from empirical disproof, no matter how mottled or motley a pea seed was, or how many brown-eyed babies were born to blue-eyed parents—is one of the ways in which sympathetic attention to Weldon's perspective can clarify Bateson's perspective too. Another neglected dimension of Bateson's work that receives thorough treatment in what follows is his emphasis, in promoting Mendelism, on its practical applications. Bateson proved an indefatigable publicist for the utility of Mendelism, not only for plant and animal breeders but also for those concerned with human breeding. Practical usefulness as a badge of the truth of theoretical principles has been a major motif in Western science since the seventeenth century. Bateson and his allies worked extremely hard, and successfully, to get that badge pinned on Mendelian principles. When, in 1909, Bateson's *Mendel's Principles of Heredity* was the subject of a review titled "The Mendelian Revelation" in the *Pall Mall Gazette*, the reviewer especially commended the final chapters, "culminating in a statement of the Eugenic ideal."

Success brings with it an air of inevitability. If histories of genetics include Weldon at all, he comes across as tiresomely obstructive—a nitpicker and a naysayer, with no larger vision of his own.[7] In fairness, Weldon expressed his views most far-reachingly in letters and manuscripts that never became public. We owe the possibility of the resurrection of his alternative science of heredity to the archive, above all to the collection assembled after his death by his heartbroken friend Pearson. That science can be summarized in three overlapping emphases and an upshot.

1. Variation matters. It matters that real pea seeds are not always just yellow or green, or just round or wrinkled. Actual variability should not be fictionalized or idealized away by the use of simple categories such as "yellow" or "round." Statistics is the descriptive language appropriate to biology because biological populations are variable and statistics capture that variability with quantitative exactness. (Hence Weldon's dismissal of the Mendelians as "unstatistical folk.") Because that variability might itself turn out to hold indispensable clues to how inheritance really works, a science of inheritance that deliberately disguises variability is one that is not moving in the right direction.

2. Ancestry matters. The particular ancestry, the particular lineage that *these* peas come from, as distinguished from *those* peas, can matter. From Weldon's perspective, the trouble with Mendelism—Mendel's paper unjustifiably blown up into a biological world view—is that it encourages ignorance and incuriosity, in this case about the deep ancestry behind any particular individual's characters. According to the Mendelian, as long as true-breeding green-seeded peas are producing green-seeded peas, that is all you need to know about them. Weldon thought that complacency was a mistake, one connected to the Mendelian mistake about variability. For Weldon, behind all of that variability might lie different kinds of ancestry. And ancestry matters because what is inherited is not just this or that individual "factor for," but a context.

3. Environment matters. From within a lineage, an individual inherits not just this or that factor but a whole suite of them. Crucially, these factors together constitute a context—an environment—conditioning the visible effects of any particular factor. Change the environment, and you can change the effect. In "Theory of Inheritance," the manuscript that Weldon worked on throughout 1904–5, the conditioning role of environments, not just germinal (specifically, for Weldon, chromosomal) but more generally physiological and physicochemical, was a major theme. Weldon took that theme to be one of the main lessons of the experimental embryology in its glory period toward the end of the nineteenth century, and so, surely, something that any science of inheritance worth having had to take into account. A character was neither all inherited nor all acquired, but always a joint product of what was inherited and what was around it.

Upshot: The science of inheritance should treat the modifying influence of context on a hereditary character not as exceptional, as Mendelism does, but as exemplary. In Weldon's view, to run a Mendelian experiment is deliberately to exclude all of the variability that would otherwise give you different kinds of patterns. If you were so minded, he thought, you could probably get a race of peas going in which greenness was dominant to yellowness. It all depends on the choices that you make, the contexts that you build. To declare one pattern the natural one, and other patterns somehow deviant, is just arbitrary. What biologists need is a concept of dominance that sees it as context dependent—a concept owing less to Mendel than to Galton. With context dependence accorded a central conceptual role, Mendelian patterns take their place not as generalizations upon which to organize all knowledge of heredity but as special cases: interesting for limited purposes, but in no way exemplary of how heredity works on the whole.

Weldon never lived to make this case. The "Beyond" chapters (10–13) look back on his quarrel with Mendelism to consider, respectively, why Mendelism went on to succeed on the scale that it did, whether Weldon's alternative could have been as successful had events worked out differently around 1906, what the legacies of Mendelism's success have been, and what to make of Weldon's legacies, as they have actually been but also as they could have been, and even might still be.

Much in these chapters is straightforwardly historical. In analyzing the expansion of Mendelism as a research program, for example, I offer a new interpretation of the Morgan fly room as the place where Mendelism became not merely chromosomal but, in the close attention paid to how context can modify the effects of chromosomes on characters, Weldonized. In showing that molecular genetics owed little to Mendelism, and could well have arisen without it, I reexamine some pioneering work in molecular genetics as well as later attempts by the pioneers to depict their achievements as continuous with Mendel's. In following the trail of the eugenic ideal from Batesonian enthusiasm to brutal enactment, I quote the popular American press of the early 1920s on Mendel's status as the scientific brains behind eugenics, and I document the role of Mendelian-Calvinist expertise in smoothing the path later that decade to more severe immigration restrictions as well as to the US Supreme Court decision that opened the way to legal sterilization of the "unfit." (Eugenics propagandists in Germany went on to emulate, and then exceed, the American model, Mendelism and all.) In suggesting that research on what came to be called "norms of reaction"—graphs comparing responses of different gene-variant combinations to different environments—traces back to Weldon, I identify hitherto unnoticed affinities between the German paper remembered as inaugurating the norm-of-reaction tradition and an earlier, unpublished lecture of Weldon's. And in trying to make sense of the curious, belated notoriety of Weldon's discovery that Mendel's experimental data are improbably close to what his theory predicted ("too good to be true"), I guide the reader through some of the subtler cultural dynamics of science during the Cold War, on both sides of the Iron Curtain.

But, as mentioned above, there are un-straightforwardly historical elements in the mix too. For all that I aspire to cast the debate over Mendel in a new light, I also want that light to shine on more general themes in the study of human knowledge. Three themes in particular wend their way through the final chapters. One theme has to do with the organization of a body of knowledge, and the cascading consequences—for everything from individual cognition to scientific advance to social justice—of some items of knowledge coming to be treated as central, exemplary, subordinating, and others as peripheral, exceptional, subordinated. Part of what makes the Bateson-Weldon debate worth thinking about historically is the complex way in which Weldon's emphases have become both thoroughly integrated and thoroughly marginalized. I try to show how that has worked in practice as well as in principle. (To my mind, a minor but stubborn tendency to dismiss the debate as much ado about nothing, since the two sides can be so easily "reconciled," betrays indifference to this issue of cognitive priority, especially as manifest in the handbooks, textbooks, encyclopedias, and popular-science books and articles from which most people—including most scientists—learn about a science.) In the first of three postscripts that round out the book, I revisit the theme via some remarks on two terms that run throughout the book: "genetic determinism"—a term Bateson used in Calvinistic mood—and "interaction."

Another theme has to do with explanation: What is it that biologists are doing when they call upon genes as explainers? What is it that historians are doing when they try to explain biologists' explaining? What lies behind judgments of some candidate explanations as better or worse than others? A facet of Bateson's artfulness as a Mendelian advocate was his extracting from Mendel's pea-hybrid paper the version of it immortalized ever since in our textbooks. I look at the choices Bateson made, and why he made them, in the course of articulating my own explanation for Mendelism's success, which I put down to its combination of teachable principles, tractable problems, and technological promise. From Bateson's time to our own, when students encounter the Mendelian explanation of the basic pattern that Mendel discovered in his hybrid peas, that explanation seems satisfying not least because, on the face of it, nothing else could explain that pattern so well. The possibility of an alternative explanation in which, on assumptions that were reasonable in the early twentieth century, the recessive organisms making up the 1 in the famous 3-to-1 ratio could turn out to harbor the dominant-character factor never comes up. In the second postscript, I draw on Weldon's work to set out just such an alternative. I hope that the experience of following his argument gives you—as it gave me when I first grasped it—that familiar pop of pleasure, though now deriving as much from having critical faculties honed as from having learned something new about how the world might work.

To say that A explains B is—often if not always—to imply that, if not for A, then B would not have been or would not have happened. This concern with explanation thus brings with it a concern with what are called "counterfactuals." I think the term is unfortunate because, at its best, the assessment of the possible—the third of my trio of themes—is a richly evidential enterprise. In trying to answer the question of whether Weldon's alternative science, had he lived to complete his "Theory of Inheritance," might have been as successful as Mendelism actually became, because potentially carried in teachable principles, tractable problems, and technological promise, I end up looking afresh at major episodes in the actual history of biology. Indeed, my pursuit of evidence on the matter of teachability led me to become an experimentalist myself. What, I wondered, would happen if students began not with Mendel and his peas but, in Weldonian spirit, with examples where context conspicuously affects the influence that a gene has? And what if those students went on to receive a whole introductory course as if it had emerged from a Weldonian past, where the organizing emphases were on causal interaction and character variability? A course along these lines actually ran at the University of Leeds in autumn 2013. We found that, whereas students taking an orthodoxly Mendelian course were, on average, just as determinist about genes at the end of the course as they had been at the start—and were, if anything, more determinist—students taking our Weldonian, interaction-emphasizing course were, on average, less determinist at the end. In other words, the Weldonian curriculum seemed to do better than a Mendelian one in enabling a basic grasp of genetics without inadvertently imparting what is, by the lights of current biology, an exaggerated notion of the power of genes.

The Weldonian curriculum has begun to inspire changes in the teaching of introductory genetics among teachers concerned to improve students' understanding of real-world variability and the interleaving of genes and contexts, internal and external, behind it. What excites these teachers is the prospect of students leaving the classroom with knowledge that is more accurate scientifically and, for that reason, gives them a larger, livelier sense of the possible. In the future, when they evaluate genetic claims wrapped up in binary categories, they will be primed to ask about variation beyond the binary. When advised that DNA sequencing has revealed a worrying mutation, they will not be satisfied with knowing the headline outcome, but will insist on finding out about the range of outcomes, and how different genetic backgrounds and wider environments—some perhaps yet to be investigated—might make a difference. When told that, say, girls set themselves up for failure by aiming for a career in science because the female brain is by nature not well equipped for scientific thinking, or that there is nothing to be done about persistent social inequalities between different races because those inequalities are rooted in the genes, they will roll their eyes, mutter to themselves about scientific Calvinism, and start scrutinizing.

Mendelism's role as a bridge between science education and social prejudice has, for too long, been hiding in plain sight. As I was finishing this book, I came across a magisterial essay by the American legal scholar and social critic Patricia J. Williams, reviewing a new facing-the-facts book on racial disparities in the United States. According to Williams, the concept of race that the author presupposed amounted to "a category of unyielding genetic difference, a sealed box of capability, disposition and destiny." Her dismantling of the case for that concept, including its entanglement in eugenics, is unsparing. To illustrate her essay, the editors chose, aptly enough, a photograph from a circa 1930 educational exhibit from the London-based Eugenics Society. The displays look very much like the ones you can see in figure 12.3, right down to the heavy didacticism about something that, in Williams's extensive historical survey, never comes up: "Mendel's law."[8]

It is time for the simplifications of Mendelian storytelling to become history. I hope that this book helps to speed them along. I hope, too, that the Weldonian reforms now under way in biology teaching—and maybe even in biology itself, as today's students become tomorrow's researchers and teachers—pique the interest of historians of science in the value of counterfactual inquiries. By way of tickling imaginations, I close the book, in my third and final postscript, with an extract from an edition of the *Dictionary of Scientific Biography* that never was, belonging to a history in which Weldon lived beyond spring 1906. Whether or not my speculations convince, there is nothing idle about them, for they are already busily opening up new options. The scientific past is over, but its possibilities remain potent.

Part 1

BEFORE

1

Who Needs a Science of Heredity?

Somebody rummaging among your papers half a century hence will find Pangenesis & say "See this wonderful anticipation of our modern Theories—and that stupid ass, Huxley, prevented his publishing them."

T. H. HUXLEY TO CHARLES DARWIN, 16 JULY 1865

Your last note made us all laugh.—The future rummager of my papers will I fear, make widely opposite remarks.

CHARLES DARWIN TO T. H. HUXLEY, 17 JULY 1865

On heredity, Gregor Mendel is our culture's greatest scientific hero, and Charles Darwin, to put it kindly, is not. Mendel's status can be checked in any biology textbook. For Darwin's, consider the physicist Mario Livio's 2013 bestseller *Brilliant Blunders*, about the biggest goofs of famous scientists. Livio places Darwin's messing up on heredity with Lord Kelvin's grotesque underestimate of the age of the Earth, Linus Pauling's erroneous structure for DNA, Fred Hoyle's theory that the universe had no beginning (though Hoyle's derisive label for the rival "big bang" theory has stuck), and Einstein's mistaken conjecture about a cosmic repulsive force counterbalancing gravitational attraction. According to Livio, Darwin not only got heredity wrong—plumbing the depths with his embarrassing "hypothesis of pangenesis"—but came wrenchingly close to the needed correction. On the bookshelves of Down House, in a German volume on plant hybrids, was a summary of Mendel's "Versuche über Pflanzen-Hybriden" (Experiments on Plant Hybrids). Alas, a photograph shows, the relevant pages were never cut.[1]

To pair up Mendel and Darwin in this way—and it is the standard way—is to assume that what Mendel succeeded in doing with his pea experiments, and what Darwin failed to do with pangenesis, was to establish the basis for a science of heredity. But in 1865, the year when Mendel presented his experimental results as lectures and Darwin first wrote up pangenesis, almost no one, Mendel and Darwin included, aspired to that goal. A felt need for a science of heredity, with its own principles and procedures and place in the scheme of knowledge, became widespread only later, over the latter decades of the nineteenth century—in no small measure thanks to debates stimulated and sharpened by Darwin's theory of evolution by natural selection, published in 1859 in *On the Origin of Species*. As new priorities emerged, new meanings got read into old methods, and eventually into an old paper of Mendel's. If we want to understand how and why that happened, we need, first of all, to recover what Mendel and Darwin circa 1865 thought they were doing.

That is the burden of this chapter. We will see that, where Mendel sought something considerably smaller in scope than a science of heredity, Darwin sought something considerably larger, even grander. For both, moreover, their real projects were, by the lights of later science, thoroughly alien. Giving that alien quality its due will mean grappling with a range of questions about heredity historically viewed, from the surprisingly late incorporation into English of the word "heredity" to the endless mischief caused by a distinction between "particulate inheritance" and "blending inheritance."

1

What, exactly, is Mendel's famous paper about? Peas, of course—and, to a lesser extent, beans. But the complex program of research it recounts is directed at a particular problem about peas and beans. Here is what readers of *Brilliant Blunders* learn about the paper:

> The modern theory of genetics originated from the mind of an unlikely explorer: a nineteenth-century Moravian priest named Gregor Mendel. He performed a series of seemingly simple experiments in which he cross-pollinated thousands of pea plants that produce only green seeds with plants that produce only yellow seeds. To his surprise, the first offspring generation had only yellow seeds. The next generation, however, had a 3:1 ratio of yellow to green seeds. From these puzzling results, Mendel was able to distill a *particulate*, or *atomistic*, theory of heredity. In categorical contrast to blending, Mendel's theory states that genes (which he called "factors") are discrete entities that are not only preserved during development but also passed on *absolutely unchanged* to the next generation. Mendel further added that every offspring inherits one such gene ("factor") from each parent, and that a given characteristic may not manifest itself in an offspring but can still be passed on to the following generations. These deductions, like Mendel's experiments themselves, were nothing short of brilliant. Nobody had reached similar conclusions in almost ten thousand years of agriculture. Mendel's results at once disposed of the

FIG. 1.1. Gregor Mendel (*standing, second from right*) with other members of his monastery in Brünn (Brno), in about 1862.

> notion of blending, since already in the very first offspring generation, all the seeds were not an average of the two parents.[2]

Those who read around a little further will soon enough encounter a compelling backstory for the image of Mendel as the discoverer of genes. The main protagonist is not Mendel but his great patron in the Abbey of St. Thomas, Cyrill Napp, who served as abbot from the mid-1820s until his death in 1867. In a playful photograph taken around the time that Mendel completed his pea experiments in the abbey's garden, Napp is seated just to Mendel's right (fig. 1.1). As the photograph suggests, even by the relaxed standards of the worldly Augustinian order, St. Thomas under Napp more closely resembled a research-institute-cum-salon than a conventional monastery. The sciences thrived, above all the practical sciences of agricultural improvement from which Brünn and the surrounding region might benefit economically. A man of broad learning and huge energy, Napp wrote a breeding manual, set up a stock nursery in the monastery grounds, and held positions in breeders' groups devoted to everything from apples and bees to sheep and vines. No one concerned with breeding could fail to take an interest in inheritance—"like engend'ring like," in the old phrase—and Napp, with customary zeal, made it his mission to put inheritance on a sound footing. When Mendel entered the monastery in 1843, mainly for the chance to pursue a scientific education (otherwise out of reach for his farming family), Napp found his instrument. On Napp's recommendation, Mendel studied at the University of Vienna between 1851 and 1853. When, on his return, Mendel—now securely launched in his career as a school science teacher—devised a plan to crack open the problem of inheritance, Napp gave him a plot in the garden.[3]

Much in the story above is not merely right but worth underscoring. St. Thomas under Napp really was an extraordinarily lively scientific hub through which ideas and visitors flowed, and Mendel's "Versuche" is very much its product.[4] Furthermore, Napp really did show a lively appreciation for ideas about inheritance. In the published transcript from an 1837 meeting of sheep breeders, for example, he is recorded at one point as saying—by way of trying to refocus a drifting discussion about "inheritance capacity" in sheep—"The question is: what is inherited and how?"[5] Looking beyond the immediate world of the monastery, we can see too that Napp posed his question in an era when, across Europe, inheritance was, for the first time, coming into its own scientifically. From the early decades of the nineteenth century, disease transmission within families became something of a specialty among medical writers, especially French ones, whose field of inquiry came to be known by a distinctive label: *hérédité*. Their efforts climaxed in the physician and alienist Prosper Lucas's two-volume *Traité Philosophique et Physiologique de L'Hérédité Naturelle* (1847–50). His title continued: *dans les états de Santé et de Maladie du Système Nerveux*—"in the states of health and of disease of the nervous system." Thus did *hérédité*, previously understood as embracing hereditary disease only, come to encompass transmission from parents to offspring generally. The expanded category made an impression, in France—where, from the late 1860s, Émile Zola devoted a famous series of Lucas-influenced novels to dramatizing how, in Zola's words, "heredity has its laws, just like gravity"—and farther afield. In the breeding-focused first chapter of Darwin's *Origin of Species*, Darwin affirmed the authoritative status of Lucas's volumes and then summarized the reasoning behind the recent expansion: "Every one must have heard of cases of albinism, prickly skin, hairy bodies, &c., appearing in several members of the same family. If strange and rare deviations of structure are truly inherited, less strange and commoner deviations may be freely admitted to be inheritable. Perhaps the correct way of viewing the whole subject, would be, to look at the inheritance of every character whatever as the rule, and non-inheritance as the anomaly." Not long after, other English writers on inheritance began to use an anglicized form of the French term, "heredity"—a word that had barely existed in English up to that point.[6] The German counterpart term is *Vererbung*. In the German translation of the *Origin* that Mendel read, "inheritance," "inheritable," and "non-inheritance" in the passage quoted were translated using *vererben* and variants on its root, *erben*. When Napp put his question to the sheep breeders, he asked, according to the transcript, "*Was vererbt und wie?*"[7]

So, it seems, at a moment when inheritance was emerging on the scientific map, Napp asked: what is inherited and how? Decades later, Mendel answered: genes are inherited, and they are transmitted unchanged down the generations, like atoms of heredity.[8]

The trouble with taking Napp's question as the Beginning in this way is that it stands out from that old transcript only because, a century and a half later, admirers of Mendel qua discoverer of genes went hunting through the documentary record in search of a "genes as the answer" backstory. The backstory, in other words, presupposes that Mendel's paper is about heredity, indeed, about heredity along at least roughly the lines familiar to the age of genic biology. If, instead, we ask what someone taking an interest in the subject around 1837 might have wanted to know about, we rapidly find ourselves in surprising territory. For example, after Napp's intervention at the sheep breeders' meeting, the discussion turned to how the power of hereditary transmission changes over the lifetime of a ram.[9] Or consider Darwin's reflections in the same period, when he was living in London and filling notebook after notebook with theorizing about the mutability of species. In Darwin's emerging theory, a new adaptive variation arising in an individual organism became consequential for the species only if the organism's offspring inherited the novelty. Accordingly, Darwin obsessed about hereditary phenomena, including what he called "Yarrell's law," after the London naturalist William Yarrell. Notebook C, filled between February and July 1838, opens with a long entry that begins: "Mr Yarrell 'Give it as his theory' tells me. he has no doubt that oldest variety, takes greatest effect on offspring." In other words, when parents come from different varieties, offspring take after the parents unequally, tending to resemble the older, more established variety much more than the younger one. In answer to the question "What is inherited?" we thus have "characters from both parents";

and in answer to the question "How is it inherited?" we have "unequally, since characters from the older variety tend to win out."[10]

Mendel's "Versuche" goes nowhere near these topics. Even in its vocabulary, it seems to be about something else: an *erben* word shows up just once, and in passing.[11] Consistent with that lack of visible interest are Mendel's annotations on his copy of the *Origin*, in the German translation that came out while he was nearing the end of his pea hybridization work. (Mendel's copy still survives in the Abbey.) Darwin's Lucas-inspired reflections elicited no responses at all. What got Mendel's pencil going was a claim of Darwin's not much noticed nowadays: that, upon being cultivated, a plant species becomes highly variable, to a degree well beyond what would be the case in a state of nature.[12] Mendel seems to have taken that claim as a challenge to the very idea that cultivated plants might be governed by natural law—whereas his "Versuche" can be seen as a demonstration of the lawfulness that reigns among at least one kind of cultivated plant (with some tart comments toward the end wondering at how anyone could have suggested otherwise).[13]

Heredity, then, was not Mendel's main concern. The clue to his subject lies in his title. "Versuche über Pflanzen-Hybriden" is about what Mendel said it was about: hybrids.[14]

2

In the opening lines of the paper, Mendel tells us that the idea for his experiments began with something first noticed through the use of artificial fertilization in the search for new color variants in ornamental flowers: that repeated crossings of the same two species reliably yield the same hybrid form. At the start of a hybrid lineage, then, there is lawful regularity. Mendel's questions bear on the subsequent fate of the hybrid form: Is there a law governing its appearance and disappearance? If so, what is that law?[15]

Hugo Iltis, Mendel's earliest biographer, reported that Mendel was esteemed in and around Brünn for the quality of the hybrid flowers he bred. One creation was even on sale under the name "Mendel fuchsia." Perhaps, in figure 1.1, Mendel is inspecting an example. We need look no further than Mendel's immersion in the hands-on culture of practical science to account for his absorption in the problem of understanding hybrid forms.[16] But to account for what Mendel *did* with that problem, we need to look more closely at the training Napp made possible for him at the University of Vienna. There Mendel had a number of distinguished teachers, among them the physicist Christian Doppler, of the Doppler effect. But the most important was Franz Unger, a botanist whose work brought together three specifically Austro-German botanical traditions about as far removed from botany-as-plant-collecting as could be. One was the biogeography pioneered by the explorer Alexander von Humboldt, devoted to counting species and, with the aid of an array of instruments, measuring temperature, magnetic intensity, atmospheric composition, and other physical variables in order to find out how the conditions in a place determine the range of species found there. Another, related tradition was comparative morphology, which, since its founding by the great Johann Wolfgang von Goethe (to whom Humboldt dedicated a key early work on plant geography), had aimed to illuminate the diversity of living forms by relating them systematically to more basic underlying forms. Both traditions sought the laws of life, ideally quantitative laws such as, from morphology, Schimper's law, which states that the leaves spiraling up a stem are spaced at points characterized by ratios of alternate Fibonacci numbers—at ½ of a turn, or ⅓, or ⅖, or ⅜, and so on. The third tradition was cell theory, according to which all organisms were made up of cells, themselves increasingly understood as always arising from other cells. Unger worked at the intersection of these traditions, searching for the mathematical laws governing the changing distribution of cell-constituted plant forms as climate succeeded climate over geological time. "A physics of the plant organism" was his phrase for what he sought. Mendel's project was Ungerian science translated to a botanical domain—plant hybrids—previously considered too unruly to be governed by law, much less by a law as elegant as the one Mendel announced in his paper.[17]

For anyone who knows Mendel's paper from the science-classroom rendition, the beginning is reassuringly familiar. Mendel sets out his reasons for choosing the garden pea, *Pisum sativum*, for experiments on crossbreeding. *Pisum sativum*, he explains, grows quickly and with minimal fuss; the arrangement of stamens and pistil in the flower makes the brushing of pollen from one plant onto another straightforward while also protecting against natural contamination; and for any given character, there are different versions that retain their distinctiveness down the generations and so are easy to track. Most useful for tracking, Mendel reckoned, are the characters that come in two cleanly differentiated versions: seed shape, in round or wrinkled versions; seed color, in yellow or green versions; seed-coat color, which is actually a bit messy but correlates with flower color, in purple or white versions; and so on. In all, he identified seven such either/or characters for tracking. He also took care to assess the stability of his stock plants, whittling down an initial supply of 34 to 22 that, upon self-fertilization (the default mode of reproduction for pea plants), reliably retained their forms, so that, for example, his round-seeded stock plants only ever produced round-seeded progeny, and his wrinkled-seeded stock plants only ever produced wrinkled-seeded progeny. At that point (in 1856), he was ready to experiment.[18]

But as the paper proceeds through Mendel's discovery of dominance and recessiveness in his hybrids, then the discovery of the 3:1 ratio of dominant to recessive character versions among the offspring of those hybrids, modern readers wait in vain for him to frame what our textbooks refer to as "Mendel's laws." In typical presentations, Mendel's first law, also known as the principle of segregation, states—to quote from the textbook recommended at my university—that "the two members of a gene pair (alleles) segregate (separate) from each other in the formation of gametes," with the result that "half the gametes carry one allele, and the other half carry the other allele." Then there is Mendel's second law, or the principle of independent assortment, according to which—from the textbook again—"the factors for different traits assort independently of one another," or, in updated terminology also provided, "genes on different chromosomes behave independently in gamete production."[19] In Mendel's paper, however, just one law is enunciated, and it does not look like either of the above. What Mendel refers to as "the law valid for *Pisum*" is the 3:1 ratio decomposed into 1:2:1 and expressed algebraically as $A + 2Aa + a$, with A standing for the dominant-character version, a the recessive version, and Aa the hybrid (outwardly, again, indistinguishable from A).[20]

For Mendel, all else follows from the recognition of this law. Over the rest of the paper, he spells out the implications, in four directions:

1. *Generalizing the law over the long run.* Keep on breeding along these lines, Mendel writes, having done it himself up to the sixth generation, and the results will continue to conform to the $A + 2Aa + a$ law. But that does not make the long-run future of the hybrid form, Aa, uninteresting. With help from some simplifying assumptions, a little arithmetic suffices for Mendel to work out the distributions of A plants, Aa plants, and a plants in every generation following a hybrid plant's self-fertilization. A table provided spells out the calculated results, summed up at the bottom in an abstract formula: $2^n - 1 : 2 : 2^n - 1$, where n is the post-hybrid generation number, the first $2^n - 1$ is the proportion of A plants in that generation, 2 is the proportion of Aa plants, and the final $2^n - 1$ is the proportion of a plants.[21] Here, Mendel notes, something observed by previous investigators, most illustriously the German plant hybridizers J. G. Kölreuter and C. F. von Gärtner, is confirmed: hybrid lineages have a tendency to revert to the parental forms. But we also find something new: that though the relative proportion of the hybrid form, Aa, regularly diminishes over time, it never disappears. The initial act of crossing leaves a permanent, if increasingly minuscule, legacy.

2. *Generalizing the law over multiple characters.* Having generalized his results in one dimension, Mendel now—bringing the first half of his paper (corresponding to the first of his two 1865 lectures) to a close—generalizes it in another. What if we track more than one either/or character at a time? Not just seed shape in its two versions, round (dominant) and wrinkled (recessive), but also, say, seed color in its two versions, yellow (dominant) and green (recessive)? Now, Mendel shows, the algebraic form of his basic ratio comes into its own, because, on the supposition that the characters behave independently, we can represent the union of a round-and-yellow-seeded plant and a wrinkled-and-green-seeded plant mathematically by multiplying $A + 2Aa + a$ by $B + 2Bb + b$. The result is a nine-term series, $AB + Ab + aB + ab + 2ABb + 2aBb + 2AaB + 2Aab + 4AaBb$. According to Mendel, plants conforming to all the terms of the series were found in the predicted proportions. And yes, he adds, when you mix in a third character-version pair, it still works out, empirically and mathematically. No matter how many pairs are united in the hybrid, they behave as if independent of one another.[22]

3. *Explaining the law.* What, Mendel asks, must the pollen cells and egg cells of a *Pisum* hybrid be like in order for their unions to produce offspring in the "law for *Pisum*" kinds and proportions? An important clue, he tells us, is the appearance among the hybrid offspring of nonhybrid individuals—that is, the *A* plants and *a* plants, whose characters resemble those of the originating stock plants and are, like them, true-breeding. Since, Mendel reasons, an *A* plant can arise only from the union of an *A* gamete with another *A* gamete, and an *a* plant can arise only from the union of an *a* gamete with another *a* gamete, it seems to follow that the hybrid produces not *Aa* pollen but *A* and *a* pollen, and not *Aa* egg cells but *A* and *a* egg cells. In other words, the gametes of a hybrid plant are not—cannot be—hybrid, but must be pure for the parental-character versions. And since, on average, *A* plants and *a* plants figure equally among the hybrid offspring, *A* gametes and *a* gametes must likewise, on average, be produced in equal numbers.[23] To check that reasoning, Mendel describes undertaking a suite of separate breeding experiments, and in every case, his predictions proved correct. And if, he goes on, among the gametes of a hybrid plant, it is a matter of chance which pollen grain meets with which egg cell, then there are four equally probable combinations:

Pollen *A* with Egg *A* = constant plant *A*
Pollen *A* with Egg *a* = hybrid plant *Aa*
Pollen *a* with Egg *A* = hybrid plant *Aa*
Pollen *a* with Egg *a* = constant plant *a*

or, in line with the law for *Pisum*, $A + 2Aa + a$. Mendel is duly triumphant, declaring that the law here finds its "rationale and explanation."[24]

4. Extending the law. How far does this law hold beyond *Pisum*? Mendel now turns to some supplementary experiments that he did with another legume, the bean *Phaseolus*. One experiment, in which the species crossed were not very different from each other, produced broadly *Pisum*-ish results. Another, however, in which the species were very different, produced results that ranged from the easily assimilated to the seemingly unassimilable. Especially egregious were results to do with flower color, where Mendel found not purple flowers and white flowers in a ratio of 3 to 1, but flowers across a purply spectrum, from pale violet to purply red, plus the occasional white flower, in a ratio closer to 30 to 1. His ingenious solution is to treat the cross as a multi-character cross, on the hypothesis that purpleness is actually a compound of independent color components. Assume there are two component colors making up purpleness, and his mathematics gives, for every 15 individuals of varying purply colors, just 1 all-white individual. Assume there are three component colors, and the ratio jumps to 63 to 1. The bean flower results thus begin to make sense, as does more generally, Mendel stresses, "the extraordinary diversity in *the coloration of our ornamental flowers*."[25]

At the end, a paper that is already a marvel of agility and ambition gets even better. In his concluding section, Mendel enlarges his focus to reveal that he regards plant hybrids as falling into two kinds: ones where the hybrid form remains constant down the generations; and ones where, as in *Pisum*, the hybrid form does not remain constant but reappears in a variable way, along with parental forms. In constant hybrids, Mendel reckons, the cellular elements in egg and pollen are in such concord that their union remains permanent. By contrast, in variable hybrids, the cellular elements are in conflict, albeit in a contained way sufficient to permit the emergence of an organism (which thus represents a kind of temporary compromise). The moment of the production of gametes by hybrids of the variable kind is thus a moment of liberation for the contents-under-pressure elements, which, accordingly, distribute themselves equally across the egg and pollen cells.[26]

With the law for variable hybrids thus elucidated so comprehensively, Mendel moved next, in the mid-1860s, to the constant hybrids. He did not, however, get very far, in part because he encountered difficulties—his choice of experimental plant, hawkweed, proved inauspicious—and in part because, from 1868, he took over from Napp as abbot.[27] He did manage, however, to make time for reading Darwin's *Variation of Animals and Plants under Domestication*, published that year in the original English as well as in the German translation that Mendel read. A two-volume expansion of the *Origin*'s first chapter, Darwin's new book included a penultimate chapter setting out his hypothesis of pangenesis. Next to the passage where Darwin introduced the hypothesis, Mendel added an exclamation mark, admonishing at the bottom of the page against "indulg[ing] in an impression without reflection."[28]

3

Before we look in more detail at pangenesis, we could do worse than take Mendel's advice and reflect a little on a widely indulged impression: that what Mendel showed in the "Versuche" is that heredity is particulate, whereas Darwin, lumbered with the belief that heredity is blending, was forced into the brilliant blunder of pangenesis in order to save his theory of natural selection from a seemingly fatal problem.

That problem was laid bare, it is often said, in a June 1867 essay in the *North British Review* by the Scottish engineer Fleeming Jenkin. For Jenkin, a hitherto unidentified weakness in Darwin's theory was its assumption that the winners in the struggle for life pass their rare advantages of structure, habit, and so on to their offspring *undiluted*. To have offspring, of course, a winner needs to find a mate. But if, overwhelmingly, the mates available lack the winning advantages (for, again, these are rare), then, because of the effects of blending, any offspring will be only half as intelligent, or half as fast, or whatever. Another turn of the generational wheel yields grandchildren only a quarter as gifted. Within a few generations, the original advantages will be swamped into near-invisible negligibility. Thus does blending inheritance nullify natural selection. And thus Darwin's need of Mendel, for Mendel, in his hybrid pea experiments, discovered that the inheritance system is really particulate and so capable of transmitting advantageous variations intact. When, in the twentieth century, Mendelism and Darwinism were at last brought together, the result was a "modern synthesis" that still defines mainstream evolutionary biology. All that came too late for Darwin, however. He published what he had, which was the deeply Lamarckian pangenesis.[29]

This story itself belongs to the modern synthesis, above all to the work of the British mathematical geneticist Ronald Fisher. His foundational book, *The Genetical Theory of Natural Selection* (1930), began with a historical chapter titled "The Nature of Inheritance," all about the bane of blending inheritance for Darwin and the (sadly postponed) balm of particulate inheritance. That dichotomy of opposed theories, "blending inheritance" versus "particulate inheritance," comes from Fisher.[30] Although the terms can be found earlier, and Francis Galton used them, neither Darwin nor Mendel did. Unsurprisingly, their actual views fail to slot neatly into their assigned sides, whether we ask about visible characters or about the invisible somethings in cells that, somehow or other, bring about those characters.[31] Darwin was the least categorical of biological thinkers, and his general view, about inherited characters as about more or less everything, is that more than one sort of thing happens. At the level of visible characters, Darwin recognized that some offspring characters are intermediate—"blended"—between the parental characters, but some are not. Sex, for Darwin, was a ubiquitous instance of an inherited character that does not blend in the offspring.[32] From the earliest days of his theorizing on species, he collected examples of quirky habits transmitted in undiluted form to descendants, even when those descendants had never witnessed the ancestral quirk in action.[33] In *Variation*, Darwin reported a crossing experiment he had done with two snapdragon varieties, in which all the offspring took after just one of the parental forms—he describes its characters as "preponderant"—but in which, among the offspring of the offspring, the forms of both parents were represented intact.[34] In that book's pangenesis chapter, Darwin brought up the idea of invisible "gemmules" behind the inheritance of visible characters (and, as we shall see, much else). According to Darwin, preponderance comes about when the gemmules from the relevant part of the favored parent overwhelm the gemmules from that part of the other parent. Otherwise, if the gemmules are evenly matched, there is blending; and if some sort of antagonism keeps them apart, there is blotching, striping, or some other motley form. But the gemmules themselves do not blend.[35]

In the "Versuche," Mendel never denied that some characters are more/less rather than either/or (recall his treatment of flower color in his bean plants), or that sometimes hybrid characters are intermediate between the parental characters. He reports that, even in the garden pea, there are some characters, such as form and size of the leaves, where in the hybrid "the formation of intermediates is indeed almost always evident." If anything, his stance in the paper is that of someone drawing attention to nonblending hybrid characters in a world convinced that hybrid blending is an exceptionless rule.[36] And his decision to attend to these exceptional cases, he makes plain, flowed from his decision to try to get beyond Kölreuter, Gärtner, and other predecessors by being as fully quantitative as possible in tracking the fate of the hybrid form in plants subject to parental-form reversion. For those purposes, he needed to operate, he reckoned, with parental-character versions that were as cleanly differentiated from each other as possible, the better to inventory the frequency of their appearance in later generations. Thus his concentration on seed shape and color rather than leaf form and size. Indeed, it might have suited Mendel's purposes even better if seed shape, color, and the rest had turned out to have distinctive hybrid forms—think squareness and redness, or better still, robust, "you know it unmistakably when you see it" intermediacy—so that, in the first generation bred from the hybrids, he could have known just by looking which plants were hybrid and which constant for both parental-character versions. In the garden pea, which he chose to work with because of the low risk of contaminating pollen, the hybrid form for each of the seven characters he alighted upon was indistinguishable from the dominant-character version: a situation that demanded a lot of extra experimentation to sort the hybrid dominants from the constant dominants.[37]

And at the level of the invisible somethings? Remember that, at the end of the paper, Mendel distinguished the class of variable-progeny hybrids, like the garden pea, from the class of constant-progeny hybrids, like the hawkweeds he would experimentally hybridize next. For Mendel, in variable-progeny hybrids, the coming together of the invisible somethings is but a temporary, unstable arrangement. That is what makes them variable. But in the constant-progeny hybrids, the coming together is permanent, stable. One could say there is, at the cellular level, nonblending in the variable-progeny class and blending in the constant-progeny class. As for the material nature of what thus combines, whether temporarily or permanently, Mendel is studiously cagey. Note, however, that when he represents the union of the same sort of material from the two parents—of, let us say, *a* from the pollen parent and *a* from the seed parent—he does so with a single letter, *a*. By contrast, a post-1900 Mendelian would use *aa*: notation that makes straightforward sense to anyone who regards *A* as a bit of chromosome, with one bit carried on a paternal chromosome and one bit carried on a counterpart maternal chromosome. But chromosomes were unknown in Mendel's day, and it is anachronistic reading-backward to see proto-chromosomal notions in his paper. Indeed, there is no sign that Mendel thought of the character material as packaged up into countable, let alone atomic, entities of any kind. Perhaps, when he thought about the nature of the character material, he regarded it as some kind of fluid. Sometimes, after all, two fluids blend perfectly, like black and white paint: the constant-progeny class. And sometimes such mixing proves only temporary, as the two fluids are immiscible, like oil and water: the variable-progeny class.[38]

So Mendel did not show that inheritance is, in any meaningful sense, particulate. Nor did Darwin believe that inheritance was, across the board, blending. And to the extent that Darwin did acknowledge, like Mendel, that inheritance was sometimes (indeed, more often than not) blending at the visible-character level, Darwin was not nearly as vulnerable to Jenkin's swamping argument as later rumored. The apparent threat depends, on inspection, on there being no potential mates around who are even a little bit advantaged along the lines of the superior competitor; or, if those potential mates do exist, on there being no serious chance that the superior competitor will find them and mate with them. But Darwin could, and did, easily bat those assumptions away—though he nevertheless regarded the argument as pushing him to greater clarity in his own thinking on these matters. Yes, sometimes variant individuals are one-offs—"sports," in the language of the time. But equally, where conditions are such that one individual is born with a slight, inheritable advantage relative to the local norm, chances are that other individuals will be born varying likewise. (We will see in the next chapter that Galton, at around the time of Jenkin's review, began reconceiving species norms in general as distributions composed of slightly differing individuals.) As to whether the superior competitor might find similarly advantaged mates: for Darwin, the fierceness of the struggle for existence meant that advantaged individuals were the only ones who survive long enough to reproduce. In a Darwinian world, as Darwin understood it, the best are the only ones left for mating with.[39]

Even if Darwin had felt in need of rescue, however, it is doubtful that the "Versuche"—had he somehow got hold of it—would have struck him as helpful.[40] In the concluding section of his paper, Mendel presented his work as *explaining* swamping, indeed, as showing that individuals bearing no influence at all from a crossed-in parent could be expected within a single generation—far sooner than Kölreuter and Gärtner had supposed. In any case, Darwin knew all about the work of the great German hybridizers on swamping or "absorption," as he called it in *Variation*, and had even guessed that it could take place much more rapidly than reputed.[41] In other respects, too, Mendel's paper was not obviously calculated to excite or delight Darwin. As already mentioned, Darwin's views on the effects of cultivation on plants came in for severe and, as he would have seen it, misguided criticism from Mendel.[42] And while the offspring of Darwin's hybrid snapdragons nearly divided into a 3:1 ratio—he gave the numbers—he regarded the results as just another example of what he called "prepotency": the transmission of one parent's character far more strongly than the other parent's. It was something that happens some of the time, from causes so diverse, and with such diverse effects, that it was "not surprising that every one hitherto has been baffled in drawing up general rules on the subject."[43]

What is more, by 1866, Darwin was reading and corresponding with another experimental hybridizer of garden-pea varieties whose results looked nothing like Mendel's. In "Observations on the Variations Effected by Crossing in the Colour and Character of the Seed of Peas," delivered in May of that year in London at the week-long International Horticultural Exhibition and Botanical Congress, and subsequently published in the proceedings, the Stamford nurseryman Thomas Laxton reported the most untidy irregularity in the fate of hybrid seed color and shape. As Laxton's testimony would go on to be marshaled in the post-1900 debate over Mendel, it is worth quoting at length:

> The results of experiments in crossing the Pea tend to show that the colour of the immediate offspring or second generation sometimes follows that of the female parent, is sometimes intermediate between that and the male parent, and is sometimes distinct from both; and although at times it partakes of the colour of the male, it has not been ascertained by the experimenter ever to follow the exact colour of the male parent. In shape, the seed frequently has an intermediate character, but as often follows that of either parent. In the second generation, in a single pod, the result of a cross of Peas different in shape and colour, the seeds are sometimes all intermediate, sometimes represent either or both parents in shape or colour, and sometimes both colours and characters, with their intermediates, appear. The results also seem to show that the third generation, that is to say, seed produced from the second generation or the immediate offspring of a cross, frequently varies from its parents in a limited manner—usually in one direction only, but that the fourth generation produces numerous and wider variations; the seed often reverting partly to the colour and character of its ancestors of the first generation, partly partaking of the various intermediate colours and characters, and partly sporting quite away from any of its ancestry. These sports appear to become fixed and permanent in the next and succeeding generations; and the tendency to revert and sport thenceforth seems to become checked if not absolutely stopped.[44]

Note Laxton's own repeated reverting to the words "sometimes" and "partly" in describing his results. All sorts of things happened under his watch, leading to all sorts of resemblances between offspring peas and their varying ancestors, going back several generations.

4

Despite Thomas Huxley's less than encouraging remarks when he saw Darwin's first write-up on pangenesis in summer 1865, Darwin had pressed on. He had been well aware of the theory's deficiencies, owning up to the most glaring in the title of his 1865 manuscript, "Hypothesis of

Pangenesis." In mid-Victorian Britain, a hypothesis was a B-grade theory. It explained one or, ideally, several classes of facts, but—crucially—without offering compelling evidence for the real-world existence of the cause invoked in the explanation. By contrast, Darwin never labeled the theory of natural selection a hypothesis and reacted indignantly when anybody else did. Natural selection was A-grade, in his judgment. The *Origin*'s early chapters set out a richly evidenced case for believing that natural selection is real, and powerful enough to produce not just new and better-adapted varieties but ultimately new species. Only at that point did the book turn to the large and diverse classes of facts—geological, biogeographical, and so on—that, as Darwin showed, natural selection explained, and in a unifying way.[45] Alas, with pangenesis, as Darwin understood, the A-grade route was closed off, since nobody had ever seen a gemmule, much less established its capacity to do what Darwin conjectured it could do. The only reason to believe Darwinian gemmules existed was that, on the assumption of their existence, so much could be explained so economically.

Might that, however, be enough of a reason? Darwin knew that the wave theory of light, although a hypothesis, explained so much, and so well, that the existence of light waves had been accepted as something more than merely hypothetical. Perhaps gemmules, too, would ultimately enjoy that fate. In any case, he struggled to see how pangenesis, like natural selection, could be wrong and yet so powerful explanatorily.[46] Compounding his difficulties was the fact that, by the mid-1860s, he had lived with pangenesis for a very long time—at least since the early 1840s, when his reading of the great German physiologist Johannes Müller's *Elements of Physiology*, recently published in an English translation, had triggered new theorizing. Müller was the first authority cited in the 1865 manuscript, on how the difference between a fertilized ovum and a plant bud, though they belong respectively to the domains of sexual and asexual reproduction, was not an essential difference. Pangenesis would unite not merely sexual and asexual reproduction, but the healing of wounds, the regrowth of amputated limbs, the seeming independence of body parts at all levels, the various forms of development, and much else—inheritance included—as so many instances of budding forth, ultimately due to the micro-buddings from each part of the body of gemmules capable in their turn of reproducing that same part. But even when pangenesis was new for Darwin, it drew on ideas, insights, and themes that were as old as anything in his scientific work, going back to his days as Robert Grant's student in natural history at Edinburgh University in the mid-1820s. The term "gemmules" traces to Grant, who invoked microscopic ones to explain reproduction in sponges.[47]

Huxley's criticisms, whatever they were (the letter is lost), never stood a chance against a theory that, for decades, thus represented Darwin's best shot at explaining the nature of organisms.[48] And there were other sources of encouragement. Thomas Laxton first came to Darwin's attention with a note in the *Gardeners' Chronicle* of September 1866 on a pangenesis-friendly finding from his experimental crosses of pea varieties. After fertilizing a non-purple variety with pollen from a purple variety, Laxton observed that not only did the offspring show the purply influence of the pollen parent, but so did the seed parent, which, well before seeds formed, grew purply pods and purply other parts.[49] If anything like that had ever happened in Mendel's garden, he made no mention of it.[50] But Darwin expected it, seeing in it a vindication of his general view that all parts of organisms are generatively active, not just the reproductive organs (hence "pangenesis"). "I wrote to [Laxton] & he sent me the specimens & they are wonderful & not to be mistaken," crowed Darwin in a letter that October to Joseph Hooker. "This is a grand physiological fact & delightful for my pangenesis." Laxton's peas, classified as an example of what Darwin called "the direct action of the male element on the female," accordingly found their place in the monumental *Variation*, where they were treated expansively in the body of the book and more summarily in the much-expanded version of his pangenesis manuscript, with which Darwin brought the book to a climax.[51]

Perhaps in deference to Huxley, that final chapter of *Variation* bore an even more self-deprecating title than the 1865 manuscript: "Provisional Hypothesis of Pangenesis." It is comparable to Mendel's pea-hybrid paper in size and even in shape, having two halves, the first establishing the facts to be explained, the second doing the explaining and then some. But where Mendel's facts concern, in the main, his character-tracking crossing experiments with a single species of plant, Darwin's facts concern more or less everything to do with how all plants and animals, in all their diversity, come into being, change over their lifetimes, and reproduce. As he passes from one class of facts to another—the various kinds of reproduction, regrowth of amputated parts, graft hybrids (plant hybrids created by the union of bodily tissues, bypassing the reproductive organs), the direct action of the male element on the female, development, "the functional independence of the elements or units of the body" (such that separated parts can sometimes keep growing), variability, inheritance, and reversion (the reappearance of characters, whether vanished for one generation or ten thousand)—he not only sets out what is known but, often, seeks some overall pattern that will sharpen or simplify the explanatory job ahead. So, in discussing development, he notes that, across all the immense diversity of the changes that maturing individuals undergo, whether slowly and smoothly (humans) or abruptly and in stages (insects), and whether a given structure arises via transformation of an old structure or de novo, what is remarkable is the *independence* of grown structures from what precedes and what follows them developmentally. And this insight, Darwin goes on, will turn out to be crucial for explaining an important law of inheritance: that these modifications reappear in the offspring at a corresponding age.[52]

About inheritance: in Darwin's classes-of-facts list, it is but one of nine, and is in no way picked out as especially important. Indeed, for Darwin, understanding inheritance depends precisely on not treating it as a fundamentally distinctive problem requiring its own special science, but rather on appreciating its connectedness with reproduction, regrowth, and the rest. Reversion, for example, he treats as one of several possible causes of the non-inheritance of a character (non-inheritance, again, being anomalous for Darwin, with inheritance the rule). But inheritance does present its puzzles. Besides the reappearance of a character in the offspring at the age, as well as in the season, in the sex, and so on, in which it first appeared, Darwin dwells on the inherited effects of the use and disuse of particular organs and limbs, along with the inheritance of associated habits. Darwin not only accepted Lamarckian inheritance as among the facts to be explained, but in *Variation* reported extensive—and, for Darwin, unusually quantitative—supporting studies of his own, comparing the bones of domesticated versus wild ducks, rabbits and other animals. As domesticated ducks flew less and walked more, one would expect them over time, or so Darwin reckoned, to have evolved smaller, weaker wing bones and larger, stronger leg bones; and so it proved. "Nothing in the whole circuit of physiology is more wonderful. How can the use or disuse of a particular limb or of the brain affect a small aggregate of reproductive cells, seated in a distant part of the body, in such a manner that the being developed from these cells inherits the characters of either one or both parents?" And a puzzle within the puzzle, related to Yarrell's law, was why it sometimes took several generations of exposure to the new conditions before the adaptively modified habits and limbs became permanently hereditary.[53]

The answer, of course, to this question and all the others Darwin poses to himself in the first half of the chapter, is gemmules: minuscule granules thrown off by all units of the body, at every stage of development (though not necessarily incessantly). According to Darwin's hypothesis, set out at the start of the second half and then put to explanatory work in the remainder, each gemmule has the ability to develop into the part whence it came. Consider his Lamarckian ducks. As the parents' legs get stronger, the cells in their leg bones cast off modified gemmules that are capable collectively, in the body of a son or daughter, of reproducing those stronger leg bones. Those modified gemmules get into the next generation by undergoing self-division, so that in time the parent ducks' systems are swarming with them, as they are with all the gemmules from every other unit. By mutual affinity, gemmules from all the units will aggregate in the male's sperm and the female's ova. After mating, mutual affinity will again ensure, other things being equal, that as embryonic structures emerge, the right gemmules will unite with the right cells at the right times, producing the next stage of structures, which will in turn, thanks to mutual affinity, be united with the right gemmules, and so on. Whether or not the developing duckling grows up to have its parents' strong legs depends largely on whether enough of the parents' modified leg-bone gemmules were transmitted to the fertilized ovum to compete with all of the unmodified gemmules that the parents inherited and, indeed, generated themselves, up to the point where their legs grew stronger. That is why, even after the conditions of domestication have altered duck habits and, in consequence, duck limbs, the modifications can take a while to become inherited, since it can take a while for all the unmodified gemmules to become sufficiently reduced in number. As to modified characters appearing at a corresponding age in the offspring, that happens because during development, the modified gemmules have an affinity for nascent cells that appear only at the preceding age.[54]

Armed with pangenesis, Darwin marches again through his classes of facts, showing one by one how fully his assumptions—not independently verifiable, but not wildly improbable either, given what was known about, among other things, the sizes and powers of disease germs (an example he gives)—can explain them.[55] Reproduction by budding? Regrowth? Graft hybrids? Laxton's purply pods? All are straightforward entailments of pangenesis, according to which all the tissues are reproductively active. The whole is a virtuoso performance that reaches a crescendo with the final class, reversion. The key to reversion at all scales, from the return of the grandfather's character after skipping a generation to the return of now-monstrous characters after geological epochs, is to see that transmission and development are distinct powers, though often conjoined. Just as seeds can remain in the soil for years, so, according to Darwin, can gemmule lineages maintain themselves for untold generations without conditions being ripe for their development. And just as the proportion of modified to unmodified leg-bone gemmules can explain the development or otherwise of stronger leg bones in descendant ducks, comparable points about gemmule proportions can explain why, with hybrids, after a hybrid generation showing an intermediate character, we often find that, as Darwin reports, "in the next generation the offspring generally revert to one or both of their grandparents, and occasionally to more remote ancestors." Under pangenesis, the cells of the intermediate-character hybrids throw off gemmules that Darwin calls, somewhat misleadingly, "hybridised gemmules." But these hybrids have also inherited lots of parental-character gemmules, which he calls "pure gemmules." So, when two hybrids are crossed, there are several options, depending on the particular mix of gemmules in the fertilized germ, with results more Laxtonian than Mendelian. If the union brings together enough of the pure gemmules from one or the other parent, complete reversion to one or the other of the parental characters will follow. If it brings together pure gemmules with hybridized gemmules, the result will be partial reversion. And if it brings together enough hybridized gemmules, the result will be the intermediate character of the hybrid.[56]

As with the *Origin*, Darwin closed on an unexpectedly lyrical and cosmic note: "Each living creature must be looked at as a microcosm—a little universe, formed of a host of self-propagating organisms, inconceivably minute and as numerous as the stars in heaven."[57]

5

So, who needed a science of heredity? Not Mendel. His "law valid for *Pisum*" had nothing to do with the inheritance of acquired characters, the reappearance of characters at corresponding ages in offspring, Yarrell's law, or the myriad other phenomena that Darwin thought about when he thought about inheritance. The apparent exception was reversion: the return of characters that had vanished from the offspring. But Mendel's sole concern when it came to reversion was with whatever character version or versions went missing from the pair or pairs that he tracked in his hybrids—the "recessive" ones. He never asked about reversion in a general way, as encompassing a wider and possibly longer-range set of possibilities, in and out of hybrid lineages. As a reader of the *Origin*, if for no other reason, Mendel would have been familiar with the return, not just of grandparental characters, but of long-lost ancestral characters, as when a black domesticated pigeon and a white domesticated pigeon produced a blue offspring—the color of the rock dove, the species ancestral to all the domesticated pigeons, Darwin reckoned.[58] Indeed, Darwin's personal pea-hybrid expert, Laxton, reported finding similar out-of-the-frame "sporting away" in his experimental lineages. Any sporty peas born into Mendel's lineages would have been thrown away as aberrant (perhaps because of pea-beetle infection) and therefore, *given his purpose*, uninstructive.[59] And that purpose, true to his Ungerian training, was to uncover the mathematically expressed law, explained by cell physiology, governing the fate of hybrid forms in plants in which the hybrid form does not remain constant. This point is worth stressing in a book that will dwell sympathetically among the critics of Mendelism. Not only was Mendel no Mendelian, with no notion of genes *avant la lettre*, but he was not seeking the laws of inheritance. All of that, as we shall see, was read back into Mendel's paper well after his death in 1884. What little Mendel required to be true of inheritance per se—that inheritable characters got transmitted—he took for granted.

And Darwin? He no more saw himself as launching a science of heredity with pangenesis than Mendel saw himself as doing so with his pea hybrids, though for very different reasons. Inheritance was one of Darwin's abiding and explicit concerns, from his earliest private notebook theorizing through his final years. One of his last short papers, published in *Nature* in 1881, the year before he died, was titled "Inheritance," and related new, confirming facts from his correspondents about inheritance at corresponding ages and the inheritance of mutilations.[60] But for Darwin, the facts of inheritance, diverse as they were, could be made sense of only when placed within a much wider grouping of facts about bodily change, from reproduction to reversion; the beauty of his pangenesis hypothesis was that it revealed so much to be the upshot of so little. Reversion in pigeons? When the gemmules making for blackness and whiteness met in the offspring, their mutual antagonism made room for the appearance of the effects of the dormant blueness gemmules. The inheritance of mutilated parts, which seemed to happen only if, after mutilation, the part became diseased? In those cases, the gemmules from the pre-mutilation part got used up in the course of healing the wound, and were thus too reduced in number to overwhelm, in the offspring, the gemmules cast off from the mutilated part.[61]

"Inheritance," wrote Darwin in the conclusion of the pangenesis chapter in *Variation*, "must be looked at as merely a form of growth, like the self-division of a lowly-organised unicellular plant." He did not mean that transmission and development were the same thing, or indissolubly connected; reversion showed plainly, in his view, that they were not. What he meant was that offspring being like their parents—"like engend'ring like"—no more needed special explanation than a budded-off microparticle from a protozoan needed special explanation for being like the bit of protozoan it budded from, because organism-level inheritance was ultimately the product of such micro-buddings. Or so Darwin hypothesized; and hypotheses, as he quoted the great Cambridge historian-philosopher of the sciences William Whewell as saying, "may often be of service to science, when they involve a certain portion of incompleteness, and even of error."[62]

2

The Meaning of the Quincunx

I don't believe anybody would have appreciated your work more than Mendel himself had he been alive. Dear old man; my heart always warms at the thought of him, so painstaking, so unappreciated, so scientifically solitary in his monastery. And his face is so nice.

FRANCIS GALTON TO W. F. R. WELDON, 31 MARCH 1905

If a future Auguste Comte arises who makes a calendar in which the days are devoted to the memory of those who have been the beneficent intellects of mankind, I feel sure that this day, the 1st of July, will not be the least brilliant.

FRANCIS GALTON, 1 JULY 1908, COMMEMORATING THE FIFTIETH ANNIVERSARY OF THE READING OF THE DARWIN-WALLACE PAPERS AT THE LINNEAN SOCIETY

Among the objects held in Leeds University's history of science museum is a mathematical demonstration device. Small enough to fit into a briefcase, the device has a wooden frame, a plate-glass front, a peg-studded interior, and chambers along the bottom, filled with hundreds of little pellets. To use it, you first turn it upside down and then, with some angling and shaking, position the pellets in a layer. Turn the device right side up, and the pellets array themselves on an upper shelf, around a central stopper. Pull the stopper, and the cascade that results, though brief, is mesmerizing. As the bouncing pellets come to rest, they pile up in the shape of a bell, every time. It seems like magic: order out of chaos. But it is just billiard-ball physics. No wonder the quincunx, as the device is known, is a classroom and museum staple the world over. There is no better way of showing that when chanciness, or something approximating to it, reigns over a large number of independent events, outcomes get distributed according to the famous bell-shaped or normal curve.[1]

The quincunx's other names include the "bean machine" and the "Galton board," after its inventor, the English polymath Francis Galton. He had the first one made in 1873 for use the next year in a public lecture on heredity. A science of heredity, Galton thought, was something that everyone needed—and urgently. He made his debut on the subject with a two-part article, "Hereditary Talent and Character," in *Macmillan's Magazine* in summer 1865, a few months after Mendel's lectures on pea hybrids and around the time that Darwin and Huxley had their exchange on pangenesis. Galton depicted the statistical study of the laws of inheritance as the key to staving off civilization's slide back into barbarism. Understand those laws, Galton believed, and we will understand how to breed into being the improved version of ourselves that we require so desperately. By the time of his 1889 masterpiece, *Natural Inheritance*, Galton had coined a name for that improvement scheme: "eugenics."[2]

This chapter charts the emergence and expansion of Galton's science of heredity between 1865 and 1889, when it began to capture the imaginations of others, W. F. R. Weldon among them. It is not an easy science to get to know, in part because some of it was not easy science; in part because there grew to be so much of it, including not just new data, explanations, and experiments but new kinds of data, new kinds of explanations, and new kinds of experiments; and in part because ideas about what it all means became fixed long ago. Most prominent among these ideas is the identification of Galton with the doctrine that heredity is destiny, imperturbable and inescapable—"hereditarianism," to give it a name. An encyclopedia entry titled "Heredity and Hereditarianism" declares that Galton "may be considered the father of modern hereditarianism."[3] Who but a hereditarian, after all, could have invented eugenics? Where Mendel is our culture's hero on heredity, Galton is our zero.

But there was once another reading of Galton—and it was the one that mattered for Weldon, inspiring him, during the debate over Mendelism, to develop what amounted to an immensely sophisticated critique of hereditarianism. The quincunx is its enduring emblem.

1

In coming to grips with Galton's work on heredity up to *Natural Inheritance*, it helps to consider the years 1872 to 1874, with the quincunx's beginnings in the middle, as marking a sort of break. On the one side of this break, from 1865 to 1871, we are dealing with a science whose agenda was largely set by Galton's revered older cousin Darwin, and where the big issues concerned selection, natural and artificial, and then pangenesis. On the other side, from 1875 to 1889, we are dealing with a closely related but nevertheless rather different science whose agenda was largely set by Galton himself, and where the big issues concerned the dynamic stability of populations, together with allied statistical and evolutionary concepts. In between, from 1872 to 1874, we find Galton engaged in the clarification of what he thought about heredity, under the double stimulus of a disagreement with Darwin about pangenesis and a dawning awareness that he had come to be seen, unfairly he thought, as dogmatically hereditarian. (The word "hereditarian" first appeared in print in 1873 in an article criticizing Galton and others for exaggerating heredity's influence.)[4] Needless to say, the innovations of 1865–71 and 1872–74 did not fossilize in those years, but continued to evolve alongside the innovations of 1875–89, eventually making up a body of writings susceptible to alternative readings and selective appropriation.

Before proceeding, we should dwell a little on Galton's wanting to devote himself to the science of heredity at all. That ambition was, again, uncommon in the 1860s, even among investigators whose questions, like Darwin's and Mendel's, intersected with inheritance. And Galton came to heredity not from natural history or some other biological science, where inheritance in some form or other might be expected to come up, but from geography. He had acquired esteem and even some fame as an African explorer with a flair for the quantitative. It was his geographical contributions that got him elected a fellow of the Royal Society of London in 1860, at the age of thirty-eight. (He came to savor the coincidence of sharing his birth year, 1822, with Mendel.)[5] As late as 1863 he seemed to be heading, with some success, in the direction of meteorology.

What, then, accounts for his turn toward heredity—and not just as something to study, but as something to build a new science around? Galton (fig. 2.1) gave two different, though not conflicting, answers. Variations on a third, rather disobliging answer have been circulating since his lifetime. None is the whole story, but each belongs in it.

There was, first of all, his interest in racial difference. Race had been on a growing number of European scientific minds since the mid-eighteenth century, when imperial venturing—bound up, in Britain and elsewhere, with the slave trade—began to sharpen questions about the nature, origins, and control of human diversity. Galton's African writings from the 1850s are full of descriptions of the characters of native tribes. Not until 1864, however, did he begin serious research on the subject—what he later characterized as "a purely ethnological inquiry, into the mental peculiarities of different races." At the time, race was seen (as it still is) as belonging to anthropology, and Galton had begun attending meetings of the London-based Ethnological Society, recently reinvigorated with the arrival of Huxley and other Darwinians. Ethnology gave way to heredity, however, as it became clear to Galton that each race was made up of distinctive family lines.[6] His race-level interests nevertheless survived in his subsequent work. In "Hereditary Talent and Character," his discussion of character relies heavily on cringe-inducing racial examples: "The Red man has great patience, great reticence, great dignity,

FIG. 2.1. Francis Galton in 1864, when he wrote "Hereditary Talent and Character."

and no passion; the Negro has strong impulsive passions, and neither patience, reticence, nor dignity." In *Hereditary Genius*, a follow-up book published in 1869, a chapter titled "The Comparative Worth of Different Races" upholds the conventional Victorian racial hierarchy, from the ancient Athenians at the top to the aboriginal Australians at the bottom, albeit while emphasizing the range of hereditary abilities to be found within each of the races.[7] When Galton took up composite photography in the 1870s and '80s, racial types were, as we shall see, among the images he tried to capture.

The second answer was Galton's Darwinism or, more precisely, his Darwinian iconoclasm. To the end of his long life, Galton testified to the huge, liberating impact on him of the *Origin*. In 1908, the year of the anniversary tribute quoted at the start of this chapter, the elderly Galton published an autobiography, in which he recalled that the book's "effect was to demolish a multitude of dogmatic barriers by a single stroke, and to arouse a spirit of rebellion against all ancient authorities":

> I was encouraged by the new views to pursue many inquiries which had long interested me, and which clustered round the central topics of Heredity and the possible improvement of the Human Race. The current views on Heredity were at that time so vague and contradictory that it is difficult to summarise them briefly. Speaking generally, most authors agreed that all bodily and some mental qualities were inherited by brutes, but they refused to believe the same of man. Moreover, theologians made a sharp distinction between the body and mind of man, on purely dogmatic grounds.[8]

The very first sentence of "Hereditary Talent and Character" is a paraphrase of the *Origin*'s first chapter: "The power of man over animal life, in producing whatever varieties of form he pleases, is enormously great." From there, Galton heads straight for those smashed barriers, arguing that this power extends as much over the mental as over the physical qualities of animals and, moreover, that the mental qualities of humans are no less inheritable than those of animals. Hence humans, too, can be bred for better minds as well as bodies—and Galton even imagines the speech to be given at the ceremony where, after the selection process, the best young men in their year are married off to the best young women.[9] In *Hereditary Genius*, after a discussion of the statistical law governing human stature, we learn that what is true statistically for stature surely "will be true as regards every other physical feature—as circumference of head, size of brain, weight of grey matter, number of brain fibres, &c.; and thence, by a step on which no physiologist will hesitate, as regards mental capacity."[10] When, in December 1869, Darwin sent Galton a kind note about *Hereditary Genius*, Galton wrote back, "I always think of you in the same way as converts from barbarism think of the teacher who first relieved them from the intolerable burden of their superstition."[11]

The third answer was Galton's pride in his own pedigree—pride laced with anxiety. In 1889, the Scottish biologist Patrick Geddes initiated a still-living tradition of speculation concerning Galton's membership in an eminent family. To Geddes, *Hereditary Genius* exuded "the conscious pride of an intellectual patrician, himself sprung of the mighty races of Darwin and Wedgwood."[12] Later historians have made the same insight the basis for an ideological interpretation of Galton's science, construing it as an attempt to naturalize the inequalities from which his class benefited.[13] Others have noted Galton's many failures up to the mid-1860s—his medical studies abandoned midway through; his mathematical studies at Cambridge ending in poor exam results; his years of rich-boy inactivity in his twenties and thirties; a marriage but no (and never any) children; a long-standing and unfulfilled sense of noblesse oblige toward a world he judged as troubled; a place at the Royal Society but none in the Darwinian circle he craved to join—and have seen in his efforts with heredity a mixture of self-soothing and self-advancement.[14] Certainly Galton reckoned he had done rather well when it came to hereditary qualities. From the fortune-making Galton (not, contra Geddes, Wedgwood) side he traced his statistical bent, and from the Darwin side, his penchant for noticing more than others did and then generalizing expansively on what he noticed. It was that heritage, Galton guessed, that predisposed him to absorb the perspective of the *Origin* so fully and swiftly.[15] As for patrician lineages generally, he was, however, very far from venerating them, for reasons we will come to shortly.

A final word on "heredity": The passage above from Galton's autobiography continues by noting that when he began his post-geographical work, its novelty was such that "even the word heredity was then considered fanciful and unusual. I was chaffed by a cultured friend for adopting it from the French." Another, apparently less cultured friend, Herbert Spencer, used "heredity" in the titles of two chapters of his book *The Principles of Biology*, published in 1864. "Heredity" does not actually appear in "Hereditary Talent and Character," although, looking back on it in *Hereditary Genius*, Galton observed that "the arguments I then used have been since accepted, to my great gratification, by many of the highest authorities on heredity." From that point onward, Galton used, as we do, a mixed vocabulary. In *Natural Inheritance*, for example, on the first page of its first chapter, we find "the science of heredity," along with "inheritance" and "hereditary."[16]

2

Darwin may have set the agenda between 1865 and 1871, but the signature moves were Galton's. At the core of "Hereditary Talent and Character" is Galton's analysis of published lists of (mostly British) men who, over time spans stretching from decades to centuries, had made it to the very tops of their intellectually demanding fields, as indicated either by the general esteem that got them onto lists of greats in the first place, or by their winning whatever highest prizes their fields offered a given cohort of competitors—the Chancellorship for lawyers, Senior Classic for classics students at Cambridge, and so on. What the remarkable proportion of fathers, sons, and brothers on such lists revealed, Galton claimed, was that the capacity for eminence ran in families. He allowed that this capacity depended on more than intellect; the right sort of character was also needed. But character, too, he urged, was inherited, no less than the physical features whose hereditary transmission no one doubted. And, he went on, just as animal breeders take advantage of inheritance to improve their stock by selectively mating the best males with the best females, generation after generation, so now, armed with an understanding of the inherited nature of intellect and character, humans could begin exercising far greater quality control over their descendants, to the benefit of all humankind. Furthermore, there was, in his view, no time to lose, given both the growing demands of a complex civilization and the quality decline that civilized life, by sheltering weaker individuals from natural selection, had set in motion in the human stock. To Galton, this prospect betokened something far grander than mere damage limitation. Just imagine, he wrote, the "galaxy of genius" that could be created if only humans attended to their own breeding with even a fraction of the attention they give to animal breeding. "Men and women of the present day," he prophesied, "are, to those we might hope to bring into existence, what the pariah dogs of the streets of an Eastern town are to our own highly-bred varieties."[17]

We saw in the last chapter that, under various guises, the regular diminishing of an individual's contribution to a lineage was a common theme for both Mendel and Darwin, as indeed for their predecessors (Kölreuter and Gärtner) and contemporaries (Jenkin). Yet it was Galton who not only put his permanent stamp on the idea but, with his first statement of it in "Hereditary Talent and Character," spelled out its subversive sociopolitical implications:

> The share that a man retains in the constitution of his remote descendants is inconceivably small. The father transmits, on an average, one-

> half of his nature, the grandfather one-fourth, the great-grandfather one-eighth; the share decreasing step by step, in a geometrical ratio, with great rapidity. Thus the man who claims descent from a Norman baron, who accompanied William the Conqueror twenty-six generations ago, has so minute a share of that baron's influence in his constitution, that, if he weighs fourteen stone, the part of him which may be ascribed to the baron (supposing, of course, there have been no additional lines of relationship) is only one-fiftieth of a grain in weight—an amount ludicrously disproportioned to the value popularly ascribed to ancient descent.

What eventually became known as Galton's law of ancestral heredity would go on to play a large—though in many ways largely symbolic—role in the Mendelism debate.[18]

Galton hinted at having examined vastly more biographical material than he could include in "Hereditary Talent and Character." With *Hereditary Genius*, he made that material available, and then some. The largest part of the book consisted of (sometimes lengthy) chapters devoted to particular domains and the men who had achieved eminence there: the Chancellors and Senior Classics, but also statesmen, military commanders, men of letters and science, musicians and painters, Evangelical Christians, even—just to check that physical prowess was no different from mental prowess—oarsmen and wrestlers from the north of England. Over and over again, he found the same pattern: to a much greater extent than ordinary people, eminent men have near relatives who are themselves eminent men. And this pattern held, Galton again concluded, because outstanding mental or physical talent is hereditary, along with the capacity for hard work and other eminence-making character traits. By his lights, children born with these innate gifts will rise to become the leading figures in their fields, provided, at any rate, that those fields are fully open to talent (as, he maintained, the ones he selected for study were—hence his leaving out parliamentary politics). At the same time, children lacking those gifts will not so rise, no matter how much they push themselves, or are pushed by others, or are given the sorts of opportunities that, Galton acknowledged, the having of eminent near relatives tends to bring. So, in Galton's view, the son of an esteemed novelist, say, will no doubt have a clearer path to first publication than the son of a working-class father. But if, from lack of talent or from laziness, the books of the famous author's son are not very good, few people will buy them, and his reputation will never become what his father's was. By the same token, for all the obstructions in his way, the working-class boy, if born with literary gifts, and—in common with the great writers of the past—compelled from within to write, and to push himself to improve his writing, will eventually publish the books that will make his reputation. Innate talent and character will out; so, too, their absence.[19]

In "Hereditary Talent and Character," Galton had described the kinds of data he was examining as "statistics," and the facts disclosed through his analysis of those data as "statistical."[20] Such usage was familiar enough at that time, as it is today, and as it had been increasingly since the rise in the eighteenth century of the first modern states, whose rational management demanded more and better numbers on everyone and everything, from steamships to suicides (thus the "stat-" in "statistics"). Nowadays, of course, the word has a second, related but distinct meaning, as the name for the specialized mathematical concepts, techniques, and reasoning used to draw conclusions from such data. This is what science, medical, and business students the world over study when they study, often groaningly, statistics. Beginning with the early chapters of *Hereditary Genius*, Galton's writings played no small role in the development of what is today a ubiquitous branch of mathematics.[21] Before setting out his biographical data, he took pains to clarify that by "eminence" he meant the level of achievement that got someone an obituary in the *Times*—on his calculation, something managed by just 250 men in a million. Next, and going well beyond what he had done in his 1865 articles, he raised the question of how ability is distributed overall, up and down the ability range. One might guess, he went on, that if English men could be grouped into ability classes, with each class separated from the next by the same number of degrees, then each class would be found to hold the same number of men—in other words, that ability is equally distributed. But that is not remotely the case, wrote Galton. On the contrary, the distribution is most uneven. The extreme classes, marking out the very lowest and the very highest abilities, are thinly occupied. The rest gradually get more crowded as they near the middle of the range, with maximum occupancy right at the middle. Average ability, in other words, is the lot of the largest number of people.[22]

And that is exactly the distribution Galton discovered, as he showed, in the scores attained in entry exams at the Sandhurst military academy. It was consistent with eminence being as rare as it seems to be (a consideration providing further support, in his view, for his use of reputation as a stand-in for ability). And it conformed strikingly with what he called "the very curious theoretical law of 'deviation from an average,'" proposed by the Belgian astronomer Adolphe Quetelet. As Galton explained, astronomers had known for decades that, due to unavoidable human error, repeated observations of the position of a stationary object in the night sky tended to deliver not the same value again and again, but different values. These values, however, showed a characteristic clustering around an average, which, once identified, could safely be taken as the correct value—just as, if one were to examine the bullet holes on a target, the holes would cluster nearest to where the shooter was aiming. (Hence one of the law's other common names: the "law of error." The name "normal curve," for the bell-shaped graphical representation of data conforming to the law, would eventually displace "error curve.") It was Quetelet's brilliant idea to see whether the same pattern might be found in measurements of human populations, as though, with each individual human, Nature were taking her best shot to hit the bull's-eye of the balanced middle. He used data from an Edinburgh medical journal on the chest circumferences of Scottish soldiers. And there it was: the symmetrically mounting cluster. Now Galton presented his work on hereditary genius as extending Quetelet's finding from the physical to the mental.[23]

3

The law of deviation from an average was one of two major theoretical borrowings in *Hereditary Genius*. The other was from Darwin: his new, though, as we have seen, long-contemplated, hypothesis of pangenesis. In the book where it had appeared, *The Variation of Animals and Plants under Domestication* (1868), Darwin had praised "Hereditary Talent and Character" for having put paid to doubts about the hereditary nature of the mental qualities underpinning eminence. Galton returned the compliment in his new book by concluding it with a glowing endorsement, and mathematical elaboration, of pangenesis.[24] In Galton's view, the deviation from an average distribution of hereditary genius in humans was yet another fact about inheritance that could be counted in support of Darwin's hypothesis, since that distribution, Galton wrote, "is identical with the special case in which only two forms of gemmules had to be considered [basically, eminence-making and obscurity-making gemmules], and in which they existed in equal numbers in both parents."[25]

No less exciting to Galton were the new, potentially answerable questions the hypothesis suggested. For example, Darwin thought that individuals of a species differ from one another in part because some portion of the gemmules transmitted to each of them gets modified in ways peculiar to each individual. Mathematically, it was, Galton showed, straightforward to derive a general formula for calculating the proportion of gemmules that pass unchanged into an individual from any previous generation. But for the formula to become predictive for a given species—and therefore useful for control of its inheritance—it was necessary to find out, by observing, how many gemmules typically undergo modification in the course of reproduction in that species. Similarly, it should be possible through the study of, in Galton's words, "the patent characteristics of many previous generations" to identify the nature of dormant or "latent" gemmules in a species, again to the benefit of predictive power. Far from the reappearance of long-vanished characters being a perpetual surprise, as at present, "the tendency to reversion into any ancient form ought also to admit of being calculated."[26]

Readers of *Hereditary Genius* responded with the usual mix of bouquets and brickbats, although reviewers for scientific journals tended to find the book rather more bouquet-worthy than did reviewers for religious journals.[27] Among scientific admirers, none was more effusive in his praise than cousin Charles. His reaction is famous and, even if one makes allowances for family friendliness, seemingly genuine: "I do not think I ever in all my life read anything more interesting & original," he wrote in a letter to Galton just before Christmas 1869, explaining that, though only up to the beginning of the biographical part, he just had to "exhale" his enthusiasm, "else something will go wrong in my inside"—though his son George had assured him the later chapters were even more interesting. "You have," Darwin added, "made a convert of an opponent in one sense, for I have always maintained that, excepting fools, men did not differ much in intellect, only in zeal & hard work; & I still think this is an eminently important difference." The extent of Darwin's enthusiasm for the book and for Galton's earlier articles—including his reflections on the baleful effects of the less talented out-reproducing the more talented, and the need to counteract those effects through better management of human breeding—would shine through in the numerous positive citations in Darwin's *The Descent of Man, and Selection in Relation to Sex*, published in February 1871.[28]

In the meantime came a collaboration by correspondence on the pangenesis hypothesis, though in a direction that no reader of Galton's work up to then would have anticipated, and with an outcome, as reported by Galton at the Royal Society in March 1871, unwelcome to both men. With help from staff in the Zoological Gardens in London, and with—he thought—Darwin's blessing, Galton had conducted blood transfusion experiments on rabbits with the aim of testing whether, as the hypothesis seemed to him to predict, blood taken from "mongrel" rabbits and introduced into purebred ones would, via the gemmules carried in the alien blood, influence the character of the offspring when the purebred rabbits mated. He tried several transfusion methods, but none produced any effect, suggesting that the hypothesis was false. His announcement prompted a defensive letter from Darwin to the letters column of *Nature* (which had already become, and would long remain, the preferred print forum for British scientific controversies). For all Darwin's encouragements before the results were known, he now maintained that Galton's experiments were no fair trial of his hypothesis, since he had never specified that the gemmules circulate through the blood. Galton—appalled and aggrieved in equal measure—immediately and publicly accepted the rebuke. He nevertheless carried on experimenting, though the results remained the same.[29]

His ties to Darwin's version of pangenesis now decisively loosened, Galton began formulating his own version. Rather than circulate, Galton's particles would, in germinal terms, stay put, shaping bodies without being shaped by them, in line with doubts about Lamarckian inheritance he had expressed back in "Hereditary Talent and Character." There indeed we find a remarkable vision of parents as little more than temporary containers for the hereditary material coursing through their lineage and ultimately—for, Galton held, even the noblest lineages were massively "mongrel" if traced back even a few generations—the race as a whole. "We shall therefore take an approximately correct view of the origin of our life," he had written,

> if we consider our own embryos to have sprung immediately from those embryos whence our parents were developed, and these from the embryos of *their* parents, and so on for ever. We should in this way look on the nature of mankind, and perhaps on that of the whole animated creation, as one continuous system, ever pushing out new branches in all directions, that variously interlace, and that bud into separate lives at every point of interlacement.[30]

And yet, for all the seeming imperviousness to experience of that genealogy of embryos, Galton had also always stressed that no one should be surprised when two gifted parents fail to produce gifted offspring. First of all, prior mongrelization will have made the parents transmitters of low as well as high ability. Second, even when high ability chances to be passed on, its becoming manifest depends on the fortuitous conjunction of innumerable favorable conditions, physiological and environmental.[31] Outcomes, in other words, were always the upshot not just of germinal underpinnings but of their contexts: another conviction Galton would carry over into the next phase of his work.

4

We have arrived at those years of clarification, 1872–74. Three projects, with broadly the same thrust, define these years: from 1872, a new theory of the physiology of inheritance; from 1873, a new inquiry into how "nature and nurture"—Galton's coinage—combined in the lives of distinguished English men of science; and from 1874, a new inquiry into how nature and nurture combined in the lives of a group of twins. In the middle of all of this activity, Galton commissioned the building of the first quincunx (fig. 2.2).

Recall that the quincunx shows how the normal curve comes about, but that Galton introduced it to help him make a point about how heredity works. That was in February 1874, at a Royal Institution lecture previewing his nature-and-nurture findings on English men of science.[32] What point, exactly, was Galton trying to make? If we suppose, reasonably enough, that there was a connection between his new demonstration device—which was not yet called a "quincunx," or indeed anything—and the lecture theme of nature and nurture, how, for Galton, did the one relate to the other?[33] Here is a suggestion. On the quincunx's debut, Galton understood the pellets to represent characters such as stature or intelligence, and the pegs to represent the internal and external influences encountered in the course of a character's development, from embryonic state to full maturity. Just before falling, a centrally positioned pellet thus represented a pre-development character, bound for averageness absent the slings and arrows of developmental fortune. As the pellet fell, its bouncing trajectory represented one of the many possible paths for the development of that character, so that, depending on how the collisions with the pegs (interactions with encountered influences) went, the falling pellet (developing character) ended up closer or farther away, in either direction, from averageness. The pellet at rest in its chamber represented the fully developed character. With resting pellets as with developed characters, averageness was the most probable outcome, with ever more extreme outcomes increasingly less probable, but still possible. Hence that bell curve.

This suggestion does not chime at all well with our received image of Galton as hereditarian.[34] But bearing it in mind, as a candidate meaning of the quincunx when Galton first invented it, will help us take seriously the concern with context that runs through these three mid-period, clarifying projects. At the same time, we will see that a closer look at the projects strengthens the case for interpreting the quincunx in this nonhereditarian way. Let us consider each project in turn.

The physiology of inheritance. In "On Blood-Relationship," published in

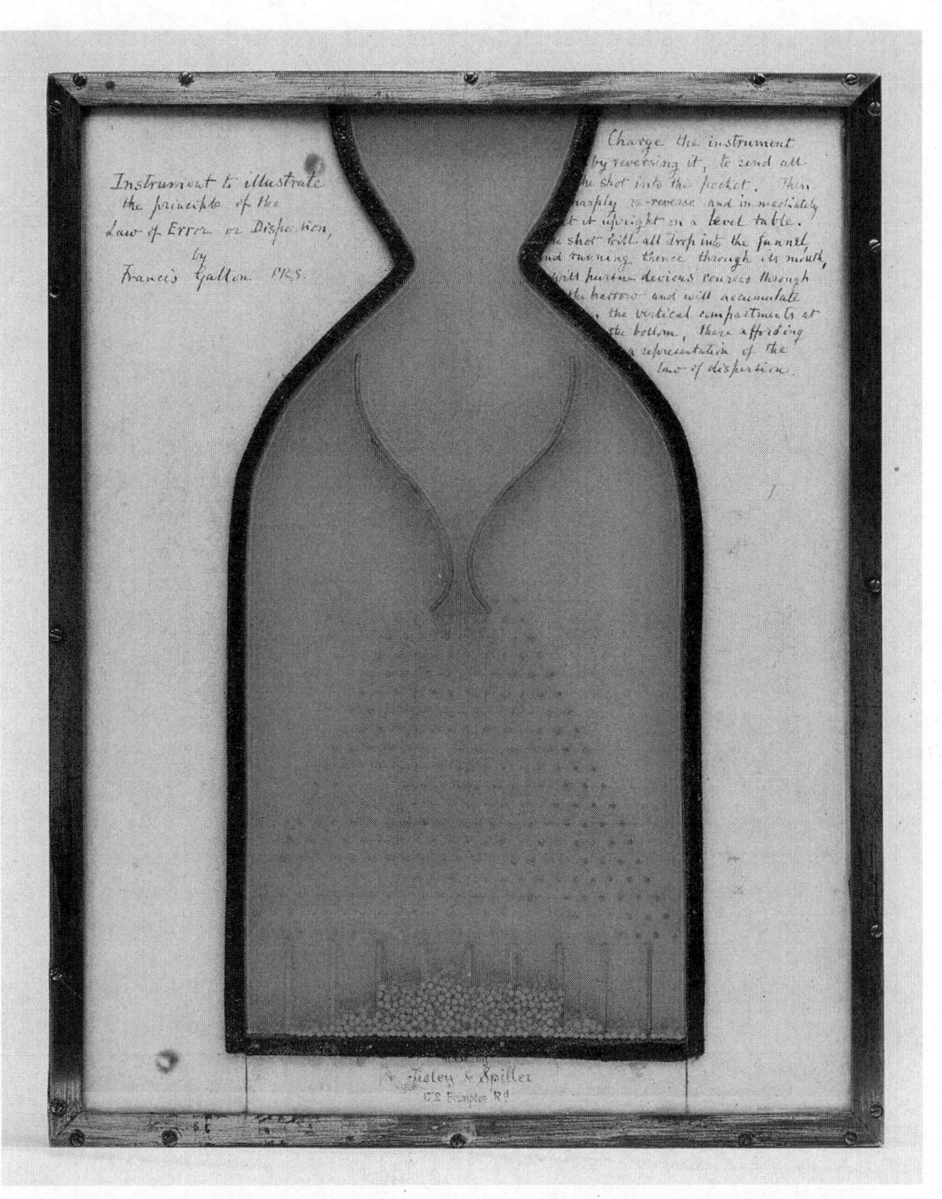

FIG. 2.2. Galton's original quincunx, "to illustrate the principle of the Law of Error or Dispersion."

1872 in the *Proceedings of the Royal Society*, Galton offered the following reflections on development and its role in producing variability:

> The embryonic elements are *developed* into the adult person. "Development" is a word whose meaning is quite as distinct in respect to form, and as vague in respect to detail, as [can be] . . . ; it embraces the combined effects of growth and multiplication, as well as those of modification in quality and proportion, under both internal and external influences. If we were to obtain an approximate knowledge of the original elements, statistical experiences would no doubt enable us to predict the average value of the form into which they would become developed, just as a knowledge of the seeds that were sown would enable us to predict in a general way the appearance of the garden when the plant had grown up; but the individual variation of each case would of course be great, owing to the large number of variable influences concerned in the process of development.

The paper was the first in what became a series of papers setting out Galton's version of pangenesis. He reckoned that the developmental process described above takes place only after a competitive struggle of sorts among an individual's inherited elements or, collectively, "stirp." Elements of the same broad class—the ones conferring intellectual power, say, or a disposition to suffer from gout—vie with one another to be promoted above the crowd and become the patent element in that class, with the losers consigned to latency. Think, Galton recommended, of the patent elements as a representative government, with each element having got there via success in an election, in which electoral success is due, no less than in political elections, not just to the victor's nature but to its interactions with the numerous competitors it happens to be up against, and to effects on the process of conditions that are themselves a matter of happenstance and susceptible to change along the way. In bodies as in politics, the complexities are such that small differences at the start can lead to large differences in outcome. Again, it is no wonder at all that offspring often do not resemble their parents. Nor is it a wonder that the offspring of those offspring sometimes resemble a grandparent or some more remote ancestor. What is patent in a parent, in Galton's view, has no bearing on what makes it into a male's spermatozoa or a female's ova. In the production of these gametes, moreover, which element ends up where (if anywhere) is effectively a matter of chance, just as much as which man ends up in the army when a nation starts conscripting indiscriminately, or which colored balls end up in your hand after you reach into a ball-filled urn. Given the necessary halving of the size of the originating stirp when eggs and sperm are made (for otherwise stirps would double in size with each generation, as clearly they do not), the odds are that some hereditary material will simply be marooned forever within an individual.[35]

Nature, nurture, and English men of science. Toward the end of 1872, Galton received an advance copy of a lengthy book by the Swiss botanist Alphonse de Candolle, claiming to demonstrate, in patterns of scientific achievement internationally, the influence of environment over heredity. The case against heredity was presented as a case against Galton. Yet Galton found almost nothing in the book that he disagreed with. "I literally cannot see that your conclusions, so far as heredity is concerned, differ in any marked way from mine," he protested to Candolle. "I never said, nor thought, that special aptitudes were inherited so strongly as to be irresistible, which seems to be a dogma you are pleased to ascribe to me and then to repudiate."[36] Candolle turned out to be far from alone in making that ascription, however. Early the next year, in February 1873, addressing a philosophy club in Cambridge, the physicist James Clerk Maxwell read out an essay on free will, in which he wittily observed that, if Galton were to be believed, whether one thinks free will exists or not is itself hereditarily determined. Maxwell went on to stress that the great lesson from recent science was that determinism is a much looser sort of thing than previously thought, because tiny causes can have out-of-scale consequences. Giving examples, he mentioned "the little gemmule which makes us philosophers or idiots"—neatly implying that Galton was not merely a determinist about heredity but a self-refuting one (while making exactly Galton's points about how small differences can lead to large ones).[37]

By the end of 1873, Galton, unhappily alerted to the reputation he was plainly earning, launched the project that culminated in his 1874 book, *English Men of Science: Their Nature and Nurture*. Meeting Candolle on his ground, it reported the results of questionnaires sent to 180 fellows of the Royal Society, including Maxwell, Darwin, Huxley, and Jenkin. The new results, Galton concluded, reinforced the old: intellectual ability and the desire to exercise it in a scientific direction were innate; but the latter depended for its expression on encouragement at home, fostering at school, and—as he emphasized in a passionately political final chapter—the continued shift of national funding and prestige away from the clerical sphere toward the scientific one, where newly created professorships and other paid positions would, he hoped, "give rise to the establishment of a sort of scientific priesthood throughout the kingdom."[38] At one point he noted that the heights of the parents of his scientific correspondents had shown the error-curve distribution. Discussing what might have brought that distribution about, he went beyond his previous remarks about the errors to which Nature as marksman was prone, or the way that gemmule inheritance probabilities in averagely gifted families align with the distribution of giftedness in a population as a whole. What we see in the height distribution, he wrote, is "what would have been the case supposing stature to be due to the *aggregate action of many small and independent variable causes*" (his emphasis). The quincunx was that statement made wood-and-metal.[39]

The phrase "nature and nurture," given title billing both at his quincunx-assisted Royal Institution lecture and in his book, was yet another new means of highlighting how seriously he took the modifying effects of the non-hereditary. To be sure, when all else was equal, nature (not the same as the hereditary, Galton insisted, since environment and chance shape us before birth) beat nurture (which he took to include all the influences acting upon an individual after birth, not just during childhood). But he also insisted, as he put it in *English Men of Science*, that "neither is self-sufficient," and indeed, that the "effects of education and circumstances are so interwoven with those of natural character in determining a man's position among his contemporaries, that I find it impossible to treat them wholly apart." If anything, he went on, those difficulties in separating them are even worse when one considers how heredity combines with the non-hereditary in making up someone's inborn nature: "Heredity and many other co-operating causes must therefore be considered in connection."[40] If he felt a sense of exasperation at having to spell all this out, it was with some justice. Even as he was corresponding with Candolle, an article of Galton's then in the press, "Hereditary Improvement," announced on its first page that "nothing in what I am about to say . . . shall underrate the sterling value of nurture, including all kinds of sanitary improvements; nay, I wish to claim them as powerful auxiliaries to my cause." At that time, when Galton wanted a word to express what nurture interacted with, he used "race." It was reading Candolle's impressive discussion of the word "nature" that led Galton to use that word instead.[41]

Nature, nurture, and twins. Twins as windows onto the nature-nurture relationship were already on Galton's mind when he wrote *English Men of Science*. The section headed "Nature and Nurture" moves from a general treatment of the complex reality captured by that "convenient jingle of words," as Galton put it, to the case of twins, real and Shakespearean. Their often amazingly close resemblance when young, he wrote, "necessarily gives way under the gradually accumulated influences of differences of nurture, but it often lasts till manhood."[42] Accordingly, when Galton began research on twins toward the end of 1874, he wrote up a questionnaire reflecting not merely his interest in nature and nurture but his expectation that nurture's role would turn out to be sizable. To his surprise, that was not what he found. The testimonials he collected, from twins who shared an upbringing but then went their separate ways as adults, showed, in his evaluation anyway, that very similar twins largely stayed similar in countless ways throughout their lives, while very different twins largely stayed different. Again, he supplied an analogy: our lives are like sticks in a flowing stream, encountering all sorts of objects en route, but in the grand scheme going with the predetermined flow. Here he stated the message bluntly: equalize for social rank and country, and "nature prevails enormously over nurture."[43] Little wonder that Darwin, on reading the published paper, wrote to Galton, "It is enough to make one a Fatalist."[44]

5

What drew younger admirers to Galton's science of heredity was not these various nature-and-nurture projects, nor his earlier studies of pedigrees and pangenesis, but the work of his third phase, 1875–89, culminating in *Natural Inheritance*. In these years his great theme was dynamic stability, statistically and evolutionarily analyzed.[45] A Friday evening lecture at the Royal Institution in February 1877, titled "Typical Laws of Heredity," staked out the new territory. In Galton's view by this time, the hereditary processes operating within a population functioned to maintain the stability of the overall species type. He posited two such population-level processes, each balancing out the effects of the other. There was the production of the bell-curve variability well modeled by the quincunx. And there was a counteracting tendency "to 'revert' towards mediocrity," as he put it. He expended some effort tinkering with the quincunx to illustrate this second process too, the better to show his audience—and, later and less effectively, readers of *Nature*—the causal churn beneath the apparently unchanging surface equilibrium. For a given character in a given population, Galton argued, parents who, between them, have the average form of the character typically produce some offspring who are above or below average for that character (the law of deviation), while parents who, between them, have the character in above-average or below-average form typically produce offspring closer to the average for that character (reversion to mediocrity). The result was the stasis that, Galton stressed, was now a documented fact of life not just for plants and animals of the present day but for extinct ones too. Compare existing forms with their fossil counterparts, and, though they may be separated by thousands of generations, you will "seek in vain for peculiarities which will distinguish one generation taken as a whole from another, the different sizes, marks and variations of every kind, occurring with equal frequency in both."[46]

The now-aged Darwin had played a part in helping Galton with the experimental work that lay behind the latter's new framing of reversion to mediocrity not merely as an unfortunate fate that often befell families of high talent and character, but as a law of heredity in its own right. Darwin was one of a number of people who, at Galton's behest, undertook to plant seventy sweet-pea seeds provided by him and divided up into seven weight classes, from three units below the average weight to three units above, with each set containing ten seeds of equal weight. On examining the weights of the seeds harvested from the grown plants, Galton found that, though the weights of the offspring seeds conformed to the law of error in their overall distribution, their mean weight was typically closer to the overall sweet-pea average than was the weight of the parental seed. Whereas, in his 1865 articles, he had treated natural selection as the driver of evolution, he now treated it as yet another force keeping populations right where they were, their characters stably distributed around means representing the optimally adaptive values under the existing external conditions.[47]

This dissent from a standard Darwinian picture of evolution had been in the works for some while. In the final chapter of *Hereditary Genius*, Galton had built up to his bravura statistical reading of the pangenesis hypothesis by way of reflections on gemmule affinities and interactions as sometimes making for stable types and characters, sometimes not, and sometimes propelling a species from one type-defining stable equilibrium to a new one altogether. As he sought to explain himself, the analogies heaped up. Types were a fact of organic life, he suggested, because harmoniously meshing gemmules tended to come together into functional wholes, in just the way that, in the absence of any central direction, harmoniously meshing people tended to come together to make different types of places—holiday villages, fishing encampments, whatever—so much of a generic muchness. In both biology and society, a local feature of some kind attracts a few individuals, whose presence creates opportunities seized upon by other individuals, and so on. And just as hybrid places may be unstable, depending on the particular mix of elements, the offspring of naturally talented parents, even when both descend from stock as pure as could be for that talent, may sometimes fail to inherit a harmoniously success-making constitution. Conversely, naturally untalented parents can produce prodigiously talented offspring when the latter inherit just the right combination of success-making elements that had lain latent in the parents' constitutions, much as, in political elections, the combining of two boroughs into a single borough can suddenly turn what had previously been two election-losing minorities into an election-winning majority. ("The character of a man," Galton wrote, "is wholly formed through those gemmules that have succeeded in attaching themselves; the remainder that have been overpowered by their antagonists, count for nothing; just as the policy of a democracy is formed by that of the majority of its citizens, or as the parliamentary voice of any place is determined by the dominant political views of the electors: in both instances, the dissentient minority is powerless.")[48]

Turning to the question of how to reconcile "the stability of types" with Darwin's *Origin of Species*, Galton in *Hereditary Genius* drew an analogy with the nonliving world in order to illustrate how discontinuous changes—what he called "changes in jerks"—can result from continuous changes at a lower level:

> The mechanical conception would be that of a rough stone, having, in consequence of its roughness, a vast number of natural facets, on any one of which it might rest in "stable" equilibrium. That is to say, when pushed it would somewhat yield, when pushed much harder it would again yield, but in a less degree; in either case, on the pressure being withdrawn it would fall back in to its first position. But, if by a powerful effort the stone is compelled to overpass the limits of the facet on which it has hitherto found rest, it will tumble over into a new position of stability, whence just the same proceedings must be gone through as before, before it can be dislodged and rolled another step onwards. The various positions of stable equilibrium may be looked upon as so many typical attitudes of the stone, the type being more durable as the limits of its stability are wider. We also see clearly that there is no violation of the law of continuity in the movements of the stone, though it can only repose in certain widely separated positions.

Or think, he continued, about the stop-start movement of a crowd through a narrow door. Most of the time, though everyone in the crowd is in agitated motion, with some even aggressively so, the crowd as a whole is stationary. Then, quite suddenly and unexpectedly, "by some accidental unison of effort," the deadlock is broken and there is forward motion. Then equilibrium rapidly reestablishes itself, and stays established until fortuitously and fleetingly broken, and so on.[49]

6

As the quincunx served to visualize the chance processes that produce the normal curve, another technique, first proposed by Galton in 1877 a few months after he published "Typical Laws of Heredity," helped him make visible the stable, type-defining averages. Photography had been around and advancing for decades when it occurred to him that if photos of the faces of individuals of a particular type could be combined in the right way, what was common to them—and so defining of the type—would stand out. His first attempt at making such a composite portrait involved training a camera on a pile of uniformly sized full-face photos, which he then began to leaf through, placing and removing the lens cap between leafings so that the same part of the sensitized plate got repeatedly exposed. He was pleased enough with the results that he continued refining the method, publishing the composites most accessibly in 1883 in *Inquiries into Human Faculty and Its Development*. Of the composite face depicted under the heading "Health" (fig. 2.3), he wrote, "This face and the qualities it connotes probably gives a clue to the direction in which the stock of the English race might most easily be improved."[50]

The book is best remembered now for the introduction there, in a footnote, of the word "eugenics," as a more apt name for "the science

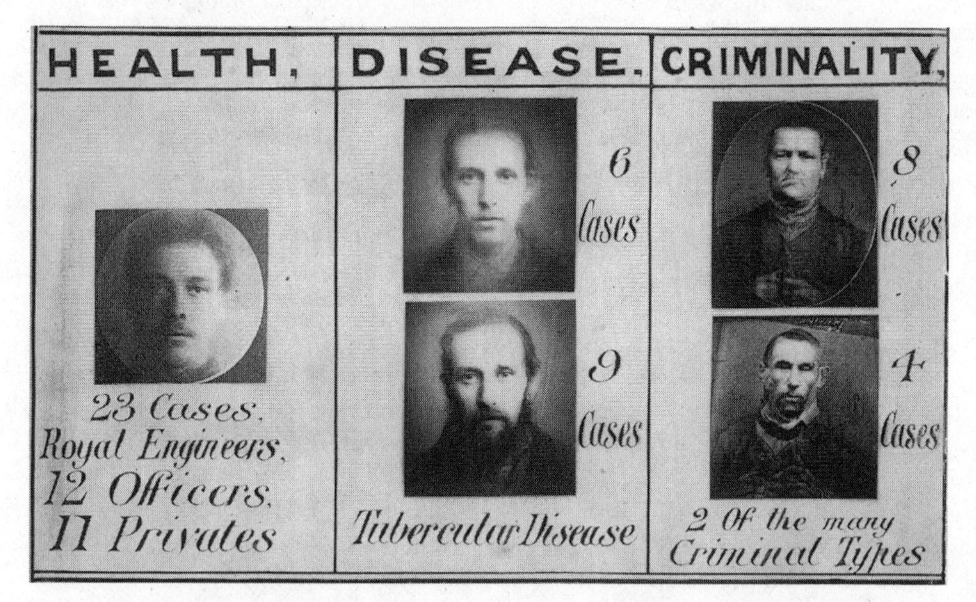

FIG. 2.3. Galton understood his composite photographs to show that a civilized race like the English has an ideal typical form (*left*) but multiple, and distinct, criminal breeds (*right*).

of improving stock" (which is, Galton insisted, "by no means confined to questions of judicious mating") than the one Galton had previously favored, "viriculture."[51] What is not well remembered is how much less sanguine Galton had by then become about the prospects for humanly engineered improvement of the human stock, in line with his deepening appreciation of the stubbornly stable nature of types. He now insisted that lasting change could come about only with the chance coming into being of individuals who represent not exceptional deviations from an old type—for their descendants would just revert to mediocrity—but "the formation of a new strain having its own typical centre." To the extent that natural selection awaits the arrival of such type-shifting individuals to effect improvements, it is responsible for those improvements only in an indirect way. Likewise, Galton went on, human improvers are very limited in what they can achieve. Returning to an image from *Hereditary Genius*, he now compared the improver to Sisyphus, doomed to an eternity of pushing the multifaceted stone of the human race up Mount Improvement, only to have it roll right back again. The only hope was the emergence of a new facet—"a new typical centre," "a temporary sticking point in the forward progress of evolution"—that might provide a new point of stability a little farther up the hill than before. That was true for human improvers of animal and plant varieties, who knew all too well how fragile an overbred variety could become. And it was true for would-be practitioners of eugenics.[52]

Not that Galton offered a counsel of despair. Eugenics in his eyes stood at its best for what he termed a more "merciful" alternative to the misery-inducing path of exposing civilization-cushioned humans to the most demanding conditions of life, such that only a handful in each generation would survive to reproduce. No, much better to keep a watch out for new-type individuals and, when they arrive—which, Galton thought, will be "no infrequent occurrence"—to ensure that they propagate their kind disproportionately, so that a new, improved race will eventually, and painlessly, supplant the old one. And again, judicious mating was not enough, since, as he put it elsewhere in the book, the "interaction of nature and circumstance is very close, and it is impossible to separate them with precision. Nurture acts before birth, during every stage of embryonic and pre-embryonic existence, causing the potential faculties at the time of birth to be in some degree the effect of nurture."[53]

Six years later, Galton considered the theme of types and their stability sufficiently significant to devote a whole chapter to it in *Natural Inheritance*. It is a curious and demanding chapter in a curious and demanding book. Galton's main concern in the book was to show that the complex calculational machinery behind what he now called "regression"—

that tendency for offspring to be less extreme in a character than their parents—was sufficiently refined that he could predict, sometimes with remarkable accuracy, the amount by which extreme values for such hereditary characters as stature, eye color, and artistic ability (to use the examples he used) would, on average, be less extreme in the next generation. He did discuss normal variability, in its own right and as preserved by natural selection's favoring of the optimal. But his comments were very much incidental to his larger, regression-oriented purposes. And on natural selection, the impression generated was that its evolutionary importance had been overrated. As for environmental conditions, he stressed, near the beginning and the end of the book, and very much in the spirit of his physiological theorizing of the early to mid-1870s, how large a role they played in determining visible characters. For the bulk of the book, however, he ignored them, suggesting that their effects would tend, at the level of analysis that concerned him, to cancel out. His concluding reflections were on how best to apply the lessons of the book to ensure "the well-being of future generations" through judicious interbreeding.[54]

The chapter titled "Organic Stability" bears attention because it displays in miniature the main features of the whole of the Galtonian science of heredity circa 1889. Galton begins by once again insisting that, between the operations of chance and conditions, far more structural potential gets transmitted from parents to offspring than is manifest in either, so that fully developed bodies in a lineage should be viewed less as links in a chain of inheritance than as pendants on that chain. "This is why," he adds, "it is so important in hereditary inquiry to deal with fraternities rather than with individuals, and with large fraternities rather than small ones." Next he turns to stability as something that can come about whenever there are large numbers of elements jostling together under myriad influences, whatever the nature of the elements and the influences. He begins with hereditary elements or, as he also calls them here, "particles," citing a recent volume on cell theory and heredity as affirming his long-held view of the fertilized egg as a place where there are "segregations as well as aggregations," no doubt due to lower-level repulsions and affinities. At this point, Galton furnishes instance after instance to illustrate how common is the spontaneous emergence of stable forms. Governments, on whatever scale, generally take one of only a handful of forms, all others being unstable; crowds at ceremonies, whatever the occasion and whoever is there, behave in the same ways; and so on. He goes on, however, to introduce a new subtlety: the recognition that stability is not, in fact, all or nothing, but can come in degrees, to such an extent that it makes sense—in biology as much as in the apparently distant realm of patented improvements to the public carriages in

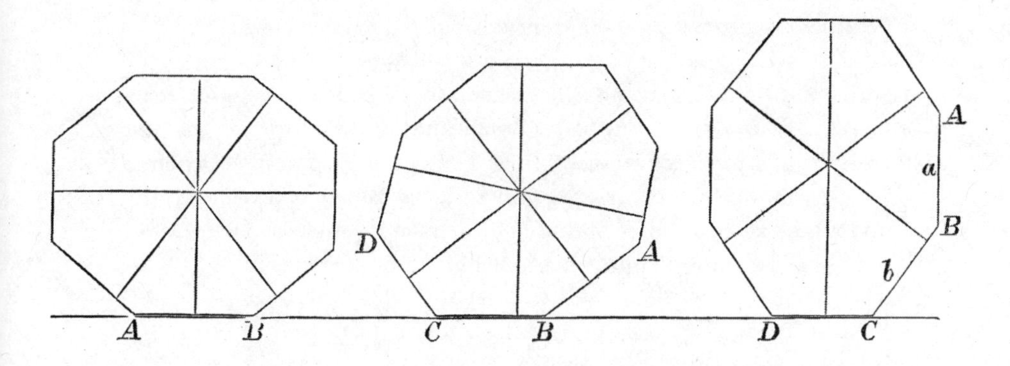

FIG. 2.4. For Galton, a lineage's evolution from one stable type to another was like the successive tumbles of a rolling polyhedron. Depicted here are three instants in a counterclockwise roll of his model, from a stable position (*left*), via a not-very-stable position (*center*), to a new stable position (*right*). (The letter symbols on the left diagram are reversed.)

London—to distinguish between primary types, subordinate types, and deviations from these.[55]

This new set of distinctions then becomes integrated with the old multifaceted stone analogy. The word Galton uses here is not "analogy," but "model," referring to an actual, physical model he had constructed: "a polygonal slab that can be made to stand on any one of its edges when set upon a level table." An accompanying drawing (fig. 2.4) showed the slab in tumbling motion, from a very stable position (primary stability), to a not-very-stable position (subordinate stability), to a new very stable position (primary stability again). As before, Galton explained that, in the model as in an organic structure, a shift takes place only if a disturbance is great enough to overcome reversion back to the old fixed point. He worked through the evolutionary consequences—and their meaning for Darwinism—in detail:

> The ultimate point to be illustrated is this. Though a long established race habitually breeds true to its kind, subject to small unstable deviations, yet every now and then the offspring of these deviations do not tend to revert, but possess some small stability of their own. They therefore have the character of sub-types, always, however, with a reserved tendency under strained conditions, to revert to the earlier type. The model further illustrates the fact that sometimes a sport may occur of such marked peculiarity and stability as to rank as a new type, capable of becoming the origin of a new race with very little assistance on the part of natural selection.

> Also, that a new type may be reached without any large single stride, but through a fortunate and rapid succession of many small ones.[56]

So, as he emphasized again near the end of the chapter, evolution need not take place only in tiny steps. Sometimes it proceeds that way. But sometimes it proceeds by medium-sized steps, and sometimes by large ones, as shown by the existence of hereditarily stable types, and by the undoubted fact that some of those types first appeared not as the last in a series of ever closer approximations but out of the blue.[57]

7

The notion that anyone might have read Galton as an interactionist on heredity comes as a surprise in an era that associates Galton with hereditarianism even more fully, and damningly, than Candolle and Maxwell did. The quincunx, though now cherished for other reasons, formed part of Galton's clarifying reply to Candolle. His reply to Maxwell came about a decade later, when, as Galton recalled in his autobiography, he became "so harassed with the old question of Determinism, which would leave every human action under the control of Heredity and Environment," that he decided to try and settle the question empirically. His conclusion, reported in *Mind* in 1884, was that the more closely he inquired into his own acts of apparent free will, the more clearly he saw them as bound within ordinary causal laws. Ever the Darwinian iconoclast, he summed up his results as supporting "the views of those who hold that man is little more than a conscious machine, the slave of heredity and environment, the larger part, perhaps all, of whose actions are therefore predictable."[58] No consolation for the Christian, then; but none for the hereditarian either.

Behind *Natural Inheritance* stood an oeuvre heterogeneous enough to license multiple, and even opposed, interpretations. Depending on where one looked, one could find support for the importance of natural selection as well as skepticism about it; an understanding of organisms as members of statistically characterized populations as well as an understanding of them as more or less faithful instantiations of types; a quincuncial stress on heredity-environment interaction as well as a thumping insistence that nature beats nurture. Along the way, Galton managed, by his direct involvement with it, to turn pangenesis into a theory of heredity, deeply flawed but worth fixing. Darwin's gemmules lived on in *Natural Inheritance* in modified but recognizable form, as in this passage, which is also full of that mid-period Galtonian concern with developmental contingency and the variability it brings about:

> We appear, then, to be severally built up out of a host of minute particles of whose nature we know nothing, any one of which may be derived from any one progenitor, but which are usually transmitted in aggregates, considerable groups being derived from the same progenitor. It would seem that while the embryo is developing itself, the particles more or less qualified for each new post wait as it were in competition, to obtain it. Also that the particle that succeeds, must owe its success partly to accident of position and partly to being better qualified than any equally well placed competitor to gain a lodgment. Thus the step by step development of the embryo cannot fail to be influenced by an incalculable number of small and mostly unknown circumstances.[59]

Indirectly, through the impact of *Natural Inheritance* on the next generation, Galton would likewise play a role in Mendel's "Versuche" coming to be seen as the basis for a theory of heredity—indeed, a particulate theory of heredity. In that generation, the book's earliest and most eager readers included two young morphologists en route to becoming ex-morphologists and, not at all coincidentally, ex-friends: William Bateson and W. F. R. Weldon.

3

Biology for the Steam Age

Five years hence no one will think anything of that kind of [morphological] work, which will be very properly despised. . . . It came to me at a lucky moment and was sold at the top of the market; presently steam will be introduced into Biology and wooden ships of this class won't sell well.

WILLIAM BATESON, LETTER HOME FROM KAZALINSK, 10/22 NOVEMBER 1886

For the purposes of comparative morphology, which endeavours to present the main points of animal structure in an intelligible way, it is not only useful but necessary to neglect many variations, and to consider species as if every individual were "typical." . . . For other purposes, such a conception of a species is altogether inadequate.

W. F. R. WELDON, UNPUBLISHED MS ON STATISTICS AND EVOLUTION, CIRCA 1900–1901

We met comparative morphology briefly in the first chapter, as one of several sciences that, as distilled in Unger's teaching at Vienna, led Mendel to search for just the sort of law that he later found for the hybrid form in *Pisum*. (*Morph–* is from the Greek word for form; "morphology," the study of form, was Goethe's coinage.) In the *Origin*, Darwin praised morphology as "the most interesting department of natural history," even "its very soul":

> What can be more curious than that the hand of a man, formed for grasping, that of a mole for digging, the leg of the horse, the paddle of the porpoise, and the wing of the bat, should all be constructed on the same pattern, and should include the same bones, in the same relative positions? . . . We never find, for instance, the bones of the arm and forearm, or of the thigh and leg, transposed. Hence the same names can be given to the homologous bones in widely different animals. We see the same great law in the construction of the mouths of insects. . . . Analogous laws govern the construction of the mouths and limbs of crustaceans. So it is with the flowers of plants.

Such unity-of-plan patterns, Darwin went on, appeared in a new light under his theory of natural selection. According to that theory, when a species gives rise to descendant species, they will resemble each other thanks to the conservation of shared ancestral structures, albeit adaptively modified to new environments. Conversely, some structural commonalities among otherwise diverse groups can be traced back to a common ancestor, located at a branching point farther down the evolutionary tree of life. (Think of the famous Galápagos finches, with their beaks adapted to the conditions in that rocky archipelago, but their bodies overall bearing the distinctive stamp of finches on the lushly tropical South American mainland—because, Darwin argued, the Galápagos birds descended from mainland species.) Thus did comparative morphology in the latter nineteenth century become a Darwinian science. If you were young, and serious about science, and excited about reconstructing the evolutionary past, comparative morphology beckoned.[1]

At the start of the 1880s, Walter Frank Raphael Weldon and William Bateson were all these things. Weldon, a star pupil in the science at Cambridge, wrote home in March 1880 to say that his travel back would be delayed because, after hearing someone talk "what I believed to be great nonsense about the evolution theory" at a meeting, he spontaneously gave such a stout defense that he was asked to read a paper about it—which he now needed to prepare.[2] Bateson, also studying zoology at Cambridge at that time, remembered the ambience vividly: "Morphology was studied because it was the material believed to be most favourable for the elucidation of the problems of evolution, and we all thought that in embryology the quintessence of morphological truth was most palpably presented. Therefore every aspiring zoologist was an embryologist, and the one topic of professional conversation was evolution."[3] This chapter tracks Weldon's and Bateson's intertwining trajectories over the course of the decade, as they diverged from morphology and, increasingly, from each other. We shall see that the ultimate roots of Weldon's anti-hereditarian reading of Galton's work lie in this period, in a forgotten history of unsuccessful attempts by Weldon to build a positive research program from his morphologist teacher Francis Balfour's self-critical ideas on heredity, environment, and Darwinism. When Weldon began, later in the decade, to take a more quantitative approach, and then to embrace *Natural Inheritance* as his guide, he did so as a refurbisher of Balfour's legacy.

Viewing Weldon as a loyal but frustrated Balfourian in turn brings into focus an unfamiliar image of Bateson as an ever more disloyal and frustrated Weldonian. Bateson's otherwise odd decision in the mid-1880s to travel all the way to the Central Asian steppe for a field study on how environments shape animal forms makes sense once we appreciate how fully Weldon's interests at this time shaped Bateson's—and equally, how much Bateson chafed at the dependency. Although open hostilities between the two did not flare up until later, the potential for conflict, and some of its sources, were already visible in these years. The contrasting attitudes to morphology expressed in this chapter's epigraphs—Bateson's wanting to shut it down and start afresh; Weldon's balanced assessment of its achievements and limitations—were wholly in character.[4]

Contra Bateson, biology in the mid-1880s was already in the steam age, in the sense that the post-Darwinian generation of comparative morphologists were forever traveling by steamer to marine zoological stations, in Naples and elsewhere, in search of abundant specimens, good working conditions, and the camaraderie of international colleagues. In 1885, Weldon received a multilingual hello from Naples—"the Patent Neapolitan Polyglot Postcard," he called it—and wrote back that "when I see so much and such varied greeting condensed into so small a space, I feel that far away though they be there are indeed some friends that stick closer than a brother."[5] Within a few years British biologists had a splendid new laboratory of their own, in Plymouth. Alongside this institutional expansion, however, came a measure of intellectual contraction. The rising influence in Britain of the work of the German biologist August Weismann meant that the flexibilities that Bateson and Weldon (fig. 3.1) early on took for granted in the linking of heredity to environment, variation, development, and evolution came under increasing pressure.

1

When Darwin and Galton studied at Cambridge, it was in many ways like an Anglican version of the Abbey of St. Thomas. Only from 1870, with the arrival of James Clerk Maxwell in physics and Michael Foster in physiology, did the natural sciences cease to be mainly the part-time business of men who answered to "Reverend." Foster—the son of an Evangelical preacher—worked hard and fast to modernize. He introduced lab work into the biological curriculum, itself no longer divided between zoology and botany but organized into a single body of facts, unified by cell theory and evolutionary theory. In short order, Cambridge biology boasted an ambitious, technically capable, internationally networked research community, specializing in physiology as well as in the other great lab biology of the era, comparative morphology (by now, as Bateson recalled, indistinguishable from the evolutionary study of embryos). What became celebrated as "the Cambridge school of animal morphology" thrived un-

FIG. 3.1. William Bateson and Walter Frank Raphael Weldon, about 1890.

der the leadership of the most brilliant of Foster's early students, the aristocratically born Balfour.[6]

Elected a fellow of the Royal Society in 1878, at the age of twenty-seven, Balfour was the most proficient and productive morphologist of his generation. He was also the most thoughtful about its methods, above all the remarkable way in which Darwinian theory had given with the one hand while taking with the other. With its giving hand, it had lifted comparative morphology into a leading role in the grand reconstruction of the genealogy of life, since the similarities and near-similarities morphologists found in developing forms could now be interpreted as shared inheritances from common ancestors. Yet—the taking hand waggled an admonitory finger—because of natural selection, no inference from similarity to shared ancestry could be entirely safe, since two separate, unrelated lineages passing through similar environments might well end up similarly adapted to them. As Balfour appreciated, there was no sure way to tell, without a great deal of extra effort anyway, whether a given similarity was a conserved ancestral legacy or an innovation modified by selection.[7] To be a Balfourian morphologist was to live this tension, carrying out the job of embryological comparison while keeping before one's mind not just the species branchings that were the product of Darwinian evolution, but the process behind the product—a process coordinating heredity and environment, as Balfour stressed in an August 1880 lecture on the impact of the *Origin* on embryology:

> The law of variation is in a certain sense opposed to the law of heredity. It asserts that the resemblance which offspring bear to their parents is never exact. The contradiction between the two laws is only apparent. All variations and modifications in an organism are directly or indirectly due to its environments; that is to say, they are either produced by some direct influence acting upon the organism itself, or by some more subtle and mysterious action on its parents; and the law of heredity really asserts that the offspring and parent would resemble each other if their environments were the same. Since, however, this is never the case, the offspring always differ to some extent from the parents. Now, according to the law of heredity, every [newly arisen] variation tends to be inherited, so that, by a summation of small changes, the animals may come to differ from their parent stock to an indefinite extent.[8]

Darwin, reading the lecture in *Nature* a few days later, loved it, and wrote to Balfour to tell him so.[9] Weldon almost certainly heard the lecture live, as he was there in Swansea, at the annual meeting of the British Association for the Advancement of Science (BAAS), when Balfour delivered it. Born in 1860, the beneficiary of a well-funded, wide-ranging, and ever more scientifically focused education, including preparatory university studies in London (where he grew up), Weldon—"Raphael" to his family and friends—had arrived in Cambridge in spring 1878, taking rooms over a small cabinetmaker's shop a short distance from his new college, St. John's (probably chosen on the advice of a London lecturer, Alfred Garrod, who had studied there and was himself an early product of the Fosterian reforms at Cambridge). A year later, Weldon's ties to home, and to his family's Swedenborgian religion, remained sufficiently tight that he wrote to his mother about a notice he had seen announcing "the formation of a society (among members of the University) for the investigation of modern psychical phenomena by the light of ancient Eastern superstitions," and to his brother about having two Swedenborgian books with him, though recently turning down a chance to buy a volume of Swedenborg's *True Christian Religion*. The remainder of Weldon's surviving correspondence home, however, makes no mention of such matters. The main subjects are his biological studies and his desire-cum-anxiety to outcompete his peers for Balfour's good opinion. In a letter to his mother in late May 1879, near the end of an exams period about which he was generally not optimistic, he reported, "Mr. Balfour has asked me to lunch on

Sunday. I suppose he has asked all his class. As he is examiner this afternoon, I hope I shall do a good paper. The subject is Comparative Anatomy, so I have hopes." He added, "The Natural Science[s] Tripos exam begins on Monday, so all men who are going in are getting pale and excited: lights are at their windows till all sorts of times in the morning."[10] In the event, Weldon did extremely well: a First Class, one of only two. By the following spring, Balfour was making encouraging noises about Weldon's scientific future and suggesting that, as original research was such good training, Weldon might take up some outstanding questions about the structure and development of beetles. In April 1880 Weldon told his mother proudly that he had been asked to substitute for a month for the regular demonstrator, Adam Sedgwick, in Balfour's lectures—a paid position that, Sedgwick let on, showed that Balfour reckoned Weldon the most able student in his year. "If I do it well," Weldon wrote, "it will help me very much to get work after the Tripos."[11]

In the same letter, Weldon described other, extracurricular doings involving a new friend. "There is an immense deal of excitement here over the general election," he reported. "Bateson, of John's, and I went about on the polling day for the borough in a trap, with a big blue and yellow flag, hunting up liberal voters. Bateson has worked awfully hard at the election, and has had a good big share in the production of the result (you probably know that we returned two liberals instead of two conservatives)."[12] A year younger than Weldon, Bateson had barely scraped through the preliminary exams needed to start the Cambridge natural sciences degree, and was still finding his feet scientifically. In later life, he would credit Weldon's influence and example as decisive. "To Weldon I owe the chief awakening of my life," Bateson wrote to a friend. "It was through him that I first learnt that there was work in the world which I could do. Failure and uselessness had been my accepted destiny before."[13] And indeed, a consistent complaint about Bateson by his teachers at Rugby School was that he lacked the self-discipline that might allow him to make more of the cleverness that fitfully revealed itself, notably in natural science. His gloomy letters home in this period likewise suggest a boy whose sense of himself was of someone not fitting in and getting along.[14]

How, then, to account for the zestful confidence that propelled Bateson, with an admiring Weldon alongside, through the streets of Cambridge, drumming up support for the return of Gladstone's Liberals? It could not have hurt that, for Bateson, the move from school to university had been a move back home. His father, also called William, was a Cambridge-trained classicist turned clergyman who, from before his son's birth, had been serving as the master of St. John's. Cambridge in general, and St.

John's in particular, were thus not merely familiar territory, in a quite literal sense, but territory where to be a Bateson was to enjoy the esteem and even glamour that came with powerful connections. As for Rugby School, which he hated while he was there, it had also left its mark, for he spent the rest of his life upholding the ideals distinctively identified with the school and its master, Thomas Arnold—ideals of seriousness, service to others, and suspicion of commerce and utilitarian values as threatening to what was best in human culture (the Greek and Latin classics) and character (the high-mindedness required to live a life dedicated to truth, goodness, and beauty). Someone coming out of Rugby School into the Cambridge college where his father was master was not obviously destined for "failure and uselessness." And indeed, the young Bateson's dashing intervention in the 1880 general election was an early sign that he intended to live up to the Arnoldian ideals.[15]

As close as they were becoming, Bateson did not accompany Weldon to Swansea a few months later for the BAAS meeting. Weldon did not lack for company, however. As noted, Balfour was there, as was Sedgwick. Weldon's entire family was there too; going to the "Bass" every summer was family tradition. Weldon's father, a self-made chemical entrepreneur and member of the BAAS General Committee (along with Balfour), gave a talk that year on his latest ideas about atomic structure. Also present, at the invitation of Weldon's parents, was the young woman who would go on to be his wife and close collaborator, Florence Tebb—"Florrie" to him and everyone else (fig. 3.2). The daughter of a family friend, she was now at Cambridge too, at women's-only Girton College, studying mathematics: an expertise that would come in handy.[16]

2

Although weighed down by family tragedy—the deaths of his brother and mother came in merciless succession—Weldon nevertheless got a First in the Tripos exam. By February 1882, he was writing to Anton Dohrn, the German boss at the Stazione Zoologica in Naples, to set up a first research visit, having been "appointed by the University of Cambridge to occupy a table in your Zoological Station at Naples." By June that year, his research was under way: an attempt to take further a paper of Balfour's on the embryology of the wall lizard by looking at earlier stages of its development than Balfour had managed.[17] But then, the next month, came another tragic death: Balfour's. He was mountain climbing in Switzerland when he fell. Britain's biological elite, already in a state of mourning after Charles Darwin's death in April that year, were devastated. The response of Balfour's London counterpart, E. Ray Lankester, was typical.

FIG. 3.2. Florence and Raphael Weldon in New York—on a side trip from a collecting expedition to the Bahamas—in 1886, three years after their marriage.

On learning the terrible news, he wrote in a letter that "when I remember [Balfour's] earnest eager way, and his benevolent beautiful eyes, as I have so often seen them as we discussed late into the night in my study or at Cambridge, and then remember that he has been cruelly smashed, and that I shall never see him again—I can only cry." As for Cambridge, where Balfour had recently been elected to a personal chair in animal morphology, Lankester could see no way forward: "All is over there with the study of animal morphology."[18]

If that epitaph proved premature, it was in large part thanks to Balfour's Cambridge legacies. The goodwill, largesse, and revived scientific ambition behind his chair soon extended to the continued maintenance of the laboratory and associated demonstratorship (now held by Weldon) and beyond. In 1882, a research studentship in morphology was established; in 1883, a new lecturership in animal morphology; and in 1884, two lectureships in advanced morphology, one dedicated to the vertebrates and the other to the invertebrates, as well as a new Balfour Biological Laboratory for Women. The invertebrate lectureship went to Weldon, who held it through the rest of the decade, in the course of which he taught what he loved to lots of excellent students, some of whom went on to become dear colleagues, in Cambridge and elsewhere.[19] More so than any other Balfour student, Weldon carried on the tradition of thoughtfulness that Balfour had brought to the Darwinian topics of variation, heredity, environment, adaptation, and natural selection. When the Royal Institution invited Weldon to give a Friday evening lecture in early May 1885, he seized the opportunity to try out his ideas—fully aware that in doing so he was risking unpopularity. "The approach of the Royal Institution business is bothering me," Weldon wrote at the end of April that year to D'Arcy Thompson, a former fellow morphology student at Cambridge then working at the University of Dundee. "I am ruining myself in diagrams: and as I am going to talk my own private and particular heresy, I am afraid of 'washing.'" There is, alas, no record of the heresy Weldon talked that evening. But his title gives an indication: "On Adaptation to Surroundings as a Factor in Animal Development."[20]

The death of Balfour in the middle of Bateson's undergraduate education—and the traumatized state of the students and colleagues he left behind—may go some way to explaining why Bateson does not seem to have emerged from Cambridge with anything like the same depth of allegiance to Balfour's science. Later in life, when Bateson recalled the lectures he attended at Cambridge, it was not Balfour's but Foster's he remembered. Bateson once recalled that the "best answer in few words" that he had ever heard to the question "What is a living thing?"

> is one which my old teacher, Michael Foster, used to give in his lectures introductory to biology. "A living thing is a vortex of chemical and molecular change." This description gives much, if not all, that is the essence of life. The living thing is unlike ordinary matter in the fact that, through it, matter is always passing. Matter is essential to it; but provided that the flow in

> and out is unimpeded, the life-process can go on so far as we know indefinitely. Yet the living "vortex" differs from all others in the fact that it can divide and throw off other "vortices," through which again matter swirls.[21]

To have been impressed with Foster's vortex definition of an organism was, at a minimum, to have been open to biological notions that had no obvious place in the Balfourian program. Be that as it may, with Sedgwick and Weldon as his teachers, Bateson got trained up in the skills of embryo dissection, description, and evolutionary inference needed to advance the program. His chance to do so began with a visit in July 1883 to the Chesapeake Zoological Laboratory in Hampton, Virginia, to work with an animal that, a few years later, would cause Weldon much embarrassment: the acorn worm, *Balanoglossus*.

It was Weldon who first got Bateson working on *Balanoglossus* and who, as Bateson's studies of the creature deepened, supplied him with specimens from the Naples station. Along with the free-swimming *Tornaria* larval form through which some species passed (though not, as it happened, the Chesapeake species), the sand-burrowing *Balanoglossus* was, by the 1880s, something of a star of comparative morphology, not least because of a sense that getting *Balanoglossus* right could throw light on one of the biggest morphological puzzles of all: the evolutionary origin of the vertebrates.[22] During Bateson's final undergraduate year at Cambridge, a notice appeared about the animal having been found at the Hampton site where the peripatetic Chesapeake lab, run by the recently founded Johns Hopkins University in Baltimore, was due to settle for the summer. At Sedgwick's suggestion, Bateson wrote to the lab's chief, the Hopkins morphologist William K. Brooks, to ask about visiting, and Brooks wrote back positively. Thus Bateson, having finished with a First, found himself, in the sweltering Virginia summer of 1883, in the best possible circumstances not merely for making original observations, of potentially wide professional interest, but for doing so in the company of the master teacher that, for Bateson, Balfour had never had the chance to be. For hours on end through those weeks, when Bateson and Brooks were not at the shore collecting or bent over microscopes, Bateson was learning firsthand about Brooks's unorthodox ideas, back at the boardinghouse where Bateson was staying along with Brooks's family. Bateson's abiding memory of that summer was of sitting spellbound as Brooks, lying on his bed in his shirt-sleeves, smoking the tobacco that served to lubricate his mind, unhurriedly held forth.[23]

Brooks's intellect was restless, and the topics occupying it that summer, though they would obsess Bateson for the rest of his career, were, for Brooks, ephemeral. He was just then in between two rather different and very demanding writing projects. One, recently published, was a long article for the magazine *Popular Science Monthly*. Titled "Speculative Zoology," it attempted to defend the morphologists' way with evidence from the criticisms of his former teacher in marine zoology at Harvard, Alexander Agassiz. The conviction that grew in Bateson after that summer, however, was not that Brooks had answered Agassiz, but that Agassiz's criticisms were unanswerable.[24] Figuring especially heavily in those boardinghouse conversations was Brooks's forthcoming book, *The Law of Heredity*, due to be published that autumn. He gave it the subtitle *A Study of the Cause of Variation, and the Origin of Living Organisms*. A more descriptive one would have been *An Homage to Darwin in the Form of a Creative Update of His Book on Variation, Excerpted at Sometimes Surprising Length*. The new (if often old) book aimed not merely to serve as a reminder of the largely untapped potential for empirical—above all experimental—research that the recently deceased Darwin had opened up in his *Variation*, but to introduce theoretical modifications that, as Brooks saw them, would help keep the pangenesis hypothesis and natural selection theory current. Among the modifications needed, he reckoned, was allowance for—and better still, a pangenetical account of—the kind of evolution illustrated by Galton's multifaceted stone: "saltatory evolution," as Brooks enduringly called it. Only if it was assumed that some complex structures come into being very rapidly, thanks to processes acting to destabilize and then restabilize organic form after equilibrium-disrupting environmental change, could natural selection possibly accomplish what Darwin supposed it did in anything less than infinite time. If that meant that some variations look more like speciations, and Darwinism itself more like Lamarckism, then so be it.[25]

3

Between 1883 and 1886, working mostly at the Morphological Laboratory in Cambridge but with a further summer stint in 1884 back at the Chesapeake lab, Bateson mapped the development of the Hampton form of *Balanoglossus* in meticulous detail. For the scientific equivalent of a grand finale, he concluded a series of outstanding papers with a sweeping synthesis, titled "The Ancestry of the Chordata." It concluded, conventionally enough, with a tree diagram relating the vertebrates and the invertebrates. But it began with a terse statement disparaging all such reconstructions as largely deserving the ridicule they had lately attracted.[26] From that disorienting, Agassiz-inspired preface—which reads as if tacked on, grudgingly and peevishly, in response to an unsympathetic referee report—Bateson proceeded to a confident, commanding, and comprehensive essay on segmentation in the animal kingdom. Picking up on one of Brooks's saltationist themes in *The Law of Heredity*, Bateson now argued against the received wisdom that whatever was ancestral to the vertebrates must itself, like the most primitive vertebrates, have been segmented, only less so. For Bateson, segmentation was not a morphological feature like any other, evolving by degrees up the animal scale. Rather, it was "the full expression of a tendency which is almost universally present": a tendency toward the serial repetition or reproduction of parts, where the repeated, reproduced parts can be anything from whole organs up to whole organisms.[27]

Of his relationship with Weldon in this period, Bateson left two rather different impressions. To his wife Beatrice, in a 1906 letter responding to the news of Weldon's death, Bateson wrote, "Until the time—about 16 years ago—when his mind began to embitter itself against me, I was more intimate with him than I have ever been with anyone but you." Yet decades later, Reginald Punnett, one of Bateson's great allies in the fight over Mendelism, recalled that, for all the undoubted closeness of Weldon and Bateson in the 1880s, "in referring to those days Bateson said that he was often made to feel like Weldon's bottle-washer." The apparent contradiction is not difficult to resolve, however. Weldon may well have been, in Punnett's phrase, "a dominant personality," and to that extent, in matters scientific, able to impose his sense of the topics that needed engaging on his more junior friend and sometimes pupil. (Bateson was among the many Cambridge students coached by Weldon before his appointment to a lectureship relieved him of the need for such work; thereafter, Weldon was the one in full-time morphological employment.) But Bateson, eager to strike out on his own, was the more ready to engage those topics on his own terms—and to take inspiration from elsewhere in doing so.[28]

With the Weldon-Bateson relationship in the frame, Bateson's next, seemingly inexplicable move becomes intelligible. In the late spring of 1886, having completed work on his final *Balanoglossus* papers, and with some funding from the Royal Society as well as from Cambridge, he set out on a year-and-a-half-long expedition to the Central Asian steppe, to a region of lakes northeast of the Aral Sea, in and around what is today Kazakhstan. Nothing he had done up to that point would have led anyone to predict such an extraordinary step. To abandon genealogical reconstruction, in line with his grousing remarks against it, was one thing. But to abandon it not for rigorously descriptive embryology in a well-equipped marine zoological station by the sea somewhere, or even for theorizing about segmentation and heredity from an armchair in his rooms at St. John's (Bateson was now, like Weldon, a fellow), but for a one-man trek to a landlocked region of harsh climates and unpromising civilization, was quite another.[29] Recall, however, Weldon's controversy-courting Royal Institution address from the year before: "On Adaptation to Surroundings as a Factor in Animal Development." That topic, in brief, was Bateson's object of study in his new venture. The animals whose development and surroundings he planned to examine were the invertebrates of the steppe lakes—which, he surmised, were attractive for the purpose, both because the lakes were known to vary in their saltiness and because their fauna probably descended from a common ancestral group, inhabiting an "Asiatic Mediterranean" that, it was conjectured, once spread over the whole of the steppe region and beyond. By looking at how the invertebrates differed from lake to lake, and seeing how far those structural differences tracked different levels of salinity, Bateson hoped to throw light on the causal relationship, if any, between variation in environment and variation in development.[30]

As for why he alighted upon the salty steppe lakes, of all places, as the home for such an inquiry, the answer seems to lie, again, with Brooks. In *The Law of Heredity*, he devoted a few pages to the role that a constant environment can play in producing species characters that, though naturalists treat them as typical, are by no means part of a fixed, hereditary, species-defining endowment. By way of example, Brooks mentioned an intriguing Russian case reported a few years back and much discussed since. After the breaking and then repairing of a dam separating parts of a lake near Odessa, a local schoolteacher, M. W. J. Schmankewitsch, had observed the effects of year-to-year increases in water salinity on brine shrimp, *Artemia salina*. Within just a few years, he claimed, the shrimp had become so transformed as to be a different species, *Artemia milhausenii*. In follow-up experiments, he found not only that he was able to induce the same transformation artificially, and in both directions, but that if he gradually reduced the salinity of the water, *Artemia salina* transformed into yet another species—indeed, one classified as belong to a different genus altogether, *Branchipus*. Morphology in these animals thus seemed indeed to track salt levels, over a span conventionally marked by three different species. And, if Schmankewitsch was right, that span could be traversed very rapidly.[31] For Bateson, Schmankewitsch's work was what made a steppe expedition worth trying—and the Royal Society agreed, awarding him the money.[32] In early May 1886, Bateson arrived in St. Petersburg, to answer a question out of Weldon's "private and particular heresy" as filtered through Brooks's *The Law of Heredity*.

Things got off to a good start. Over the summer months, Bateson managed to find some fossils that seemed to support the hypothesis of an Asiatic Mediterranean. He also collected enough brine shrimp to begin addressing the Schmankewitsch puzzle whose solution was his main aim. "I am beginning to get my *Branchipus* into order," he reported to his sister Anna, "and I think that they make for Schmankewitsch's view, but they don't vary so much as his did."[33] Better still, he learned not only that cockles were abundant in the Aral Sea region, but that those arrayed in one cluster of recently dried-up lakes he had stumbled upon in mid-September showed every sign of varying regularly in their shell forms as salt in the surrounding water became more concentrated. On the day of this discovery, he wrote a triumphant letter home detailing it; and it would go on to form the basis of his main publications from the expedition.[34] Further searching made him doubt, however, that the same pattern could be found in the shells of other species. But even as he expressed that doubt to his mother, he shared his ambition to rethink the nature of biological patterning along the motion-and-equilibrium lines he had glimpsed in Foster's classroom and then in Brooks's company:

> I have an idea that in winter I will make an attempt to analyse possibility of treating the evolution, development, progress, or whatever one calls it (meaning thereby the passage of races across the earth, the succession of forms, and so on), as if it was motion or no. One is accustomed to metaphors which assume such possibility (as, for example, that protoplasm goes on lines of least resistance, etc.), and I think it would be worth while to make a rigid examination if there is any truth in this feeling. If there is, then, can biological forces be represented by lines and treated geometrically—and if not, why not? I have often wished to think about this seriously, and I think a winter at Kazalinsk should give complete and suitable, if not exactly academic, leisure for such a purpose.[35]

So high was Bateson's confidence as he settled in for winter (fig. 3.3) that he took in his stride the news that his application for the Balfour studentship had been unsuccessful. In another letter home (from which this chapter's opening epigraph comes), he dismissed the sort of research that the Cambridge morphologists admired, his own *Balanoglossus* work included, as destined for science's dustbin.[36]

4

By no means had Weldon abandoned the risky themes of his Royal Institution lecture. He addressed them again in a short paper communicated to the Royal Society on his behalf by Foster in March 1887. On a research trip to the Bahamas with Florrie, Weldon had found himself intrigued

FIG. 3.3. Some of Bateson's Kazakh associates on the Central Asian steppe, winter 1886–87.

by larvae that kept appearing in his tow-nets—larvae that he recognized, in light of Bateson's recent papers, as belonging to *Balanoglossus*. So distinct were the larval and adult forms of this animal that the larvae had been given a name of their own: *Tornaria*. Weldon reported that once his Bahamian larvae had developed to the point where they had formed gill slits, unexpected variability followed. In some individuals, development proceeded as normal. But in others—the majority, according to Weldon—there was loss of structure, or degeneration, accompanied by an increase in size. What fascinated Weldon was less the degeneration (a major interest for Lankester) than the apparent role of the environment in determining whether an individual larva achieved its ordinary hereditary destiny by developing into an adult, or whether it instead developed along different, variable, sometimes retrograde lines. To Weldon, the fact that almost all of his abnormally developed larvae were caught in deep water, probably driven there by the currents and winds, was a major clue.

Might it be that, in some species at least, the developmental sequence characterizing normal maturation beyond the larval stage requires the presence of the right kind of environmental stimuli? Suppose that natural selection had made the larvae of these species into sensors, capable of detecting the physical or chemical indicators of the standard environment to which the adult form had been adapted. When the standard environment obtained, maturation proceeded through the normal, adaptive-for-that-environment stages, from a normal, adaptive-for-that-environment larva. And when the standard environment did not obtain, maturation did not so proceed. Instead, the late-stage larval features—provided, that is, the environment permitted the organism's continued survival at all—became highly variable, though potentially in well-defined directions. Weldon concluded the paper with what seemed to him four inescapable conclusions:

> *First*, that, at least in some cases, the transmission by a larva of hereditary changes is only possible on the application of the stimuli afforded by particular surroundings; *secondly*, that some larvae, in the absence of these stimuli, but in conditions otherwise favourable, are highly variable; *thirdly*, that the variations produced by a given change in the environment may be of an uniform and definite character; and lastly, that these changes may result, not in the modification of ancestral organs, but in the hypertrophy of those which are purely larval.[37]

Thus did the Bahamian *Tornaria* appear to bear out Balfour's teaching that larvae can be remodeled via the interaction of heredity and environment, however inconveniently for the reconstruction-minded morphologist.[38] If there was heresy here, it lay in treating that possibility not as an obstacle to morphological study but as one of its objects.

The riposte from orthodoxy came a few months later, in the form of letters from a German zoologist, J. W. Spengel. He was, he told Weldon, about to publish a monograph on *Tornaria*, which would show that what Weldon had interpreted as later, aberrant, degenerate developmental stages actually belonged to earlier, normal, complexifying stages. Requesting help in May 1887 from the Naples station's assistant, Paul Mayer, the Cambridge-based Weldon displayed typical good humor in the face of professional calamity. "In a week or two, when I have finished teaching the unhappy students of this university what I hope is morphology, and not something else umgekehrt," Weldon wrote to Mayer, using the German for "other way around" (which, according to Spengel, was how Weldon had gotten his *Tornaria*), "I am going to beard Prof. Dr. Spengel in his den. . . . I venture to think even Dr. Spengel may be umgekehrt in his interpretation of my beast."[39] In the end, however, it was Spengel who did the convincing and Weldon who, thanking Spengel for his efforts, and apologizing to the Royal Society, acknowledged the error in print, via Foster again, in June.[40]

Even without the retraction, however, Weldon's *Tornaria* paper was not timed for glory. The mid-1880s saw the beginnings in Britain of a new polarization on evolutionary, developmental, and environmental matters, provoked mainly by the work of the Darwinian physician-turned-zoologist August Weismann, based at the University of Freiburg. Weismann held that the hereditary substance, whatever its molecular makeup, was sequestered from the rest of the body in the germ cells, and indeed, in the nuclei of those cells. His thinking had stabilized into two related doctrines: first, what Weismann called the "continuity of the germ-plasm," according to which the hereditary substance passed continuously from the germ cells of the parents to the germ cells of the offspring, without being affected by the fates that befell the surrounding bodily tissues; and second, the total non-inheritance of acquired characters.[41] In September 1887, the month before Bateson's return to Cambridge, the latter doctrine got a session unto itself at the BAAS meeting in Manchester, under the title "Are Acquired Characters Hereditary?," and with contributions from Weismann as well as from Lankester and others. Later celebrated as a turning point in British biology for forcing the issue of Lamarckism as it had never been forced before, the session also did a great deal to cement Weismann's association with ideas that, as we have seen, closely resembled ones that Galton had earlier endorsed.[42]

Increasingly, it seemed, the choice was either to back a boldly Lamarckian interpretation of any links between structural variation and environmental conditions or, with equal boldness, to dismiss any such Lamarckian view as utterly discredited.[43] At the Manchester meeting, Weldon gave a couple of talks, so he was almost certainly present for the discussion of "Weismannism" (as it would soon be known), and thus able to give Bateson a firsthand report when he got back to Cambridge the next month.[44] Even while he was in the steppe, Bateson had objected to the way this new dichotomizing rigged the discussion on variation and environment so as to make a vast middle ground suddenly uninhabitable. "By the way," he asked, in the letter where he saluted the advent of post-morphological biology,

> whoever originated that ridiculous piece of bad logic about variations due to environmental change seeming not to be "permanent"? How the deuce should they be on any hypothesis, which supposes that they result from change, which when reversed, or withdrawn, leads naturally to a return to [a] former state? If iron in soil make hydrangeas blue, why is this to be regarded as a false variation? Because the same hydrangea without iron is not blue?[45]

The frustration that Bateson expressed here was that of someone inspired to his labors not because he understood Schmankewitsch's example as a case study of Lamarckian evolution in action, but because he saw it as a window onto the role that constant environments can play in the maintenance of the attributes of certain species. For such attributes to be considered "not permanent" was, Bateson recognized, for them to be downgraded as uninteresting because they were so easily, and boringly, assimilated to the non-Lamarckian side of the emerging debate. Yet they counted as "not permanent" only if one accepted the terms of that debate—terms that, to the likes of Bateson, seemed maddeningly arbitrary. Why, after all, should something that immediately changes back to a previous state with the return of a previous environment be considered wholly different from something that likewise reverts upon a similar environmental change, but does so gradually? Only from a perspective already structured around the question of Lamarckism as the be-all and end-all did the difference of time scale loom large. Otherwise, it faded into the background—where Bateson thought it belonged.[46]

Despite Bateson's long absence from Cambridge, his conversation with Weldon had never really stopped. Nor had the emotional complexity on Bateson's side: references in his correspondence home about Weldon are by turns warm and waspish, playful and prickly. "Weldon seems to have done fairly—getting 'some new' beasts," Bateson had written to Anna in December 1886 from Kazalinsk, "but I greatly fear that he is lost to any work that I think anything of now." At least Weldon had come through with good local gossip: "I wish I understood the Balfour Studentship Election—I fear from Weldon's account it must have been a thorough job [i.e., a stitch-up]."[47] When, early the next year, Weldon sent more news about his recent discovery, Bateson wrote of his delight: "Isn't it funny? Weldon's new beast turns out to be a sort of *Balanoglossus* larva in disguise! Vivant les hemichordates! They are the true friends of the rising naturalist."[48] For his part, Weldon, while preparing his ill-fated Royal Society note on his beastly *Balanoglossus*, managed to send Bateson a pipe through the post, as requested—though the Russian officials unfortunately confiscated it en route. Such was Weldon's generosity that Bateson advised his mother to be ready to nag for the bill.[49] In telling his family in early summer 1887 that he would not be going for the Balfour studentship next time around, however, Bateson begged, "Please don't discuss this with Weldon . . . I wrote him my feelings about it—he doesn't agree with me & I did not expect he would. Our views of life differ more & more widely as time goes on."[50]

By contrast, Weldon seems to have borne nothing but sunny goodwill toward Bateson. In March 1888, as Bateson, newly flush with the Balfour funding he eventually did apply for, prepared to make a brief visit to Egypt for research among its cockles and salt lakes (perhaps in response to criticism at a talk he gave on his steppe research at the Cambridge Philosophical Society in February), Weldon wrote from London that he had just bought, and urged Bateson to buy, a new book-length attack on Weismann by the German naturalist Theodor Eimer. "I see from my casual look through the book, that he treats of such things as the effects of climate on butterflies; and there is a section among those treating of variation . . . with the suggestive title 'Ist Alles Angepasst?' [Is everything adapted?] This is emphatically a question applying to your cockles." And would Bateson's departure be by steamer from Plymouth, Weldon wondered? If so, he would come aboard to make his farewell in person.[51] A few months later, near the end of June, with Bateson now back in Britain, and the Laboratory of the Marine Biological Association in Plymouth up and running, Weldon sent a note from his London club, the Savile, in Piccadilly. He quoted at length from an item in the *St James's Gazette* about the modified habits of English rabbits in Australia and local suspicions that the rabbits had changed due to interbreeding with marsupials. "You see the merit of reading a good old Tory [paper] instead of the P. M. G. [the then-Liberal *Pall Mall Gazette*]," joked Weldon—for, he went on, here was a case far better explained by the view that adaptive-for-Australia changes in habit and associated structure originated with the rabbits themselves. And it did not much matter, he might have added, whether one glossed those changes in Lamarckian terms (by supposing that offspring inherited the altered habits and anatomical structures of their parents), or in Darwinian terms (by supposing that the rabbits who happened to be born with more suitable habits and structures out-survived and out-reproduced the rest), or supposed instead, in the manner Bateson had been exploring recently, that each individual rabbit of each new generation acquired the new habits and structures afresh. Whatever one's favorite style of evolutionary explanation, here was a lovely example of a new environment bringing about an adaptive change in species character.[52]

5

Between the stimulus of dialogue with Weldon and continued immersion in the business of analyzing the steppe specimens, Bateson's own theorizing bubbled away. "My brain boils with Evolution," he wrote to Anna in September 1888. "It is becoming a perfect nightmare to me." What haunted him above all, it seems, was the question of how to understand a curious pattern in his data on cockle variation. He had stressed in the Cambridge talk that while the textures and colors of the shells making up the individual terraces in the Central Asian lakes were remarkably, almost perfectly, uniform, the shapes showed much greater gradation and diversity, with round shells being found even among what were generally the longest shells. Now, as he explained to Anna, he was beginning to think that the key lay in seeing an organism as a "system"—that is to say, as an integrated whole, whose parts are such that if one part should begin to vary in response to environmental change, then so will all the others, and in such a way that these secondarily variable parts could well become much more dramatically altered (and so conspicuous as variations) than the change-initiating, originally variable parts. Only if variation works along such system-wide, ramifying, "correlated" lines, Bateson suggested, could new environments bring forth new, viable versions of old species.[53]

A few days later, the annual BAAS meeting began, in Bath. Weismann dominated at a distance. When it came to matters Darwinian, "the biological world now looks to Professor Weismann as occupying the most prominent position," declared the Darwinian botanist and Kew Gardens chief William Turner Thiselton-Dyer in his address as president of the biology section. Although Thiselton-Dyer confessed to finding aspects of the theory of the continuity of the germ-plasm obscure, he welcomed the impetus it had given to empirical studies of the causes of those variations on which natural selection depended.[54] It was a prophetic speech: that same month, Weismann announced the experimental results for which he is still best remembered. Despite the repeated lopping off the tails of mice, generation after generation, in his laboratory at Freiburg, he found that tailless mice persisted in producing offspring with tails.[55] Experiment and theory seemed to be unanimous: the Lamarckian inheritance of acquired characters did not occur.

Again, Lamarckism for-or-against was not Bateson's issue, nor Weldon's. Even so, they began to calibrate their own positions on variation, heredity, and environment against what they took Weismann's to be. "My dear Bateson," began Weldon in a long letter sent at the end of that same September, "I have not written to you for a long time, because I have the spirit of polemic upon me; and I have wished to consider carefully the words I should say to you." The main thrust of what followed was an insistence on the complexity of the relationship between variation and environments, and the need, in making sense of ubiquitous nonadaptive variation, not to ignore the fundamental Darwinian insight—recall again the Galápagos finches—that whatever cannot be explained as adaptive to present environmental conditions can often be explained by attention to ancestry, expressed as persistent hereditary tendencies. Three further points about this letter, and the light it throws on author and recipient, bear emphasis. First, the tone of friendly challenge that Weldon announced at the start is sustained throughout. "I will tell you three sets of things which ought as it seems to me to annoy you," was how he introduced three examples of extensive nonadaptive variation that made sense once ancestry was brought into the picture. And at the end of the letter, signing off with a series of friendly questions about the progress of Bateson's research, Weldon asked, "And when are you coming to crush one???" Second, in driving home the inadequacy of any ahistorical correlated-variations story such as the one Bateson had sketched in his letter to Anna, Weldon dwelt on the differences between larval characters and adult characters, and how poorly a common environment accounted for those differences—something on his mind in the wake of his *Tornaria* studies. Weldon's reflections on the issue, including the tentative framing of laws governing the partly separate and partly linked evolution of larval and adult characters throughout the animal kingdom, seem energized by Bateson's dismissive attitude to the whole topic. (Wrote Weldon, "You have often, when occupied in scoffing, remarked to me that all pelagic [oceanic] larvae in a given area, must be sensibly under the same conditions"—an attitude that neglected the possibility that outwardly similar larvae could have very different ancestries.) Third, for all the spirited polemic, Weldon aimed not to crush Bateson but to convince him that his steppe findings did not require him to go quite so far out on a limb theoretically as he seemed to think. Yes, some slight structural modifications were undoubtedly brought about by different environments, with the modifications as they were because of particular features of those environments. But none of that required one to deny that, across the potential range of environmental differences and associated structural modifications, underlying hereditary traditions, stable and unchanged, persisted. Such a recognition, wrote Weldon, "would allow of the occurrence of all your variations . . . without touching the question of general influence of conditions on a species from generation to generation. . . . In this way one could understand both your contentions, and those of Weismann?"[56]

South Front of the Laboratory of the Marine Biological Association, on the Citadel Hill, Plymouth.

FIG. 3.4. The magnificent new Laboratory of the Marine Biological Association, on Citadel Hill in Plymouth, in 1888.

Weldon wrote this letter from the new Plymouth marine biological laboratory. Sited on a hill by the sea and complete with aquaria in the basement, abundant lab space on the ground floor, and amenities including a library and dining room above, this splendid stone structure (fig. 3.4) had opened for scientific business at the end of June 1888.[57] Even before the official launch of the laboratory, Weldon—a backer of the scheme from the start—had become a Plymouth regular, indeed the first recipient of funding dedicated to the laboratory's work from the Royal Society.[58] Not, it should be stressed, that he needed the money. Between the wealth inherited at his father's death and, it seems, the income brought in by the family stake in the chlorine industry, he was, as an acquaintance later put it, "never hampered by want of means."[59] Institutional provision of assistants, materials, and laboratory space was simply something the new professionals took for granted as desirable. What the Plymouth laboratory represented for Weldon was thus an irresistibly well-appointed and conveniently located place to pursue the ever-expanding lines of inquiry that interested him. By 1885, those interests had come to include the crustaceans.[60] By 1887, he had begun recording observations of different species of decapod (ten-legged) crustaceans—mainly shrimp, prawns, lobsters, and crabs of various kinds—encountered in and around Plymouth Sound, as preparation for the study for which he received the Royal Society grant, on variation in the decapods.[61] And in early November 1888, a week after sending that long letter on variation to Bateson, Weldon began the new variation project in earnest. "The Book of the Variations of Palaemon serratus [the common prawn]," he wrote on the title page of the first of what would become a set of notebooks titled "Measurements of Crustacea."[62]

Note that Weldon began his study several months *before* the publication of Galton's *Natural Inheritance* in early 1889. Independently of Galton's influence, anyone who, for whatever reason, wished to get serious about variation in animals and plants already knew that being serious meant being quantitative.[63] Yet *Natural Inheritance* made a deep and lasting impression when Weldon read it in Cambridge in spring 1889, as he and Florrie were making preparations for a planned year of research at Plymouth. The then-director at Plymouth, Gilbert Bourne, remembered Weldon that summer as being increasingly "dissatisfied with the methods he had hitherto employed" and, by way of remedy, turning to "the works of Francis Galton," which he now read "more attentively than he had done before." Bourne recalled Weldon "coming into the Plymouth Laboratory one morning, armed with Galton's 'Natural Inheritance,' the contents of which he forthwith proceeded to expound with his accustomed eagerness." Weldon's message, according to Bourne's testimony, was that further progress in biology depended on biologists' embracing measurement and mathematical methods.[64] But again, Weldon was already exploring a more quantitative style of research. What he found in Galton's pages, in the interstices of Galton's own program, were the elements of a solution to the problem of how to recast the big but vague Balfourian questions about heredity, environment, variation, and natural selection that had long fascinated him—and that had most recently led him to start measuring variability—into an answerable form.

Weldon's first attempt at *Natural Inheritance*–inspired research was something he could start right away: a breeding experiment. In the book, Galton had sought volunteers willing to breed pedigreed moths using the methods and, if desired, the moths of an entomologist collaborator. By such experiments, Galton explained, he hoped to accumulate the data that would allow a more precise determination than otherwise possible of the influence of ancestors on the physical characters of their descendants.[65] In June 1889, Weldon wrote to Galton to put himself forward. It proved, however, an inauspicious start: many of the larvae succumbed to mold; and then came, as Weldon put it in a note of apology sent in November, "an unfortunate accident." Somehow—Weldon did not explain how, and he was, he said, away when it had happened—the covers had got left off the bottles holding the caterpillars, and many escaped.[66]

Weldon made no mention of a rather different but no less Galtonian project that had already been launched and, indeed, was nearing completion: an inquiry into whether populations of wild organisms exhibit the same normal-curve distribution of characters that Galton had famously found—and, in *Natural Inheritance*, magisterially analyzed—in humans. That question was of interest, Weldon suggested, because of what seemed, on the face of it, a major difference between wild species and domesticated ones; namely, the exposure of the former, but not the latter, to the full force of the Darwinian struggle for life, and so to natural selection. The organism Weldon had chosen to study was one plentiful in Plymouth Bay: the common gray shrimp, *Crangon vulgaris*. Using a pair of compasses and, as needed, the crosshairs of his microscope, Weldon had measured the lengths of several of the exterior parts on each shrimp, including the carapace (upper-body covering) as a whole; a distinct section of the carapace, the last-but-one segment on the lower-body covering; and the last segment, nearest the tail. He did this for two hundred shrimps from Plymouth, and also for a large number of spirit-preserved specimens sent by friends working at the seaside elsewhere in Britain. In each case, he had found, after due plotting, calculating, and comparing, the same result: the distribution curve was abnormal. In a wild population, it seemed, magnitudes did not clump around an average and then slope off symmetrically to either side. Natural selection, Weldon concluded, must have a skewing effect—which meant that the normal-curve regularity of the populations studied by Galton (mostly humans, but also sweet peas and moths) might well be an artifact of domestication.[67]

Recall that natural selection and how to study it were not exactly major themes of *Natural Inheritance*. There were, however, a few pages where Galton laid out his expectation that the normal curve would be found in populations whether or not they were under the influence of natural selection. Other readers would not, and did not, see much in these pages to get excited about. Viewed through a Balfourian prism, however, they gave the search for patterns in Weldon's shrimp data something extraordinarily valuable: an evolutionary—indeed better still, Darwinian—significance.[68] Weldon's paper, for which Galton was the referee, went on to be rejected. But it initiated a scientific relationship between them that would last the rest of Weldon's life—a life devoted to exploring the immense potential Weldon found in Galton's work for advancing the perspective on Darwinism that had been so lively in Balfour's teaching. As we will see, Weldon made the choices he did—concentrating on some aspects of Galton's biology while challenging or ignoring others—thanks to a training against which Weismannism, and later Mendelism, looked hopelessly impoverished and impoverishing.

6

Plymouth was also where, from April to October 1889, Bateson was based, conducting research on the sense perception of fishes and shellfish, much of the time alongside Weldon, who was then in the midst of his own pure and applied research on crustaceans. For Bateson, these months proved productive and provocative.[69] Just as he finished work on the final, much-expanded version of his Aral Sea and Egyptian cockles paper, submitting it to the Royal Society in mid-May, Weldon was beginning to preach the virtues of Galton's *Natural Inheritance* to all and sundry.

Given Galton's virtual but vivid presence at Plymouth as the apostle of a vigorously quantitative biological science of variation, the copious measurements included in Bateson's new paper on that topic, and his steppe-tested esteem for Galton's classic how-to guide from his African exploration period, *The Art of Travel*, it was utterly unsurprising that, when Bateson's paper came out in early autumn 1889, he sent Galton a copy. A brief but friendly bout of correspondence ensued. If Bateson did not quite succeed in answering Galton's query as to whether there might be shellfish that—perhaps rather better than moths and plants—might be suited to an experimental partitioning of the effects of ancestry and environment on variation, he nevertheless had now, like Weldon, brought himself to Galton's attention as a promising young biological collaborator.[70]

A significant and enduring change in Bateson's overall outlook as a biologist was now fully in swing: a shift away from environmentally adjusted variations as a major focus for research, and a dropping with it of any lingering allegiances he might have had to the theories vying to explain such adjustments. He had come to see that his cockles study, for all its evidence of the regularity with which shells and saltiness varied, was not remotely representative of his findings from the steppe. Such regularity was the exception, not the rule. Schmankewitsch was a bust. Shortly before departing for Plymouth, Bateson wrote to an old Cambridge teacher, Alfred Newton, that for the most part, and notwithstanding the cockles paper soon to be published, "no variations can be found which can be shown to be correlated with the constitution of the waters."[71] Liberating himself from the questions that had ever made him think otherwise—questions that were, ultimately, Weldon's—Bateson found himself able to turn to different ones, more securely his own. He went back to the Brooksian issue that framed his chordate ancestry paper, the repetition of parts, but approached not as a source of insight into the reconstructed tree of life (he was done with that) but as a problem in the biology of variation. Now he began a systematic program of cataloguing and collecting. At Plymouth, when someone found crabs with an extra pincer or two, or other examples of freakish growth, they went to Bateson. He surveyed his initial results under the title "On Some Cases of Abnormal Repetition of Parts in Animals," hinting at both a large-scale work to come and a spelling out in that work of momentous theoretical implications.[72]

The course Bateson was starting to chart for biology could hardly have been less congenial to Weldon. For where Weldon was seeking to measure differences so small that less careful investigators could easily miss them, Bateson wanted to concentrate on whole parts whose unusual repetition was unmissable. And where Weldon was eager to press home statistical inquiries as a means of understanding the interplay of development, environment, incremental variation, and natural selection in the wild, Bateson—who had not long before traveled all the way to Siberia to study that interplay—now passed over these topics almost entirely, stressing instead how far the facts of repeated parts supported a view of variation as proceeding by "integral steps," such that, in his words, "the control which the symmetry of the body exercises over variations" matters more than anything else in propelling evolutionary change along the directions it takes.[73] What was, then, already a pronounced divergence between these two would grow over the next decade into open—and increasingly bad-tempered—disagreement.

4

Royal Entrances (and Exits)

What have you done about proposing Weldon for the Royal? Sedgwick thinks that Bateson ought to be put up also—and I am inclined to agree. I don't quite approve of holding men back because there are two or three good enough to come in. It seems to me that there must sometimes be more good men in one subject than in another—and that unless the Royal Society definitely limits its election to 10 mathematico-physicists, 2 chemists, 2 doctors and one *biologist, we have simply to propose those men whom we think worthy of election.*

E. RAY LANKESTER TO ALFRED NEWTON, 4 FEBRUARY 1889

The most dangerous thing about the discussion seemed to me the deliberate antithesis which these men sought to establish between what they called "Natural History" on the one hand, and any sort of statistical enquiry . . . on the other. Several members of the Ctee., of whom Bateson is the leader, have the idea of such an antithesis, and are strongly influenced by it. However sincere and able such men may be, I think they do harm, and I think it would be a great misfortune if a Committee of the Royal Society were to officially adopt their view.

W. F. R. WELDON TO FRANCIS GALTON, 28 NOVEMBER 1899

I want your Husband's help & sympathy inexpressibly just now. I see that . . . these Cambridge biologists took the opportunity of your Husband's death to get a brilliant paper rejected at the R.S. & this within a few weeks of the last letter to W.F.R.W. asking whether in view of referees' reports it should or should not be printed. But I am very likely to make some horrid slip for want of R.'s aid, whereas the whole of Cambridge is in arms against me; it is Saint Biometrika contra mundum!

KARL PEARSON TO FLORENCE WELDON, 19 OCTOBER 1906

Is the Royal Society good for science? What, exactly, is the point of a self-constituted, self-congratulatory, luxuriously accommodated scientific elite, so powerful that it can dictate to the rest of the community—that great mass of the aspirant-FRSs—yet so weak that the contents of the dictates get decided by the scientific version of office politics? And just how intellectually awesome are the decision-makers anyhow? Such questions were not at all idle in Britain in the nineteenth century. Then as now, the seventeenth century was seen as the society's golden period. Victorian counterparts to Boyle, Newton, Hooke, and company tended to be more closely identified with other bodies, of more recent vintage: Faraday with the Royal Institution, founded at the end of the eighteenth century; Darwin with the Geological Society, which was even younger; Huxley with the BAAS, which was younger still. Nor were the questions restricted to backstage grumbling. In 1830, Charles Babbage published a rebuking pamphlet, *Reflections on the Decline of Science in England and on Some of Its Causes*—the major cause, in Babbage's view, being the Royal Society. What science needed was a cross-channel Académie des Sciences. What it had was a club for posh dilettantes and patent-medicine hustlers.[1]

The history of the society over the rest of the century is a history of standard-raising reform, in step with the wider entrenchment of the sciences' position in the nation's culture and institutions. The reason that Lankester in 1889 had to ask whether Weldon should be proposed alone or with Bateson, for example, was a cap on numbers introduced in 1847 in order to make election more competitive—which it did. Attempts were also made to improve the caliber of the society's publications and discussions. Alongside the venerable *Philosophical Transactions* came a new journal, *Proceedings of the Royal Society of London*, publishing abstracts and sometimes full texts of the papers read at the weekly meetings, principally as a means of engaging the fellows more fully in the scientific life of the society. A move to new premises in Burlington House in bustling Piccadilly, around the corner from the Royal Institution, was undertaken partly in the hopes that the setting would promote livelier meetings (fig. 4.1). As we shall see, a very public, and ultimately very consequential, falling out between Weldon and Bateson in 1895 was the result of an experiment in the stage management of debate initiated by the Royal Society's council. Medals and committees proliferated as the society became more ambitious for science and—thanks in part to that ambition—had more funding to disburse.[2]

Throughout the 1890s, Weldon and Bateson dealt with each other most often and most directly not in Cambridge or in Plymouth but in Piccadilly. Weldon was elected in 1890, Bateson in 1894. The Evolution Committee they served on so unharmoniously—along with Galton and a new collaborator of Weldon's and Galton's, the London mathematical physicist and socialist thinker Karl Pearson (elected in 1896, awarded the Royal Society's Darwin Medal in 1898)—funded the hybridization research that would eventually lead Bateson to the "Versuche." For Weldon's part, when he wanted to see for himself what the growing fuss over Mendel was all about, he borrowed the relevant volume of the Brünn Association's an-

FIG. 4.1. The Royal Society's meeting room at Burlington House. The portrait behind the president's chair was of Isaac Newton.

nual proceedings from the Royal Society library. It has been easy to miss or minimize the role of the society in the making and unfolding of the Mendelism debate. This chapter puts "the temple of science at Burlington House," and the business of getting in and then getting on, at the center.[3]

1

To be elected a fellow of the Royal Society, you first needed your proposers to fill out and submit a form listing your publications, along with the signatures of all the fellows prepared, from personal or general knowledge, to support your candidacy. At that point, you entered the candidate pool, from which the council once a year fished out what it judged to be the fifteen most eligible candidates, who were then presented to the fellows for election. Weldon's form arrived at the society in February 1889 with signatures from Lankester, Newton, Sedgwick, and thirteen others, among them the aged Huxley. A few months later, when the council drew up its slate of candidates, Weldon was not among them. Just a year later, however, he was. Elected in early June 1890, he duly took part a couple of weeks later in the society's annual "ladies'" conversazione (meaning that it was for women and men, unlike an earlier, men-only event), exhibiting larval and adult inhabitants of Plymouth Sound on behalf of the laboratory.[4]

Why the change of fortune? It was down to a single new paper, initially rejected by the society after Weldon submitted it at the end of 1889. As noted, the paper used newly collected measurements of various and varying parts of the bodies of wild shrimp to throw doubt on Galton's assumption in *Natural Inheritance* that the normal-curve distributions he had found among humans and domesticated lineages of moths and sweet peas would also be found in the wild, where natural selection operates.[5] Predictably enough, the society sent the paper for review to Galton, recipient of its Royal Medal in 1886 for his services to the new statistical biology. Predictably enough, he found it unpersuasive, especially given what he reckoned were sets of measurements too small to support such large claims. But Weldon took the rejection well, and Galton responded encouragingly to Weldon's letter in reply. Thus, in early January 1890, there began an extensive correspondence between them on Weldon's measurements and how to interpret them. Before long Weldon came to accept that his ever-fuller sets of measurements indeed showed what Galton had expected: that variations in wild shrimp follow the normal curve.[6]

The conclusion Weldon drew—and here, too, he affirmed Galton's brief treatment in *Natural Inheritance*—was that, for a certain body part in a particular group of organisms in a particular place, natural selection affects not the error-curve normality of the variations but the average size, and the width of deviations from the average size. That was what made shrimp from geographically distant Plymouth, Southport, and Sheerness biologically distinctive, though all belonged to the same species, and no individual shrimp was in all respects identical to any other. If, as Galton wrote in *Natural Inheritance*, "the mediocre members of a population are those that are most nearly in harmony with their circumstances," then the average size of each race of shrimp must be the size most conducive to survival under the conditions in that locality. Likewise, the width of deviation would go up or down depending on the strength of natural selection acting on that character in a particular locality. Where selection was strong, size would be expected to deviate little from the average. Where selection was weak, size could be expected to deviate a great deal. The construction of frequency curves thus functioned as a new kind of selection detector, revealing, as Weldon put it in the revised paper that he sent back to the society in mid-March, "the selective action of the surrounding conditions—an action which must vary in intensity in different places."[7]

Disappointingly for Weldon, the paper sparked no discussion at all when it was read at a meeting in mid-April. But Lankester, at least, was kind about it, Weldon reported to Galton; and its subsequent printing in the *Proceedings* gave it a much wider audience. Before May was out, Galton was congratulating Weldon on his imminent election.[8]

"The Variations Occurring in Certain Decapod Crustacea. I. *Crangon vulgaris*" nevertheless fell short of Galton's hopes for it. After sending *Natural Inheritance* to the press, he had realized that, mathematically speaking, regression—the most mathematically developed idea in the book—was but a particular, cross-generational form of a relationship that could be framed much more generally. Given a set of measurements of stature, or exam scores, or whatever from one population, and another set of measurements from another *or even the same* population, one could ask to what degree abnormality in the first set is, on average, accompanied by abnormality in the second. Galton called this measure of linked abnormality "co-relation," indicating, in the *Proceedings* paper where he introduced it, that he saw himself as giving newly quantitative expression to the old biological notion of the "correlation of structure": that when certain parts of organisms vary, other parts tend to vary too.[9] In a letter to Galton in mid-February 1890, Weldon had reported himself busily learning "your test of co-relation," promising it "shall be tried upon the shrimps as soon as possible."[10] But as the correlational work proceeded, Weldon soon realized that the task ahead was very considerable. Writing from Plymouth in early March, he asked for Galton's blessing to publish the paper as it stood—before the Royal Society began to raise awkward questions:

> Would the curves, without any co-relation, be worth publishing soon? I do not ask from impatience but for this reason: I am sent down here by a joint committee of the Royal Society and the Marine Biological Association, to report upon the Crustacea here, and especially upon the Lobster fishery. It was understood that I should do what I liked, except with regard to the Lobsters, provided I did something. The glorious thing which I did at Christmas [i.e., the rejected paper] must have disgusted this committee rather much: and I am rather anxious to remove as much of their disgust as may be,—or at least to show that I have not simply been idle. This is a very paltry reason for wanting to print something, but perhaps you will see that it has some justification.[11]

Blessing secured, Weldon spent the next few years making good on two promissory notes at the end of the paper: to apply the new correlation test to the shrimp measurements, and to extend his measurement-and-mathematics program to "races" (that is, local populations) of other organisms.[12] His next paper, presented to the society in spring 1892 and duly published in the *Proceedings*, summed up the correlation testing work carried out on measurements gathered from still larger numbers of shrimp, now from five places (thanks to colleagues from Helder in Holland and Roscoff in France). The remarkable upshot was one that he had already glimpsed in 1890: for all the selection-induced variational diversity of the local races of *Crangon vulgaris*, the correlation measure for any two parts held constant across the races. These correlation constants, in other words, were species characters, previously unknown to science. In autumn 1893, again at a meeting and then in the *Proceedings*, he presented the results from redoing the whole exercise with a new species, the shore crab *Carcinus maenas*, using eleven parts from races collected in Plymouth and Naples. Once again, the measurements clustered with normal-curve symmetry and the correlations proved constant. The sole apparent exception to the normal-curve rule was the "frontal breadth" (the crab equivalent of a forehead) of the Naples race, for which the measurements gave an asymmetrical curve. But Weldon showed that this curve could itself be decomposed into two overlapping normal curves, representing two races-within-the-race. "It cannot be too strongly urged," Weldon wrote in triumphant, bordering on triumphalist, conclusion, "that the problem of animal evolution is essentially a statistical problem."[13]

For checking his analysis of the frontal-breadth curve, Weldon thanked Pearson. The two men were colleagues at University College in London, where Pearson, a few years older, was professor of applied mathematics (from 1884) and Weldon the Jodrell Professor of Zoology (from 1891). Initially they had joined forces to oppose a plan to remake the College, and the wider University, into a teaching-only institution. But they soon enough discovered shared interests and complementary expertise. Pearson's published research in applied mathematics up to that point had been concerned mainly with aspects of ether theory. As a thinker, writer, and lecturer of enormous energy and wide interests, however, he ranged over the whole of knowledge, including philosophy and history, often in directions connected with his deep commitment to socialism. The 1880s and '90s were a period when Darwin vied with Marx as a socialist hero, especially in London, which teemed with socialist luminaries, lectures, and discussion groups. Pearson (fig. 4.2) belonged to this world, and introduced *Natural Inheritance* into it. In March 1889, shortly after the book's

FIG. 4.2. Karl Pearson in 1890. He changed from "Carl" to "Karl" in late 1880, shortly before he volunteered to translate *Das Kapital* into English. (The other Karl turned him down.)

publication, he made it the subject of a superb lecture to fellow members of one of those groups, the Men and Women's Club. He apologized at the start for returning them yet again to heredity and its implications—a topic from which they might reasonably have hoped for a break.[14]

Frequency curves soon joined sexual selection and natural selection in Pearson's intellectual repertoire. But it was his growing acquaintance with Weldon that turned these enthusiasms into a new line of biological-mathematical research.[15] At the Royal Society meeting on 16 November 1893, the paper read after Weldon's on his crab curves and correlations was an abbreviated version of a new memoir from Pearson, setting out the general mathematics of curve decomposition or "dissection." Published in full in the *Philosophical Transactions* under the title "Contributions to the Mathematical Theory of Evolution," it became the first installment in what eventually grew, over the next twenty-three years, into a twenty-six-part series.[16] And the flow of stimulation was two-way. Early in 1894, around the time that Pearson published a popular essay on the roulette wheel at Monte Carlo as not always conforming to the expectations of probability theory, Weldon, with help from his wife and a clerk, undertook a monumental sequence of dice rolls—over 26,000 rolls of 12 dice—putting that theory to empirical test.[17]

But biology, of a recognizably Balfourian, adaptive-development-in-environments cast, remained as central to Weldon as ever, though it was more visible in his correspondence than in his publications. Consider three passages from his letters to Galton in 1890. While working on the shrimp variation paper in February, Weldon wrote to Galton with a challenge. Suppose that among newly born shrimp there is a certain extent of variation, and that later on, among the adults, the variation has narrowed. How can we tell, by statistics alone, whether that narrowing was due to natural selection, as Galton presumed, or instead, in Weldon's words, to "the modification of each individual in such a direction that its deviation from the median is reduced"? Did the extreme individuals die off, or did they converge developmentally on the forms of the more average individuals?[18] To raise the latter possibility was to touch on what Weldon, in his 1885 Royal Institution lecture, had called "adaptation." In another letter a few days later, he invoked that terminology explicitly in pursuing a similar, and similarly skeptical, line of questioning, this time about Galton's grounds for transferring the findings from pedigreed moth lineages to wild races. From breeding experiments, Weldon explained, it was clear that moth body size could be increased by selection, but also that it could be increased by raising the moths in what turned out to be more favorable conditions, not always themselves easily analyzed. So, asked Weldon, "in any given case in which [the moth] Selenia is found to vary in size, in a wild condition, how is one to assign the proper share of responsibility for the change to Selection on the one hand, and to Adaptation on the other"?[19]

What the Galtonian student of selection needed, Weldon saw, was an organism that showed normal-curve variation yet was developmentally insensitive to changed conditions. By September of that year, Weldon was writing excitedly about a potential candidate: the ditch shrimp *Palaemonetes varians.* ("I think I have the beast you want to do all kinds of things with . . . [It] behaves in a very strange manner.") In experiments at Plymouth strongly reminiscent of the work quasi-assigned to Bateson in the steppes, Weldon found that, though there were distinctive saltwater and freshwater forms, gradual diminution of the salt concentration in water inhabited by the saltwater form produced no change in the normal curve of variation in the offspring. Here, it seemed, was a species ripe for selection experiments that really would provide a safe basis for theorizing about selection in the wild.[20]

For whatever reason, it was not ditch shrimp, however, but shore crabs that would become the objects of Weldon's most extensive studies of natural selection through the rest of the decade. They would prove a troublesome choice.

2

Throughout 1890, Bateson endured one professional setback after another. His Royal Society candidacy form, submitted in February, was discussed in May but not selected. In June he applied for the position of deputy to the Linacre Professorship in Comparative Anatomy at Oxford, furnishing the electors with a lengthy manifesto for his new vision for a steam-age biological study of variation. But the post—and indeed, the professorship—went, as everyone expected, to Lankester, who duly resigned his professorship at University College, thus making room for, as it turned out, Weldon. A brief moment of triumph came in the summer, in the form of a book contract. By the autumn, however, it was clear that there would be no renewal of Bateson's Balfour Studentship at Cambridge, and little official encouragement there for him to apply for Weldon's morphological lectureship.[21]

Whatever persecution-complex tendencies Bateson harbored now began to get the better of him. He wrote to Sedgwick in October offering to give up his room at the Morphological Laboratory, as the new, distinctly non-Balfourian direction of Bateson's research was so plainly out of favor. Sedgwick—whose signature had topped the list on Bateson's Royal Society candidacy form—wrote back to tell him he was mistaken, chiding him for being "a trifle touchy on this point, perhaps a little morbidly so."[22] Bateson never lost that morbid touchiness. But he kept the room, and stayed on at Cambridge, soon eking out a living as a college steward at St. John's—not the worst of appointments (he looked after the kitchen garden), but a far cry from the academic position that he craved, and that new FRSs were expected to hold. At the Royal Society, meanwhile, the council continued to decline to put him on the ballot. But they changed their minds in 1894, following the publication, at last, of his book. For the rest of his life, Bateson saved press clippings reporting his election—what one newspaper called "the blue ribbon of science."[23]

The immense scale of the modest-sounding *Materials for the Study of Variation*—nearly six hundred pages—caught everyone by surprise. But there had been plentiful signs, in the run-up years of institutional failure, of a developing research program of unusual energy, range, and originality, as well as contrariness. In 1890–91, in a paper read at the Cambridge Philosophical Society and then at the Linnean Society, Bateson presented evidence of irregularities in the numbers and symmetries of petals in two wild and two cultivated flower species, some examined by him around Cambridge and some by his sister Anna in Wales. He contrasted the slight variations familiar from the work of Galton on stature and Weldon on shrimp with variations of this kind, where neighboring individuals in a series differ by having more or fewer parts—each perfect in itself, and with the parts around it often reorganized into an accommodating new symmetry (as opposed to staying where they were, as per the old symmetry, but with new squeezings-in or subtractions). This second, neglected class of "integral" variations ("which, if they occur at all, occur in their complete form always") was important, Bateson held, because the more that knowledge of its extent grew, the less the student of evolution needed to rely on the theory of natural selection and its increasingly doubtful assumption about adaptive utility. Not, Bateson hastened to add, that Darwin himself was guilty of ignoring integral variation: see the pages, he advised, in *Variation* on non-blending character versions, notably in the snapdragon varieties Darwin crossbred.[24]

The symmetry of perfect-from-the-start parts was the theme again when Bateson exhibited—as non-FRSs often did—in the Royal Society Ladies' Conversazione in June 1891, displaying a mechanical model illustrating his new theory of the complex mirror-image arrangements assumed when supernumerary legs arose around normal ones in beetles (fig. 4.3).[25] At the Entomological Society as well as in the pages of *Nature*, he spent much of 1891–92 engaged in public criticism of the evidence-light claims of the Oxford Darwinian E. B. Poulton—one of the "apologists of Adaptation," Bateson called him—about the supposed adaptive value of certain insect characters.[26] At the Zoological Society in February 1892, in a paper describing animal skulls with more or fewer teeth than usual, he blasted Darwinian teachings on wild versus domesticated variation (there was no difference, said Bateson) and on homologous forms (which should be thought of, he said, less like successive states of a modified piece of wax, and more like the results of independent meltings-down and recastings).[27] At the same venue in November that year, he

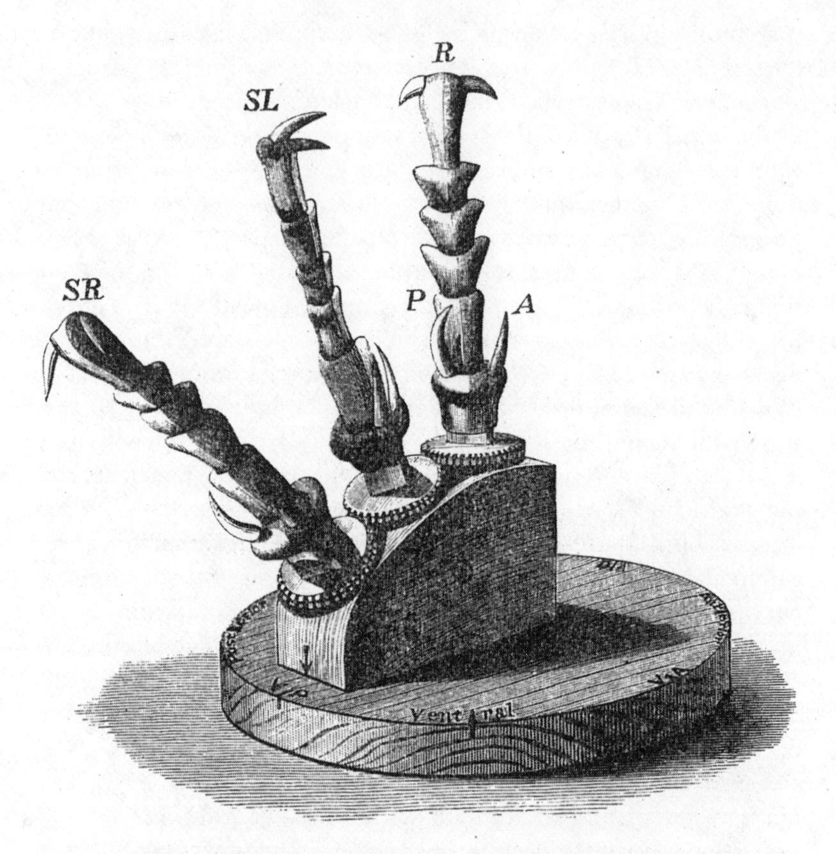

FIG. 4.3. A machine devised by Bateson to show, by two sets of rotations, how extra or "supernumerary" left (SL) and right (SR) legs on the body of a beetle are oriented in space relative to each other and to the normal leg that they duplicate—here, a normal right leg (R).

presented new research on size variation in wild populations, à la Weldon. The most striking thing that Bateson found among his earwig tail-forceps and beetle horns, however, was clustering into two widely separated error curves—what Bateson called "dimorphism." "To those who are acquainted with the chapter on Organic Stability in Galton's *Natural Inheritance*," he wrote in the published paper, "this will be recognised as an instance of Variation about two positions of stability, the intermediate position being one of less stability."[28]

All this time, Bateson was engaged in a vast correspondence with naturalists around the world, while quarrying libraries and museums closer to home for information on integral variation in animals in all its diversity, and alarming his publisher, Macmillan, with projections of an ever larger, ever more copiously illustrated book.[29] Completed in late December 1893 and published in early February 1894, *Materials for the Study of Variation Treated with Especial Regard to Discontinuity in the Origin of Species*—to give it its full title—offered detailed commentary on nearly nine hundred cases of what Bateson now called "meristic" variation (from *meros*, Greek for part) in animals. His ideas about symmetry, adaptation, homology, dimorphism, organic stability, and much else appeared embedded in a lengthy programmatic introduction representing the study of discontinuous variation as an ineluctable next move for students of evolution. Their problem, as he saw it, was that, whether glossed in a Darwinian or a Lamarckian fashion, the conventional view of evolution assumed that changes in the forms of living things track changes in environments. But that, according to Bateson, had turned out to be incorrect, for whereas environments form continuously grading series, species form discontinuous series (hence their classification as distinct species in the first place). It was time for a fresh start: for an end to lazy speculations about might-have-been phylogenies and could-be utilities; for gathering evidence without prejudice as to the possibility of sudden, large variations in wild species; for developing and testing theories aiming to explain those variations by appeal to internal rather than to external dynamics. Bateson meant to relaunch the whole evolutionary project, putting discontinuous change at the center.[30]

Recall that letter he sent from the steppes in 1886 about wanting to turn biology into a geometrized science of forces and motion. In *Materials*, he affirmed that ambition as well as the congruence he saw between the goal of geometrized, dynamic biology and the book's emphasis on the repetition of parts and other kinds of symmetrical patterning:

> Looking at simple cases . . . there is, I think, a fair suggestion that the definiteness of these variations is determined *mechanically*, and that the patterns into which the tissues of animals are divided represent positions in which the forces that effect the division are in equilibrium. On this view, the lines or planes of division would be regarded as lines or planes at right angles to the directions of the dividing forces; and in the lines of Meristic Division we are perhaps actually presented with a map of the lines of those forces of attraction and repulsion which determine the number and positions of the repeated parts, and from which Symmetry results.[31]

In the physics of the day, and above all at Maxwellian Cambridge, "mechanically" conjured up not clockwork but vibrations. At a time when waves and whirls were overtaking matter as the ultimate explanations for regularity in nature, Foster had taught, and Bateson had learned, that a

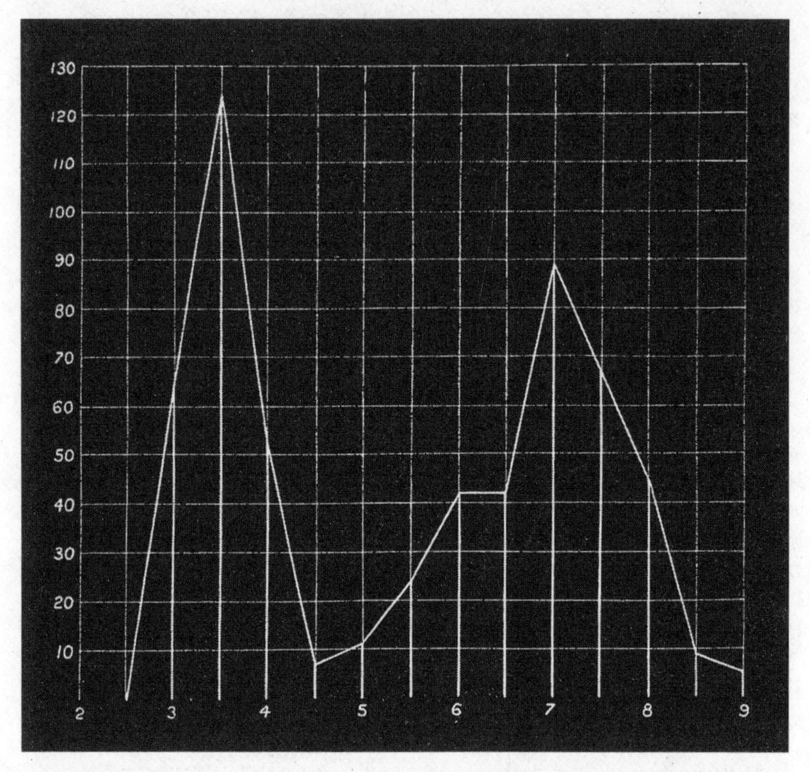

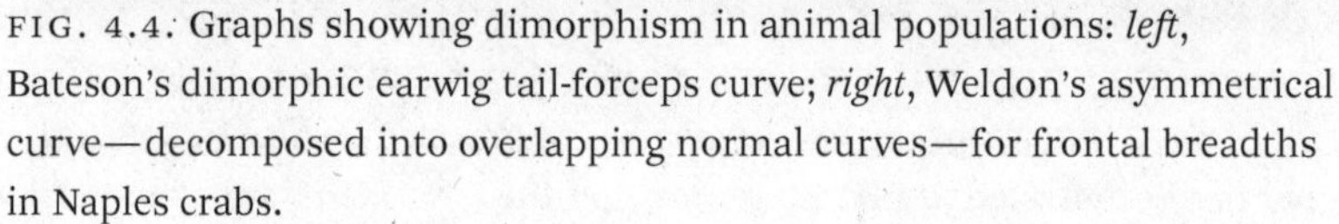
FIG. 4.4. Graphs showing dimorphism in animal populations: *left*, Bateson's dimorphic earwig tail-forceps curve; *right*, Weldon's asymmetrical curve—decomposed into overlapping normal curves—for frontal breadths in Naples crabs.

living thing is a vortex. This theme—so important for understanding why Bateson became a Mendelian, and what kind of Mendelian he became—will be one we revisit in the next chapter.[32]

After the publication of *Materials*, Macmillan sent out eighteen presentation copies. Foster and Brooks were on Bateson's list. So was Weldon.[33] Although Bateson and Weldon were no longer in regular correspondence, the still-just-about-friends had kept tabs on each other's research. As mentioned, Bateson had cited Weldon's 1890 shrimp paper, by way of distinguishing the different sorts of variation they studied. Weldon sent the odd (in both senses) bit of animal-parts information Bateson's way.[34] And there are small but telling signs that Weldon knew some of Bateson's papers. Within two weeks of the reading of Bateson's paper on dimorphism in insects at the Zoological Society, Weldon, working at his crab data, managed to convert what had been a stubbornly asymmetrical curve into double-humped normality (fig. 4.4), and wrote to Pearson with

The only case in which an undoubtedly asymmetrical result was obtained is that of the frontal breadth of the Naples specimens. From an inspection of the curve of distribution of these magnitudes, I was led to hope that the result obtained might arise from the presence, in the sample measured, of two races of individuals, clustered symmetrically about separate mean magnitudes. Professor Karl Pearson has been kind enough to test this supposition for me: he finds that the observed distribution corresponds fairly well with that resulting from the grouping of two series of individuals, one with a mean frontal breadth of 630·62 thousandths, and a probable error of 12·06 thousandths; the other with a mean breadth of 654·66 thousandths, and a probable error of 8·41 thousandths. Of the first race, Professor Pearson's calculation gives 414·5 individuals, of the second, 585·5. The degree of accuracy with which this hypothesis fits the observations may be gathered from fig. 3.

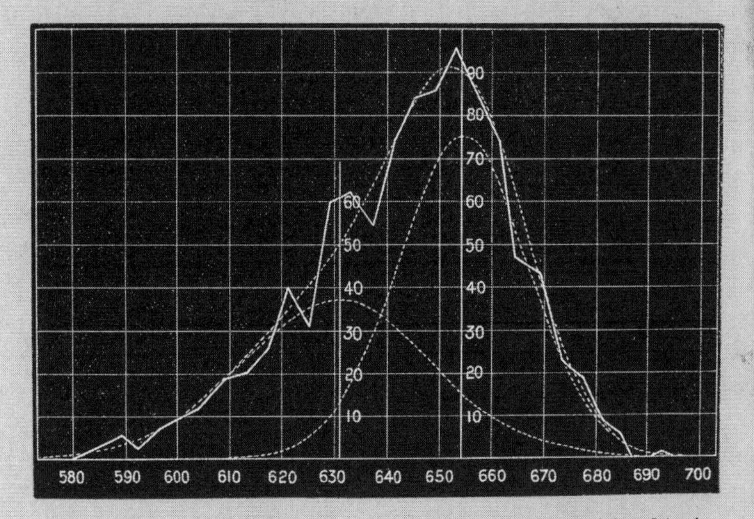

Fig. 3.—Diagram to show the distribution of all observed frontal breadths in the Naples specimens. The horizontal scale represents thousandths of the carapace length, the vertical scale numbers of individuals. Each ordinate of the upper dotted curve is the sum of the corresponding ordinates of the two component curves.

We may, therefore, assume that the female *Carcinus mœnas* is slightly dimorphic in Naples with respect to its frontal breadth; and that the individuals belonging to the two types are distributed in the proportion of nearly two to three.

the good news.[35] (Again, Pearson's whole career in biological statistics started with the mathematizing of that conversion; so there is a sense in which he owed that career at least as much to Bateson as to Weldon.) The reflections with which Weldon ended the 1893 paper where his double-hump analysis appeared—concluding that "the problem of animal evolution is essentially a statistical problem," so that henceforth (to quote from the final line) "the only legitimate basis for speculations as to [the] past history and future fate" of species will be numerical answers to questions about the distribution of degrees of abnormality, associated death rates, and so on—are jarring in both their stridency and their loose connection to what precedes them. But they make sense as a rejoinder to a comparable statement in the Batesons' flower symmetry paper: "In the absence of some knowledge of the mode in which variations occur, it is useless to guess at the relationships or past descent of existing forms; while conjecture as to the developments which may in the future be possible to these forms is still more hopeless."[36]

Weldon teased Bateson for his book's "fin de siècle Arianism," as Weldon put it in a letter. But he generally declined to play Defender of the Faith to Bateson's Heretic. Indeed, Weldon thought Bateson had done biologists a favor by drawing attention to some neglected classes of variation, and said so, most effusively in a review in *Nature* in May 1894. Weldon especially admired the treatment of symmetry in supernumerary legs and the coverage of what Bateson enduringly dubbed "homeosis": the growing of a complete and perfect part somewhere other than in its usual place (e.g., a foot where an antenna should be).[37] And Weldon was already in print as accepting the fact that sometimes a single population could harbor two types, with their own mean values and bell-curve distributions. In a letter to Galton on the same day as the one to Pearson about the double-humped curve, Weldon wrote, "Either Naples is the meeting point of two distinct races of crabs, or a 'sport' is in process of establishment," adding, "You have so often spoken of this kind of curve as certain to occur, that I am glad to send you the first case which I have found."[38]

But Weldon had criticisms too, even severe ones, putting some in the review and more in letters to Bateson in February and March. The conflict Bateson saw between continuous environments and discontinuous species disappeared, Weldon pointed out, once one took seriously, as Darwin always insisted one should, the overwhelmingly greater importance of cohabiting organisms than of physical conditions in the struggle for life and so in the shaping of animal and plant forms. To the extent, furthermore, that physical conditions might have form-shaping power—as when an abnormally hot or cold or wet or dry summer, for example, leads to the development of abnormal morphologies more frequently than usual—Bateson's method of willy-nilly classing together specimens collected across different times and places made the detection of such effects impossible. That collecting style suffered, too, from its tendency to reflect the bias of museum curators and traditional naturalists toward remarkable specimens, whose appearance of being on the opposite side of an unbridged gap from typical forms was at least sometimes due to less remarkable bridging forms going uncollected or unnoticed. As for the Galtonian notion of discontinuous positions of organic stability, Weldon pressed at a couple of points of potentially fatal unclarity. He brought Bateson up to date on the recent dice-rolling experiment by way of illustrating how, with no breaking of the ordinary laws of chance, low-probability deviations from the norm (e.g., 11 out of 12 dice showing a 5 or a 6) can, in a given sample, happen more often than higher-probability deviations (e.g., 10 out of 12). How many observations did one have to make before being entitled to say that the pattern of variation in a race or species was truly unexpected? It was difficult to say. It was also difficult to say how large a deviation had to be in order to count as a discontinuity. In the dimorphism Weldon had found in his Naples crabs, he noted (and he now sent Bateson the paper), the stable means were quite close. At the limit, two positions of stability could be right next to each other: at which point discontinuity collapsed into continuity.[39]

Others' responses to *Materials*—which was very widely reviewed, nationally and internationally, though it never sold especially well—were on the whole much more positive, in line with the Royal Society's verdict. "That it will have an influence upon the biological work of the future we have not the slightest doubt, and we believe also that its influence will be great, permanent, and beneficial," declared the *Westminster Review*. In *Science*, the book was predicted "to be one of the few valuable and lasting additions to the literature on the general subject of the evolution of organic nature."[40] The most uniformly glowing review came from an FRS. In the July 1894 issue of *Mind*, under the title "Discontinuity in Evolution," Francis Galton puffed the book as offering brilliant support for a perspective on evolution that, though he was more convinced than ever of its truth, had seemed doomed never to interest anyone else, let alone persuade them. Galton began by stressing, as ever, the "complexity of circumstance under which each germinal element is placed, and the multitude of interacting elements," and the need, in the face of that complexity, for patience in awaiting the day when we understand in detail the causal chains behind any particular variation. Next Galton reviewed the arguments and evidence, from his own work—most recently on fingerprints—as well as from Bateson's, for the general case for evolution proceeding by leaps between types. True to form, Galton took the opportunity to introduce some new vocabulary, distinguishing "divergent" variation ("a mere bend or divergence from the parent form, towards which the offspring in the next generation will tend to regress") from mean-shifting "transilient" variation (the sports that go on to spawn new types). Equally true to form, he ended with a eugenic flourish, urging readers who shared his concern with how low our species had fallen since antiquity to take courage from the fact that humankind throws up the occasional mental or moral sport. "It is reasonable to hope that when the power of heredity and the importance of preserving valuable 'transiliencies' shall have become generally recognised, effective efforts will be made to preserve them."[41]

3

After "Discontinuity in Evolution" was published, Galton sent Weldon a copy. In reply, Weldon rehearsed some of his objections to Bateson's museum-stroll methods before pointedly making his excuses: "I can't write any more now.—I have measured 160 crabs today; and in measuring by turning a screw one's hand becomes so tired that it is a great effort to write, or indeed to use one's hand for any reasonably human purpose."[42]

Biological measurement for statistical ends had recently brought Weldon and Galton into closer collaboration. Inspired by the discussion of the crabs-and-curves papers at the Royal Society the previous November, Galton had suggested to Weldon that a society-sponsored committee might usefully be set up to promote further research in the same vein. After a preliminary meeting at Weldon's club in Piccadilly, the first official meeting of the new Royal Society Committee for Conducting Statistical Inquiries into the Measurable Characteristics of Plants and Animals had taken place in early January 1894. Participating along with Galton (the chair) and Weldon (the secretary) were Charles Darwin's botanist son Francis and Raphael Meldola, an industrial chemist whose passion for natural history had led him to Darwinism. Thanks to start-up funding from the council, several projects, including one under Weldon's direction on dimorphism in herring, were soon under way. But the herring work dragged on unexpectedly, and soon Weldon's own research over the last year, on natural selection in Plymouth crabs, was being counted as a committee project.[43]

With the committee's authorization, Weldon submitted his new results to the society in November 1894 in the form of a first report from the committee. Subtitled "An Attempt to Measure the Death-Rate due to the Selective Destruction of *Carcinus maenas* with Respect to a Particular Dimension," it in fact considered two crabby anatomical dimen-

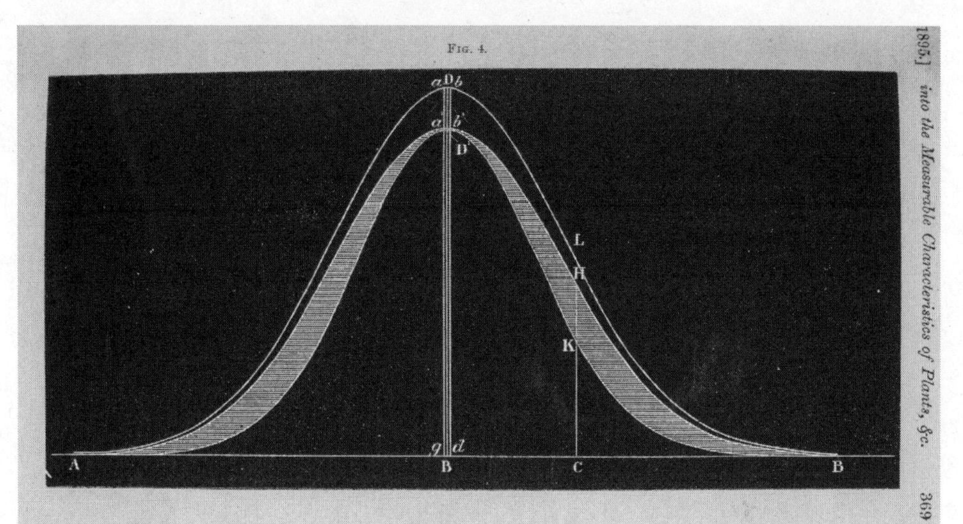

FIG. 4.5. Selection in action, from Weldon's 1894–95 report. The middle curve represents a crab population after unselective thinning out and before selective narrowing (the shaded area).

sions, the lengths of the frontal breadth and the right dentary margin (a region of toothlike serrations on the carapace, just to the side of the frontal breadth). He had measured these lengths in over eight thousand female crabs, whose size he took to be a surrogate for age. To his surprise, he found that, for both dimensions, the range of variation *increased* as the crabs got older, with a subsequent decrease occurring only in frontal-breadth length. Elegantly transforming puzzles into predictions, he pointed out that the late onset of many variations had been one of Darwin's great themes (explained by him, as we saw, by hypothesized gemmules), and furthermore, that the dimorphism among the Naples crabs suggested that, whatever the function of frontal-breadth length, or of some other part varying with it, it was important to survival—more so than the right dentary margin, which, like human height in selection-shielding civilizations, just got more and more varied. Weldon went on to calculate, by way of mathematical arguments well beyond anything in his previous papers, that in Plymouth Sound, female crabs with too-short and too-long frontal breadths were, relatively late in life, being destroyed at a rate of seventy-seven per thousand (fig. 4.5).[44]

Weldon stressed that what mattered were less his numbers—which depended on too many assumptions and limitations to be entirely trustworthy—than the forms of the reasoning by which he reached them. He had shown how to catch natural selection in the act, and even how to measure the intensity of its action. The paper marked a milestone worth making a fuss over.[45] And it arrived at the Royal Society at a moment when, as never before, the council was on the lookout for papers to make a fuss over. In summer 1894, it had accepted, in a provisional way, recommendations made in response to complaints from that inveterate modernizer, Michael Foster, about the meager intellectual yield of too many of the society's meetings and publications. The new thinking was that, four times a year, there would be meetings organized around single topics, with papers programmed to advance the discussion on those topics. The pioneer attempt in this direction took place in late January 1895, gathering the latest work on the greatest achievement of recent science, certainly in Britain, arguably anywhere: the discovery of a new element, dubbed "argon." So many people wanted to attend that the meeting had to be moved out of Burlington House to a larger lecture theater at the University of London.[46]

How to follow it up? Foster asked Weldon to let his paper be the centerpiece of the next discussion meeting, and he agreed.[47] Here is what happened, according to the *Athenaeum*'s correspondent:

> The second special meeting for discussion was held at the Royal Society on the 28th [February], and the new experiment bids fair to be a decided improvement over the old-fashioned "reading of papers," which it is designed, if not to supersede, at least to supplement. The subject on this occasion was "Variation in Animals and Plants," and was introduced by Prof. Weldon, F.R.S. The printed documents in the hands of the meeting were (1) The Report of a Committee of the Royal Society for conducting Statistical Enquiries into the Measurable Characteristics of Plants and Animals; (2) some remarks upon variation by Prof. Weldon; (3) a contribution by Mr. H. M. Vernon on the effect of environment on the development of echinoderm larvae. Prof. Weldon's opening statement, which was illustrated by lantern slides, was followed by speeches from Mr. Thiselton Dyer, Prof. E. Ray Lankester, Prof. Alexander Agassiz (the newly admitted Foreign Member of the Society), and Mr. Bateson. Mr. Dyer illustrated his remarks by specimen plants from Kew, a wild cineraria from the Canaries being placed side by side with the latest cultivated variety, and specimens of variation in the Chinese primrose being also exhibited.[48]

Of these documents, (1) was Weldon's paper. (2) was a brief big-picture commentary by him, written for the occasion, and setting the paper in the context of the post-*Materials* debate over variation and its role in the modification of species.[49] (3), from an Oxford-trained chemical physiologist based at the Naples zoological station, reported experimental studies of how slight differences in temperature, salinity, density of metabolic products, and so on, of a kind that developing sea urchin larvae might actually experience in nature, affect variation in the sizes of various parts of those larvae. Weldon was sent it to referee, and loved it, declaring it a "model" study. (He had spent the latter half of 1894 getting excited and then despondent about an experimental variation study of his own using jar-bound water fleas, so he was primed to appreciate Vernon's success.)[50] Of the speakers beyond Weldon, Thiselton-Dyer was a Darwinian to his boots, as was Lankester. Balancing them, as it were, were Agassiz—the ultimate source, recall, of Bateson's broadsides against Darwinian speculation—and Bateson himself, at whose *Materials* Weldon had made passing but provocative digs in (1) and (2).[51]

What set off controversy was the subsequent publication in *Nature* of a version of Thiselton-Dyer's remarks. He included a few sentences about a diverse group of ornamental flowering plants, the *Cineraria*, which, as he had explained at the meeting, he took to show the power of "human selection"—and thus the "gradual accumulation of small variations"—to produce very large differences.[52] A couple of weeks later, Bateson sent in a long response that dealt exclusively with the *Cineraria* example. The history of its cultivation, he wrote, revealed that selection was used only in a secondary way, after hybridization and sporting had produced the big novelties.[53] Three months of often acrimonious debate on the Letters page followed, with Weldon weighing in on Thiselton-Dyer's side, and with no clear winner at the end. The disagreements were various and detailed, but broadly speaking, they clustered around three topics: first, what *Cineraria* cultivators and commentators had themselves reported about the origins of new kinds of *Cineraria*, with some stressing hybridization, some selection, some hybridization here but selection there, or vice versa; second, how to interpret those reports, gleaned from sources including *Paxton's Magazine of Botany* and the *Ladies' Magazine of Gardening*—whether, for example, "seed" should be taken to mean seed (Weldon), or only the female element (Bateson), and likewise whether what the *Cineraria* cultivators referred to as "hybrids" could be taken with confidence to refer to what botanists today call "hybrids"; and third, how to relate the historical reports to what could be observed now—for example, in making up one's mind about the claim of hybrid parentage for cultivated *Cineraria*, did it matter that most of their characters could be found, albeit in smaller dimensions, in the feral form? To Bateson, it did not; for sometimes hybridization left no trace, and we know of hybrid origination only because it was documented.[54]

In the course of this "silly row," as Weldon called it, his relationship with Bateson became still more fraught. Their private correspondence in these months is a record of deteriorating goodwill, with Weldon trying and failing to answer Bateson's ever-expanding charge sheet of bad faith and bad behavior, and Bateson oscillating between giving Weldon another chance and condemning him as a toady to the loathsome Thiselton-Dyer. It was a bad ending; but it was not necessarily the end, as shown by a brief conciliatory exchange more than a year later, when Bateson assured Weldon that, contrary to the impression that Weldon had apparently formed (as Bateson had belatedly discovered), there had been no deliberate snubbing of him back at the Royal Society meeting. A relieved Weldon wrote of hoping they would meet on better terms in the future, and Bateson agreed.[55]

4

Weldon and Bateson became fellows by defining new directions within an established science that was already theirs. Pearson did it by inventing a new science—call it "mathematics for statistical biology"—for which his previous research in physics and engineering had little prepared him. In *The Grammar of Science*, a popular (and eventually classic) introduction to scientific method that he published in early 1892, his sole discussion of his own work concerned the possibility that, in step with the era's vortex physics and non-Euclidean geometry, atoms might be "ether-squirts" in higher-than-three-dimensional space. His book used the basic concepts of physics to show how, in science properly pursued and understood, the raw material of sense-impressions becomes economical, prediction-enabling law through respect for the facts, eagerness for consensus, vigilance against the distorting effects of bias, and other habits of mind valuable to all citizens. Yet Pearson's interest in biology, and above all Darwinian biology, shines throughout. He regarded science itself in a Darwinian light, as the fullest expression of the powers of reasoning by which humans had become planetary masters. Among Darwin's successors—and notwithstanding the "metaphysical" muddle, as Pearson saw it, of the germ-plasm theory—Weismann stood tallest, for demonstrating that acquired characters are not inherited. Thanks to Weismann, it was now clear, concluded Pearson, that "the stern processes of natural law" in civilized society needed to be replaced not with better sanitation or education, but with "milder methods of eliminating the unfit." (If Pearson's socialist comrades had become bored of hearing about the social implications of heredity before *Natural Inheritance*, Weismann was to blame.)[56]

Pearson's Weldon-inspired memoir of 1893–94 was the somewhat awkward start of something new. At its heart was a mathematical procedure for deciding between two biological interpretations of a frequency curve like the frontal-breadth curve from the Naples crabs, where the overall asymmetry was known to arise as a secondary effect from the summing of symmetrical component curves. (Pearson also cited Bateson's earwig paper as another example of such double-humpedness in natural populations.) One interpretation was that a single population was in the process of diverging—a "homogeneity" scenario, in Pearson's terminology. Alternatively, two distinct populations might have become mixed together, perhaps as an artifact of the collecting method—a "heterogeneity" scenario. By applying his procedure to the Naples crab data, Pearson was able to affirm that Weldon's results were no artifact: "Professor Weldon's material *is homogeneous*, and the asymmetry of the 'forehead' curve points to a real differentiation in that organ, and not to a mixture of two families having been dredged up."[57] Welcome though Weldon found this news, the issue was not something he had actually worried about. Nor had the Royal Society's referees—the physicist George Darwin (son of Charles) and Galton— who registered unease with what struck them as a blend of surplus-to-requirements biology and not-very-interesting mathematics. "It would hardly occur to the mathematician to undertake such a piece of work," wrote Darwin, who went on to wonder whether biologists on the whole really understood homogeneity and so on in the ways presumed.[58] Galton, too, doubted the usefulness of Pearson's method to biologists, but hoped the memoir—"a vigorous and original attempt to supply a statistical want"—would spur others to simplify its mathematics and generally improve on its efforts.[59]

Already Pearson was at work on that front himself.[60] Throughout 1894, in time snatched from heavy teaching commitments, voluminous popular-polemical writing for the *Fortnightly Review* and other periodicals on scientific and political topics (a distinction he gloried in trampling), and a busy correspondence with Weldon and Galton among many others, he wrote what became his second "Contributions" memoir, submitted to the Royal Society in late December that year. Again the subject was asymmetrical curves and their biological significance. But now he took up cases in which the asymmetry, rather than being resolvable into symmetrical components, had to be taken as genuine and primary. Considered causally, he argued, the situation could be thought of on the model of pellets dropping through a quincunx where the probability of a bounce in one direction was greater than that of a bounce in the other. (Pearson had such a "probability machine" built.) Mathematically, that model, in turn, could be described by modifying the standard equation for the normal curve—that graphical upshot of infinite pellets streaming through stream-bisecting pegs—so that the left-or-right probabilities were treated not as fixed and equal (½, ½) but as variable quantities (greater than ½, less than ½). With this more general form of the equation, a normal curve became the limiting case of a skew curve—one for which the degree of skew was zero—while a huge family of nonzero skew curves could be easily generated. Better still, as Pearson demonstrated at length, these curves actually fitted statistical data, in domains from biology and medicine to psychology and economics, far more closely than the one-size-fits-all normal curve ever did. Even the most symmetrical of Weldon's crab curves turned out to be better matched by one of Pearson's skew curves.[61]

This time Galton, again refereeing, was impressed, so much so that he dashed off an enthusing article for *Nature*, under the title "A New Step in Statistical Science." "There can be no doubt," declared Galton, "that the descriptive efficiency of Prof. Karl Pearson's method is of the highest order."[62] Galton's advertisement came out in late January 1895, a week after the abstract of Pearson's second memoir was read at the Royal Society and—on that same day—his certificate of election to the society was submitted, with Galton a signatory.[63]

But there was room for doubt, expressed most pungently, in public and in private, by Weldon. Alerted to the criticism in Pearson's new memoir, Weldon—who had let Pearson analyze not just his published data but also his unpublished data from his ditch-shrimp studies—surprised Pearson by mounting a slide-illustrated riposte during the discussion at Burlington House, claiming that, on Pearson's misguidedly exacting criteria, there could be no real-world examples of the normal curve, not even among dice.[64] Suspicions that Pearson was too much mathematician and not enough biologist were confirmed for Weldon when, in early February, and with the Royal Society discussion of his committee report just weeks away, Weldon received a lengthy commentary on the report from Pearson, "attacking violently every point in the paper," as Weldon told Galton. For Weldon, these new criticisms again betrayed an Olympian unconcern with whether the deliverances of abstract reason were actually reasonable in light of the concrete facts—for instance, about the propensities for crab limbs to break off and regenerate—in ways that, as Weldon saw it, easily accounted for small departures from strict normal-curviness. "Here, as always when he emerges from his cloud of mathematical symbols," complained Weldon to Galton, "Pearson seems to me to reason loosely, and not to take any care to understand his data."[65]

Throughout the spring of 1895, as Weldon's scientific disagreements with Bateson grew irremediably personal, those with Pearson never did. In part that was because Pearson thought so highly of Weldon, with whom he continued to enjoy friendly knockabout lunchtime discussions. They even began to collaborate on a new human-measurements project.[66] For Weldon's part, though he never held back from criticizing Pearson's work (and vice versa), he also began to admire it, and even to champion it. "What a pretty thing Pearson's general expression for heredity is!" Weldon remarked to Galton, after Pearson previewed his third memoir at a Royal Society meeting in June.[67] In October, in a speech at University College summarizing the college's recent scientific achievements, Weldon was generous in his praise.[68] A few months later, asked by Galton to support the election of another statistically engaged mathematician, Weldon agreed, but added that he hoped in doing so not to hurt Pearson's chances.[69]

There was no need to worry. Pearson's new memoir, published in full in spring 1896, was his most successful attempt yet at showing what his new science could do. Taking *Natural Inheritance* as a starting point, he offered brilliantly rigorous explorations of a number of Galtonian themes, including variation, correlation, selection, heredity (whose deviation patterns he had captured in just two formulas, linking multiple coefficients of correlation), regression, and "panmixia," a Weismannian term recently adopted by Galton and more or less everyone else. Literally meaning "universal crossing," it was taken to be what happens when, as a result of natural selection ceasing to operate, individuals from anywhere on the frequency curve interbreed.[70] Talk of panmixia, with sometimes explicit links to broader anxieties about biological and moral decline—a degeneration "boom," it was called at the time—was then in vogue (fig. 4.6).[71] Weldon and Pearson had earlier argued that, as Galton had taught, selection's cessation just allowed the span of variation in a population to increase; it did not cause the mean to slide backward limitlessly (and in case of a backward slide, Pearson was wont to add, thoughtful socialists embraced struggle with other nations, along with internal management of mating, as progressive).[72] Now Pearson reinforced and extended the Galtonian message on panmixia in the context of a creatively wide-ranging, mathematically sophisticated response to a book he hailed as "epoch-making." When Galton read the final version in the *Philosophical Transactions* in May 1896, he was moved by the tribute. Within weeks, Pearson was elected to the society.[73]

Almost immediately, at the Ladies' Conversazione in mid-June, Pearson and his growing team of assistants, led by his right-hand man G. Udny Yule, were out in force. Their exhibit comprised seven mini-exhibits, including the skew probability machine and a diagram illustrating a newly identified form of selection: what Pearson called "reproductive selection," in which a relatively small proportion of the population produces a relatively large proportion of the next generation—as seemed to be happen-

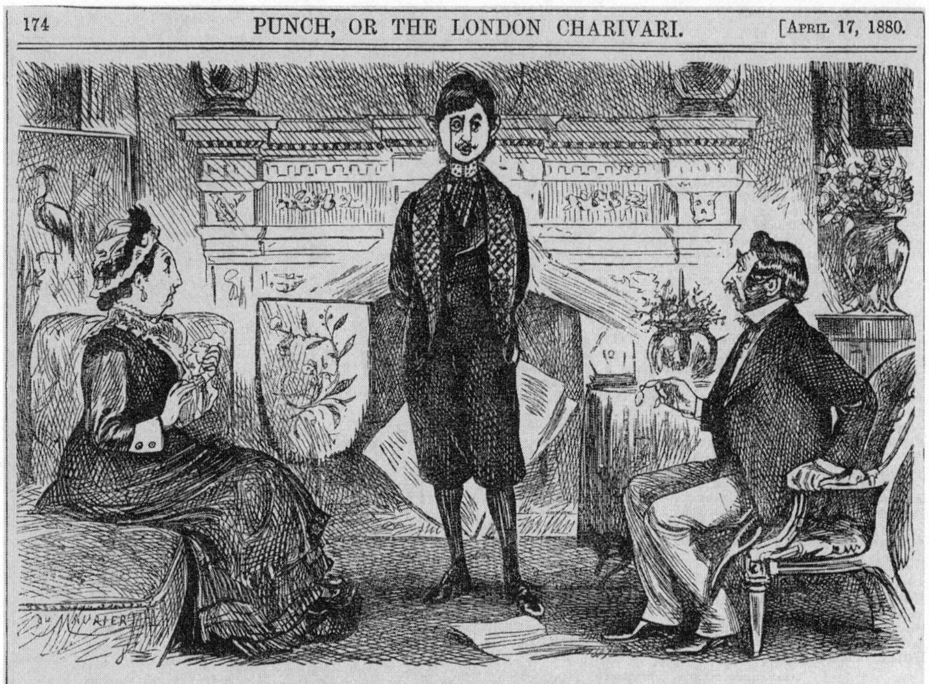

174 PUNCH, OR THE LONDON CHARIVARI. [APRIL 17, 1880.

RESPONSIBILITIES OF HEREDITY.

Son and Heir (suddenly dissatisfied with his Stature, his Personal Appearance, and the Quality of his Intellect). "AW—WHAT ON EARTH EVAH COULD HAVE INDUCED YOU TWO PEOPLE TO MAWWY?"

Sir Wobert and Lady Mawiah. "THE OLD, OLD STOHWY, MY DEAR BOY! WE FELL IN LOVE WITH ONE ANOTHAH—AW—AW——"

Son and Heir. "AW—WELL—YOU 'RE BOTH SUCH AWF'LY GOOD OLD DEAWS, THAT I FORGIVE YOU. BUT YOU WEALLY SHOULD HAVE HAD BETTAH TASTE, YOU KNOW, AND EACH HAVE FALLEN IN LOVE WITH A DIFFEWENT KIND OF PERSON ALTOGETHAH, AND GIVEN A FELLAH A CHANCE! YOU SEE, IT 'S ALL OWIN' TO YOUR JOINT INTERFEAWENCE IN MY AFFAAWS THAT I 'M UNDER FIVE FOOT ONE, AND CAN'T SAY BOH TO A GOOSE, AND—A—JUSTLY PASS FOR BEING THE GWEATEST GUY IN THE WHOLE COUNTY—AW! JUST *LOOK* AT ME, CONFOUND IT!" [*They look at him, and then at each other—and haven't a word to say.*

FIG. 4.6. A weedy young man holds his parents to account for becoming parents in *Punch*, 1880—an example from a vein of degenerationist satire tapped in H. G. Wells's *The Time Machine* (1895).

ing with humans ("the battle is to the most fertile, not the most fit," as Pearson put it in a letter to Yule). Mathematics for biological statistics had arrived.[74]

5

Suppose, in Pearsonian spirit, we graph the nineteenth-century life of the Royal Society committee that Galton and Weldon established to promote the new statistical biology. Let the *y*-axis represent quantity of business transacted, as measured by meetings held and associated communications, public and private. Let the *x*-axis represent time in years, from 1894, when the committee met for the first time, to 1900, when—following Galton's lead—Weldon and Pearson submitted their resignations. Appropriately enough, what emerges is a broadly symmetrical, bell-shaped curve. Between 1894 and autumn 1896, the committee putted along, meeting infrequently for uneventful catch-up discussions. Weldon's work on shore crabs and natural selection, partly funded by the committee and published in its name, was the main attraction. Then, from autumn 1896 to spring 1897, there was a Galton-led surge in activity, set off by the prospect of buying Down House and turning it into a "Darwinian Institute" devoted to long-term, continuously observed breeding experiments. That dream died quickly, but not before it transformed the committee through the influx of new, breeding-oriented members. Chief among them was Bateson, whose research on discontinuity had shifted, post-*Materials*, to studying what he called "substantive" variation—that is, variation in a part's size or makeup (as distinct from its repetition or location)—via the tracking of non-blending character versions in experimental crosses. From spring 1897, the Evolution Committee, as it was now called, relaxed back into smallness, in its core membership as well as in the frequency and fruitfulness of its meetings. Weldon continued dutifully in his role as secretary. But his crab research disappeared from the meeting agendas, which became, like the committee members, increasingly aligned with Bateson's vision.[75]

It was as a critic of the 1895 report on the crab research that Bateson first insinuated himself into the committee's workings—not that year (the minutes make no mention of him, despite his role in the Royal Society discussion and the *Cineraria* controversy), but the next. Throughout summer 1896 and into the autumn, *Nature* hosted a lively correspondence about the evidence in the report and what it showed, touching on everything from whether any selective destruction, from whatever cause, had been demonstrated—as distinct from adaptive convergence as the crabs aged—to the lessons of Hume, Kant, and Mill on how to distinguish causal relationships from mere correlations.[76] For Weldon, holed up that summer (as he had been the previous one) at Plymouth with hundreds of bottle-inhabiting crabs, in order to work out their "law of growth" (fig. 4.7) and so, he hoped, rule out the adaptive convergence possibility, it was all an unproductive distraction, though he found reading Kant's *Critique of Pure Reason* unexpectedly rewarding. "I have read a great lot of it," Weldon wrote to Pearson in early August, "although after ten hours of bottle-washing it is not very easy to read anything. But it is a good book! I had never seriously tried it before: and I now feel properly ashamed." In a subsequent letter he added, "I think it is the most magnificent piece of human ingenuity I have ever had the pleasure of examining: and I don't enjoy it less because of its wholly irrelevant character. It

FIG. 4.7. Weldon's setup at Plymouth for determining how the curve of variation for the local shore crabs changed as the crabs—safe in their bottles from destructive agencies—matured.

seems to me quite like the sort of mathematics you are so fond of abusing: the magnificent development of a possibility which can never have any bearing on anything in particular."[77]

In mid-October, Bateson joined the ranks of the critics. He sent his letter not to *Nature*, however, but to Galton, as chair of the committee. What safeguards, Bateson now asked, had been put in place to ensure that, in the samples of crabs Weldon had used to assess breadth of variation at a given stage of maturity, the individual crabs really were of broadly similar age? After all, two crabs could go through the same number of molts and end up very different sizes; likewise, two crabs that are the same size may have gone through different numbers of molts.[78] Pearson had raised the same "size as an index of age" issue in his commentary on the draft report. But it was one thing coming from Pearson, who knew nothing about real crabs, and so nothing about how easy it was for anyone who did know them to tell older and younger apart. It was something else coming from Bateson. "I am sorry he thinks me such an idiot as to neglect this very obvious point," Weldon wrote to Galton on receiving Bateson's letter. Of course, Weldon went on (as he had written in the report), the law of growth extrapolated from his samples needed experimental verification, of exactly the kind he had recently been laboring to supply. The results so far were encouraging: captive crabs that grew through their first molt—and so were known to be at the same molt—showed the expected increase in breadth of variation. Even so, Weldon outlined additional, arithmetically involved tests he could apply to his data to check that he had not accidentally created mixed-age samples.[79]

As the to-and-fro between Bateson and Weldon intensified, Galton made a couple of decisions that turned out to be highly consequential. One was to recommend that the committee cease publishing its own reports, effectively cutting Weldon loose to publish his crab research as he saw fit, and the committee loose from having to answer for it (and the chair from having to engage in time-consuming mediation).[80] The other was to invite Bateson to join the committee. Weldon was understandably uneasy about the latter.[81] And when he saw the eagerness with which Bateson voted for the former—"he probably felt like Dogberry, and thanked God that he was rid of a knave," Weldon glumly wrote afterward to Galton—it prompted Weldon to worry that the Royal Society would deem the crab research rejected on grounds of "inherent futility."[82]

But soon enough Weldon made the most of his new freedom. Even as he had been extolling the virtues of Kant the previous summer, he had glimpsed an exciting new direction for his research. A University College student working at Plymouth, Herbert Thompson, had discovered that, over the handful of years when crab frontal breadth had been under scrutiny there, its mean value had not—as Weldon, following Galton, presumed—stayed constant, but had diminished, at least in male shore crabs. It seemed, in other words, that natural selection was not holding the mean value fixed by trimming away the extremes, but was actually driving it downward, at a pace too slow to be detected visually but unmistakable statistically. And already Weldon had an idea as to why that might be so. Casting aside his former caution about functional hypotheses, Weldon guessed that shorter frontal breadths might help to protect crabs, male and female, from the effects of the increasingly muddy and polluted conditions of Plymouth Bay.[83] Throughout the summers of 1897 and 1898, he threw himself into new measurement, experimental, and observational studies to tie it all down. He presented the results as the centerpiece of a general defense of the theory of natural selection in early September 1898 in Bristol, in his address as president of the zoological section of the BAAS.[84] Along the way, and as he made plain in the address, he had come to a new appreciation of the power and importance of Pearson's mathematical work for dealing, conceptually and operationally, with the asymmetrical curves that Weldon now saw everywhere in his data.[85] Accordingly, Weldon—a member of the Royal Society's council from 1896 to 1898—had proposed Pearson for the recently established Darwin Medal. Previous awardees included Hooker and Huxley. In November 1898, the medal went to Pearson.[86]

A gap now yawned between the increasingly celebrated work of these two FRSs on evolution and the ambitions of the Royal Society Evolution Committee of which they were members. (Pearson joined at the same expansionist, top-of-the-curve moment as Bateson, but hardly ever attended.) In the wake of the doomed Darwin Institute proposal, and with Bateson now supplying the programmatic energy previously supplied by Weldon—and also, from late 1897, replacing Weldon as recipient of the program-defining research funding, for hybridization experiments with butterflies, plants, and poultry at Cambridge—the committee mainly became an occasional talking shop for people interested in what biologists could learn from practical breeders and vice versa.[87] Weldon never overcame his misgivings about this mission, and especially about the potential for mischief due to the biologists, with the imprimatur of the Royal Society, pretending to greater knowledge than they actually had.[88] When, early in summer 1899, the committee's funding was reduced, Weldon suggested to Galton that maybe it was time to wind down the committee. The matter was discussed at the June meeting, but a decision was put off to the next meeting.[89]

That meeting took place in November. Galton was not there, but had asked that a letter from him, recommending an end to the committee, be read out. As Weldon reported afterward to Galton, the meeting proved deeply uncomfortable. A quotation from what Weldon wrote appears at the start of this chapter. Here is a fuller extract:

> After I had read your letter, the others passed a resolution urging the reappointment of the Committee.
>
> The discussion seemed to me to show more than ever the futility of the Committee. The view taken was that the Ctee. should not attempt to direct, or to discuss investigations, but that it should exist in order to confer upon its members some sort of authority in their dealings with breeders and such people.
>
> It was distinctly said that in this way the study of "Natural History" might be encouraged, and that those who did not care for "Natural History" ought to leave the Ctee.
>
> I do not see that it is possible to go on passing votes of confidence year

> after year in work of which one knows practically nothing. That kind of relation between a Ctee. and the work of its members is absurd.
>
> The "authority" lent by the Ctee. to its members is to me detestable. The whole idea of such a thing seems a piece of vulgar snobbishness; and I do not believe that competent people feel the need of such assistance when they seek for information.
>
> But the most dangerous thing about the discussion seemed to me the deliberate antithesis which these men sought to establish between what they called "Natural History" on the one hand, and any sort of statistical enquiry, leading to numerical results, on the other.
>
> Several members of the Cttee., of whom Bateson is the leader, have the idea of such an antithesis, and are strongly influenced by it. However sincere and able such men may be, I think they do harm, and I think it would be a great misfortune if a Committee of the Royal Society were to officially adopt their view.
>
> I cannot go on acting as secretary to this Committee. Will you advise me what to do?[90]

Soon after, Galton submitted his resignation, as did Weldon, Meldola, Pearson, and Thiselton-Dyer (another top-of-the-curve joiner).[91]

In line with the committee's own majority-vote decision, however, the council renewed its appointment. Its first meeting of the new year (and new century) was held on 25 January 1900. At the end, Bateson took over from Weldon as secretary.[92] Thus was the way cleared for Bateson to transform the committee into the institutional headquarters of what was shortly to become his career-defining passion: Mendelism.

Part 2

BATTLE

5

Between Boers and Basset Hounds

Looked at from the social standpoint, we see how exceptional families, by careful marriages, can within even a few generations obtain an exceptional stock. . . . On the other hand, the exceptionally degenerate isolated in the slums of our modern cities can easily produce permanent stock also; a stock which no change of environment will permanently elevate, and which nothing but mixture with better blood will improve. But this is an improvement of the bad by a social waste of the better. We do not want to eliminate bad stock by watering it with good, but by placing it under conditions where it is relatively or absolutely infertile.

KARL PEARSON, *THE GRAMMAR OF SCIENCE*, 2ND EDITION, 1900

As Heine said half bitterly, half laughingly, "A man should be very careful in the selection of his parents." On the other hand, although the human organism changes slowly in its heritable organisation, it is very modifiable individually, and "nature" can be bettered by "nurture."

J. ARTHUR THOMSON, "FACTS OF INHERITANCE," 1900

There were no Mendelians in Mendel's own lifetime. But that was not because his work on plant hybrids was hard to find or, in a getting-the-gist way, hard to follow. From Philadelphia to St. Petersburg, from Uppsala to Venice, anyone who wanted to read the "Versuche" or Mendel's companion report on hawkweed hybrids could, without too much trouble or travel, pluck the volumes of the Brünn Naturalist Association's *Proceedings* from library shelves. In Britain, as mentioned, the Royal Society had them. So did the Linnean Society, the Greenwich Royal Observatory, and Cambridge University. Beyond the bound versions, there were offprint copies circulating internationally, among people who had either received theirs directly from Mendel—he had ordered forty copies of the "Versuche"—or acquired a hand-me-down from someone else. Typically, what kindled interest in reading Mendel was a précis or citation encountered in an article or book surveying the scientific literature on plant hybrids, like the German volume whose uncut pages now lie exposed in that shaming photograph in *Brilliant Blunders*. By the end of the nineteenth century, botanists who had become serious about the fate of hybrid characters in plants were bound to bump into a reference to the "Versuche" or, if they somehow managed not to do so, to have the paper thrust upon them. Thus, early in 1900, did Hugo de Vries, professor of botany at the University of Amsterdam, receive a copy from a Delft colleague, Martinus Beijerinck.[1]

Between them, Beijerinck and de Vries illustrate a striking shift in how botanists and others had come to think about plant hybridization. A little younger than de Vries, Beijerinck was involved in hybrid plant research earlier on, throughout the 1880s. Like Mendel, he saw the subject as an end unto itself, important because of its relevance to improving the range and quality of varieties for the garden and the farm. He was especially concerned with hybrid wheat, and how low fertility in modern strains might be illuminated by using crossing experiments to identify wild parent forms.[2] By contrast, when de Vries, who had made his name in plant physiology, started to become interested in hybridization, it was mainly as a means to the end of understanding something more fundamental: the nature of heredity. In 1889 he published a manifesto in German, under the arresting title *Intracellulare Pangenesis*. He argued that, for all its flaws, Darwin's pangenesis hypothesis had been on the right track, above all in its picture of organisms as mosaics, composed of elements inherited and functioning independently of one another. One challenge was to bring that picture up to date in light of more recent discoveries, notably the central role of the cell nucleus in inheritance and, after Weismann, the non-inheritance of acquired characters. Here de Vries had a suggestion: recast the gemmules—renamed "pangens"—as undergoing transmission only *within* cell lineages, and indeed within cells, moving from inside the nucleus, where the pangens all reside in a latent state, to the cytoplasm, where a selection of them become active (hence "intracellular pangenesis"). Another challenge was to use experimental hybridization in order to clarify the mixing-without-merging, atom-like independence of hereditary characters.[3]

In the 1890s, while Beijerinck retooled as a microbiologist specializing in plant diseases, de Vries increasingly concentrated on his experimental crosses, along the lines of what Darwin had done with snapdragons. In 1892, de Vries crossed two varieties of the flowering plant *Lychnis*, one with the hairiness typical of the species, another without it. Could he transfer "hairlessness" from the one to other? The hybrids all turned out hairy, albeit less abundantly so than their hairy parents. The hybrid character, in other words, was intermediate. As for the offspring of the hybrids, he found that a conspicuous minority—between a third and a quarter—were fully hairless. He got broadly similar results when, from

1893, he began crossing a red-and-black-flowered poppy variety with a fully white-flowered variety. Here the hybrids had flowers that were not pinkishly-grayishly intermediate, but uniformly red and black; whereas among the offspring of the hybrids, around a quarter showed full whiteness. In a lecture slide he used when teaching, he summarized the poppy generational pattern, labeling black-and-redness—in an echo of his pangenesis reforms—the "active" version of the character, and whiteness the "latent" version.[4]

To anyone familiar with the "Versuche," the resemblance to Mendel's findings—down to the analysis of the active-version hybrid offspring as dividing into roughly one-third that breed true and two-thirds whose offspring show the hybrid pattern—leaps to the eye. Certainly Beijerinck knew in a general way what de Vries was up to as a hybridist. For de Vries's part, he had kept up with Beijerinck's new ideas about infectious living fluids in plants: witness the title of a lecture that de Vries gave in London in July 1899, "Hybridisation as a Means of Pangenetical Infection."[5] Whatever prompted Beijerinck, in the early months of 1900, to send his "Versuche" offprint (a hand-me-down, from a botanical relative) to de Vries, it prompted de Vries in turn to reassess his experiments through a Mendelian lens.[6] In mid-March 1900, he completed two versions of a new paper, setting out this new view of hybrids: a longer one in German, submitted to the German Botanical Society in Berlin; and a shorter one in French, submitted to the Académie des Sciences in Paris.[7]

Read at the Académie's meeting in the last week of March, the French version, titled "Sur la loi de disjonction des hybrides" (On the Law of Disjunction in Hybrids), began and ended with de Vries's signature themes of organismal characters as hereditary units and the usefulness of experimental hybridization in making their independence plain. In between, he offered—without naming Mendel or citing the "Versuche"—a remarkably Mendelized rendition of the past decade's work. Now he wrote of the character versions in each antagonistic pair not as "active" and "latent" but as "dominant" and "recessive." Much more boldly than before, he picked out a simple 3:1 ratio in the hybrid offspring as what a lot of his data were converging on. And he explained this pattern as a result of the dominant and recessive character versions joined in the hybrid dis-joining during the formation of gametes, so that each pollen grain and each ovule received one or the other version in its pure, pre-hybrid state.[8]

Mendelism's story begins with Bateson's encounter with this paper. But before looking at what excited him about it, we will do well to stay a little while longer, and stray a little more widely, in that last week of

March 1900. Besides the Paris reading of de Vries's paper, the week saw several other events:

- In Oxford, where Weldon now lived, the sowing of a large number of poppy seeds, in connection with a new project of Pearson's on the statistically reckoned likeness of like parts, was completed.
- In London, at the Royal Institution, a popular evening discourse on the "facts of heredity" ended with an introduction to Galton's recently announced "law of ancestral heredity."
- Also in London, around the corner at Burlington House, a meeting of the Royal Society brought together Weldon, Pearson, Bateson, and others for a discussion of new work on variation, experimentally and mathematically studied.

A brief stop at each of these destinations will set the scene for the immediate impact of the 1900 "rediscovery" (a term used in quotation marks even then) of Mendel's work in the world where it mattered first and most. Here lie clues to why Bateson's initial response to Mendel's law was to annex it to the new Galtonian law; why Weldon began his study of the "Versuche" that autumn in a mood of rueful appreciation for what Mendel seemed to have achieved; and why, by the end of the year, the cordial relations on display at Burlington House in March were history.

1

When Weldon's mind turned to biological research, it was research on animals, not plants. For a few months in 1895, he had just about held his own in a public squabble over wild and cultivated *Cineraria*. But that had been under duress, and he had dropped the subject the moment he could. It was not merely that his training in seaside-station embryology had biased him toward animals, or that the bias became permanent as one animal-focused project gave way to another, from wall lizards to acorn worms to common shrimp to shore crabs and so on. Bateson, after all, was a product of the same background, and he moved with seeming ease between animal research and plant research. No, for Weldon, the preference was principled, harking back to his teacher Balfour's emphasis on the need to take seriously the interplay of development and environment. As Weldon once explained to Galton, the problem with plants was that, because a plant typically stays in just one place, whatever conditions happen to prevail there go unbalanced, whereas an animal typically moves around so much that no one set of conditions exercises undue influence on it. Weldon was thus venturing into the unknown in more ways than one when, on Sunday, 25 March 1900, he began planting poppy seeds en masse.[9]

By then, he and his wife were living in Oxford. Weldon had long been the obvious candidate to succeed Lankester as Linacre Professor at Oxford, having previously succeeded him as Jodrell Professor at University College. Already in August 1898, when Lankester was confirmed as the new director of the Natural History Museum in South Kensington, Pearson had been sure the Oxford post would be Weldon's—and was morose in advance at the prospect. "I suppose Lankester in London, you will go to Oxford," he wrote sighingly to Weldon. "That is the worst aspect to me of his appointment. It will be chaos at U.C. and a personal loss to me that I won't describe for fear of its effect on your vanity."[10] A few months later, the award of the Darwin Medal brought home to Pearson the more keenly what he would be losing with Weldon's almost certain move. "Any mathematician could have done what I have done, a dozen of the better men far better," Pearson confessed to Galton, "especially if they had had the suggestive Weldon almost daily at lunch for four or five years."[11] When Pearson's fears—and Weldon's hopes—were finally realized, in late February 1899, Pearson's note of congratulations mixed appreciation with anguish and not a little self-abasement:

> What vexes me just now most is that this break down [in Pearson's health: he was often ill] leaves me little chance of grasping the few lunch hours which remain, and when you reach Oxford—amen!—I am afraid our friendship at Univ. Coll. has been a very one-sided one. I have received all—given little. Perhaps that is the reason I shall be punished more at your removal. At any rate you have changed the whole drift of my work and left a far deeper impression on my life than I on yours.[12]

In the event, Weldon's move to Oxford brought the two men even closer together. In lieu of those lunchtime discussions, they now engaged in an even more voluminous correspondence, exchanging ever longer and more frequent letters, some of them masterpieces of learning and wit, all shining with admiration and friendship. There was still occasional teasing on Weldon's part. "You are like H. G. Wells's martians," he wrote to Pearson in mid-October 1899, "you stalk upon stilts over plane space, and discharge a poisonous cloud of formulae every time you come to four cross roads."[13] But Weldon was in no doubt at all about the importance of the work Pearson was doing and the need for biologists and others to absorb it. For several months, Weldon devoted himself to preparing and delivering a set of lectures-cum-textbook-chapters setting out the new statistical Darwinism at a level appropriate for those who found Pearson's popular writings unsatisfying but his technical publications too difficult.[14] When, early in 1900, Pearson brought out a second edition of *The Grammar of Science*, expanded to include two fat new chapters on evolution and heredity, Weldon was delighted, especially by the challenge it presented to the biology-ignoring tradition in Oxford philosophy. "Pearson's new edition is published; and his new portion about biology is splendid," Weldon crowed to Galton in mid-February. "Such of the Oxford philosophers who condescend to read him are, I believe, very angry; but he will do them (and others) good!"[15]

The poppy seeds Weldon planted the next month were a part of a Pearsonian research program too recent to make it into the book. Recall that in Pearson's philosophy of science there was no higher goal than the framing of ever more economical descriptions, encompassing ever wider domains of experience. Up to the end of the 1890s, Pearson's own greatest—indeed medal-winning—achievement in this direction had been his subsuming of normal curves within the more general mathematics of frequency distributions. From 1899, he was on his way to something at least as grand, but with a more distinctively biological cast: the subsuming of likeness between related organisms (e.g., two sons of the same father resembling each other sufficiently, despite differences between them, to be picked out as brothers) within a more general description of biological likeness, as between the parts of an organism (e.g., two leaves from the same tree resembling each other sufficiently, despite differences, to be picked out as from that particular tree—or two blood corpuscles from the same frog being identifiably from that frog—or two petals from the same flower being identifiably from that flower). Pearson would shortly give this distinction-erasing topic a new name, "homotyposis"—roughly, "being of the same type." Mathematically, he was ringing changes on the correlation mathematics of which he was the undisputed master. Biologically, the idea went back at least to Darwin, whose pangenesis hypothesis likewise netted together the production of likeness within and across individuals. Pearson, like de Vries, was a pangenesis admirer. But no one before Pearson was so convinced that there was nothing special to be said about variability in heredity as distinct from other likeness-making biological processes, nor anyone so determined to prove the point empirically. And since no one before had sought to measure like parts with the precision that Pearson needed for his purpose, he undertook the job himself, recruiting allies as he could.[16]

That was how Weldon (fig. 5.1) came to be outside his house in late March, filling over twelve hundred 5-inch pots with a soil mixture and then sowing Pearson-supplied seeds from a British variety of poppy, the

FIG. 5.1. Weldon in the garden.

Shirley poppy. The part of the mature poppy that interested Pearson was not the petal or the leaf but the starfish-like stigma to be found on top of the bulbous seed capsule—in particular, the ray-like "stigmatic bands" radiating from the center of each stigma. To count the stigmatic bands on each capsule—and a single plant could produce many capsules—was the goal. For now, with all the sown pots resting on a gravel path along a south-facing, wind-sheltered wall, Weldon's aim was to ensure that as far as possible, each pot got uniform treatment, receiving the same amount of water as the others, cycling regularly through the same spots in the garden and house every day or two, and so on. "I feel like a hen in charge of a lot of duck's eggs," Weldon reported to Pearson, after the sowing was finished that Monday. "I have not the remotest idea what will happen next!"[17]

2

What, in this moment and milieu, was known about heredity? What counted as the "facts of inheritance"? Fortunately for us, as well as for his audience, one of the most gifted and best-informed popularizers of the era, the zoologist J. Arthur Thomson, Regius Professor of Natural History at the University of Aberdeen, delivered an evening discourse with that title at the Royal Institution on Friday, 30 March.[18]

He opened with the observation that two general features of scientific progress during the nineteenth century held for heredity. First, Occam's razor had sliced away at obscuring excess, so that no biologist now spoke of "heredity" as if it were an active principle, a force that somehow made offspring resemble their parents. Rather, "heredity" was understood, as Thomson put it, as "but a convenient term for the relation of organic or genetic [in the sense of "genealogical"] continuity which binds generation to generation." It was a name for what needed explaining, not the explanation itself. Second, knowledge of heredity was increasingly arrived at not just through experimentation but through precision measurements subjected to mathematical analyses. The leading figures here, according to Thomson, were Galton and Pearson.[19]

The Physical Basis of Inheritance. Under this heading—the first of five Thomson used—he now skillfully built toward the following picture. The physical basis of inheritance is in the fertilized egg cell. Within the egg cell, the "true bearers of the hereditary qualities" are, mainly but not exclusively, in the cell nucleus. And within the nucleus, they are probably in the rod-shaped chromosomes (literally, "colored bodies," because they became visible under the microscope with the use of certain stains). Thomson noted the challenge of understanding how so much complexity could reside in such a tiny space, but advised that, since so much of what an organism becomes is the upshot of development interacting with the environment, the challenge was less than might appear, for "we are not forced to stock the microscopic germ-cells with more than initiatives." As to how those initiatives travel from parents to offspring, the best-supported theory, he reported, was the Galton-Weismann theory of germinal continuity, according to which a bit of germinal material ("germ-plasm" for Weismann) gets hived off early on and kept in isolation, eventually producing sperm cells or egg cells.[20]

Dual Nature of Inheritance. Sticking with Thomson's headings: "dual" here refers to the fact that, parthenogenesis aside, a multicellular plant or animal has two parents and so two inheritances. Thomson nevertheless regarded parthenogenesis—increasingly inducible in the laboratory—as instructive, because it showed that egg cells are complete unto themselves when it comes to hereditary qualities. He guessed that sperm cells probably are too. Another, rather different point he made was that dual inheritance does not at all mean that each inheritance is equally visible, or "patent," in the offspring. On the contrary, "hereditary resemblance is often strangely unilateral, the characters of one parent being 'prepotent,' as we say, over those of another."[21]

Different Degrees of Hereditary Resemblance. From here, Thomson's coverage amounted to a selective introduction to the new—and forbiddingly demanding—chapters in Pearson's revised *Grammar of Science*. Crediting Pearson with proving what had merely been suspected before, such as the inheritance of fertility and longevity, Thomson dwelt at length on a new three-way classification of inheritance introduced in the book. First, and familiarly enough, there was "blended" inheritance, when the offspring character is intermediate between the parental character versions. Second, there was "exclusive" inheritance, when the offspring character is either wholly the same as the paternal version or wholly the same as the maternal version—though, Thomson again stressed, such either/or, non-blending characters can be inherited without being visible, remaining "latent, neutralised, silenced (we can only use metaphors) by other characters, or else unexpressed because of the absence of the appropriate stimulus." Third, there was the confusingly labeled "particulate inheritance," when both parental versions are present, as when a dark horse and a light mare produce a piebald foal. Thomson then drew on that staple Darwinian idea, the struggle for existence, to integrate these varieties of inheritance with the latest cellular understanding: "Perhaps a unified view will be found in the theoretical conception of a germinal struggle in the arcana of the fertilised ovum, a struggle in which the maternal and paternal contributions may blend and harmonise, or may neutralise one another, or in which one may conquer the other, or in which both may persist without combining."[22]

Regression. As Galton taught, other things being equal, the average values in a population stay put, however much up-and-down there is in the particular family lines making up that population. That is because of regression: the tendency for individuals on average to be more mediocre than their parents. And the explanation of that tendency? Thomson followed Galton and Pearson in blaming the fact that inheritance, though dual, is also multiple: for parents had parents, who had parents, who had parents, who had parents . . . "and unless very careful selection has taken place," wrote Pearson (quoted by Thomson), "the mean of that ancestry is probably not far from that of the general population." Here Thomson introduced another of Pearson's new three-way classifications. Yes, there was regression, when the tendency to mediocrity stabilizes a population. But there was also "reversion," when, within a lineage, an ancestral character returns, as in some of the pigeon hybridization experiments conducted by Thomson's former Edinburgh colleague James Cossar Ewart. And then there was "atavism," a throwback to a character lying deep in the evolutionary heritage of a species. But Thomson allowed that in practice it would be just about impossible to tell one of these from the others, or from arrested or otherwise abnormal development, with any confidence.[23]

Galton's Law. "The most important general conclusion which has yet been reached in regard to inheritance," Thomson announced, "is formulated in Galton's Law." In 1896–97, Galton's study of some new data on the coat colors of pedigreed basset hounds had prompted him to revisit an old insight about ancestry's multiple nature: that with each generation, ancestral influence halves without ever fully disappearing. Now he restated the point as an infinite series of diminishing fractions summing to 1: ½ + ¼ + ⅛ + 1⁄16 + . . . , where ½ represented the average contribution of the two parents, ¼ the average contribution of the four grandparents, ⅛ the average contribution of the eight great-grandparents, and so on. In 1898, Galton drew attention to this new "law of heredity" in the form of a diagram (fig. 5.2). Although Pearson at first was skeptical—his own work on heredity had persuaded him that ancestry beyond the parental generation could be ignored—he rapidly adopted, and began modifying, what he rebranded "Galton's Law of Ancestral Heredity." In a popular book from 1899, and again at the Royal Institution, Thomson quoted Pearson's encomium:

> The law of ancestral heredity is likely to prove one of the most brilliant of Mr. Galton's discoveries; it is highly probable that it is the simple descriptive statement which brings into a single focus all the complex lines of hereditary influence. If Darwinian evolution be natural selection combined with *heredity*, then the single statement which embraces the whole field of heredity must prove almost as epoch-making to the biologist as the law of gravitation to the astronomer.[24]

FIG. 5.2. Galton's law of ancestral heredity, as diagrammed by the American horse breeder A. J. Meston. Squares 2 and 3 represent the average parental contributions, squares 4 through 7 the average grandparental ones, and so on.

Having led his audience to the frontier of knowledge about heredity, Thomson brought proceedings to a close. Some listeners might, he acknowledged, have expected to hear about the inheritance of acquired characters; but as his theme was the facts, and that did not appear to be among them, he had been silent. In conclusion, he warned them to beware the exaggerated "fatalistic impression" that tended to accrete around the topic of heredity. Between, on the side of nature, the scope for making discerning choices about mates, and on the side of nurture, the limitless possibilities for improving the world that children were born into, there were plenty of faithful-to-the-facts reasons to be cheerful.

3

"What between the Boers & the Basset Hounds I don't get much sleep o' nights!" Pearson had confessed to Galton in mid-November 1899. The war with the Boers would remain a largely private concern for another year, until, as we shall see, the opportunity of a public lecture gave Pearson an occasion for pulling together his thoughts on Britain's military faltering in South Africa and the lessons that, with science's help, the nation should draw. His more immediate problems were with basset hounds, and the law of ancestral heredity generally. Even now, deep into the third round of proof corrections on the new edition of his *Grammar of Science*, he was rethinking and rewriting his coverage of the topic.[25]

He was beginning to suspect that different laws apply to blended and exclusive characters. For blended characters, such as human stature, the law of ancestral heredity seemed to apply more or less straightforwardly. On average, the stature of an individual was predicted to be the sum of half of the average statures of his or her parents (modified to allow for the lower average statures of women), plus a quarter of the average statures of his or her grandparents (likewise modified), and so on, the theoretical sum getting ever closer to the actual average the further back one took one's measurements from within the family tree. Again, the final sum was just an average, which meant that, even if one knew the measurements for all the relevant ancestors going back centuries, there was no expectation that an individual's actual stature would equal the calculated one. Rather, that individual would be expected to have a stature falling somewhere along the population's bell-curve distribution for stature—albeit, in the usual way, with averageness more likely than extremity.

But what about exclusive characters, such as coat color in basset hounds or, it seemed, eye color in humans? As Galton appreciated, a very different interpretation had to be given to the terms in the ancestral heredity series. They could not represent the averaged, modified, and summed values of the character in different past generations. But maybe they could represent the average *proportions* of individuals showing this, that, or the other version of an exclusive character. In that case, the law could perhaps be understood as predicting that, although any given family line would see all sorts of vanishings and returns down the generations, on average, half of the individuals overall would have the version most characteristic of the generation before (say, brown eyes), a quarter would have the version most characteristic of the generation before that (say, blue), an eighth, the version most characteristic of the generation before that (say, hazel), and so on.

In Pearson's new *Grammar of Science*, as well as in a new Royal Society paper, presented just as the new edition was being published, he diplomatically dissented from Galton's view that, differently interpreted though it needed to be for the two kinds of characters, the same law applied in both cases. Pearson allowed that for both, the applicable law was *a* law of ancestral heredity. But he argued that only for the blended case did the law known as *the* law of ancestral heredity apply. For the exclusive case, the applicable law was what Pearson now called "the law of reversion." The nomenclature mattered, in his view, because it acknowledged what he saw as the nontrivial difference between regression, at work behind the bell-curve distribution of the values of a blended character, and reversion, at work behind the proportional distribution of the versions of an exclusive character. He stressed, however, that it was early days, that much was still unclear—it appeared, for example, that not all exclusive characters were covered by the law of reversion, and even those that were proved hard to square with the formula—and that what was needed above all was more research.[26]

It was a message repeated even more emphatically a couple of months later, in a paper Pearson presented at the Royal Society on Thursday, 29 March 1900, on the evening before Thomson gave his Royal Institution lecture.[27] Pearson generally had low expectations for Royal Society meetings. In 1895, he had declined the invitation to speak at the presentation of the crab research committee report because, as he explained to Galton at the time, "points are always missed in discussion & do not reach the real students of the paper."[28] Two years later, at the sole committee meeting he attended, this most articulate and voluble of men found himself tongue-tied around biologists who, he was sure, would only have taken wounded offense at anything he had managed to say. "I always succeed in creating hostility without getting others to see my views—infelicity of expression is I expect to blame," he reflected afterward, again to Galton.[29] Nor was the flipped situation, when Pearson's own work was under scrutiny, any better, since in his experience, the only FRS who could ever really be counted on to understand it enough to comment usefully was Galton—and in spring 1900, as increasingly often these days, Galton was abroad, enjoying an extended health-bolstering holiday from cold, damp, foggy London.[30] So the omens were not good for a productive time at Burlington House.

And yet the discussion proved to be excellent. Whether by accident or design, the program comprised, after two papers on the nerve system of frogs, four varied papers on variation. The first was from Ewart, whose most recent animal-hybridizing investigations fatally undermined belief in what Darwin had called "the direct action of the male element on the female" but was now—in the age of the telegraph and, increasingly, the telephone—called "telegony."[31] The next paper was from Horace Vernon, now at Oxford, but at a different college from Weldon (Merton for Weldon, Magdalen for Vernon). Once again, Vernon reported quantitative results from lab studies of how environmental conditions affect developing larvae, now showing—as might have been expected—that the earlier the experience of changed conditions, the larger and more enduring the influence.[32] Finally, there were two papers from Pearson. The shorter one was a critical note, aimed at Vernon's claim in recent publications that there might be circumstances under which a species could split apart without natural selection (namely, when like individuals mate with like individuals, and when the fertility of such unions is maximal).[33] The longer one was an overview of a massive memoir on hereditary characters and laws, written with Alice Lee, a University College student turned collaborator. First read in mid-November 1899, but then withdrawn and rewritten—the "Boers and basset hounds" letter was sent the day after its underwhelming debut at the Royal Society—the memoir was basically a data-rich restatement of Pearson's latest thinking on blended versus exclusive characters and how they fitted or failed to fit with the law of ancestral heredity. In closing, he returned to the notion that, thanks to Galtonian statistics (and in spite of Weismannian metaphysics), biology was set to enter its Newtonian period.[34]

4

We know Pearson found the 29 March 1900 Royal Society meeting worthwhile because he said so, in a letter he sent afterward to Yule, who had recently left University College for a better-paying position with an exams board (another personal-professional loss Pearson felt grievously). "We had a rather good R. S. meeting & talk about variation on the last Thursday, the best I have been at," Pearson wrote. Then he listed those involved: the three speakers, interspersed with two FRSs who contributed informally—"Ewart, Bateson[,] Vernon, Weldon & your humble servant."[35] The presence of Bateson—on good form and best behavior—when Pearson aired his latest thoughts on Galton's law and the challenge of extending it to exclusive characters deserves underscoring.

Ever since the publication of *Materials*, exactly those either/or, nonblending, non-regressing characters had been Bateson's main concern. As he saw it, with the job of cataloguing the facts about whole-part repeats now finished, it was time to do the same for that other great class of discontinuous variations: the characters whose versions do not blend away when combined but—like tail-forceps sizes in earwigs—stably coexist, defining subtypes or varieties within a species. What was needed for these characters, he reckoned, were field studies and crossbreeding experiments, ideally done in tandem.[36]

A field study was Bateson's own starting point: a statistical description of a beetle species whose amazingly colorful and durably persisting varieties he had encountered while holidaying in Spain.[37] From the mid-1890s, however, he concentrated on the raising and crossbreeding of captive varieties at home in Cambridge, first using butterflies (from 1896) and then chickens (from 1898).[38] His new domesticity—marriage to his fiancée, Beatrice Durham, in 1896, and the birth of a son, John, in 1898—may have tipped the balance toward a more family-friendly style of research.[39] To be sure, he kept up with, and promoted, work on discontinuity-illuminating field studies, including research showing that dark forms of the peppered moth seemed to be replacing light forms throughout the north of England. (The case of the peppered moth is remembered nowadays as a famous illustration of Darwinian adaptation to a changing environment: darker won out, it was said, because darker individuals were better camouflaged against the pollution-darkened trees of the factoried north and so out-survived lighter individuals. Characteristically, Bateson did not even mention that possibility. For him, what was instructive about the example was that, far from all moths gradually getting darker, as per Darwinian expectations, fully dark forms were replacing the light forms.)[40] But increasingly it was crossbreeding that fascinated him, especially the prospect of explaining why one character version rather than another predominated in a cross. His suspicion was that the more inbred a variety was, the greater its power to transmit its character versions; and likewise, that the failure to take degrees of inbreeding into account was what made for all the seemingly inconsistent data on prepotency.[41]

In Cambridge, Bateson soon found a new research ally: Edith Rebecca Saunders, a botany lecturer at the university's other women's-only college, Newnham. At his suggestion, and with materials he supplied, she crossed a hairy variety of an Alpine mustard plant with a smooth variety living alongside it in the wild, finding that, when fully grown, the offspring were either hairy or smooth, with no intermediates.[42] By 1898, they were a team—Saunders responsible for plants, Bateson for animals—with funding from the Evolution Committee and, thanks to Bateson's taking an allotment next to the Cambridge University Botanic Gardens, dedicated premises for their experiments (fig. 5.3).[43]

By way of extending Saunders's investigations into hairiness/smoothness, the team got in touch with de Vries, who sent them seeds of the smooth form of *Lychnis*. (The only known stock was from plants he had discovered near Amsterdam.)[44] In July 1899, when de Vries came to Britain to speak at an international conference on hybridization and plant

FIG. 5.3. Edith Rebecca (Becky) Saunders at work in the Cambridge Botanic Garden. Alongside her botanical breeding research, she served, from 1899 to 1914 (when it closed), as director of the Balfour Biological Laboratory for Women.

breeding—where he gave his "pangenetical infection" lecture—he stayed with Bateson, who wrote to Beatrice, "de V. is a really nice person, very simple & rather rough in style. . . . He is an enthusiastic Discontinuitarian & holds the new mathematical school in contempt. So we hit it off to admiration." Saunders, too, was charmed: she "talked & chattered as I never saw her do before."[45] At the conference, the title of Bateson's lecture, "Hybridisation and Cross-Breeding as a Method of Scientific Investigation," registered the new emphasis, shared with de Vries, on hybridization as means rather than end. Bateson told his audience that evolutionists had been catching up to horticulturalists, who had long known that "it is not Evolution but Revolution" that makes for new varieties, and that, far from distinctive character versions regularly getting blended away into meanness or oblivion, they were often transmitted perfectly to the next generation and beyond.[46] In toasting the hybridists at the conference banquet, he expressed his confidence in the impending discovery of "the law which governs hybridism and kindred phenomena."[47]

So Bateson was primed to take notice when, at the Royal Society on 29 March 1900, Pearson indicated that the law Bateson sought might turn out to be a modified form of Galton's new law. In a lecture Bateson gave less than six weeks later, titled "Problems of Heredity as a Subject of Horticultural Investigation," Galton's law was the star attraction.[48]

5

That was not, however, how Bateson remembered his lecture, and not why it ranks second to Mendel's 1865 lectures in the annals of Mendelism.

Bateson delivered "Problems of Heredity" on 8 May 1900, at the monthly general meeting of the venerable Royal Horticultural Society. His connection to the society was via the botanist Maxwell Masters, an FRS who chaired its scientific committee and was also, like Bateson, an 1897-vintage recruit to the Evolution Committee and a true believer in the cause of bringing practical breeders and professional biologists into harness. It was Bateson's and Masters's encouragement that led the RHS to organize the 1899 conference. Since then, however, things had gone quiet, as Masters gently reminded Bateson the following March, in a letter belatedly congratulating him on taking the reins at the Evolution Committee.[49]

What makes Bateson's 8 May lecture famous is not its role in maintaining good relations between the committee and the horticulturalists, but its status as the occasion when the news broke in Britain about Mendel's law. The subsequent published version of the lecture indeed reads just that way. As for the spoken version, Bateson later told his wife, who relayed the story in a biographical memoir, that he was en route to give the lecture when he first read the "Versuche," and was so taken with it that he instantly rewrote his ending.[50]

Probably Bateson did read a thrilling new paper on that train ride and amended his lecture. And he had surely encountered Mendel's name here and there (it came up, for instance, in one of the lectures at the 1899 conference).[51] But neither the "Versuche" nor Mendel is mentioned in a report of Bateson's lecture published in the next issue of the weekly *Gardeners' Chronicle* (of which Masters was editor). According to that report, after Bateson extolled the virtues of Galton's statistical approach to heredity, and introduced the new ancestral law framed on its basis, he discussed the prospects ahead for modifying Galton's law in order to fit a version of it to non-blending characters, citing de Vries's new paper.[52]

Bateson's coming to Mendel through de Vries's tacitly Mendelized "Sur la loi de disjonction des hybrides" helps make sense of something that has come to seem surprising, even bizarre: that anyone could have seen a vindication of Galton's law in the Mendelian pattern. Within the next year or so, not even Bateson would see it that way, instead stressing the contrast between a law in which hereditary influence gradually diminishes without ever going extinct and a law in which that influence can go to zero in a single generation. (As we have seen, when, in a Mendel-style cross, hybrids have recessive offspring, those offspring are, with respect to the character in question, constitutionally identical to their recessive grandparents, showing no vestige whatever of the intervening dominant-character parents.)[53]

It is on this point that fussiness about whether Bateson read Mendel or de Vries pays dividends. More directly than the "Versuche," "Sur la loi" offered up Mendel in a form that looked like a good match for what Bateson, thanks to Pearson, was searching for in spring 1900.

Recall that Mendel, in the "Versuche," represented his "law valid for *Pisum*" as $A + 2Aa + a$. Again, the terms stand for, respectively, the constant dominants, the hybrids (showing the dominant-character version), and the constant recessives among the offspring of hybrid parents. Altering Mendel's nomenclature so as to represent the dominant-character version with a D and the recessive-character version with an R gives

$$D + 2DR + R$$

That is the nomenclature that de Vries used in "Sur la loi." But he stated the law rather differently, as the right side of an equation:

$$(D + R)(D + R) = D^2 + 2DR + R^2$$

By the lights of the "Versuche," that makes no sense at all. One would be hard-pressed, if thinking in "Versuche" terms, to say why the expression ($D + R$) should be multiplied by itself, and what could be meant by "D^2" or, more plainly, "*DD*." Generally, when Mendel multiplied expressions, and *D* met with *D*, the product was *D*, signifying constancy for that character version.

Why did de Vries present Mendel's law so differently? Because of what de Vries brought to the "Versuche." Unlike Mendel, de Vries was a reformer of pangenesis, who thought of hereditary characters as ultimately due to stable, countable, material particles, probably carried on chromosomes. If you brought two of these particles together, you would still have two. Also unlike Mendel, de Vries became an admirer of *Natural Inheritance* and the work it inspired. His Galtonian enthusiasms led him to study the mathematics of probability—not as deeply as Weldon, but deeply enough to learn that, according to elementary probability theory, given two successive draws from an urn containing equal numbers of black (B) balls and white (W) balls, the probability of getting a BW mix is two times the probability of getting either BB or WW. (The reason is that there are two ways to get a mix—by drawing B first and then W, or by drawing W first and then B—but just one way to get all black, and just one way to get all white.) If we represent the either-black-or-white nature of each draw as (B + W), the relative probabilities of the outcomes of two draws can be calculated thus:

$$(B + W)(B + W) = BB + 2BW + WW$$

with the terms on the right-hand side thus forming a 1:2:1 ratio, with the mixed or hybrid condition in the middle. Familiarity with this result meant that when, in the mid-1890s, de Vries first noticed, among the offspring of some of his hybrids, a 1:2:1 ratio of (in his terminology at the time) constant active, hybrid, and constant latent character versions, respectively, he found it straightforward to assimilate that ratio to the probability equation and the pangenesis-friendly interpretation it supported. It was through the prism of this understanding that de Vries, in early 1900, read Mendel.[54]

As for Bateson, although he was no fan of pangenesis, and more or less innocent of probability theory, he was, again, someone who thought that inbreeding increased the transmission potency of a character version. So de Vries's representation of the constant dominants and constant recessives as *DD* and *RR*, in giving the quantitative nature of character-version purity its due, would have rung true. And as Pearson himself showed years later, once one adopts that double-dosing nomenclature, then Galton's law, lightly modified and applied to a simple case, becomes Mendel's law. All one needs to suppose is that a *DR* mother mates with a *DR* father; that each of their parents is purebred, represented as *DD* and *RR*; and that each of them in turn comes from an unbroken line of, respectively, *DD* and *RR* ancestors. Galton's law could then be said to predict that, out of four children, half of them will be like their parents (2*DR*), a quarter like the one set of grandparents (*DD*), and a quarter like the other set (*RR*).[55]

Whatever the affinities Bateson glimpsed that day in May, the months that followed brought paper after astonishing paper on this new-but-old law—from de Vries, from the German botanist Carl Correns, from the Austrian breeder Erich Tschermak—not only affirming the law but drawing attention to its discoverer, Gregor Mendel, and his 1866 paper.[56] In a letter in August, Bateson recommended that Galton get hold of the "Versuche." "Mend[e]l's work seems to me one of the most remarkable investigations yet made on heredity," Bateson declared, "and it is extraordinary that it should have got forgotten." Having now read the "Versuche" for himself, however, he belatedly recognized, and regretted, some of de Vries's alterations. "I don't like de Vries' introduction of <u>squares</u> of D and R which is much misleading, but that is a trifling point as yet."[57]

6

Bateson signed off on the proof of the printed version of "Problems of Heredity" on 31 October 1900. Published the next year in the Royal Horticultural Society's house journal, it reads very much like the lecture summarized in the *Gardeners' Chronicle*, save for the ending: an admirable, and admiring, guide to the Mendelian basics, with Mendel's law rendered without those misleading squares (albeit still with de Vries's new symbols—1*D* : 2*DR* : 1*R*), and with full references to the "Versuche" and the recent papers of de Vries, Correns, and Tschermak.[58]

Already up for grabs was the question of how to integrate Mendel's law with the emerging picture of the physical basis of heredity. To Correns, the most suggestive point of contact was the resemblance between Mendel's hypothesis that gametes in the hybrid get just one or the other of the parental-character versions and what Weismann had termed the "reduction division": the halving of nuclear material in the production of gametes.[59] Soon others would develop that idea, looking to explain Mendel's law as the upshot not merely of nuclear but of chromosomal mechanics.

Not Bateson. He became the leading Mendelian without ever accepting, in anything more than a grudging, resigned way, the identification of Mendelian genes with bits of chromosomes. Instead—and true to the vision he set out in *Materials*—he wanted Mendel's law to take its place in a comprehensive theory of biological patterning, with chromosomes treated as cellular elements that, like other such elements, come to be coordinated into pattern-generating movement. To concentrate solely on the likes of chromosomes, he thought, was to close off any possibility of understanding the coordinating forces.[60]

Perhaps the best way into Bateson's perspective is to dwell a little on a passage near the start of his "Problems of Heredity." We must, he says, observe a firm distinction between two aspects of heredity. First, there are the outward, visible, presumably law-conforming relations between parents and offspring. And then there are the inward and invisible causes constituting the "essential process" that brings about those relations. Where our knowledge of the former is progressing, our knowledge of the latter is stagnant, never mind how much has been learned about the workings of the cell. For what we really need to know cannot be found by looking down even the most powerful microscope:

> We can study the processes of fertilisation and development in the finest detail which the microscope manifests to us, and we may fairly say that we have now a thorough grasp of the visible phenomena; but of the nature of the physical basis of heredity we have no conception at all. No one has yet any suggestion, working hypothesis, or mental picture that has thus far helped in the slightest degree to penetrate beyond what we see. The process is as utterly mysterious to us as a flash of lightning is to a savage. We do not know what is the essential agent in the transmission of parental characters, not even whether it is a material agent or not. Not only is our ignorance complete, but no one has the remotest idea how to set to work on that part of the problem. *We are in the state in which the students of physical science were in the period when it was open to anyone to believe that heat was a material substance or not, as he chose.*[61]

To succeed as physicists had, Bateson implied, biologists would need to break free from properties-of-a-substance explanations. Heat had ceased to be mysterious once it was understood not as the distinctive property of a material (caloric) but as a generic property of moving particles of any material. Heat was "a mode of motion," as John Tyndall had put it in the title of an 1863 bestseller. So was the pressure of gases. Matter itself looked for a time as if it would yield to mechanical analysis. Under a theory developed and popularized in the late 1860s and 1870s by the Scottish physicists P. G. Tait, William Thomson, and Maxwell (from initial studies by the brilliant Hermann von Helmholtz), atoms were pictured not as material particles, but vortices in a universally distributed ether. Their whirling accounted for their hardness and permanence, on the model of the spinning smoke rings that, as Tait demonstrated experimentally, bounced off each other when they collided. The "encounter between two smoke rings in air," affirmed Maxwell in a Galton-challenging 1875 encyclopedia article on the atom, "gives a very lively illustration of the elasticity of vortex rings." Smoke rings became standard fare in the lecture theater as well as on the page. At the height of the vortex theory's popularity in Britain, between roughly 1875 and 1885, it promised to unify everything from gravity to chemistry. But the difficulties mounted, and enthusiasm waned. The rest of the world was aloof from start to finish. Yet even into the early twentieth century, the theory's admirers held it up as a shining example of the mechanical view of nature that British science had served so well.[62]

Educated in the natural sciences at Cambridge during the heyday of vortex physics, Bateson carried a torch for vortex biology to the end of his life.[63] A living thing, wrote Bateson in a letter of 1924, "is not matter. It is a system-vortex . . . through which matter is passing."[64] In an address that same year, he defined biology as the study of such vortices.[65] When he tried to give a sense of how differently the vortex biologist looked at a cell or a body part or an organism, he reached for simple physical phenomena showing how vibrations can organize matter into form. Chladni patterns were a particular favorite: pour sand on a plate, set it vibrating with sound or a violin bow, and the sand will gradually clear from the high-motion places and clump in the low-motion ones, forming different patterns according to the frequency of the vibration. Wind-rippled beaches, octave-apart piano wires, and twisting smoke rings were other staple analogues. Discussion of these analogues shows up in Bateson's private correspondence and published writings from the early 1890s.[66] As we have seen, he credited the first stirrings of his interest to Michael Foster, boss of laboratory physiology at Cambridge when Bateson was an undergraduate. In a 1917 lecture in Cambridge, he recalled what Foster used to say about living things, adding, "If we could in any real way identify or analyse the causation of growth, biology would become a branch of physics. Till then we are merely collecting diagrams which some day the physicist will interpret. He will I think work on the geometrical clue."[67]

7

Between late spring and early autumn 1900, as Mendel's peas became ever more intriguing for Bateson, Pearson's poppies became a source of deepening dismay for Weldon. To be sure, Weldon's research in these months embraced more than the Shirley poppies. There was still the writing of the new textbook, though his work on it—along with his confidence in its value—soon dropped away. There were, once again, breeding experiments with pedigreed moths, done in collaboration with a talented former student from University College, Ernest Warren (who had earlier managed to wring order from the chaos of Weldon's water flea breeding experiments). And from June, there were snails, after Weldon came across a German book purporting to show that, judged by their shells, snail species had been imperturbably constant across vast changes of environment. That provoked him into a sustained bout of summertime snail collecting and shell measuring. It was, in many ways, a return to his shrimp and crab research, down to the question of whether, as he suspected, natural selection maintained that constancy by eliminating individuals whose shells deviated from the optimal mean morphology.[68]

But all of this activity did not quite manage to distract him from, as he saw it, the mess he had made of the poppy research. When he started, he had planned not merely to raise plants from which decent capsule-correlation data could be gathered, but to undertake some artificial cross-fertilizing in hopes of studying the consequences for fertility. He never even got close. As the months wore on, over and over again he realized—always too late—that he had accidentally subjected his pots to different conditions, giving them different amounts of water (which seemed to have an alarmingly large effect), thinning out the seedlings to different degrees (ditto), and so on. "I told you that an ignorant ass who had never grown a flower would play the fool in this way," Weldon wrote dejectedly to Pearson in July. Of his increasingly pitiful-looking poppies, Weldon wrote in August, "They are shameful, like everything I do . . . I am very sorry you trusted me with them." Pearson was reassuring about the data, and indeed, he would eventually use them. But in early October, when Pearson submitted a long memoir on homotyposis to the Royal Society, Weldon's data were not included.[69]

It was around this time—when, as never before, he knew intimately how extraordinarily skillful one had to be in order to get meaningful correlation results from breeding experiments with plants—that Weldon read Mendel. In a letter to Pearson on 16 October, after some tut-tutting about recent shady dealings at University College, Weldon changed the subject:

> About pleasanter things, I have heard of and read a paper by one Mendel on the results of crossing peas, which I think you would like to read. It is in the Abhandlungen des Naturforschenden Vereines in Brünn for 1865—I have the R.S. copy here, but I will send it to you if you want it.

> The point seems to me to be that the results indicate an exclusive inheritance with a very high fraternal correlation.

For Pearson's benefit, Weldon then furnished an epitome of the "Versuche" that—as against later suspicions that he never really understood the paper—was at least as good as Bateson's for the Royal Horticultural Society, and in two respects went beyond it. First, Weldon turned the Mendelian pattern into an easily graspable diagram, with "y" representing the plants producing yellow seeds only, "g" the plants producing green seeds, and "m"—the Galtonian symbol for mediocrity—the plants producing both kinds of seeds (fig. 5.4). Second, he included an abbreviated version of Mendel's $2^n - 1 : 2 : 2^n - 1$ table, showing that the proportion of hybrid plants rapidly dwindles without ever disappearing utterly. Weldon closed his letter, "Of course the whole value of the thing depends upon experimental details which are not quite given: but he gets closely similar results for several other characters of the seeds and pods, and it seems a very good starting point for further work?"[70]

If Weldon's engaged and positive assessment opened up new prospects for rapprochement with Bateson, they soon vanished. "Bateson and these folk have no 'real facts': their only idea is to take isolated cases of inheritance and to worship them as marvels," Weldon complained to Pearson a little over a week later. Weldon had cause to be grouchy. In Cambridge, where Bateson was now lecturing on evolution and heredity, a dismissive attitude toward Weldon's kind of research seemed to have become part of the curriculum, as he was startled to discover. His source was a Newnham zoology student named Hilda Sollas, one of Bateson's prize pupils, but also the daughter of an Oxford colleague, W. J. Sollas, professor of geology. Where the father had heaped praise on Weldon, lauding his Plymouth crab studies for showing that Darwinian evolution was possible even within Lord Kelvin's contracted age for the Earth, the daughter had heaped scorn, explaining to Weldon "very fully that I was a damned fool," as he reported to Pearson in mid-November; "but I think before she went back to Newnham she came to the conclusion that at least there was something to be said for your methods."[71]

Weldon told Pearson about this exchange by way of accounting for the vehemence of Bateson's attack on Pearson at the Royal Society the evening before, 15 November, when the new homotyposis paper was discussed. In Weldon's reckoning, additional reasons for Bateson's becoming annoyed included "Sedgwick's interest in the Grammar of Science, and my statement that Mendel seemed to me a confirmation of your suggestion about direct parental regression." Be that as it may, Pearson

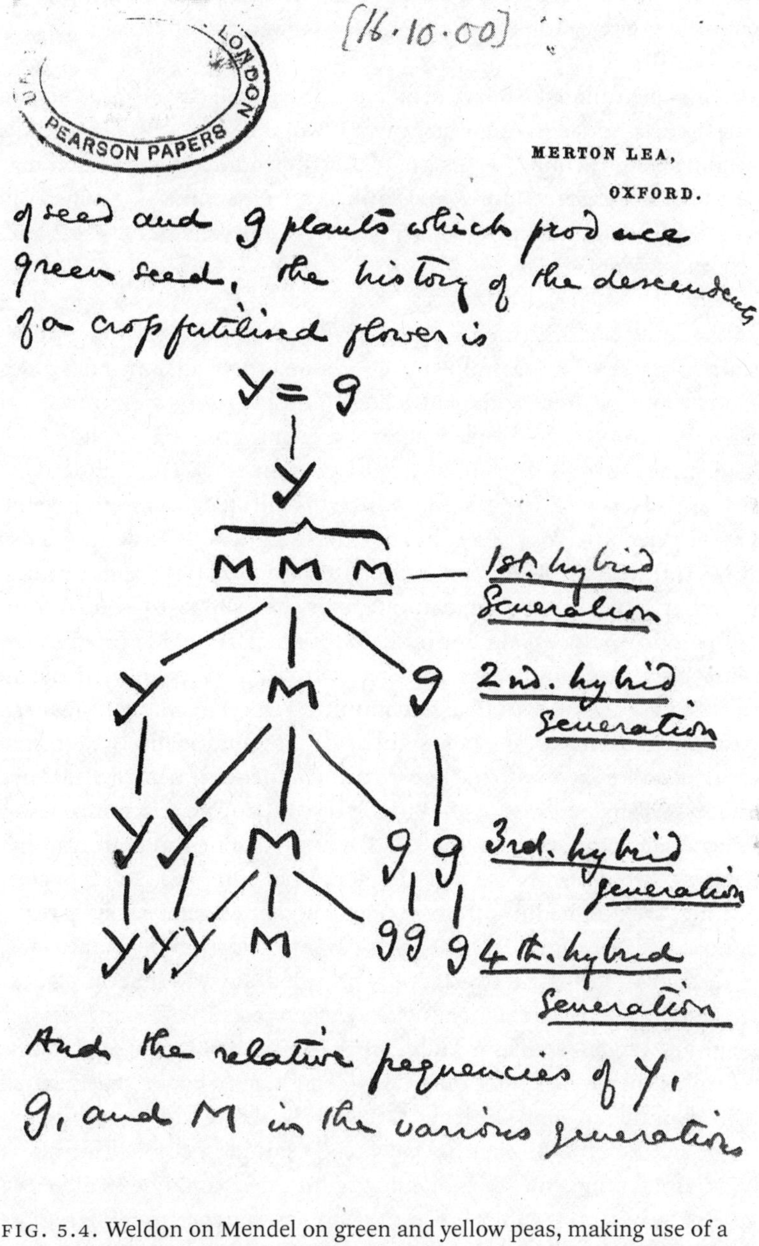

[16.10.00]

MERTON LEA,
OXFORD.

of seed and g plants which produce green seed, the history of the descendants of a cross fertilised flower is

y = g

y

M M M — 1st. hybrid generation

y M g 2nd. hybrid generation

y y M g g 3rd. hybrid generation

y y y M g g g 4th. hybrid generation

And the relative frequencies of y, g, and M in the various generations

FIG. 5.4. Weldon on Mendel on green and yellow peas, making use of a simple inverted-tree diagram showing "the history of the descendants of a cross fertilised [pea] flower," 16 October 1900.

came away from the meeting certain that Bateson would do all he could to ensure the rejection of the paper. Weldon was incandescent:

> Your paper cannot be rejected, and it is just as well that Bateson should be induced to talk rot now and then. It will the sooner bring about his ultimate destruction. . . . [His] contention that numbers "mean nothing" and "do not exist in Nature," and so on, is a very serious thing, which will have to be fought. Most other people have got beyond it, but most biologists have not.[72]

We should not take the remarks attributed here to Bateson—which Pearson repeated to Galton—at face value.[73] For all Bateson's jibes against the new "mathematical school" and his promotion instead of "natural history," he was never against a quantitative approach in biology. On the contrary, he campaigned hard, at the Royal Horticultural Society and elsewhere, to get numerical data collected, and from populations large enough for the results to be meaningful. But he wanted those efforts directed, as Mendel's had been, toward the discontinuous variations that, Bateson believed, natural history had shown to be evolutionarily potent. An understanding of these variations was what promised to transform biology into a branch of geometrized physics. To want to concentrate instead on continuous variation, he suspected, was to be in thrall to the enchantment of the new statistical mathematics invented to analyze that variation, and so to choose mathematics over nature. Pearson's homotyposis paper, with its bid to dissolve all apparent symmetries in biology into correlation tables, was that preference taken to the limit.[74]

Bateson's concerns were well founded. Even before the Royal Society meeting, Weldon had begun to wonder whether Mendel's results really represented reverting-directly-to-one-parent exclusive inheritance. Yes, for seed color, the case looked good—provided, anyway, that the character versions really were in just the two categories, "green" and "yellow." But given Mendel's confessed difficulties in establishing the seed colors, hidden as they were under differently colored seed coats, that seemed to Weldon increasingly unlikely. "I think," he wrote to Pearson, "there can be no doubt at all that his peas really varied in colour." Suppress that doubt, and maybe, Weldon allowed, the case could be treated as a "case of 'reversion (?)' to one parent, without reference to more remote ancestors?" Otherwise, those more remote ancestors would have to be acknowledged.[75]

Over the final weeks of 1900, the main topic of correspondence between Weldon and Pearson was the "Versuche" and how to interpret it. A number of questions came up, including whether hybrid plants might show a single, uniform character version as a result of injury sustained in the course of artificial fertilization. But for Weldon, the already adumbrated possibilities—of there being more to the variability of Mendelian characters than Mendelian categories suggested, and more to the bearing of ancestry on those characters than Mendelian methods disclosed—would prove to be the keepers.[76] When, a year later, keen to help a new scientific journal in need of copy, he pulled together his latest thoughts on Mendel's peas as the basis for a science of heredity, these ideas about variability and ancestry would come to the fore.

6

Two Plates of Peas

I take it that the length of a man's little finger, if undeformed has no advantage one way or the other for his fitness in walking, but if you were to measure the British Army that went & that which comes back from South Africa you would find a selection I take it of little fingers, due to the correlation of this character with other parts of the system, which were of direct service for survival.

KARL PEARSON TO W. F. R. WELDON, 30 JULY 1900

I have no vanity about this paper, because mere destructive criticism never seems to me worth much.—"If this work be of men it will come to naught; but if it be of God ye cannot prevail against it!"—But as an advertisement, I think this is the psychological moment for a thorough row about Mendel; and I think this rubbish of mine will start a right royal row.

W. F. R. WELDON TO KARL PEARSON, 15 DECEMBER 1901

[I]t would almost certainly be possible, by selecting cases of marriage between men and women of appropriate ancestry, to demonstrate for their families a law of dominance of dark over light eye-colour, or of light over dark. *Such a law might be as valid for the families of selected ancestry as Mendel's laws are for his peas and for other peas of probably similar ancestral history, but it would fail when applied to dark and light-eyed parents in general,—that is, to parents of any ancestry who happen to possess eyes of given colour.*

W. F. R. WELDON, "MENDEL'S LAWS OF ALTERNATIVE INHERITANCE IN PEAS," 1902

One of the great overlooked thought experiments in the history of science is, aptly enough, about eyes.

Imagine you have complete knowledge of every family tree within a large human population in which, from time out of mind, men and women have been choosing mates freely. Thanks to all this mixing, their hereditary constitutions are marvelously diverse, even between people outwardly the same in this or that character. You set to work identifying parental couples who fulfill the following criteria:

1. Each couple has a large number of children.
2. Each parent comes from a long line of large families.
3. In each couple, one parent has dark eyes and the other has light eyes.

Within this collection of couples you find at least one couple whose many children are uniformly dark-eyed, in conformity with a family-tree pattern in which, whenever children have a dark-eyed parent and a light-eyed parent, darkness is dominant. But you also find at least one couple whose many children are uniformly light-eyed, in conformity with a family-tree pattern in which, whenever children have a dark-eyed parent and a light-eyed parent, lightness is dominant.

So, is darkness dominant over lightness? In the first family, yes. But there is no general fact of the matter, as the second family shows—and of course, in the wider population, there will be lots of people with eye colors of every shade between extreme dark and extreme light, and other variations too, such as dark with light flecks. Different selections, involving different ancestries, yield different dominance patterns, if any. You could, if you wanted to, declare the darkness-over-lightness pattern to be exemplary and the lightness-over-darkness pattern to be exceptional. But that would be arbitrary because, whatever the actual proportions of each kind in the population, it is easy enough to imagine that chance and conditions (themselves chancy) could have made for different numbers, and might well do so in the future.

Weldon's thought experiment on human eye color appeared early in 1902, in an article titled "Mendel's Laws of Alternative Inheritance in Peas."[1] As billed, it dealt mainly with peas, whose un-Mendelian variability Weldon not only documented and described at length but depicted in two accompanying photographic plates, one illustrating seed color, the other seed shape. Weldon turned himself into a hands-on student of pea breeds only toward the end of 1901, after a year spent fitfully concerned with Mendel, but mostly occupied with other projects, including the founding with Pearson of the journal *Biometrika*, where the peas (and eyes) article appeared. Over the same period, Mendel's paper came out in an English translation, introduced by Bateson, while Galton turned himself into the spokesman for the cause that would define his final years, and ultimately his legacy: the publicizing of the need for the British people to start taking charge of their own hereditary improvement, under the banner of the new eugenic science. When Weldon spun a selective-breeding speculation about humans, he did so in the wake of Galton's doing similarly, in a controversial and widely noticed address, only weeks before.

That Weldon's speculation revolved around eye color, of all possible human characters, was due to his initial, intense correspondence on Mendel's paper with Pearson in late 1900, when Pearson—an assiduous analyst of the data on human eye-color inheritance collected by Galton—raised the eye-color data as a potential difficulty for Mendel.[2] A royal (as well as Royal) row over Mendel began, as Weldon predicted, with the publication of "Mendel's Laws of Alternative Inheritance in Peas." But the story of how that paper came to take the form it did, and to argue the case it did, begins with Pearson in late 1900, once again pitched between Boers and basset hounds.

1

For Pearson, the war with the Boers was but the latest episode in the long Darwinian struggle of nation against nation—a struggle that, on recent evidence, Britain was starting to lose. The humiliating string of defeats inflicted on the British Army in the early part of 1900 by a tiny band of South African farmers showed, in his view, that Britain, although hard to outnumber, could now be easily outsmarted. The depression into which the nation fell at the relentlessly bad news from the front had thus been entirely justified, because the weaknesses that the war exposed were fundamental. Yet the same biological perspective that illuminated the nature and scale of the problem also, he reckoned, suggested the outlines of a solution. Above all was the need to ensure that an ever-higher proportion of future citizens derived from the best stock, because there was no other way permanently to raise the quality of the average British body and—most pressingly now—mind. Education could then develop these born-superior minds still further, but it had to be the right kind of education, suited to honing their capacities for observation and foresight, so that they could adapt swiftly when encountering unaccustomed circumstances. Training in the method of science would be ideal. Furthermore, these intellectually nimble men and women must find themselves living among people as much like themselves—racially, socioeconomically, educationally—as possible, so that the instinctive moral regard all humans are born with, but which shrivels as differences accumulate, can help bind them together in a time of crisis.

Pearson laid out this vision in a popular lecture before the Newcastle Literary and Philosophical Society on 19 November 1900. He confessed to Yule beforehand how hard he was finding it "to compress my views on the war, civilisation, evolution, science & cosmos generally into an hour's discourse!" What gave the whole coherence was a repeated emphasis on heredity and its importance. Over the previous ten years, Pearson told his audience, a vague sense that heredity somehow mattered had given way to exact knowledge. "The form of a man's head, his stature, his eye-colour, his temper, the very length of his life, the coat colour of horses and dogs, the form of the capsule of the poppy, the spine of the water-flea, these and other things are all inherited, and in approximately the same manner," he declared, adding—unbowed by the clash with Bateson at the Royal Society meeting four days earlier—that his own most recent work on like-producing-like relations suggested an even greater domain of extension for the principle of heredity. Here was a law "as inevitable as the law of gravity," the opposing of which was just as pointless and self-defeating.[3]

Weldon never saw, or wrote about, heredity in quite that way. Even so, for him too, the war showed up the feebleness of Britain's ruling class, whose nice-but-dim sons ("pretty imbeciles," he once called them) had made his teaching at Oxford such a dispiriting chore.[4] Returning home from Madeira in mid-January 1901 from a snail-hunting trip with some students, Weldon found the published version of Pearson's lecture—which he had dedicated to Weldon—waiting. He was touched by the gesture, and, as he wrote to Pearson, even more convinced of the lecture's necessity after talking with invalided soldiers on Madeira (and contrasting their moral seriousness with its absence in the students). "I hope," he added, "you will go on doing work of this kind, as you can do it so well."[5]

A still more enthusiastic response came from Galton. Although he had not dealt with eugenics in more than a cursory way in decades, the lecture inspired him to dust off—and lightly imperialize—an old idea. Consider, Galton wrote to Pearson, how far the "zeal for military usefulness will cause many men to be physically examined as to fitness to serve." Might those examinations, in combination maybe with others, become the basis for a new system of certification? Those who passed could then be issued degrees confirming their "V. H. T." status: "Valid for Hereditary Transmission of Qualities Suitable to a Citizen of an Imperial Country." And if people took to the new degrees, then, over time, they might "help in forwarding marriages of the fittest & discouraging others."[6]

Pearson was thrilled, less by the proposal—he thought such degrees in themselves would not be enough to persuade the middle classes to have larger families, or to counter the influence of the workhouses and charities in encouraging breeding by the feckless poor—than by Galton's interest in making it. For in Pearson's estimation, Galton was exactly the man that the nation needed right now. "You are known," Pearson wrote back, "as one who set the whole scientific treatment of heredity going; no one has ever suspected you of being in the least a 'crank,' or having 'views' to air. You will be listened to & it will be recognised that you write out of a spirit of pure patriotism."[7] In large measure, the world beyond biology came to know about eugenics—at the same moment that it came to know about Mendel—because Pearson's plea to Galton worked.

2

By now, plans were afoot for a new journal in which Pearson, Weldon, and—they trusted—a growing band of allies could publish with impunity. "Do you think it would be too hopelessly expensive to start a journal of some kind?" Weldon had asked in his commiserating letter to Pearson after the drubbing of homotyposis at the Royal Society meeting on 15 November. Weldon reckoned that Pearson could probably fill such a journal ("an Archiv für K.P.ismus") all on his own. But for other workers too, it would be an encouragement to know that their research in the statistical study of heredity and evolution, if done well, might find a home. Certainly the *Philosophical Transactions* was not up to the job. Quite apart from the distressing new gatekeeper problem, biologists on the whole did not read it—Pearson's papers excepted—and no one abroad could afford to buy it.[8] Meanwhile, abroad was where signs of interest were most heartening, most notably in the United States, where the founding of new agricultural stations and research universities had helped to bring about a boom in biology research and teaching. The morphologically trained Charles Davenport (fig. 6.1), based at the recently established University of Chicago, was proving an especially enterprising exponent of the new biology. His *Statistical Methods with Special Reference to Biological Variation*, published in 1899, was the first textbook in the field. In Davenport's essay on the new biology's history, culled from his report to the American Association for the Advancement of Science's version of the Evolution Committee and published the next year in *Science*, Pearson, Weldon, and their students featured prominently.[9]

Soon Davenport was in the frame as coeditor of the new journal. Before long, the journal had a name: *Biometrika*—Pearson's coinage and spelling. (He was, he reminded Galton, "K. P. not C. P.")[10] Alas, the case for the new journal got a boost as *l'affaire homotyposis* rumbled on into 1901. The initial attack on the paper had taken place as and when it did because Michael Foster—keen as ever to make Royal Society meetings intellectually lively—had agreed beforehand to a request from Bateson to speak after Pearson, so that the assembled fellows could hear Bateson's criticisms right after Pearson read his abstract. In putting the request, Bateson had recognized that a breach of etiquette was involved, the more egregious because his criticisms would bear on the longer paper that he

FIG. 6.1. Charles B. Davenport around 1900—"the best American," according to Weldon.

alone, in his capacity as referee, had seen. Small wonder that Pearson became convinced not merely that his paper was going to be rejected but that it was now futile for him, or even for anyone too closely associated with him, to try to publish in a Royal Society journal. The news in January that the homotyposis paper would in fact be recommended for publication, though reassuring, was quickly followed by another, less welcome surprise. At the invitation of the controversy-stirring Foster, Bateson had written up his criticisms not in the form of an ordinary referee report but as a paper in its own right, to be read and then published in the *Proceed-*

ings. The first Pearson knew of Bateson's "Heredity, Differentiation, and Other Conceptions of Biology: A Consideration of Professor Karl Pearson's Paper 'On the Principle of Homotyposis'" was when he received a postcard in mid-February with the agenda for the next society meeting.[11]

The paper said nothing about numbers being meaningless. Bateson began by congratulating Pearson on sensing, correctly, that at bottom, the process that brings about likeness between parents and offspring (heredity) is the same as that which brings about likeness between the repeated parts of an individual (repetition of parts). But then Pearson went wrong, as Bateson saw it, in trying to capture that unity in the statistical language of correlation rather than the geometrical language of symmetry. Consider, suggested Bateson, that a part can be repeated perfectly, then repeated slightly less perfectly, then repeated . . . At a certain point, the next member in the series will cease to be a variation on the original part and instead count as something new, something differentiated. But what point is that? And how will we know it? According to Bateson, the inability to provide principled answers to these questions was not a problem for the symmetry-minded student of variation, who allowed for the possibility that outwardly identical variants might belong to types with different centers of stability, in line with the deeper rhythms of vortex biology. But for the correlation-minded student, who does not look beyond the measured values, and whose calculations are reliable only if the values compared are from genuinely like parts, the problem is fatal. So, Bateson concluded, while we should doff our caps to Pearson for the industry and ingenuity with which he purported to show, over and over again, that average correlations between all sorts of characters in all sorts of parents and offspring are well-nigh identical to average correlations between all sorts of repeated parts, the suspicion lingers that it all depends on judicious circumscribing of the data sets. To that extent, his correlational route to making sense of the fundamental unity of heredity and repetition of parts was a failure.[12]

The justice-for-Pearson party has always been small, even smaller than the justice-for-Galton party. But the extraordinary publicity the society granted to Bateson's criticisms would have made anyone on the receiving end feel shabbily treated. And the situation was even worse than it looked, as Pearson explained in correspondence with Bateson, for Pearson's own paper had not yet been accepted for publication by the council. If his paper was now rejected, all would assume that Bateson's annihilating criticisms had been the reason. And if it was accepted, then, given the immense slowness of *Phil Trans* compared with *Proceedings* publications, any improving change Pearson might later make, in the ordinary way, on the proof of his paper would be interpreted as a concessionary response to Bateson's paper, whether or not it was. (Pearson returned the proof that Bateson sent him unread.)[13]

Belatedly, Bateson agreed that Pearson had been put in an unfortunate position, and publication of Bateson's criticism was duly postponed until the homotyposis paper appeared. In the meantime, Pearson let Bateson know that his own work would be welcome for *Biometrika*. "We do not intend to be exclusive—'Nothing vital will be foreign to us'—so that if you do not aid us, we at least may find room to print and meet your future criticisms."[14]

Subscriptions rose steadily, reaching about sixty by mid-April: a good start, but not nearly enough for the journal to pay for itself. Before Oxford University Press—then the preferred publisher—would take it on, they needed a guarantee fund of £200. Might Galton help? Pearson wrote to him explaining the situation, and also asking if Galton would consider joining the editorial committee, and perhaps also contributing something to the inaugural issue.[15] From his hotel in Italy, Galton replied with a check for the full amount ("I like 'forlorn hopes' in a good cause"), permission to name him as an editor in whatever way made it clear that he would be a sort of distinguished figurehead-consultant, an offer to write something general about "Biometry," and a hope that they would try to keep the contents not too mathematically technical. "It ought to be attractive to medical men and such like; also to statisticians of the better kind."[16] Pearson and Weldon were overjoyed, but remained open-eyed, as Pearson put it to Galton, about the "uphill fight" ahead.[17]

3

To judge from surviving correspondence between Pearson and Weldon, Mendel was not at all on their minds in winter and spring 1901. The first mention of him is in early May, in a letter from Pearson bringing Weldon up to date on sundry topics. Among them was a new article by Davenport on some breeding experiments recently published by an associate of Weismann's named Georg von Guaita. Tracking the inheritance of coat color in the progeny of ordinary white mice crossed with "waltzing" piebald mice (whose scurrying was punctuated with little whirls), von Guaita had found a pattern of reversions that, in Davenport's view, little matched the law that Galton had found for basset hounds. Pearson, however, reckoned that another law might be a better match. "Have you read Guaita on crossing of white & waltzing mice?" he wrote to Weldon. "I have only seen Davenport's account of it, but it seems as if the percentages were based on too few cases of real value. Still it appears to resemble Mendel's case of pea hybrids."[18]

Weldon was underwhelmed, replying only, "No Oxford botanist has ever heard of Mendel!"[19] But word was spreading. Already it had reached Davenport, whose next article—as Pearson might have predicted—was an enthusiastic introduction to Mendel's law.[20] News of the law had also reached the readers of *Nature*, or at any rate, the ones who saw the review that spring of the second edition of *The Cell in Development and Inheritance* by the Columbia embryologist E. B. Wilson. (J. Arthur Thomson had relied on Wilson's book the previous year for his facts about the physical basis of inheritance.) *Nature*'s reviewer was J. B. Farmer, professor of botany at the Royal College of Science in South Kensington, with charge over the old Physic Garden in Chelsea. In a passage on Wilson's coverage of the specialist debate on the fate of chromosomes during the production of gametes, Farmer praised Wilson for even-handedness while yet detecting a favoring of the idea that, compared with chromosomes elsewhere in the body, gametic chromosomes were not merely halved in number—Weismann's "reduction division"—but altered in character. By way of speaking up for the other, chromosomal-persistence view, Farmer brought up Mendel. "Perhaps it hardly falls within the scope of the author's work," wrote Farmer, "but a consideration of the reversion of hybrids to the original stocks, such as indicated by Mendel's law, which has recently formed the subject of important communications by De Vries and by Correns, might have been discussed in this connection."[21]

Looking for a way to do research on this new law, Farmer contacted Weldon, who was thrilled by the prospect. "He is mad about Mendel's law, and wants to know you, and to use the Physic garden for properly devised statistical breeding work," Weldon wrote to Pearson, a day after telling him that Mendel was unknown at Oxford. "He is a good man, one of the best young botanists, and well worth keeping in this excellent frame of mind."[22] Just how widely Farmer's madness was shared was something Weldon rapidly came to appreciate thanks to his decision to supply *Biometrika* with a comprehensive guide to recent work on statistics, heredity, and hybridism. Throwing himself into systematic reading and précis-making with typical gusto (he loved new projects), Weldon soon enough discovered the Mendel-confirming papers of Correns and Tschermak, among others. "Mendel's 'law' has a huge literature growing up round it," he reported to Pearson, "and much of this seems to me important," not least for general theorizing about inheritance.[23]

For the rest of the month, as Weldon came to grips with the new Mendeliana, his correspondence with Pearson became a record of developing impressions. Three themes in particular stand out.

1. *Some crosses produce results fitting Mendel's law, but many do not.* Correns and Tschermak had both dwelt at length in their 1900 papers on how

limited the domain of application of Mendel's law seemed to be, as Weldon stressed to Pearson. "The Law is not true of all hybrid characters, even of peas," was how Weldon put the point to himself in the abstract he made of one of Correns's papers. The law did not hold universally even for the handful of hybrid pea characters that Mendel had studied. With seed shape, for example, the dominance of roundness turned out to depend on which races of *Pisum* were crossed. For some combinations it held, but not for all.[24]

2. *Variability beyond the binary may show the influence of ancestry beyond the parents.* Here the work that struck Weldon most forcefully was that of the Zürich entomologist Max Standfuss. Weldon had read Standfuss's book on experimental crosses with moth and butterfly varieties a few years earlier, following up a reference in one of Bateson's articles. What stood out for Weldon now was how much variability Standfuss had found—far more than was conveyed by Batesonian talk of "non-blending characters," with offspring inheriting either the one varietal form or the other. "Standfuss recognises clearly enough," Weldon explained to Pearson, "that each of the names he uses signifies a great range of variability: and all he means is that the curve of distribution of cross-bred offspring is in these cases bimodal." When Standfuss classified some of the descendants in his crosses as "like" one parent, and others "like" the other parent, he made it plain that likeness was to be construed statistically. Better still, Weldon went on, Standfuss sometimes knew the backstories of the varieties he worked with—how and when they had first emerged; which other varieties had been intermixed, and when—and was able to marshal this knowledge to account for the otherwise unintelligible departures from expectation in his crossbred lineages. "I think," Weldon summarized, "Standfuss has a fair case for his belief that ancestry does affect the result of a cross"—indeed, that "the whole ancestry of each parent counts."[25]

3. *Dominance can occur without segregation and vice versa.* Weldon expanded most fully on how he saw the new Mendeliana bearing on the general theory of inheritance in a letter to Pearson connecting three clusters of observations:

- the diverse possibilities for the character of a hybrid, which could be wholly like one or the other parent (dominance), or a blend or a patchwork (no dominance)
- the diverse possibilities for the hereditary constitution of the gametes produced by hybrids, given that some lineages maintain the hybrid character uniformly (no segregation), while others witness a return of the old parental characters (segregation)

- the distinctness of dominance and segregation, which sometimes occur together, but can also occur in isolation

So, yes, with Mendel's peas, there is both dominance and segregation. But Mendel himself showed with his true-breeding hawkweed hybrids that there can also be dominance without segregation. And there can be segregation without dominance, as von Guaita showed: the coats of his hybrid mice were a blended gray, but then whiteness and piebaldness came back. As an attempt to extend Mendel's law (and Mendel's explanation of that law) to encompass the apparently anomalous, it was a good start. "The whole thing is very fascinating," Weldon reflected, "but it is like all the other things. One cannot really know anything about it from books, and one cannot find out otherwise in less than several lifetimes."[26]

By the end of May, he was asking Pearson for advice on how to get hold of some waltzing mice.[27] But Weldon's many other inheritance-related projects now reclaimed his attention, and peas and mice receded from view. Along with the moths, and the snails, and the poppies, and *Biometrika*, there were human skulls, from an Oxford collection for which the Linacre Professor was responsible (fig. 6.2). A bout of lecturing with them led Weldon to notice a specimen whose features were remarkably Neanderthaloid. That in turn prompted him to begin wondering about deep-past reversions, and how to tell a true one from the chance coming together in a single individual of variations jointly suggestive of the Neanderthal type. Here, as Weldon wrote to Pearson, were questions that promised to throw as much light as anything on the mysteries of inheritance.[28]

4

After Weldon's 1895 committee report, there had been nothing in the way of an original research publication from him until his 1898 BAAS address. And after that, there was nothing until "A First Study of Natural Selection in *Clausilia laminata* (Montagu)," his snails paper, completed in summer 1901 and published that autumn as the final paper in the inaugural issue of *Biometrika*. From a distance, this looks like a steady if slow rate of production from someone who, due to a combination of high standards, habitual overextension, a preference for starting new projects to finishing old ones, and attention drift once he understood something to his satisfaction, published in a trickle rather than a Pearsonian torrent.

Yet the two gaps have very different explanations. In 1895–98, as we have seen, Weldon was dealing with the controversy over his claims about the Plymouth shore crabs, plus new information that led him to revise those

FIG. 6.2. Weldon's Oxford group, summer 1901, in a photograph organized by the graduating zoology students and showing off some of the specimens available at the university for studying the comparative anatomy of the primate skull. *Left to right, bottom:* Frank Sherlock and Arthur Darbishire; *middle:* Edgar Schuster, Edward Alfred Minchin, Weldon, John Wilfrid Jenkinson, and Robert Gurney; *top:* H. B. Gray, C. F. Ryder, and W. Hine.

claims, along with the research program supporting them, substantially. In 1898–1901, however, he was dealing mainly with a crisis of confidence that set in once that controversy subsided. He no longer trusted himself not to mess up. "I am so horribly afraid of making another foolish blunder, such as I have always made hitherto in everything I have published," he wrote to Pearson in mid-August 1901. "It won't do to spoil *Biometrika* by repeating such performances!" Pearson by then was used to reading such overwrought confessions. From a subsequent letter, he learned that Weldon that summer had even considered resigning his Oxford professorship, for "one cannot go on indefinitely feeling that all one's work is vitiated by the practical certainty that there is a mistake somewhere which may probably affect the validity of the whole thing!"[29]

The new paper seems to have restored Weldon's faith in himself. Snails turned out to have a major advantage over shore crabs for a study of natural selection's power to maintain the form of a species. The trouble with crabs, as Weldon belatedly appreciated, is that, because of molting, there is no inarguably compelling way to show that the normal individuals who survive to adulthood are the grown-up versions of *the same individuals* who, among juvenile crabs, are the normal ones. By happy contrast, in snails like *Clausilia laminata*, that identification job is straightforward, because the uppermost part of each adult's long, conical shell is that individual's juvenile shell, exactly preserved. (As Weldon put it to Pearson, "God made snails for you and me to work at, because their shells don't grow." In an aging snail, new shell material accretes from the bottom edge, leaving the older, existing shell unmodified.) When Weldon compared measurements of the radii of successive turns of the upper shell spiral in one hundred juveniles and one hundred adults, all collected from the same ancient beech wood in Germany, he found that the spread of variation—the standard deviation—was greater in the juveniles, but that the mean values were the same: the classic signature of norm-enforcing natural selection.[30]

But could something as seemingly trivial as shell size really matter so much to survival? Probably not directly, Weldon thought. Far more likely was that mean radii were being indirectly selected, on the view that—like the lengths of the little fingers of British soldiers fighting in South Africa—they were correlated with some other character that was being directly selected.[31] The paper exemplified everything that *Biometrika* stood for, down to the endorsement of Galton's law. (For Weldon, the snail data bore out two predictions from the law as generalized by Pearson: first, that races inhabiting a locality for even a short time should have a selection-fixed mean character; and second, that even in races where the mean is constant, the young will exhibit beyond-the-norm variability, which selection then prunes.)[32] Weldon also contributed an editorial on the new journal's scope and a note showing the superiority of Galton's way of calculating correlation over a clumsy rival form. But the bulk of the editorial work—as Weldon guiltily recognized—had fallen to Pearson, from finding a publisher (now the University Press at Cambridge), to filling the pages (he wrote or co-wrote or ghost-wrote over half of them), to arranging the frontispiece (a photograph of Darwin's statue at Oxford, above the motto "Ignoramus, in hoc signo laboremus"—"we are ignorant, and so we labor"—though Weldon joked that a more fitting emblem "would be a portrait of you on a pocket handkerchief, with Davenport and me as little angels holding it up").[33]

The handsomely produced first issue went out to over a hundred individual and institutional subscribers.[34] When Galton received his copy near the end of October, he was impressed but, as he explained apologetically, preoccupied with another debut, related to another aspect of his legacy. He was about to present the annual Huxley Lecture to the Anthropological Institute, and planned, as per Pearson's request to him earlier in the year, to use the occasion on behalf of eugenics.[35]

Galton's title was "The Possible Improvement of the Human Breed under the Existing Conditions of Law and Sentiment." His aim, however, was to show how much more rapidly, and even profitably, the human breed could be improved by changing existing laws and sentiments. He began by setting out the case for thinking that what he called "civic worth"—for Galton, the opposite of criminal worthlessness—is distributed in accordance with a normal curve, just like human stature. Given that distribution, he argued, then even with regression to the mean, the highest civic-worth couples could be expected to be the most efficient producers of highest civic-worth children. It was therefore in the national interest to concentrate breed-improvement efforts on this natural elite, with a view to encouraging them not merely to marry each other but to have as many children as possible as early as possible, and to raise those children in physically and morally salubrious conditions. To this end, Galton recommended, financial help should be made available, so that having lots of children and raising them well became the most prudent choice (changes in law). The nation should also cultivate what he called a "quasi-religious" belief in the duty of all to do their part in improving the breed (changes in sentiment).

And at the other end of the civic-worth scale? Galton noted that many who knew it best "do not hesitate to say that it would be an economy and a great benefit to the country if all habitual criminals were resolutely segregated under merciful surveillance and peremptorily denied opportunities for producing offspring." Yet he insisted that, when it came to "increasing the productivity of the best stock" versus "repressing the productivity of the worst," the former was more important. Best to get on with the immediate challenges it presented, such as designing a selection process that would reliably identify the most promising men and women—no mean feat, Galton reckoned, given how late some faculties develop—and calculating the amount appropriate to spend on them given the likely return on investment.[36]

H. G. Wells, always eager for news from the scientific frontier, was at the lecture. So was Pearson, who loved it. The next morning's *Times* supplied a summary, accompanied, however, by a dismissive leader. According to the leader writer, if put into practice, Galton's Utopian scheme would surely lead to misery, not only for the vast toiling majority (however "consoled by the fact that they are improving the human race") but even for the privileged few, who were bound to grow "fat, lazy, conceited, and stupid," stuck in unhappy marriages they were bribed into entering. Galton, the *Times* writer advised, should stick to his statistics. People are not pigs.[37]

Weldon caught up with the lecture the following day in *Nature*, where it was printed in full. Explaining his own reservations to Pearson, Weldon drew on the most recent Boer War headlines for illustration:

> The danger of all that kind of thing seems to me to be our utter ignorance of what we want to produce. What are you going to do with a heroic incompetent fool like [the recently fired Sir Redvers] Buller? He is a very common type, and an army of Bullers [i.e., brave dolts] commanded by a general and a staff of Moltkes would be rather a fine thing. But are you going to breed races of men for different purposes, as different as breeds of horses, or are you going on with the ideal man who is fit for anything?[38]

5

A successful scientific journal needs high-quality papers, regularly published. With the deadline for *Biometrika*'s next issue already looming, Weldon was back at work on his planned review of the ever-expanding literature around the "Versuche."

Unexpectedly, the most recent addition to that literature had come from Mendel himself, in the Royal Horticultural Society's *Journal*: an English translation of the paper, with notes and a short but powerful introduction by Bateson. It even had a Batesonian title, "Experiments in Plant Hybridisation," as though the subject under experimental investigation was not hybrids themselves, as Mendel's German straightforwardly suggests ("Versuche über Pflanzen-Hybriden" = "Experiments on/upon Plant Hybrids"), but something else to be illuminated through the act of hybridizing.[39] Bateson's commentary left no doubt that Mendel's concern was the nature of heredity and so the nature of species. More so than before, Bateson, like de Vries, now glossed Mendelian heredity as a matter of atomic units, combining and recombining down the generations. Thanks to Mendel, wrote Bateson,

> the conclusion is forced upon us that a living organism is a complex of characters, of which some, at least, are dissociable and are capable of being replaced by others. We thus reach the conception of *unit-characters*, which may be rearranged in the formation of the reproductive cells. It is hardly too much to say that the experiments which led to this advance in knowledge are worthy to rank with those that laid the foundation of the

> Atomic laws of Chemistry. . . . [H]owever much it may be found possible to limit or to extend the principle discovered by Mendel, there can be no doubt that we have in his work not only a model for future experiments of the same kind, but also a solid foundation from which the problem of Heredity may be attacked in the future.

Why did biologists have to wait so long for such a profoundly important message to get through to them? Bateson blamed a general paralysis of the will to hybridize experimentally, brought on after 1859 by premature acceptance of the Darwinian theory of species—a theory that had made the experimentalists' concern with unitary characters, and lack of concern with adaptive utility, look misdirected—and dispelled only by the triple rediscovery of Mendel's work in 1900 by de Vries, Correns, and Tschermak.[40]

Bateson concluded with a brief, bare-bones biographical paragraph about Mendel, drawing on information that Correns had learned. The fact that Mendel was a Catholic friar who had done his experiments in the garden of his monastery was news to most people, Weldon included. "I have read Bateson's translation of Mendel," Weldon wrote to Pearson in mid-November 1901, "and I am struggling to avoid a tendency to disbelieve the whole thing because Mendel was a Roman priest."[41]

Well before then, Weldon had identified other, more compelling grounds for skepticism. As we have seen, for some while, he had held doubts about the uniformity of the pea character versions Mendel described, linked with doubts about the wisdom of ignoring the ancestral histories of the initially hybridized varieties in analyzing the inheritance of those character versions in the hybrid lineages. But now Weldon began looking more closely at Mendel's data—in particular, at whether the numbers reported and the numbers predicted were not surprisingly, improbably close. What, exactly, were the odds against Mendel getting such a tight fit between data and theory?

To find out, Weldon checked the "probable error" of Mendel's results from his simplest, single-character crosses, using a standard formula to calculate expected deviations from the theoretically predicted values given the number of observations made. For example, in his seed shape cross, Mendel reported that in the offspring of the hybrid pea plants, 5,474 out of 7,324 seeds showed the dominant-character version, roundness. Theoretically, the predicted number was 5,493, with a probable error of ±24.995. In other words, if the experiment were re-done at the same scale, under the same conditions, umpteen times, half the results would be expected to fall somewhere between 5,468 and 5,518 (clustering in the middle in the usual way), and half would be expected to fall outside those boundaries. In itself, Mendel's result of 5,474—6 seeds above the lower boundary—was not suspiciously good. But the fact that almost all of his results, for all seven characters, both in the hybrid offspring and in *their* offspring, fell within the probable-error boundaries was, on the face of it, remarkable. At the other extreme of experimental complexity, Weldon found the same improbable closeness when he looked at Mendel's results for his triple-character cross, here using not the probable-error formula but a new test of Pearson's, the "chi-square test."[42]

Weldon organized his findings into table form and sent them to Pearson in late November. About Mendel, Weldon wrote, "He is either a black liar, or a wonderful man"—"wonderful" in the older, literal sense of "wonder-making." For the most part, Weldon was inclined to think neither that Mendel was lying nor that he was miraculously lucky, but that he had reported truthfully on what he had observed in the particular varieties he worked with, under the conditions that he observed them in:

> [I]f you take all Mendel's figures together, they are wonderfully good approximations to his hypothetically probable results. ~~Remembering his shaven crown, I can't help wondering if they are not too good?~~ I do not see that the results are so good as to be suspicious, so that I can see no alternative to the belief that Mendel's "laws" are absolutely true for his peas, and absolutely false for Laxton's, while those of Tschermak are intermediate. . . . But the fear of Mendel is before my eyes. Really one has never seen such perfectly devised observations, lasting over 8 years, give a result so absolutely untrustworthy. It seems to me to show an influence of conditions so great that I feel it hardly worth while to grow any thing. If only one could know whether the whole thing is not a damned lie! Segregation of hybrids into apparently pure bred offspring can't be a lie, because every one gets such a result. But the consistent dominance, and the regularity of the separate inheritances <u>only in the monastery garden at Brünn!</u> Shall I shave my crown too?[43]

Recall that Laxton was Darwin's pea expert, whose crossing work Darwin cherished for the support it seemed to provide for the "direct action of the male element on the female," and so for the pangenesis hypothesis. Weldon cherished it for a different reason: when Laxton tracked hybrid characters in peas, his results—published in 1866, the same year as Mendel's—looked, as noted, nothing like Mendel's. For where Mendel's hybrids always showed just the one parental-character version in color and shape, Laxton's were sometimes blended, sometimes wholly like the one parent, sometimes wholly like the other, and sometimes mosaically like both. Introducing Laxton's work to Pearson, Weldon wrote, "While

Mendel was making his 'laws,' Laxton, of whom Darwin speaks so often! was crossing peas and making all the main races we now eat."[44]

Tschermak has come to be remembered as the least scientifically impressive of the rediscovery trio. But Weldon got a lot out of Tschermak's papers, appreciating his care in not sweeping all the variability he observed in, say, 400 hybrid seeds into the category "yellow," but instead noticing, and recording, that 40 of them showed different kinds of not-yellow.[45] And Tschermak affirmed what Weldon suspected: that the particular varieties or "races" used mattered hugely. The more closely the conditions of a cross approximated Mendelian conditions—as was sometimes the case for Correns, Tschermak, and others—the more closely the results approximated Mendelian results, especially when the categories used were the Mendelian binaries. "There is no doubt at all," Weldon wrote to Pearson, after sending the data tables, "that the only thing Mendel or anyone else who tried to repeat his work can have done is to put each pea into one of two categories, each containing very variable elements, but not generally overlapping."[46] As to where all that left the black liar-or-wonderful man question, Weldon was a little clearer in a letter a few days later: "I believe myself, after reading the others, many of whom worked at first without knowing Mendel,—that he cooked his figures, but that he is substantially right.*" The asterisk took Pearson to a further comment at the top of the page: "I mean, right for particular races.—That is, the amount of dominance is a function of ancestry as well as individual character: and his attempt to treat parental character as a sort of chemical unit is rot."[47]

So some pea races were minimally variable like Mendel's, and, provided one was willing to categorize every seed as either "yellow" or "green" and either "round" or "wrinkled" (or otherwise to reject it as a bad seed), hybridizing those races could well lead one to think that yellowness and greenness and roundness and wrinkledness in peas were like chemical units. But other pea races were far more variable, some at a grand scale. They were less like Mendel's peas and more like humans when it came to eye color and, probably, just about everything else.

By the end of November Weldon had a favorite example, first learned about from Tschermak: a pea variety marketed under the name *Telephone*. Weldon was now getting shipments of it from several breeding firms, including Suttons, Vilmorin in Paris, and *Telephone*'s originator, Carters, which was also supplying helpful information on the family history of the breed. In Weldon's view, there was no way that belief in the chemical-unit nature of pea characters could survive contact with the sight of *Telephone* seeds in all their variable glory. Sending the first dozen that he de-coated and examined to Pearson, Weldon crowed, "It smashes

St. Gregory of Brünn pretty thoroughly . . . I think you will admit (1) that the specimens sent show a fairly good series from undoubted yellow to undoubted green; and (2) that the intermediates are very unmistakably patchy."[48] But how to give *Biometrika* readers that experience? Weldon briefly considered whether a card with peas pasted on—a version of what he sent Pearson—might go out with each copy of the second issue. Color photography was in its infancy, and though sometimes it worked well (as could be seen, for example, in one of Correns's papers), even the best color photograph of pea seeds would be no match for the real thing. There might even, Weldon joked, be scope for using the pea cards to help increase the journal's popular appeal, thus meeting concerns Galton had recently raised with Pearson about its forbidding technicality. "There is a nice popular suggestion.—A Sample of Sutton's Guaranteed Seeds Presented Gratis with Each Copy! Given away with a Pound of Tea!! Let us add a missing word competition?" (As his tone suggests, Weldon thought the day of the journal that could be both scientifically valuable and accessible to the nonspecialist was over. "I think," he explained to Pearson, "people now want very technical journals, plus Rudyard Kipling and the evening papers." In Weldon's view, biology had suffered from the need for evolutionary ideas to be presented popularly, with "the inevitable dominance of the windbag.")[49]

Photographic plates it would be, then: one black-and-white, for showing examples from the spectrum of wrinkledness; and one color, for showing greenness-to-yellowness. Weldon submitted the paper for publication in early December.

6

His title, "Mendel's Laws of Alternative Inheritance in Peas," referred to Mendel's laws, plural. That was in keeping with Weldon's view that, contra Mendel, dominance and segregation often come apart in practice and so need to be kept apart in theory too. As Weldon saw it, Mendel's singular "law valid for *Pisum*"—$A + 2Aa + a$—tacitly smuggled a law of dominance into a law of segregation. To see what Weldon was getting at, consider the "extracted recessives" in a Mendel-style cross: say, the green-seeded plants that appear after the yellow-seeded hybrid generation. Why, in the Mendelian formula, are these represented as *a* (Mendel) or *aa* (Mendelians)? Because, according to Mendel's analysis of dominance, any individual showing the recessive version of a character *must* be pure for that version in its hereditary constitution. Any other constitution makes for a plant showing the dominant-character version.[50] Likewise,

Mendel predicts a 3:1 ratio of dominant to recessive character versions among the hybrid offspring as a whole because, on that same analysis of dominance, the constitutionally hybrid plants, the *Aa*s, can never show an intermediate character, whether blended or piebald. To have the elements for the dominant-character version is to show that character version, and so to be indistinguishable outwardly from any pure dominants.

After briefly summarizing the already standard three-way classification of hereditary characters—the blended, the exclusive (called "alternative" by Weldon), and the piebald or mosaic—Weldon announced his intention of dealing with "some cases of alternative inheritance, which have lately excited attention." He then introduced Mendel's paper and what were, on Weldon's reading, Mendel's two laws of alternative inheritance in peas:

> *The Law of Dominance:* If peas of two races be crossed, the hybrid offspring [i.e., the cross-bred pea plant] will exhibit only the dominant characters of the parents; and it will exhibit these without (or almost without) alteration, the recessive characters being altogether absent, or present in so slight a degree that they escape notice.
>
> *The Law of Segregation:* If the hybrids of the first generation, produced by crossing two races of peas which differ in certain characters, be allowed to fertilise themselves, all possible combinations of the ancestral race-characters [i.e., *A* with *A*, *A* with *a*, *a* with *A*, and *a* with *a*] will appear in the second generation with equal frequency, and these combinations will obey the Law of Dominance, so that characters intermediate between those of the ancestral races will not occur.[51]

Weldon used the example of seed color to give a sense of the significance of these laws as they emerge from Mendel's paper, through to the point in the first half of the paper where Mendel derives his $2^n - 1 : 2 : 2^n - 1$ generalization of the segregation law. Weldon's tutorial in the Mendelian basics ended with a list of the seven characters Mendel studied and a short discussion of qualifications he noted.

What follows is a detailed inquiry into two questions: First, how fully did Mendel's own results support his conclusions? Second, how fully do the results of other observers support those conclusions? By way of answering the first question, Weldon guided the reader through the statistical testing of the numbers that Mendel reported, chiding admirers who, in their enthusiasm, had skipped this precaution. Weldon now calculated the odds of getting worse results than Mendel did from his single-character crosses if repeated at the same scale and under the same conditions as 16 to 1. For the three-character crosses, the odds of worse results were closer to 20 to 1. Plainly, then, Mendel's results were too problematic to count in support, not because they contradicted his laws, but because they confirmed them with improbable exactness.

Turning next to the bearing of others' results, Weldon was unsparing. On his survey of the evidence, hybrid peas observed by equally trustworthy workers often behaved in ways utterly unlike Mendel's peas. It appeared, then, that Mendel's laws did not hold universally even for the pea characters that Mendel had investigated. In reaching this verdict, Weldon added, he had "no wish to belittle the importance of Mendel's achievement," but wanted "only to call attention to a series of facts which seem to me to suggest fruitful lines of enquiry." Those facts were of three kinds: evidence on dominance, mainly in the color and shape of pea seeds; evidence from a multi-generation lineage of hybrid pea varieties; and evidence on segregation.[52]

Other Evidence concerning Dominance in Peas. For page after page, Weldon under this heading piled on counterexamples, from Tschermak and others: crosses where greenness was dominant to yellowness, where greenness "dirtied" the yellowness, where the wrinkled offspring went on to produce round and wrinkled progeny, where seeds were intermediate between full roundness and full wrinkledness, and so on. The lesson Weldon drew, as in his correspondence with Pearson, was that whether Mendelian dominance holds—and sometimes it does—depends on which pea races, with which ancestries, are being crossed. Dominance is not an invariable property of one side of a character-version binary; there is no sense in which yellowness and roundness are simply "dominant" across the board, and greenness and wrinkledness "recessive" across the board. Mendel's failure to appreciate the dependence of his results on his using particular races was, Weldon judged, a symptom of a more fundamental failure to take ancestry seriously. Bateson, at least, had actually known better, once upon a time, as Weldon recalled in bringing his survey on dominance in peas and beyond—the thought experiment about human eye color appears here—to a close:

> These examples, chosen from many others which might have been cited, seem to me to show that it is not possible to regard dominance as a property of any character, from a simple knowledge of its presence in one of two individual parents. The degree to which a parental character affects offspring depends not only upon its development in the individual parent, but on its degree of development in the ancestors of that parent. A collection of cases which illustrate this point is given in Bateson (No. 1).

That was a reference to a two-part paper of Bateson's from the late 1890s, when he was interested in the possibility that the "predominance" of a character in a cross was the upshot of how inbred the lineage was.[53]

The Hybrid Peas of the Telephone *Group.* Now Weldon offered up a new empirical study that illustrated how greater precision in describing variability went together with greater curiosity about ancestry. He first got hold of seeds from the pea variety *Telephone* in order to try and follow Tschermak's descriptions of it. But the more Weldon learned, the more fascinated he became, not only with these peas but also with their history. *Telephone* turned out to be one of several varieties established over the previous quarter century by Carters through selection of promising-looking variants from an original, highly variable hybrid pea stock called *Telegraph*. Although advertised as "green," *Telephone* had seeds that—as displayed in the top row of Weldon's color photographic plate, reproduced on the cover of this book (you are looking at the middle, transition-zone part of the row)—could be arranged into a continuous green-to-yellow color scale. The colors of the "green" seeds of another *Telegraph* descendant, *Stratagem* (see the next row down), formed a continuum in between the yellowy-greens and greeny-yellows of *Telephone*. "I think," commented Weldon, "there can be no question that in *Stratagem* a blend of green and yellow has been inherited, and fixed by a process of selection."[54]

Through further crossings, *Stratagem* had itself become the begetter of an extraordinarily diverse and, for Weldon's purposes, especially instructive sub-lineage. The first descendant variety was *Daisy*, a green-seeded pea. Next, *Daisy* was crossed with a yellow-seeded variety called *Lightning*. By cultivating the green seeds found among the offspring of the yellow *Daisy* × *Lightning* hybrids, Carters created still another variety, *Early Morn*. In the green-to-yellow-to-green path leading to the *Early Morn* seeds, then, it might appear that, having vanished during the hybrid generation, the recessive-character version had returned à la Mendel. But a closer study of the available evidence, according to Weldon, suggested a more complicated reality. Thanks to Carters, he had obtained seeds from the original stock of *Early Morn*, and in their coloring, they showed, as Carters' staff pointed out, traces of their yellow parentage: a Mendelian no-no. As for seed shape, Weldon directed readers to his second photographic plate (fig. 6.3). Seeds of *Early Morn* can be seen at 16–18. It is, no doubt, a "wrinkled" pea. So is *Daisy*, just above it (10–12). But are these two related varieties wrinkled in the same way, to the same degree? Mendelian reasoning does not at all invite the question. But ask it, and the answer, Weldon suggests, is an emphatic "no." If there

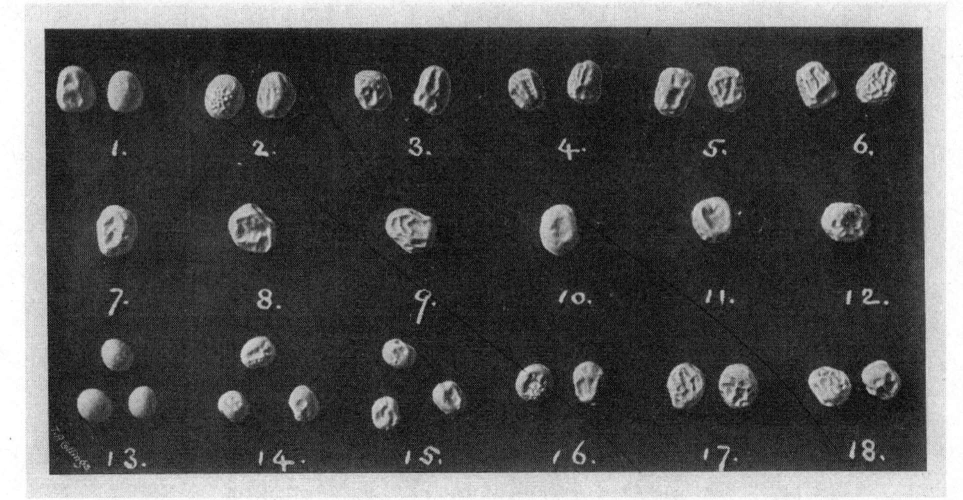

FIG. 6.3. Degrees of wrinkledness in the hybrid pea varieties *Telegraph* (1–4), *Telephone* (5–6), *Stratagem* (7–9), *Daisy* (10–12), *Lightning* (13–15), and *Early Morn* (16–18).

are seeds on the plate whose wrinkles *Early Morn*'s resemble, they are the seeds of *Telephone* (5–6), and above all, the most wrinkled seeds of the ur-pea, *Telegraph* (4).

What we have here, Weldon advised, is reversion; but it is reversion to an ancestor whose generational remoteness puts it outside of the Mendelian frame. "This," tutted Weldon, "is another example of the danger incurred by using Mendel's categories without careful examination."[55]

And what of *Telegraph* itself, then twenty-five generations on from its emergence? Surely, Weldon noted, if dominance in seed characters in peas works as per Mendel, and in particular, if the long run of hybrid lineages is truly captured in his $2^n - 1 : 2 : 2^n - 1$ formula, then, with $n = 25$, the number of *Telegraph* seeds that show anything other than pure-form colors and shapes should be virtually nil. Yet Weldon found that *Telegraph* seeds remained stubbornly variable, with a modest-sized sample yielding up seeds that ranged in shape from round to extremely wrinkled (see, again, fig. 6.3, 1–4), and colors that fell all along the green-to-yellow *Telephone* scale, with a sizable piebald remainder.[56]

Mendel's Law of Segregation. To Weldon, the history of the *Telephone* group, in which "recessive" characters in hybrid offspring again and again failed to remain constant, was as comprehensively undermining of the wider validity of Mendel's law of segregation as it was of the wider validity of his law of dominance. But the group's behavior did fit very well, Weldon reckoned, with the results of another observer of pea hybrids. Weldon now presented the long passage from Laxton's 1866 paper quoted in chapter 1 of this book, in connection with the question of the likely consequences for Darwin had he read Mendel. Laxton's statements, wrote Weldon, "show that the phenomena of inheritance in cross-bred Peas, as Laxton observed them, were far more complex than those described by Mendel; but they do not preclude the possibility of a simple segregation, such as Mendel describes, in particular cases." In closing, Weldon stressed the need in the future to give complexity-making ancestries their due:

> The fundamental mistake which vitiates all work based upon Mendel's method is the neglect of ancestry, and the attempt to regard the whole effect upon offspring, produced by a particular parent, as due to the existence in the parent of particular structural characters; while the contradictory results obtained by those who have observed the offspring of parents apparently identical in certain characters show clearly enough that not only the parents themselves, but their race, that is their ancestry, must be taken into account before the result of pairing them can be predicted.[57]

7

The color plate proved, as feared, expensive and slow to produce (it pushed the January *Biometrika* publication date into February). Given the technical constraints of the day, it was also bound to be a bit of a disappointment, as in certain ways it was.[58] But the need for something like it as part of an argument like the one mounted in the paper was brought home to Weldon not long after he submitted it, when he stopped for lunch at his club in Piccadilly, the Savile. He told the story in a letter to Pearson:

> Lankester and others were talking about Mendel's law. Lankester is getting Maize from Bateson to show in the Brit. Mus. as illustration of these laws. I had a box of peas in my bag, and demonstrated; but Lankester had not read Mendel, and did not see the point at all. . . . [S]ome other men saw the point of intermediate peas: but it came as a great shock. They are all taking Mendel as gospel.[59]

Against that gospel, Weldon was already moving on other fronts. He was breeding Japanese waltzing mice, looking at whether their waltzes indeed bred true. He was also making plans for the re-hybridizing, in the Oxford Botanical Gardens, of *Early Morn* from stocks of *Daisy* and *Lightning*.[60] When the right royal row over Mendel came, Weldon would be ready.

7

Mendel All the Way

I believe that few new & good things are ever accomplished except in the teeth of violent opposition. What does Bateson say of Weldon's paper in Biometrika*?*

FRANCIS GALTON TO KARL PEARSON, 28 FEBRUARY 1902

Bateson is a great man. Do you know:
There was a strong man on a syndicate;
The truth he determined to vindicate;
He rose to deny that his words could imply
What they seemed so distinctly to indicate!

W. F. R. WELDON TO KARL PEARSON, 28 MAY 1902

Personally I am quite prepared to find out there is truth in Mendel, only it must be done by a man who does not become vulgarly abusive in a purely scientific discussion. Bateson has a strong personal feeling about Weldon. He wrote me comparatively recently a most disgusting & fulsome letter, which amounted to a request to chuck Weldon overboard & take a certain Cambridge naturalist—Bateson—as a confidential adviser in his place! If ever a man stood in need of horse-whipping it was the writer of that letter.

KARL PEARSON TO G. UDNY YULE, 27 AUGUST 1902

On the train yesterday many of the party arrived with their "Mendel's Principles" in their hands! It has been "Mendel, Mendel all the way" and I think a boom is beginning at last.

WILLIAM BATESON TO BEATRICE BATESON, 3 OCTOBER 1902

With the publication of "Mendel's Laws of Alternative Inheritance in Peas" in early February 1902, tense relations between Bateson and his *Biometrika* antagonists grew febrile. A closer look at the "disgusting & fulsome" letter Pearson received from Bateson will help set the scene.[1]

Sent in mid-February, the letter caught Pearson very much by surprise. For one thing, it came a few days after a notably un-fulsome letter from Bateson on the latest *Biometrika*. ("Mr Bateson is very angry with Weldon's article," Pearson reported to Galton. "We have evidently got a big fight on hand with the old type of biologists.")[2] For another, Pearson at that moment was once again skirmishing with Bateson via the Royal Society over homotyposis and the future conduct of their debate, now that their homotyposis papers had at last been published.[3] Beyond these live professional hostilities, at a personal level, the two men barely knew each other. "I had only spoken to [Bateson] for ten minutes in my life!" a still-flabbergasted Pearson recalled to Yule.[4] Who would have guessed, all that time, that Bateson was admiring from afar?

In retrospect, the signs are there. A number of passages in *Materials for the Study of Variation*—on the need for wariness of scientific terms as "riddled with theory"; on Weismann as having rendered hereditary transmission less like inheritance in the familiar sense and more like trusteeship; on living matter as having an ultimately mechanical, indeed vibrational, nature—read like cribs from Pearson's *Grammar of Science*.[5] In 1895–96, the two men had a brief but friendly correspondence, initiated by Bateson, on their shared commitment to university education for women.[6] And in 1900, as we have seen, it was Pearson's call at the Royal Society meeting in late March for a modified version of Galton's law that induced Bateson shortly afterward to embrace Mendel's law as a Pearsonian prophecy fulfilled.[7]

So it was not at all unthinkable that Pearson himself might be likewise induced. Bateson first gave it a try in October 1901, sending Pearson a copy of the new "Versuche" translation and inviting him to Cambridge. Pearson declined the invitation, citing busyness. He nevertheless thanked Bateson, saying how welcome Mendel's appearance "in an English dress" was, though adding that efforts to square Mendel's results with the data on human eye color had so far come to nothing.[8] Bateson was undeterred. A lengthy and brilliant report that he and Becky Saunders completed a few months later, showing in detail what the new Mendelian program of research looked like, included a footnote reminding readers that, in dealing with non-blending characters, Pearson himself had actually proposed a separate "law of reversion," distinct from Galton's law, and different from Mendel's law only in the numbers predicted.[9]

Not even *Biometrika*'s swing against Mendel could keep Bateson, one last time, from extending an offer of comradeship. "I respect you as an honest man, and perhaps the ablest and hardest worker I have met," gushed Bateson in his letter, "and I am determined not to take up a quarrel with you if I can help it." Regrettably, the Royal Society had bungled its management of the homotyposis affair, but Bateson trusted there were no lingering hard feelings. As for the new Mendelian research, Bateson hoped he could still win Pearson round, even if Weldon, on the latest evidence, was now irredeemable: "It was to me exasperating, that Weldon should so utterly have missed the point of Mendel and mangled a great man's work beyond recognition." Alas, Bateson went on, his relationship with Weldon was beyond recovery, partly due to "faults of temperament on both sides," but more unforgivably due to Weldon's response to what, it was already clear, was a revolutionary biological discovery. Weldon knew fully well what he was doing, and would deserve history's indictment when it came. But Pearson was no biologist, and—up to now—could not be held responsible for getting Mendel wrong, in Bateson's view. It especially pained Bateson to think that perhaps he had inadvertently played a role in estranging Pearson from the Mendelian future. But if Pearson wanted to be part of it, then, with Bateson's help, there was still time to change course.[10]

Delivering his emphatic "no thanks," Pearson put Bateson on notice as well that further disobliging references to Weldon—"one of my closest and most valued friends"—would not be taken lightly.[11] For Weldon's part, he had recently reaffirmed to Pearson the readiness for that "good row" now surely looming with Bateson, seeing in his anger over the new article "a good sign. . . . It means that he has a bad case, and knows it." When Pearson passed along Bateson's failed-courtship letter, Weldon was understanding but unimpressed. "The worst of Bateson," wrote Weldon, "is that he is in earnest. . . . He is, no doubt, grieved that we do not all work together, and he thinks bitterly that it is my fault. He forgets that the only condition on which he would tolerate us would be our unhesitating acceptance of all his ideas."[12]

Bateson aside, Weldon's critique of Mendel seemed to be hitting home. Galton judged it "scholarly & timely." Thiselton-Dyer congratulated Weldon on how completely he and Pearson had "knocked the bottom out of Mendel." And anyway, Weldon observed, tongue (mostly) in cheek, a juicy controversy or two would do *Biometrika*'s circulation no harm.[13] The battle over Mendel, formerly visible only in the closed, private world of elite English science, was now open, public, declared—and, as Weldon would learn to his cost, Bateson would fight as if nothing less than the future of biology was at stake.

1

After the publication of "Mendel's Laws of Alternative Inheritance in Peas," as Weldon and Pearson awaited the impending storm, they remained busy on the usual multiple fronts. Each also found time, however, to write a new, Bateson-criticizing article for the next issue of *Biometrika*, published in early May 1902. Their common target was Bateson's claim that variation comes in two fundamentally different classes: continuous or, as he was now calling it, "normal" variation, which is evolutionarily inert; and discontinuous or "differentiant" variation, which alone was responsible for new species.[14]

Pearson's critique took the form of a grindingly thorough reply on homotyposis, wrapped in a defense of the biometric ambition to unify mathematics and biology. (Although Pearson was offered a right of response in the Royal Society's *Proceedings*, he decided *Biometrika* was a more suitable venue.) He dwelt at remorseless length on what he saw as Bateson's inability, in *Materials* and elsewhere, to give clear definitions of key terms such as "variation" and "discontinuity," or to grasp the nature of regression. It is not, Pearson insisted, a distinctively biological process, maintaining stability in bell-curvy systems during their evolutionary resting periods between episodes of mean-shifting transformation. Regression is just what happens when two sets of chance results are correlated, in dice or duckweed or whatever.[15]

Weldon repeated that lesson in his article, but without the scholasticism—and with Bateson dealt with largely by proxy. Weldon's ostensible subject was Hugo de Vries's new book, *Die Mutationstheorie*, in which de Vries defended a distinction identical to Bateson's, between—in de Vries's terminology—"fluctuating variation" and "mutation." To show that no amount of selection on fluctuating variation could produce permanent, species-changing evolution, de Vries had marshaled the evidence of selective breeding with plants. Yes, repeated sowings and growings in a new environment can eventually lead to a change in the average value of one or more characters in a plant variety, as per selection theory. But return the new seeds to the old environment, and, in the many cases that de Vries documented, the plant returns to the old characters. The relevant chapter heading in de Vries's book put the point succinctly: "Selection Alone Does Not Lead to the Origin of New Species."[16]

In response, Weldon took the opportunity to revive what was, as we have seen, a career-long theme: the difficulty, when an environment changes, of disentangling the effects of selection from the effects of developmental reshaping. In Weldon's view, not one of de Vries's examples was an unambiguous instance of selection even being tested, let alone failing to achieve its purported effects. What was more, Weldon went on, recent work in experimental embryology, including his own, had shown that even very minor changes in ordinary environmental conditions can have dramatically large effects on developing form, underscoring the scale of the disentanglement challenge. To illustrate, he directed readers to accompanying plates—not of pea seeds but of chick embryos. Under standard conditions, Weldon explained, a chick embryo

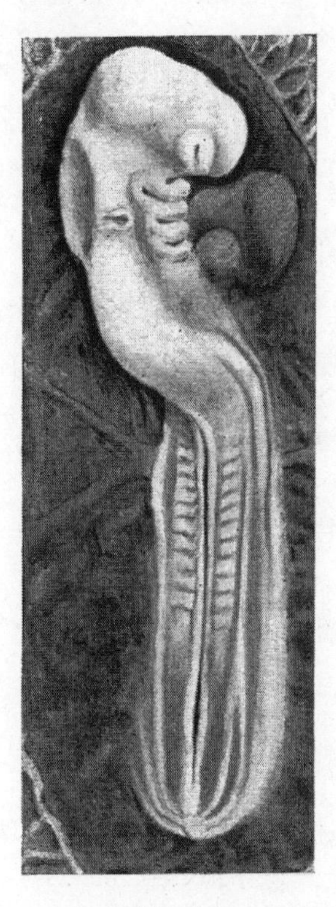

Fig. 1.
First Embryo: 72 Hours of Incubation.

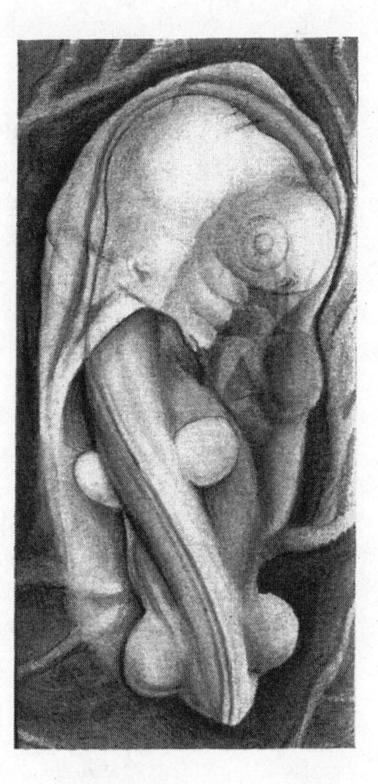

Fig. 2.
Second Embryo: 96 Hours of Incubation.

FIG. 7.1. Weldon's drawings of two abnormal chick embryos, from experiments in which he siphoned water into hens' eggs from which it was evaporating.

develops suspended in a fluid-filled sac, the amnion. But when, in experiments not previously published, Weldon had artificially replenished the water that usually evaporates from a freshly hatched egg, the amnion's growth was suppressed, first partially, then wholly (fig. 7.1). Only by such environment-manipulating interventions was an embryonic character widely assumed to be the upshot of ancestral inheritance alone revealed to be dependent, in a fundamental way, on interaction between inheritance and environment.

Lest anyone miss the overall message, Weldon stated this old Balfourian conviction with newly categorical firmness and updated references:

> Now it cannot be too strongly insisted upon that every character of an animal or of a plant, as we see it, depends upon two sets of conditions: one a set of structural or other conditions inherited by the organism from its ancestors, the other a set of environmental conditions. There is probably no race of plants or of animals which cannot be directly modified, during the life of a single generation, by a suitable change in some group of environmental conditions. The work of [Camille] Dareste, [Hans] Driesch, [Curt] Herbst, and others has shown that some of the most normal and universal phenomena of animal development are each directly dependent for their occurrence upon a certain group of external conditions. . . . Until we know far more than we know at present about the relation between an organism and its environment, it is simply useless to discuss the stability of characters, whether "variations" or "mutations," except under environmental conditions which are as constant as we can make them during the period under discussion.[17]

And Mendel? His name was absent from the issue, and barely present in the letters that survive between Weldon and Pearson that spring. A sporadic and pea-full correspondence with Tschermak nevertheless kept Mendel in view for Weldon. So did a notice he read in *Nature* on a French investigator, Lucien Cuénot, who reported getting the Mendelian pattern of uniformity followed by the 3:1 ratio in crossings of gray and white mice. "The result that white and grey mice when crossed give always grey young is contradicted by a good many people," Weldon groused to Pearson. As ever for Weldon, it was ancestries, environments, and their interactions that needed attending to if such perennial contradictions in the data on dominance were ever to be resolved. "The solid thing which ought to come out of the Mendel business," Weldon continued, "is some kind of knowledge of the conditions which determine whether inheritance shall be blended or alternative." He went on to express regret that he was not a better gardener; for if he were, he could throw himself with confidence into growing the hybrid pea seeds that Tschermak had sent him in order to see what difference, if any, the Oxonian environment—so different from the Viennese one—might make to the characters of the next-generation seeds.[18]

By late April, Weldon had heard that Bateson was preparing a book-length reply to the Mendel critique. As for Weldon's own next steps in research and writing, matters were less clear. He was through with writing little critiques, he told Pearson, dismissing the ones on Mendel and de Vries as "journalism." But he lacked for a major new project.[19]

That would change between the end of May and the beginning of June, when not one but two extraordinary books from Bateson came out in quick succession. The first—the occasion for Weldon's limerick, quoted at the start of this chapter—was coauthored with Saunders and unprepossessingly titled *Reports to the Evolution Committee of the Royal Society: Report 1*. The second was a solo effort, *Mendel's Principles of Heredity: A Defence*, which included a reply to Weldon so vitriolic as to cross the line, in the view of Pearson and others, into vulgarity.[20] The remainder of this chapter considers these books and their impacts, which reverberated into the autumn of 1902.

2

Submitted to the Evolution Committee in mid-December 1901, lightly revised at proof stage in March, *Report 1* was a tour de force of Mendelian reasoning in action. After a briefish introduction to Mendel's law and its transformation of biology ("the whole problem of heredity has undergone a complete revolution"), Bateson and Saunders summarized her experiments with plants, then his experiments with poultry, then rounded off with a long discussion titled "The Facts of Heredity in the Light of Mendel's Discovery," but which, with equal justice, could have been called "Mendel's Discovery in the Light of the Facts of Heredity."[21] For, far more comprehensively and creatively than any of the rediscovery papers, the report took Mendel's tightly focused paper on the law governing variable plant hybrids and recast it as the foundation for an endlessly resourceful—and winningly user-friendly—research program on heredity. How so? Consider some of the report's innovations:

1. *New terminology.* When a new science emerges, a specialist language often emerges too, easing communication among insiders while subtly reinforcing their sense of being a part of something new and special. As we have seen, biometry's proponents—"biometricians," as they now styled themselves—invented "regression" and reinvented "normal" and "correlation." Mendelians already had "dominance" and "recessiveness," and were well on their way with "segregation." But the report took the jargon-minting to another level. Henceforth, each of the unit-character versions making up an antagonistic pair would be known as an "allelomorph." Two similar allelomorphs united during fertilization would be deemed to form a "homozygote." A zygote comprising two different allelomorphs would be known as a "heterozygote." And to avoid ambiguity in discussing the generations in a crossbred lineage, a new labeling system would be used, broadly inspired by what Galton had used in *Hereditary Genius*: "P" for the parental generation; "F" for the filial generation (a.k.a. the hybrids); "F_2" for their offspring; and so on.[22]

2. *New "Mendel numbers."* Thanks to the report, the signature Mende-

lian ratios grew beyond the classic 3:1. One newcomer was 9:3:3:1, for the results of a cross involving two characters. (So, with the seed characters in peas, the cross gives, in the F_2, 9 yellow-and-round seeds to 3 yellow-and-wrinkled seeds to 3 green-and-round seeds to 1 green-and-wrinkled seed.) The other newcomer was 1:1, the result when hybrids are "backcrossed" with the pure recessive parent—a crossing, in the new terminology, of a heterozygote and a recessive homozygote. Again with pea seeds, half the offspring will be yellow or round, and half will be green or wrinkled. Return to Mendel's paper, and you will not find either ratio spelled out as such. With two-character crosses, as we saw, Mendel had a completely different way of articulating the outcomes, in line with his hair-raising ambition to show how these outcomes mapped onto the terms of the expanded series $(A + 2Aa + a)(B + 2Bb + b)$. Yes, anyone looking for the 9:3:3:1 ratio can spot it in Mendel's data from such crosses. But Mendel drew no attention to it. As for the 1:1 ratio, it, too, lurks inconspicuously in Mendel's paper, in data from the crosses with which Mendel tested his hypothesis about pollen and egg cells from the hybrids receiving just one or the other of the parental-character versions. In isolating these easily missed ratios from Mendel's paper, the report bequeathed to the growing corps of the Mendel-curious some new numbers to hunt for in their own data on heredity.[23]

3. *New flexibilities.* Over and over again, the report showed by example how Mendelian reasoning could be extended to accommodate seemingly all of the "facts of heredity," not just the ones conforming obviously to Mendelian ratios. A major source of this new power was Bateson's identification of Mendel's law not with Mendel's "law valid for *Pisum*," $A + 2Aa + a$, but with Mendel's explanatory hypothesis about the gametes. Thus circumscribed, Mendel's law could in principle be made compatible with any and all visible patterns of inheritance, since unions of pure gametes could, in theory, give rise to all sorts of possibilities in the developed organisms. Among other things, that meant, as the report made clear from the start and throughout, that the undoubted diversity of dominance patterns was irrelevant. An allelomorph's exclusive dominance over its antagonist in the hybrid was no more evidence for Mendel's law then a blended hybrid character, or a mosaic one, or something different from the parents entirely (the report cited what became an enduring example: the blue-gray feathers of an Andalusian fowl from a cross between a black and a white parent) was evidence against it.[24] As for facts that, even on the gametic understanding of Mendel's law, appeared to contradict that law, there was often a minor but ingenious modification at hand that could turn the anomalous into the intelligible. So relaxed were Bateson and Saunders about these facts that they grouped them un-

der the heading "non-Mendelian."[25] Take, for example, their handling of a dominant individual where only recessives are expected, as happened with both Saunders's plants and Bateson's poultry. Surely that was fatal to Mendel's law? Not necessarily, as they explained. It might be that, sometimes, a *DR* organism can show the recessive character, with the result that down the line, *D* will eventually meet with *D*. Or maybe, sometimes, the *R*s in an *RR* organism are sufficiently different from each other for their union effectively to make a heterozygote, and as such, to show the dominant character as heterozygotes generally do. Or maybe there has been contamination, or it was just one of nature's mishaps.[26]

4. New relations to Galton's law. The penultimate section of the report was titled "Galton's Law of Ancestral Heredity in Relation to the New Facts." The ground for an uncompromising dismissal of the idea of regularly diminishing ancestral influence had already been prepared: in a statement early on about how extracted recessives are not just similar to but *identical to* their recessive grandparents; then later, in a discussion of "reversions" as simply cases in which, as with the Andalusian fowl, the hybrid has a distinctive character (perhaps, the report went on, because the parental characters are themselves compound); then, more obliquely, in a brief remark on how a continuously varying character such as human stature—the Galtonian blend par excellence—could turn out to be the upshot of four or five allelomorphs plus the effects of environmental accidents.[27] Now the break was made brutally explicit. Galton's law, with its rejection of gametic purity, was declared "irreconcilable" with Mendel's law, and furthermore, the data seeming to support the former would, soon enough, all turn out to be explicable by the latter.[28]

5. New implications. Looking backward, the report indicated, biologists could now easily enough see new Mendelian meaning in puzzling old results, such as Darwin's snapdragon crosses.[29] Looking forward, they could make out practical boons in the offing, not just narrowly scientific ones. Agriculture and horticulture especially would benefit, notably from the new understanding of why valuable hybrid characters prove so difficult to stabilize, no matter how assiduously one applies selection. Relatedly, breeders could now be on the alert for the possibility that a desirable character is dominant, in which case as long as they were sure to breed from "genetically pure" forms, rather than from possibly identical-looking hybrids, crossing would maintain the purity with no need for selection. (The list of twenty-six simple allelomorphic pairs in the report included characters in wheat, barley, maize, fowl, cattle, and goats, as well as peas.)[30] But there were even signs of medical and social benefits, as explored in a long footnote on the inheritance of rare conditions. When Bateson analyzed family data collected by a London medic, Archibald Garrod, who had become interested in the cause of alkaptonuria (so-called because the main symptom is that the urine of sufferers turns black when exposed to air), it appeared, first of all, that the condition is recessive, and second, that marriage between first cousins greatly increases the chances of otherwise rare recessive allelomorphs being joined. The notion that cousin marriage might be bad for the health of offspring was, of course, nothing new. But to explain at a stroke *why* it was bad, with such precision and economy, was unprecedented.[31]

3

"Cannot you manage to use me less often as (personal) ammunition?" Saunders asked Bateson in comments she sent on his draft response to Weldon.[32] No sooner had Bateson read the *Biometrika* Mendel critique than he had sprung into action, in short order securing a book contract from Cambridge University Press; obtaining permissions from the Royal Horticultural Society to reprint (with revisions) his inaugural paper on Mendel as well as the English "Versuche"; undertaking a translation of Mendel's short paper on hawkweed hybrids; and writing what grew, in the published version, into over a hundred pages of animadversions on Weldon's critique of Mendel.[33] Saunders was not the only backstage reader of the new book to register discomfort over the element of score settling. "I cannot think that it is for the interests of science that controversy should be carried on in quite this tone," wrote Bateson's Press contact, asking him to drop or rephrase some of the over-the-top sentences in the preface. Bateson's wife Beatrice sided with the Press: "Far from the scene of action, one detects a howl sounding through the words."[34]

For the most part, Bateson succeeded, in the Weldon-directed final section of *Mendel's Principles of Heredity: A Defence*, in letting the new Mendelian machinery of the 1902 report do the talking. As we have seen, Bateson and Saunders there dispensed utterly with the dominance patterns that Mendel observed in his pea crosses as in any way defining the commitments of "the Mendelian," to use a label that Bateson was now using. In this follow-up book, Bateson was even more forthright. Yes, of course, dominance could be "very dazzling at first" when encountered as Mendel had encountered it. And how fortunate that it can be so: for, according to Bateson, only because Mendel deliberately selected the most dazzlingly dominance-exhibiting varieties for his crosses could he have made his truly important discovery: the production by the hybrids of pollen and egg cells pure for the parental characters. After Mendel had reached that essential insight, however, dominance no longer mattered in the slightest—it was a kicked-away stepladder.[35]

In Bateson's view, nothing could be more misleading about Mendel's achievement than Weldon's framing of a "law of dominance" on Mendel's behalf. Mendel recognized no such law. If anything, his contribution lay precisely in seeing that *dominance has no law*, because the character of a hybrid is the upshot not of parental transmission but of gametic combination.[36] Once grasped, Mendel's "all bets are off" approach to predicting the hybrid character, Bateson now insisted, marked a major departure from the traditional view of hybrid-character predictability, most recently and subtly defended under the aegis of the law of ancestral heredity. Where the ancestry-obsessed investigator pored over pedigrees in hopes of identifying the formulae with which to forecast average character (blending inheritance) or character proportions (alternative or exclusive inheritance), the Mendelian looked to experimental crosses à la Mendel, making no presumption that what holds for one pair of varieties or species will hold more generally, let alone universally. Not even the purity of the gametes was sacrosanct. On the contrary, there were plainly exceptions to it, and the Mendelian was as interested in the anomalies as in the cases conforming straightforwardly to the simple ratios. Dominance, and the purity of the gametes, served the Mendelian as new baseline expectations by which to order what had been the chaos of observations—nothing more and nothing less.[37]

What, then, were the "principles of heredity" being defended in Mendel's name? Bateson at one point lists seven key ideas: purity of the gametes, unit-character allelomorphs, and so on.[38] But by the end of the book, the impression conveyed is that the most important principles are methodological. For anyone interested in understanding heredity, Mendel's example showed what to do: how to design experimental crosses, what to look for in the data in the first instance, what not to be sidetracked by (principally, the influence of conditions, which, unless controlled for, could have "masking effects" on what would otherwise have been "clearly detected"). There were no dogmas save for the belief that dogmas were bound to be undermined as experiments progressed and the complexities accumulated.[39] As for ancestry, it was, for purposes of understanding what happened and why, irrelevant. The individuals making up the 1:2:1 pattern in a Mendelian F_2 would all have the same ancestry but strikingly different posterities, in a manner that Mendelian gametic purity illuminated in full. And unlike the ancestors, those individuals could be studied in the here and now, not just conjectured about.[40]

In just two years, the "Versuche" had gone from being a low-key enthusiasm among specialists in plant hybrids to the foundation for a new science of heredity that, in this latest statement, was invincibly all-conquering. No critique could have landed with much force on a terrain that had become so mobile and so malleable. Weldon's dominance-highlighting critique provided easy pickings for Bateson. All those painstakingly documented exceptions to the classic Mendelian pattern? Not only were these unsurprising to the Mendelian, but, as Bateson showed case by case, myriad explanations for the departures from simplicity lay to hand, some drawn from *Report 1*, others from the arcana of pea breeding (something Bateson had recently been learning about at Suttons' nursery). As for Laxton's 1866 report on his pea-hybrid experiments, Bateson's creative Mendelizing of it formed a bravura set piece.[41]

Not every aspect of Bateson's demolition was equally commanding. On Weldon's claim that nothing is naturally dominant but ancestral setting makes it so, Bateson offered both blustery condemnation ("one of the most remarkable examples of special pleading to be met with in scientific literature") and relaxed concession (saying that of course every variety is different when it comes to dominance, with some more sensitive to surroundings than others, and that even if breeders ended up universally favoring a genuinely exceptional variety like *Telephone*, so that dominant greenness became the rule, the use of Mendelian baselines for understanding would remain just as valuable).[42] There was barely an allusion to the "too good to be true" data problem, and not even that much, here or in the report, to Mendel's own ancestral law for *Pisum*, predicting the gradual (but never total) disappearance of the hybrid character.[43]

Bateson was unapologetic about offering readers *his* Mendel, whose service to science could be measured only now that so much work following his lead was well under way. For a frontispiece, Bateson chose a portrait of Mendel: the first glimpse of him for most people outside Brünn. Here was that man who, alone and unheeded, had pointed the way for the study of heredity and evolution at the very moment Darwin was inadvertently imposing his stultifying authority.[44] Now naturalists were at last making progress, and this new book from Bateson, proud evangelist for the new gospel (the religious language was his), aimed to ensure that no young biologist would be dissuaded by Weldon's critique from being part of the future. And a truly amazing future beckoned, as Bateson signaled in conclusion:

> In these pages I have only touched the edge of that new country which is stretching out before us, whence in ten years' time we shall look back on the present days of our captivity. Soon every science that deals with animals and plants will be teeming with discovery, made possible by Mendel's work. The breeder, whether of plants or of animals, no longer trudging in the old paths of tradition, will be second only to the chemist in resource and in foresight. Each conception of life in which heredity bears

a part—and which of them is exempt?—must change before the coming rush of facts.[45]

4

It took a while for Weldon to grasp just how profoundly *Report 1* and the *Defence* had neutralized criticism along the lines he had ventured in "Mendel's Laws of Alternative Inheritance in Peas." Every sort of departure from Mendelian simplicity—nondominance, impurity of the gametes, a dominant-character version where only recessiveness was expected, reversions back to long-vanished characters, character versions seeming to stick together down the generations rather than behaving independently (Saunders had found an instance), anything—could now be absorbed, and with no destabilizing effect on confidence in the underpinning ideas, because those ideas had been recast as guidelines for inquiry.[46] "Mendel's principles of heredity" were not statements about what is the case, even in a limited domain, but starting-point generalizations to organize experiments and the knowledge that flowed from them. For the Mendelian, the interesting challenge now was not to show that there were departures from simplicity—that was admitted—but to figure out how to modify the principles so as to explain those departures. If there was a serious objection to be lodged against a science of heredity so conceived, it was not whether it "worked," but whether the way it worked put it fundamentally at odds with a more basic biological truth, to the point of occluding it. And whatever that truth might be, it needed to have experimental credentials at least as impressive as those supporting Mendel's principles (as the Galton-Pearson ancestral law did not).

Eventually, Weldon would reframe his critique along the newly necessary lines. In the meantime, however, he carried on where "Mendel's Laws of Alternative Inheritance in Peas" left off, continuing to explore the gap between what the new Mendelians said and what was actually observed. His doing even that much was not a foregone conclusion when he got hold of a copy of the *Defence* and found to his dismay that he was a threatened "established prophet," a "vehement preacher of precision" whose critique of Mendel could now be exposed as a shameful farrago of "perverse inference," "slovenly argument," and "misuse of authorities, reiterated and grotesque."[47] (Thus the tempered version of Bateson's preface.) Pearson rejected out of hand any talk of Weldon being dropped from the editorial board of *Biometrika*, lest the journal become permanently tainted with controversy.[48] But the personal nature of Bateson's abuse was not something Weldon could shake easily, as he explained:

> I do not want to give up the fight at all, and my wife says I am very fit. But you have not read this book. It is a deliberate and very explicit charge against me of having wilfully lied and misrepresented Mendel, in order to conceal from my pupils that my occupation is now gone, because the only thing I believe in is the Law of Ancestral Heredity, and since Mendel has shown that ancestry has nothing to do with inheritance (in fact there is no such thing as inheritance) the whole of my poor attempts at teaching become futile. Of course I shall not propose to defend myself against this sort of thing but one feels that somehow one must be to blame, if it is possible for such things to be said about one, even untruly. This is the sickening part of it. As an instrument for facilitating a thorough exposure of Bateson's own attitude with regard to inheritance, [however,] one could not wish for anything better.

For Weldon, the "sickening" element was compounded, he went on, by his now understanding a puzzling remark from Francis Darwin (botanist brother of George) a little while back. Learning that Mendel's law now featured in the Tripos examination at Cambridge, Weldon had asked Darwin what he had thought of Weldon's critique, and was told that, in Darwin's view, Weldon had been unfair, but "he added in a very curious way that he did not think I had done so intentionally." Belatedly, Weldon appreciated that at Cambridge he was already thought of as yesterday's man, unable to grasp the new principles, let alone concede their truth. "I am ashamed to have let the thing get on my nerves," he admitted to Pearson, "but it has. It is not nice to know that half the undergraduates in one's old laboratory are being brought up to believe one a deliberate liar!"[49]

A new project was always the best therapy. In *Report 1*, the results of Saunders's crossing experiments with hairy-leafed and smooth ("glabrous") -leafed forms of *Lychnis* conformed most closely to Mendelian predictions, with hairiness said to dominate exceptionlessly in the hybrids, and smoothness returning in their offspring in a proportion pleasingly close to the expected quarter.[50] Now Weldon threw himself into collecting and examining the *Lychnis* varieties that carpeted the fields of Oxfordshire, soon involving a young collaborator, Miss C. B. Sanders. As with "yellowness" and "wrinkledness" in peas, so with "hairiness" in *Lychnis*: the Mendelian categories contained so much variability that, absent finer-grained descriptions, there was no reason at all to trust that what were counted as pure dominants and pure recessives among the hybrid offspring really were "identical" to the grandparent forms.[51]

Perhaps it would even be possible to make sense of the Mendelian pattern of dominance-then-3:1 on the view that (1) super-hairiness and super-smoothness occupy poles on a continuum; (2) breeding from individuals

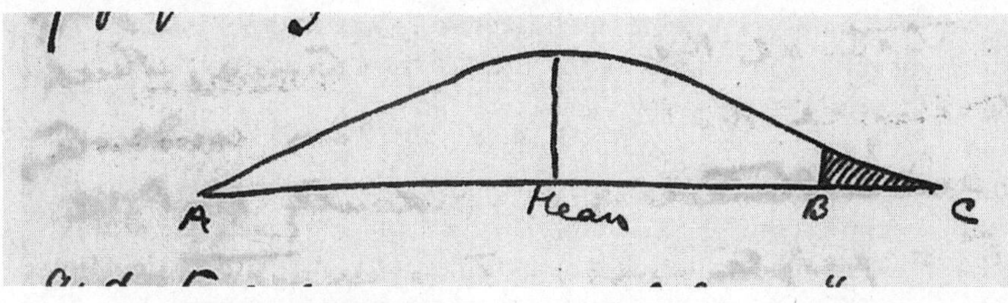

FIG. 7.2. "What all Mendelians do, is to take the diagram of frequency and to call a range AB one 'character,' and the range BC the other 'character' of a Mendelian pair": from Weldon to Pearson, 23 June 1902—source of the "battle" passage quoted as the first epigraph in this book.

near those poles would produce blended offspring that would count, for the Mendelian, as "hairy," and (3) random mating among those offspring would produce an expanded bell curve in the next generation, in which around a quarter of the individuals would count, for the Mendelian, as "smooth" (fig. 7.2). "If one interprets 'a chance = ¼' with 'neo-Mendelian' latitude," Weldon explained to Pearson,

> I think it quite likely that a large range of cases of blended inheritance could be shown from the Law of Ancestral Heredity to 'obey Mendel's Laws': and a more important point would be this: that since some races of the same species are said to obey these laws, while others do not, one has a chance of predicting in a given case, from a knowledge of parental variability, what will happen, and so smashing the 'pure gamete' once for all.[52]

As Weldon prepared his next, *Lychnis*-based move, *Report 1* and the *Defence* made their way through the biological community, and the low-blow tactics of the *Defence* inevitably drew attention. Pearson let Bateson know that, though something from him might in principle appear in the pages of *Biometrika*, nothing with the tone of his *Defence* would ever be published there.[53] Galton expressed his disappointment too, to Bateson and others. ("It is a great pity," Galton wrote to Pearson, "that Bateson can so easily be tempted into using unconciliatory language.")[54] Tschermak, asked to review the *Defence* for *Biometrika*, wrote to Weldon to register surprise that an English work of science should stoop to this level of unconstructive polemic.[55] In a review of *Report 1* and the *Defence* in *Nature* in October, the Oxford entomologist F. A. Dixey noted that, even allowing for the "zeal of an advocate," Bateson would have done better to follow Weldon's example of writing in "a critical, but, as it seems to us, not unfriendly spirit."[56] Yule, in an important review essay published the following month (and to which we will return), wrote, "Mr. Bateson may no doubt congratulate himself on a *succès de scandale*, but it is difficult to see that his 'Defence' attains any worthier goal."[57]

Yule got that wrong: it was precisely by writing the kind of book he did that Bateson dramatized the excitement of being on his side in the battle that now commenced. A Cambridge research program hitherto carried on by just Bateson and Saunders now began to swell with younger people, mainly trained up in Cambridge biology.[58] As Bateson was writing the *Defence*, he drew on the work of Hilda Blanche Killby, who was already assisting him with hybrid pea experiments.[59] When he checked the proofs, he was traveling with Robert Lock, a botany student who had attended Bateson's Mendel-promoting lectures on "the practical study of evolution" and would go on later that year to conduct Mendelian hybridization studies at the Royal Botanic Garden in Ceylon. (In 1899, Bateson had been appointed deputy to the aging Alfred Newton, whose lecturing duties Bateson took over—a step up from his kitchen steward duties.)[60] When the *Defence* was published, Bateson sent copies to two other graduates of those lectures, the agricultural botanist Rowland Biffen and the zoologist Leonard Doncaster. "I am glad to see Weldon's 'exceptions' turned to such a good use," replied Biffen in a thank-you note, adding, "Couldn't you translate the other part of Biometrika now & render it in plain English? I'm sure the majority of people are like myself ignorant of Karl-Pearsonese." Before long Biffen had Mendelian wheat breeding experiments under way on the farm of Cambridge's new Department of Agriculture, where a slightly older colleague and fellow attendee of Bateson's lectures, Thomas Wood, was doing likewise with sheep breeding.[61] When Doncaster received copies of the *Defence* as well as *Report 1*, he was based at the Naples zoological station, where he was looking at dominance in hybrids of sea urchins. He confessed to not quite seeing why the *Defence* was needed, given the excellence of *Report 1*, though he acknowledged that probably the former would reach more readers, and that Weldon's criticisms did need answering at length. Doncaster also shared some insights into Weldon gleaned from a fellow Naples researcher, Edgar Schuster (seated in the middle row, far left, in fig. 6.2):

> I have been particularly interested in the Weldon controversy because I have been living for 4 months with an Oxford man, a pupil of Weldon's, who knows him intimately, and so I have heard what is to be said on both sides. He tells me that when Mendel's work was first discovered, Weldon was exceedingly keen about it, and became a convinced Mendelian, but when he came to examine the evidence his faith was upset. It seems to me a case of "a little knowledge." . . . I hear he is vigorously breeding white

and waltzing mice; it will be interesting to see if he gets results agreeing with von Guaita's.[62]

Of these new Cambridge Mendelians, Biffen would come closest to sharing Bateson's sense of mission in the service of Mendel. Another, non-Cambridge recruit from around this time, the Leicestershire breeder Charles Chamberlain Hurst (fig. 7.3), would come closer still. Nine years younger than Bateson, Hurst was the very model of the scientifically engaged practical breeder whom Bateson had long fantasized about: a devoted experimental hybridist who saw his experiments as potentially illuminating the mysteries of variation and inheritance as well as furnishing valuable new products for market. At the 1899 hybridization conference in London, Hurst had presented a commanding survey, concentrating on his own experiments with orchid hybrids (his specialty) and how fully they had or had not conformed to Galton's law. Hurst's paper even got cited positively in Bateson and Saunders's *Report 1*, for evidence that some non-Mendelian results could be explained by supposing that sometimes crossing induces parthenogenesis in the seed plant.[63]

Yet by late March 1902, when Bateson first wrote to ask if he could visit Hurst at his family's nursery, they seemed much less like natural allies. Politically, on the matter of the Boer War, whereas Bateson had become "a fanatical pro-Boer" (as Weldon remarked to Pearson the previous month), the patriotic Hurst had actually tried to join the fighting in South Africa in 1900, settling instead—after failing to pass the medical—for vigorous volunteer work on the home front.[64] Scientifically, "Captain Hurst," as he now was, had a new paper on Mendel in press that was far more sympathetic to Weldon's position than to Bateson's. Published in the Royal Horticultural Society's *Journal*, "Mendel's 'Law' Applied to Orchid Hybrids" was to be the first of a two-paper set. This first paper described the restricted sense in which Mendel's law could be said to apply to orchid hybrids, and the second was to explain why Mendel's own theory of why his law holds should be rejected. According to Hurst, what Mendel and his rediscoverers claimed about one of two antagonistic character versions dominating in hybrids could be safely ignored as the "abnormal and exceptional" upshot of choosing to hybridize plants whose "peculiar kinds of elements" produced that anomalous pattern. Nevertheless, Hurst went on, "I have reason to believe that the application of Mendel's formula"—$A + 2Aa + a$—"is strictly limited to hybrids and crosses of *a certain ancestry*." What Hurst meant, it turns out, was that in crosses like the hundred or more that he reported between closely related and, crucially, known-to-be-true-breeding orchid species, the distribution of characters in the hybrid generation fell along a spectrum, with roughly a quarter of indi-

FIG. 7.3. Bateson's allies in Mendelian plant breeding: Rowland Biffen (*above*) and Charles Chamberlain Hurst (*right*).

viduals taking after the *A* parent, roughly a half being intermediate, and roughly a quarter taking after the *a* parent, with the whole thus "making a perfect series of intermediate forms between the two parent species." At the very end of the paper was a note, added in proof, about how at the time of writing Hurst had not seen Weldon's excellent *Biometrika* paper, where Weldon suggested, in Hurst's words, "that in former experiments sufficient attention has not been given to the important question of ancestry. With this I agree, and submit that the experiments with Orchid hybrids detailed above are not open to this objection."[65]

If the Boer War—then winding down—came up at all during Bateson's overnight stay with the Hurst family that April, it did not get in the way of Bateson and Hurst hitting it off. Over the succeeding months, a friendly correspondence gradually picked up pace.[66] Hurst never wrote his planned follow-up paper contra Mendel. Indeed, Hurst's next paper, sent to Bateson in mid-September, was a celebration of Mendel's experimental methods as constituting an astonishingly powerful tool that deserved universal adoption immediately. Whereas Hurst had earlier recounted the story of the Mendelian discovery and rediscovery as not terribly exciting background facts, he now hailed Mendel's 1866 paper as marking a "new epoch" in plant hybridization, albeit an epoch in suspended animation until the "psychological moment" for it arrived in 1900, after which "progress in certain directions has been phenomenal." And whereas Hurst had previously characterized the dependence of Mendel's results on the peculiar ancestries of the races of peas he worked with as a reason to distrust his experiments as a guide to any more general understanding, those experiments were now not only "a great advance on what has been done before" but would "probably prove a stepping-stone towards the final solution of the problems of inheritance." Indeed, Mendel's caution in spending two years ensuring that his pea races were pure in order to avoid ancestry-caused confusion in his crosses was the sign of a "master mind," as were his tracking of antagonistic versions of unit characters with a clear dominance/recessiveness relationship, his working at a large enough scale to get statistically meaningful numbers, and his not stopping his hybridizations at the first or even second generation. Here, for Hurst, were "classical experiments [that] will serve as a model for the hybridist who wishes to attack the perplexing problems of inheritance." Hurst was leading by example, noting that he already had Mendelian experiments under way not only with orchids but with a range of other flowering plants, with *Pisum*, even with poultry. For further reading, he recommended not Weldon—mentioned only as an example of a critic whose failure to take Mendel's methodological precautions no doubt accounted for discrepant results—but Bateson's *Defence*.[67]

Hurst had intended to read his paper himself at a conference taking place later that month in New York City on hybridization—a successor to the 1899 London conference. But the demands of harvest time proved too much, so he asked Bateson, also due to present, to bring the paper with him and have it read out.[68]

5

At the three-day International Conference on Plant Breeding and Hybridization, organized by the newly formed Horticultural Society of New York, Hurst's paper, "Notes on Mendel's Methods of Cross-Breeding," was second on the program, just after Bateson's opening address. Bateson's invitation had come about through his connections in the Royal Horticultural Society, whose representative he was.[69] He made the most of the opportunity.

In important ways, the horticulturalists and agriculturalists gathered that morning in late September were better positioned than anyone else in the world at that moment not only to appreciate Bateson's Mendelian message but to put it into action. Some were, like Hurst, commercial breeders with a serious interest in understanding the scientific principles behind success and failure. But most were research scientists working within a distinctively American system of publicly funded institutions aiming to help American farmers stay competitive, above all by harnessing the power of experiment to give them new products to sell. With encouragement and coordination from the ambitious United States Department of Agriculture in Washington, DC, a nationwide culture of experimental breeding, with a particular stress on hybridization, had taken hold in a growing number of state agricultural colleges and experiment stations. In 1901, the *Experiment Station Record*, the USDA's house journal, published a digest of Mendel's "Versuche," based on the Bateson-introduced English translation, that quoted Bateson on Mendel's discoveries as doing for heredity and evolution what the atomic laws had done for chemistry. Positive summaries of Mendelian papers by Tschermak and Bateson soon followed. In July 1902, the seventy-five students who attended the USDA's first Graduate School of Agriculture, held at Columbus, Ohio, received instruction in the new Mendelian work.[70]

So the ground was well prepared for Bateson's address, titled "Practical Aspects of the New Discoveries in Heredity." He began with flattery: it was a treat for anyone investigating breeding experimentally to come to the United States and meet fellow workers, since nowhere else were the problems being studied by so many, under such varied conditions of soil and climate, and in such an amply resourced way. In studies of this kind, he continued,

> we have reached a critical moment. That crisis, as it is known to many of those present, has been brought about by the rediscovery and confirmation of Mendel's work on heredity. These discoveries intimately concern the art of the practical breeder, and I propose to use the present opportunity to indicate some of the ways in which we can employ them for his purposes.[71]

In the exposition of Mendelian principles that followed, Bateson dipped once more into his fund of chemical analogies. Given a simple salt, one can swap the base for a range of other bases, or the acid radical for a range of other acid radicals, decoupling and recoupling at will. The elements that constitute a compound are understood as replaceable components. So too, Bateson explained, are the hereditary characters making up the sorts of hybrids that Mendel examined. Greenness and yellowness, hairiness and smoothness, tallness and dwarfness, are passed on to offspring in an all-or-nothing, unit-character way, and so, like the chemical elements, are susceptible to endless swapping in and out. Here was an insight, he continued, whose future significance for breeders and hybridists would be hard to overstate. We should nevertheless guard, he cautioned, against overstating how far it had *already* yielded practical men the tools needed to achieve their commercial ends. But the time was not far off when, like the chemist, the "breeder will be in a position . . . to do what he wants to do, instead of merely [to use] what happens to turn up." Those involved with hybridization had got used to hopeless confusion in their findings, and had become resigned to never knowing what the results of a cross would be. Now Mendel's work was the thread that would lead the way out of "those wonderful mazes of heredity."[72]

Above all, Bateson went on, breeders needed to absorb the lesson that plants that look exactly the same can be fundamentally different when it comes to the characters composing them and so transmitted in crosses. A pea plant grown from a yellow seed may, with respect to seed color, be the product of a union between two yellowness gametes, and so itself be the producer of yellowness gametes exclusively. Or, because yellowness is dominant and greenness recessive, this plant might instead be the product of a union between a yellowness gamete and a greenness gamete, in which case the plant will go on to produce both kinds of gametes. Failure to distinguish between these two possibilities—what Bateson was already, with minimal fuss, calling a "homozygote" and a "heterozygote"—had caused endless mischief. Here, in his diagnosis, was the source of "an immense number of the contradictions which the practical breeder experiences," not least the perennial breeders' headache of "rogues which are not true to the character which he desires to put on the market—rogues which he is unable to eliminate." The old thinking had it that to eliminate rogues, the breeder had to hoe them out, and continue to do so generation after generation. By such means the number of rogues would diminish, and the breeder would "gradually fix his type." The new Mendelism taught otherwise. Character fixity was a matter not of selection plus time, but of gametic purity. For the breeder attempting to breed plants in which the desired character was a dominant character, all that was needed to achieve rogue-free fields was to ensure that the starting materials contained no recessive-harboring hybrids. And that was just a matter of judicious, constitution-exposing breeding among individual plants exhibiting the character.[73]

Famously, after the conference, Bateson wrote a letter to his wife about the electric response to his lecture, as quoted in the final epigraph at the start of this chapter ("It has been 'Mendel, Mendel all the way'"). To judge by the published transcript of the discussion following Bateson's lecture, he did succeed in switching on mental light bulbs for a number of his listeners, above all with his discussion of rogue plants. He illustrated the abstract lesson with an agricultural example, the unwanted persistence of whiskery "bearded" wheat in fields meant to be growing only "beardless" wheat. The trouble, he explained, was that farmers had not understood that beardedness was a recessive character and was bound to turn up whenever farmers failed to exclude heterozygote plants from the beardless breeding stock.[74] William Saunders, a director of experimental farms in Canada, spoke of how illuminating he had found Bateson's explanation of bearded rogues:

> Mr. President, this paper has thrown light upon many subjects which have been somewhat dark in my mind. For instance, in the cross-fertilizing of wheats we have often found that the crossing of two beardless forms will produce a bearded form, or we have a beardless wheat as the result of the crossing of two bearded forms. This explanation that Professor Bateson has given us throws light on that point and on many similar points which have puzzled many of us who are practical workers in this very interesting field.[75]

Note that Saunders's praise was founded partly on a misunderstanding. He reported getting a beardless form (D) when he crossed two bearded ones ($R \times R$); but of course, according to the Mendelian analysis, bearded crossed with bearded should yield bearded ad infinitum. If Bateson was bothered, he kept quiet, as he generally did in his lecture about all the hard-case exceptions that occupied so many pages of *Report 1* and the *Defence*. His whole performance before his New York audience set the pattern for future Mendelian publicity, emphasizing the simplicity of the principles and the excitement of putting them into useful practice.

Saunders's comment came in between two other, ready-to-spread-the-gospel ones. Liberty Hyde Bailey, chair of General and Experimental Horticulture at Cornell's agricultural college, commended Bateson's *Defence* to all, declaring the dawning Mendelian era to be comparable to the period after the publication of Darwin's discovery: "It seems to me it is as important as that. I expect to use [Bateson's] book as a basis for all our work in plant breeding." (Bailey was among those who, before 1900, had published citations to Mendel's papers without troubling to read them.)[76] William J. Spillman, from the USDA, had spent the last year realizing that in his research on wheat hybrids, conducted while he was based at Washington State Agricultural College, he had unknowingly joined the ranks of the Mendel rediscoverers. In New York, he used the occasion of Bateson's lecture to announce that he was exhibiting figures illustrating Mendel's law applied to wheat.[77]

The last words in the session went to the chair, presumably the society's president, James Wood, a farmer and businessman who was a tireless servant of the cause of agricultural improvement. For Wood, as for Saunders, Bateson had indeed dispelled mystery:

> I could have presented from my own fields this season ten acres of illustration of Mr. Bateson's statement in regard to growing wheat. I have been growing a hybrid wheat for a number of years in a practical way as a farmer, and the seedsmen have taken the crops, and every year I have had to fight these bearded specimens of plants that came up in this field. To me it has been one of my greatest puzzles, as I was making no progress whatever; and while I never allowed one of those plants to go into my field, yet year after year I had the same result. I can see that it has been a bottomless work that I have been trying.[78]

What to make of such testimonials? It seems surprising that the likes of Saunders and Wood should have needed Bateson to tell them that, in order to take a hereditary character out of circulation, they needed to take the germinal causes of the character out of circulation, and that when unwanted characters popped up, it was because the elimination job had been botched. Nor should it have been news that a character can lie dormant, reemerging only when its germinal causes are not inhibited in their effects. Darwin and Galton had said as much. And yet, plainly, listening to Bateson, some in his audience experienced a kind of epiphany—indeed a powerful one.

Months later, Spillman wrote to Hurst about the great success of the conference, not least for putting the new Mendelian science on the map: "Of the fifty-one papers or more, no matter what the subject, every one dealt with Mendel's law more or less; for as soon as Bateson's paper was read there was an excited state of mind until the end of the meeting."[79] On the evidence of the published record, Mendel did come up a fair bit over the first day and a half, and with little assistance from Bateson. After Hurst's paper, Hugo de Vries had a paper read on his recent work exploring a theme Bateson had touched upon: how the breaking up of a compound character into its components upon crossing can occasionally, as component meets component according to Mendel's law, lead to the reappearance of the "lost" character—a process that, once understood, might even enable the deliberate re-synthesis of the compound character.[80] In the discussion of a paper on the classification of hybrids, Bailey named Mendel, along with de Vries and Bateson, as among those investigators whose work was putting pressure on the old notion that the term "hybrid" referred only to the product of crossing between distinct *species*.[81] Willet Hays, an agriculturalist with the state experiment station in St. Anthony Park, Minnesota, touched on Mendel near the end of a magisterial overview of "breeding for intrinsic qualities," which stressed the economic significance of the enterprise and the utility of the method of America's most commercially successful breeder, Luther Burbank—basically, hybridize to create novelty; grow individual plants on a massive scale; select the best one; repeat. Hays put in a plea for questions about Mendel's law, exciting though they were, not to crowd out all the other, sometimes even more important questions about breeding that cried out for answers.[82] Hays's counterpart at the Geneva, New York, experiment station, S. A. Beach, in a paper on correlation in plant characters, quoted Mendel on how seed-coat color correlated with flower color in the garden pea.[83] In "Some Cytological Aspects of Hybrids," William Cannon, a research fellow at Columbia with E. B. Wilson, pointed out that, according to the Mendelian hypothesis on how character versions are distributed in the gametes of hybrids, "we might expect that the chromatin derived from the primitive parents maintained its individuality, and was disposed in such a manner . . . that the resulting sex cells were not *hybrid*, but *pure*"—something already observed, he announced, in hybrids of roundworms. If it turned out to be true of plant sex cells as well, "a very important and forward step in explanation of the nature of this and other types of hybrids shall have been made." (The attendees were not Cannon's people: his paper evoked no discussion at all.)[84] After a paper on the breeding of florists' flowers, Spillman piped up again to wonder about the value of carefully watching offspring grown from seeds of hybrid apple trees for Mendelian splitting into types.[85]

Then, in the middle of the second day, controversy briefly flared. Reflecting on ten years of crossing experiments with pumpkins and squashes, Bailey said that breeders stood to learn a great deal by using Mendel's methods, but in doing so did not need to accept that Mendel's conclusions hold true in a general way, or to believe that practical plant breeding, where success depends so much on selecting the right individual plant, would one day be guided exclusively by an individuality-flattening statistical law like Mendel's.[86] At the start of the discussion. O. F. Cook, a USDA botanist, seconded the notion that Mendel's "so-called laws" might turn out to have limited applicability beyond Mendelian investigations. That, in turn, provoked a monumental reply from Bateson, who politely but firmly insisted that, as he summarized it near the end, "Complexity itself is no bar at all to the application of Mendel's law." Bailey conceded the point. But then, closing off the discussion, came a remarkable long comment from T. V. Munson, from the Texas Experiment Station in Denison. The problem with talking about laws of hybridization like Mendel's, said Munson, is that inevitably one ended up with considerations on both sides and not much progress made. What was needed was a perspective on plants that took seriously their dynamic chemical and environmental situations, not just their biologies narrowly construed. "It is true we want laws; we want something by which we can guide ourselves in this work; but if you try to make a law out of a mere theory which is only set up for experimentation you are wasting time."[87]

So the meeting was by no means uniform in wanting to take Mendel all the way. But there was plenty of enthusiasm for doing so. A little later that same afternoon, at the end of a discussion of a paper titled "The Improvement of Carnations," Spillman put the case for the practical value of theoretical science generally, and the cash value of Mendelian science in particular:

> I can't help saying . . . that when these practical men, who occasionally rather speak slightingly of theoretical men, learn the immense value in dollars and cents of knowing the why and the wherefore, and how to do it again, it will be worth a great deal to them. In other words, if Mendel's law is true, it is worth millions of dollars to the breeders of plants in this country. If it is not true, it is vastly important that we should know it soon.[88]

8

Damn All Controversies!

What I hate is that I want to get a definite result. I want the thing to be proved nonsense. That is a thoroughly unhealthy and immoral frame of mind, and I expect it will lead to a well-deserved smash. Damn all controversies!

W. F. R. WELDON TO KARL PEARSON, LATE JULY 1902

I was glad to hear Weldon accept the Mendelian principles in regard to albinism. This is something, at any rate, & it is quite possible soon that our good friend will evolve gradually into a Mendelian.

C. C. HURST TO WILLIAM BATESON, 27 MAY 1903

It is a horrible nuisance to have all these controversies, but it seems to me there are only two ways: one may find out what one wants to know, as well as one can, and publish occasionally for one's own amusement, not caring whether anybody reads one's stuff or not. . . . The other way is to take things fighting, and to be careful to hit hard every time.

W. F. R. WELDON TO KARL PEARSON, 31 DECEMBER 1903

We adjourned for lunch and on resuming found the room packed as tight as it could hold. Even the window sills were requisitioned. For word had gone round that there was going to be a fight. . . . [By the end,] Bateson's generalship had won all along the line and thenceforth there was no danger of Mendelism being squelched out through apathy or ignorance.

REGINALD PUNNETT, "EARLY DAYS OF GENETICS," 1950

It was no part of Weldon's plan, as he meticulously totted up the hairs on *Lychnis* leaves in summer 1902, to spend the better part of the next two years trying to squelch Mendelism. He had his own projects, after all, with the ongoing study of natural selection in the wild (about which he published one more paper, on snail shells) closest to his heart. Moreover, he was, as we have seen, and as the quotations above affirm, ambivalent about the value of scientific controversies. When the prospect of a confrontation over Mendel arose that summer, he did not relish it. "The ruffians who run the biological sections of the British Ass. want a sensational row," Weldon complained to Pearson; "they want to get Bateson & me to discuss Mendel." Weldon was in a bind, not wishing to do it ("that sort of debate leads only to vanity: one cannot do any good, and one may easily do harm by it; it would be far better to wait and say one's little say in writing!"), but not seeing how he could honorably decline, as he was going to be at the BAAS meeting in Belfast anyway, for an evening lecture on inheritance, and he could scarcely seem unwilling to defend himself.[1] Fortunately Ulster politics scuppered the debate, or so Weldon reckoned. He explained to Pearson that the section president dropped the proposal after Weldon pointed out "how awkward it would be to discuss in an Irish city the question whether Orange and Green were mutually exclusive characters."[2]

Between the 1902 BAAS debate in Belfast that never was, and the 1904 BAAS debate in Cambridge that Reginald Punnett, Bateson's then-newest protégé, recalled so vividly, Weldon's constant involvement in Mendelian controversies ensured that, for all the stress and sourness, no one in this period thought harder or in a better-informed way about what it meant for Mendelian conceptions to enter the core of the science of heredity. Even so, working out the consequences of that entrance, and then working out what to do about it, proved a tricky business. It was not just that Bateson, in *Report 1* and the *Defence*, had so adroitly positioned those conceptions as presumptive baselines; had so creatively protected them behind a range of new explanatory options (the Mendelian equivalent of the Ptolemaic astronomer's epicycles and equants); and had so bullishly set them forth as organizing the experimental knowledge now presented as the only knowledge about heredity worth having. It was also that the new science remained a moving target.

The hybridization conference in New York inaugurated a phase of remarkable spreading and strengthening for Mendelism. Before 1902 was out, the term "Mendelism" was in use, introduced by Liberty Hyde Bailey in an address in Washington, DC; a clutch of embryologists published papers advancing the case for Mendelian patterns as arising from the behavior and powers of chromosomes; and G. Udny Yule argued that, far from Mendel's law being inconsistent with the law of ancestral heredity, the former entailed the latter.[3] In January 1903, *Popular Science Monthly* alerted a wide audience to the excitement over Mendelism with a piece by William Spillman on "Mendel's Law."[4] Meanwhile, Bateson and an ever larger Cambridge group, now including Punnett, continued to extend the experimental breeding research. New international allies emerged, among them the Danish plant physiologist Wilhelm Johannsen, discoverer of selection-resistant "pure lines" in beans, and Harvard's William Castle, who was studying coat-color inheritance in mice, guinea pigs, and other animals. Outside the breeding laboratories and the experimental gardens, agriculturally applied Mendelism proceeded apace, thanks chiefly to Rowland Biffen in Cambridge and to work encouraged through a new American organization, the American Breeders' Association, led by Willet Hays (soon to be assistant secretary of agriculture).[5] In January 1904, Charles Davenport, a founding member of the ABA, published two guardedly Mendelian papers in *Science* and was appointed director of the new Carnegie-funded Cold Spring Harbor Station for Experimental Evolution, where he soon presided over a more generously resourced version of Bateson's Cambridge breeding program.[6] At that summer's BAAS meeting in Cambridge, the Mendelian work showcased included waltzing-mouse experiments begun under Weldon's supervision by a former student, Arthur Darbishire.

This chapter charts the shifts in Weldon's understanding, as these changes unfolded, of what he was up against and how to confront it. "It is easy to say Mendelism does not happen," he remarked to Pearson in March 1903, during a wrangle with Bateson over the interpretation of Darbishire's early data. "But what the deuce does happen is harder every day!"[7] As we will see, this episode disabused Weldon of faith in problematic data as argument winners in their own right. With Mendelism now powerful enough to absorb any amount of discrepancy between its predictions and the world, what was needed was an alternative theory of heredity, founded on distinctive baseline principles and tied to distinctive methods of inquiry. From the end of 1903, in letters to Galton and Pearson, Weldon started hammering out the beginnings of such a theory. At its center was an old but recently refreshed theme: The developed form of an inherited character can be highly variable because development is always and fundamentally dependent on environmental conditions. A theory of heredity had to treat that dependency not as a complication, to be invoked to explain departures from a general pattern, but as the general pattern itself.

Nature, nurture, and their interactions were just then coming to a new public prominence in Britain, thanks, belatedly, to the Boer War. In summer 1903, concerns that an alarming proportion of would-be recruits had needed to be rejected on grounds of physical inadequacy led to the launching of a cross-departmental government committee to investigate national "deterioration."[8] In a newspaper article welcoming the committee's work, Galton offered his own state-of-the-nation reflections. It was, in his view, high time for an appraisal of British hereditary stock carried out in the critical spirit with which "an authority of the Royal Agricultural Society might criticise the stock of his neighbour over the hedge." Galton was upbeat about the findings:

> The imaginary critic above mentioned could emphatically affirm with justice that the whole of a race which was able to furnish the large supply that is produced in Great Britain of men who are sound in body, capable in mind, energetic and of high character, has the capacity (speaking as a rearer of stock) of being raised to at least that same high level. How to do this is a question of both Nature and Nurture.

On the side of nature, he reiterated the recommendations he had made in his Huxley Lecture about adjusting existing laws and sentiments to maximize breeding from the best while undertaking the further anthropological studies needed to ensure that those efforts were not misdirected. On the side of nurture, he stressed the importance of finding out at what age interventions from the state would be most effective. As ever, the Boer War provided a fine illustration: "The immense improvement in the physique of previously ill-fed recruits, after a year's good feeding with an out-of-doors life, is well known, but much irremediable mischief may have been done before that age is reached."[9]

In May 1904, at a meeting of the newly formed Sociological Society, with Pearson in the chair, Galton made this pitch yet again, in a short address titled "Eugenics: Its Definition, Scope and Aims." Weldon was there, having agreed to serve as a commentator, not out of enthusiasm for eugenics—which he continued to see as premature—but for the chance afforded to criticize another commentator, Bateson. "You know my own feeling is that these things should not be put too quickly into the hands of people such as those of the Sociological society," Weldon wrote to Pearson beforehand. "However, I hope my share may only be to revile the Mendelian, which I shall be delighted to do."[10]

1

When Weldon confessed guiltily to Pearson that he wanted "the thing to be proved nonsense," he was writing about discovering, on a patch of ground thick with *Lychnis*, just a single plant with hairless or "glabrous" leaves. Surely, Weldon reasoned, on the Mendelian view that glabrousness is recessive, one should find that, where *any* glabrous plants are growing, many are growing, since self-fertilizing recessives breed true? But then again, maybe the Mendelian would just say that the lone individual Weldon had plucked was a freshly sprung-up "mutation," whose line of pure descendants had now been cut off prematurely.[11]

Weldon's article on *Lychnis* appeared in December 1902 in *Biometrika* (the November issue: it always ran late) under the title "On the Ambiguity of Mendel's Categories." Here Weldon did for "glabrousness" and "hairiness" in *Lychnis*—the star exhibit in Bateson and Saunders's *Report 1*—what he had previously done for seed characters in *Pisum*: he showed by example how much less patly Mendelian everything looks when one inquires into the backstories of the stocks being crossed and describes the character versions under study without the Mendelian binaries.

But first, and more innovatively, came a preface on why there was room for doubt about whether Mendel himself had ever endorsed the supposedly essential doctrine of the purity of the gametes, much less the ancestry-annulling consequences claimed for it. Weldon pointed out, correctly, that Mendel had never asserted that, for instance, all of the green-seeded pea plants among his hybrid offspring had seeds of *exactly* the same shade of green as their respective green-seeded grandparents. He had reported merely that the grandparents were "green" and that roughly a quarter of their grandchildren were "green." Furthermore, when it came to the hybrid-generated gametes whose unions brought those grandchildren into existence, Mendel had hypothesized that each gamete corresponds in its inner constitution to one or the other of the constant forms among the grandchildren, not to the grandparental forms. So Mendel's own statement of his theory left open the possibility not only that true-breeding grandchildren could show differences from their true-breeding grandparents, but that these differences would be built into the hybrid's gametes. Only Bateson, as self-appointed spokesman for Mendel, but with no sanction from Mendel's own words, asserted hybrid-overleaping identity between the forms of the grandparents and grandchildren.[12]

So Mendel was potentially off the hook when it came to that vitiating Mendelian indifference to differences in ancestry as accounting for differences in character. But he remained very much to blame for the binary categorizations that made that indifference such a natural attitude for anyone following Mendel's methods. "The confusion between resemblance to a race and resemblance to an individual involved in Mr Bateson's treatment of Mendel's work," Weldon summarized, "is one of the many unfortunate results which follow when Mendel's system of dividing a set of variable characters into two categories, and of using those categories as statistical units, is carried too far."[13]

In the remainder of the article, Weldon drove these points home via his most recent enthusiasm: the spectrum-spanning hairiness of *Lychnis*, past and present. Turning first to the past, he noted that, against Mendelian expectations about how a recessive variety comes into being (perfectly and completely in one generation, with no selection required, just the matching of like gametes), the first glabrous *Lychnis* emerged, on its originator de Vries's own account, only thanks to repeated selections—and even then, years later, traces of the old hairiness could still be seen.

Pre-1900, de Vries had reported a couple of other results that, Weldon noted, sat uncomfortably with post-1900 Mendelian thinking. One concerned a cross de Vries had made between his glabrous *Lychnis* race and a hairy but non-*Lychnis* race. The offspring were hairy, as expected. But the hairs were distinctively *Lychnis*-y: a striking instance of ancestral reversion. The other result concerned glabrous × hairy *Lychnis* crosses that de Vries went on to cite as Mendel-supporting because exhibiting uniform hairiness in the hybrids, followed by 3:1 hairiness-to-glabrousness among their offspring. Back before de Vries knew what ratio Mendelian theory predicted among the hybrid offspring, however, his numbers looked considerably closer to 2:1.[14]

The Mendelism-undermining lessons Weldon drew from his reading of the literature on *Lychnis* were corroborated by the hair-counting research he now reviewed. As with seed color and shape in peas, so, Weldon found, with density of hairs on a patch of leaf on a *Lychnis*: "It is possible to pass by a series of small steps from the glabrous condition through individuals with various numbers of hairs per square centimetre of leaf-surface, up to a condition of very great hairiness." Given that enormous range, it was not enough, for purposes of establishing the dominance of "hairiness," simply to check, as Bateson and Saunders had done, whether hybrids were hairy or not. It was necessary to say *how hairy* they were—and more generally, to dispense with Mendel's ambiguous categories.[15] *Lychnis* crosses using quantitative descriptions of hairiness would undoubtedly reveal greater variability than would be picked up otherwise, and in so doing would prompt greater attention to ancestral backgrounds as a potential source of that variability.

Between the awkward new data it reported, the awkward old data it extracted from de Vries's work, and the imaginative yet scholarly readings of the "Versuche" on gametic purity that it offered, "On the Ambiguity of Mendel's Categories" was more than just an updating of the earlier peas paper. But as an answer to what Bateson had wrought in *Report 1* and the *Defence*, Weldon's latest fell a long way short of what was needed—and he knew it, dismissing it in a letter to Pearson as a "foolish paper," "very inadequate," with "shamefully little work in it."[16] Unsurprisingly, Bateson was no more impressed, commenting to Hurst in late January 1903, "Weldon's *Lychnis* is of course mere dust-throwing, apparently he is not going to defend his own whackers!"[17]

2

Far more exciting to Weldon were the experimental crosses that Darbishire was making with Japanese waltzing mice, albino mice, and their descendants. Although Darbishire's Oxford college, Balliol, had kept him on after graduation, appointing him to a demonstratorship in zoology and comparative anatomy, his position at Weldon's feet in that group photograph from summer 1901 (fig. 8.1, *left*) was indicative. Under Weldon's direction, Darbishire made his initial crosses at the end of the following summer, after Weldon judged that his lab's stock of waltzers had been true-breeding for long enough to count as pure. For a stock of albino purebreds, Weldon turned to a commercial dealer, who, he was intrigued to note, was also selling mice that, though born of albino parents and showing the customary white fur, had dark rather than pink eyes. With Cuénot and now Bateson claiming albinism in mice as a Mendelian recessive, Weldon was intrigued. "But the 'recessive' albino should never, on Mendel's view, throw any but albino young!" he pointed out to Pearson. "This is rather a nasty one for M. Cuénot, if it can be verified?"[18]

Initially, coat color, not eye color, was Darbishire's main concern. Weldon's waltzers had piebald coats of blotchy light brown ("fawn") against a white background. The Mendelian literature predicted that, barring impurity in the stocks, the crossbred mice would have coats of uniform gray, à la wild mice—an instance once again, for Bateson, of a heterozygote character spuriously resembling something bygone. By late September, however, Darbishire's crosses had yielded twenty-seven offspring, and "all," Weldon wrote to Pearson, showed "a demonstrable patchwork of white and something wild-colour."[19] Far from the albino parent's whiteness receding into invisibility behind uniform wild-mouse grayness, the whiteness always persisted, and in varying amounts, in the crossbreds' coats, which were usually mostly gray but occasionally mostly fawn (fig. 8.1, *right*). In late October, Leonard Doncaster, Edgar Schuster's friend from Naples, visited from Batesonian Cambridge. (Schuster was also doing mouse crossings.) "We showed him the mice," Weldon reported to Pearson about Doncaster, "and he became sad."[20]

In addition to the purebred albinos, Darbishire was making crosses with another, homegrown stock of albinos, descended from non-waltzer piebald mice that Weldon kept in the lab as material for embryological studies. If Mendelism taught anything, Weldon reckoned, it taught that the ancestry of a recessive should make no difference to the character of its progeny. From whatever generation in a lineage one plucked a wrinkled pea or an albino mouse to use in a cross, the range of characters in the offspring from that cross should be the same, since recessive character versions can be manifest only where there is gametic purity for them, all traces of everything else having been extruded. Furthermore, purity was all or nothing: the Mendelian scheme furnished no grounds for expecting that, other things being equal, different recessive-exhibiting

FIG. 8.1. Arthur Darbishire (*above*) and drawings of his mice (*right*), showing the four coat-color patterns he found among the crossbred mice in his first experiments, autumn 1902.

individuals might be more or less powerful transmitters of recessiveness. Yet when Darbishire crossed Weldon's waltzers with the piebald-descended albinos, the offspring tended to have *more* white in their coats than when purebred albinos were used. It was as if there was some kind of trade-off between an albino's power to transmit albinism and the degree of inbreeding in the race whence the albino derived. At a minimum, what Darbishire had shown, Weldon explained to Pearson in December 1902, was that "the colours of the first generation [the crossbreds] depend very considerably upon the pedigrees of the recessive white forms used in making the crosses." He went on, "I think the whole thing shows so far only that ancestry matters, and that no Mendelian hypothesis can be considered."[21]

A brief note from Darbishire describing this first stage of the research came out that month in *Biometrika*, in the same issue carrying Weldon's *Lychnis* article.[22] A more substantial follow-up appeared in the next issue, published in March 1903. Now Darbishire had data on over two hundred crossbred mice as well as on *their* offspring, from three sorts of matings: crossbreds with crossbreds; crossbreds with purebred albinos; and crossbreds with piebald-descended albinos. Overall, the earlier conclusions about coat color were corroborated. Ancestry continued to matter, in that the more complex the ancestries of the individuals used in a cross, the more heterogeneous were the characters of their offspring. Thus, in Darbishire's new crosses, the most heterogeneous offspring came when both parents were crossbreds, and the least heterogeneous from crosses in which one parent was a purebred albino. As with hybrid peas, so with hybrid mice: variability waxed and waned depending on the pedigrees of the stocks used.[23]

Another parallel with the peas case was that the ambiguous categories that Mendelians favored tended to obscure this waxing and waning.

If, for example, the description of coat color in mice was restricted to a choice between "albino" and "not-albino," then, yes, offspring of waltzer × albino crosses had "not-albino" coats only, while the offspring of those offspring had both kinds of coats, roughly in the proportion 1 "albino" to 3 "not-albino." But that Mendelian result stood only because "not-albino" here embraced six different patterns and six different colors, including a pale blue-gray one that mouse fanciers called "lilac," seen only in the coats of the offspring of the offspring.[24]

What held for coat-color heterogeneity as the upshot of complexity in ancestral backgrounds seemed to hold even more dramatically for a new character Darbishire was tracking: eye color. Waltzers had pink eyes, and so did both sets of albino mice. The offspring of waltzer × albino crosses uniformly had dark eyes: not what elementary Mendelism predicted, but not fatal to it either, nor an especially obvious "win" for a theory of heredity rooted in Galton's rather than Mendel's law. More telling, on Darbishire's Weldonian analysis, was what happened next to the character versions newly conjoined in the crossbreds, darkness of eye and coloring of coat. When these crossbreds were mated with albinos—and the complexity of ancestral backgrounds thus minimized—darkness of eye and coloring of coat stayed conjoined. In other words, when the offspring from these crosses were dark-eyed (and about half were), they were a perfect bet for having non-albino coats, and vice versa. However, when two crossbreds were mated together—and the complexity of ancestral backgrounds thus maximized—this new conjunction fragmented, resulting in a sizable proportion of non-albino offspring with pink eyes. Among this generation, although a dark-eyed individual was still a perfect bet to have coloring of coat, there was only a 2 in 3 chance of the reverse being true. Here was evidence of ancestry effects that were invisible as well as unintelligible under Mendelism, with its indifference to the complexity of pedigrees. "The behaviour of eye-colour," the article concluded, "is thus in every respect discordant with Mendel's results."[25]

It was a confidently contrarian conclusion. Even so, over the previous months, Weldon's letters to Pearson had shown a growing awareness that, given Mendelism's remarkable elasticity, taking it to the breaking point with failed predictions was not going to be easy. For all that Weldon was inclined to invoke ancestry to explain the variability in the mouse progeny, "such a view," he wrote to Pearson, "would be very difficult to test, and the upholder of 'compound allelomorphs' will be rather difficult to confute, if it is true!"[26] The willy-nilly postulation of compound allelomorphs in the face of messy data was just one of the devices by which, as Weldon put it in a letter on Darbishire's eye-color results, "these people wriggle out of every test." There were, in addition, those ambiguous binaries, and "the way in which the adoption of a plausible category—albino and not albino brings about an appearance of Mendelian segregation in the 2nd generation."[27] Even among the likes of Correns, there was also an exasperating willingness to excuse any and all exceptions, on the view that Mendel's laws, in some foundational sense, had now been proved valid, and so the only legitimate question was how best to reconcile apparent exceptions with the laws. In that spirit, Correns had recently raised the possibility that aberrant frequencies among hybrid forms might be due to selection operating on the gametes. "Really the thing is becoming too comic," Weldon groused to Pearson. "With that and 'irregular or imperfect dominance' one can get a good way towards Mendel, with any given data whatever?"[28]

3

Weldon was shrewd about the flexibility of Mendelian reasoning. But Bateson was a shrewdly flexible Mendelian reasoner. He set out his ingenious solution to the puzzle of Darbishire's eye-color data in a letter to *Nature* in its 19 March 1903 issue. According to Bateson, their Mendelian structure became clear with just two suppositions. First, eye color and coat color had to be treated as constituting a single unit character. The version of this two-for-one character in the albinos comprised pinkness of eye and an albino coat. Bateson labeled this allelomorph *G*. Its antagonistic partner, *G′*, comprised pinkness of eye and a non-albino coat, as in the waltzers. Second, when the unlike gametes *G* and *G′* united upon crossing, the character of the heterozygote, *GG′*, had to be acknowledged as a distinctive third version of the character, comprising darkness of eye and a non-albino coat. As with the Andalusian fowl, so with waltzer-albino mice: dominance and recessiveness did not occur; just purity of the gametes.

From here, Bateson showed, elementary Mendelism applied straightforwardly. According to the "purity of the gametes" doctrine, when a *GG′* mouse forms gametes, half of them will be pure *G* gametes (pinkness of eye plus an albino coat), and the other half pure *G′* gametes (pinkness of eye plus a non-albino coat). Now consider what happens when a mouse of this sort mates with a *GG* mouse, that is, an albino. Half of the offspring will develop from the union of *G* and *G* gametes, and so will be pink-eyed and albino-coated; while the other half will develop from the union of *G′* and *G* gametes, and so, like the similarly constituted crossbred parent, will be dark-eyed and non-albino-coated. Checking Darbishire's numbers, Bateson found that indeed, out of 205 offspring, 111 were pink-eyed albinos and 94 dark-eyed non-albinos: tantalizingly close to the predicted

1:1 ratio. As for the other kind of crossing, of *GG′* mice with each other, these unite one set of *G* and *G′* gametes with another set of *G* and *G′* gametes. According to the familiar Mendelian combinatorial math, the expected ratio among the offspring is 1 *GG* (pink-eyed, albino) mouse to 2 *GG′* (dark-eyed, non-albino) mice to 1 *G′G′* (pink-eyed, non-albino) mouse. Again, the actual numbers were tolerably close: out of 66 mice, there were 13 pink-eyed albinos, 36 dark-eyed non-albinos, and 17 pink-eyed non-albinos.[29]

Weldon had not seen that coming. But he immediately spotted what he was sure was a problem. As Weldon explained to Pearson, Bateson, with his usual rhetorical savvy, had postponed discussion of the other great difficulty in Darbishire's data, the heterogeneity of the coat colors. And when that discussion came, Bateson was bound to suggest that coat color in waltzers is a compound character, to be resolved into more basic components, as per Mendel on the purply spectrum of flower colors in *Phaseolus*. But upon any such analysis, according to Weldon's understanding, the proportion of albinos among the offspring of the hybrids should be much smaller than a quarter. The Mendelian could not have it both ways, with coat color regarded both as part of a simple character (which gave the right number of albinos but left coat-color heterogeneity unexplained) and as a compound character made up of two or more component characters (which allowed for the right degree of variability in the coat colors, but at the cost of predicting far too few albinos).[30]

Another bout of letters to *Nature* ensued, in the course of which Bateson indeed sketched out a compound-character explanation for coat color, but without admitting the looming contradiction. The exchange ended, inconclusively, in mid-May, when the editor called a halt, giving the last word, however, to Weldon, who complained that the ceaseless ad hoc modifications of Mendel's ideas meant that the name "Mendel" had been "made to shelter almost any hypothesis," and so comprehensively that "almost any experimental test is evaded."[31]

In the meantime, Weldon had started writing an article for *Biometrika* itemizing the modifications introduced so far, including the insistence that constant characters in the grandchildren and the grandparents are identical, and the use of that asserted identity as evidence for "the great fact of gametic purity" (Bateson's phrase). There was, Weldon insisted, no warrant in the "Versuche" for these and other proposals whereby the Mendelians deflected attention from ancestry. Nevertheless, that deflection ultimately depended on the Mendelians' keeping faith with Mendel's example in the use of categories "so vaguely defined," wrote Weldon, "that they convey wholly inadequate information; and with a little skill such categories may be found to fit almost any series of results." Weldon illustrated the point with a discussion of Darbishire's latest mouse-breeding experiments, now extended to a further generation. When Darbishire bred the albino grandchildren from his initial crosses (the "extracted recessives," in Mendelian parlance) with standard purebred waltzers, the offspring had non-albino coats. So far, so Mendelian. But, Weldon went on, by exchanging "non-albino" as a category for a more exact description of the colors in the coats of these offspring, and by expanding the frame of observation to include the parents of the albinos, a strikingly non-Mendelian pattern could be seen within and beyond the Mendelian one. It turned out, for example, that in cases where the parents of an albino had some yellowness in their coats, the coats of that albino's offspring were twice as likely to have yellowness too. Similarly, wild-grayness in the coats of an albino's parents nearly guaranteed it in the albino's offspring. Such evidence, sifted through the more refined categories Darbishire applied to it, suggested that his extracted-recessive albinos were not, after all, identical in their hereditary constitutions to the albinos used in the originating crosses, but differed from them in directions set by the hereditary constitutions of more immediate ancestors. Here was predictive accuracy undreamt of in "Mr. Bateson's Revisions of Mendel's Theory of Heredity" (to quote the eventual title of Weldon's article) because that theory ruled out the possibility in advance.[32]

Would these latest difficulties prove any more damaging to Bateson's cause than the previous ones Weldon had flung at it? Hope sprang eternal. "There is no plausibly Mendelian way round it!" he wrote to Pearson in mid-April about Darbishire's new color-correlation observations. Weldon conceded that, in the generation from which Darbishire got his extracted-recessive albinos, Mendelian expectations had been met in an approximate way, in that "there seems always to be a segregation into albino and not albino, which is nearly in a Mendelian proportion." But "that is the only Mendelian thing." What needed publicizing was the not-even-remotely-Mendelian thing: the apparent transmission by the albinos of the distinctive coat color in the generation before them.[33] A subsequent letter to Pearson spelled out its significance more boldly:

> At present, the purity of "extracted" albinos, and the correlation in coat colour between their parents and the young they produce when crossed with pure waltzers [i.e., with purebred, color-coated mice], seems to me the most easily settled and the most conclusive point against the Mendelian. The one fixed point in the whole shifting quicksand of hypotheses is the purity of recessives, whenever they appear.[34]

A few weeks later, Weldon was looking forward to a visit from Galton. By way of a preview of what he would see when he inspected Darbishire's mice, Weldon summarized the new work, in passing noting how a seemingly Mendelian result of Darbishire's—that when albinos were mated together, their offspring so far were uniformly albino (and thus not so nasty for M. Cuénot)—now took on a Mendelism-undermining meaning:

> These facts seem to me fatal to any theory of "gametic purity." The fact that albinos of coloured ancestry, although they produce albino young when paired together, yet show traces of coloured ancestors when paired with a coloured mouse.—this kind of fact [viz., an ancestral character returning when conditions permit] is surely direct disproof of the purity of albino gametes and is exactly the phenomenon Darwin called atavism?[35]

Weldon wrote that on 22 May. Four days later, he sat fuming as he listened to Bateson at the Zoological Society presenting a paper titled "The Present State of Knowledge of Colour-Heredity in Mice and Rats." Beforehand, Weldon had expected to flay the opposition. "We will all take our largest and sharpest carving knives and have a row," he wrote to Pearson in anticipation. "Get some chain mail and a long sword."[36] But though Darbishire made it to the meeting, squeezing in the rail journeys to and from London to meet his new responsibilities as a demonstrator at Owens College in Manchester, Pearson did not. Worse, weak chairing, coupled with Bateson's volubility—he provided a detailed *tour-d'horizon* of all the published work on the subject plus unpublished work from his Cambridge group, where his wife's sister Florence Durham was conducting microscope studies of the pigments making up mouse coat colors—meant that neither Weldon nor Darbishire nor anyone but Bateson got much time to speak.[37]

Afterward, Weldon rued what little he had managed to say as woefully ineffective, writing to Pearson, "I am ashamed to have made such an ass of myself last night, making no attack worth any thing at all. I was positively afraid. I do not see any weapon which is effective against a man who can talk words for an hour and a half as Bateson did, with such a magnificent air of transmitting a special revelation."[38] Other friends were reassuring, as friends will be, about Weldon's performance.[39] But Hurst, a friend of Bateson's, was disappointed by Weldon in the flesh, finding him not only less impressive than expected but also, with the concessions he was prepared to make on the inheritance of albinism, more Mendelian than expected—hence the crack to Bateson, quoted at the start of this chapter, about Weldon's gradual evolution.[40]

To read the published version of Bateson's paper is to experience something of what Weldon found so unnerving. Nowhere is there any hint that anything discovered so far, whether in Weldon's lab or anywhere else, was threatening to the Mendelian program in the slightest. In generating that impression, it helped that Bateson had quietly dropped his $GG + GG' + G'G'$ formula for the inheritance of eye color and coat color. But what helped even more was Bateson's calm insistence, as if he were reporting a witnessed event, that the program's validity was no longer up for grabs:

> Though for convenience we may still speak of inheritance as being "Mendelian" or "non-Mendelian," we are rapidly passing out of the initial phase of the inquiry in which such expressions are demanded. In our further investigations we are concerned not so much with the question of the applicability of the simplest Mendelian hypothesis to special cases, as with the formulation of the specific laws followed by the several characters of various animals and plants. . . . As in chemistry, these laws must be worked out separately case by case, and each as it is determined has the value rather of fact than of hypothesis.[41]

The paper included a postscript, added in July after Bateson had seen the latest *Biometrika*. Responding to Darbishire's new correlation data took no more than the briefest rummage through the existing kit bag. The asserted identity of the originating and extracted recessives was not an assertion that they were identical *in every respect*. Indeed, those other, non-identical respects might sometimes make for differences so great that the union of an originating and an extracted recessive could be a heterozygous union, with the usual unpredictable consequences. As for the color correlations revealed in Darbishire's latest crosses, they could well turn out to be the upshot of imperfect segregation in the gametes in the races being used.[42] Whatever the ultimate explanation, there would be no budging of Mendel's principles from their position at the center of the knowledge that, thanks to that position, was now exploding.[43]

4

"Take things fighting, and be careful to hit hard every time." Between the summers of 1903 and 1904, Weldon lived up to that credo fully. On the page and in person, he remained every inch the critic-as-naysayer, concerned less with saying what the deuce does happen than with exposing overreach on the part of opponents.

As ever, *Biometrika* offered a ready platform. In a short commentary, drafted in August 1903 and published later that year, Weldon and Pearson took the fight to Wilhelm Johannsen. In support of Bateson's view of the discontinuous nature of heredity and evolution, Johannsen had announced the discovery that populations are actually aggregates of stable types ("pure lines"). A pure line could be selectively filtered out of a population but, he believed, no amount of selection could shift a pure line's mean value. That conclusion was bound to provoke Weldon and Pearson, who duly reanalyzed Johannsen's data and found, on the contrary, that the bean plants Johannsen studied had transmitted their distinctive variations with enough fidelity for selection to work its gradual, mean-shifting magic after all.[44]

The next issue of the journal, out in early 1904, carried a solo note from Weldon in belated response to the Mendelians' house physician, Archibald Garrod. According to an emerging Mendelian consensus, albinism behaved as a Mendelian recessive in humans just as much as in mice. Mendelism's newest American recruit, the Harvard zoologist William Castle, took that view.[45] Bateson thought it probable.[46] So did Garrod, who had found, in family records unearthed from an old Italian anthropological article on albinism in Sicily, a pattern of incidence uncannily like the one he had found for alkaptonuria.[47] Unsurprisingly, when Weldon consulted the same article, what he saw did not look quite so Mendelian. Yes, albino offspring had come from albino parents—but so had pigmented offspring. And yes, albinism had disappeared for a generation and then reappeared in the next one. But the proportion of albinos among the latter was not anywhere near the Mendelian quarter.[48]

Beyond *Biometrika*, London scientific life afforded many opportunities for confrontation—so many that Weldon could be discriminating. After the Zoological Society debacle in May 1903, the Royal Society approached him about setting up another Weldon-versus-Bateson "discussion" evening, along the lines of the one organized around Weldon's first report on crabs. The Royal Society rematch never happened.[49] But Weldon went along when, in February 1904, Bateson and Rowland Biffen were scheduled to give Mendelian papers at the Linnean Society, with Bateson talking about primulas and Biffen about wheat. On this occasion, victory in debate went to Weldon.[50] He left so energized that he briefly threw himself into his own investigations of primulas, spending two days at Suttons' nursery in Reading, where Bateson had got all of his Mendelism-supporting specimens, and where Weldon now found all the usual Mendelism-undermining complications. ("Suttons are good folk," he wrote to Pearson in March, "only they were so surprised, after five years of Bateson, to meet a 'man of science' who was not a Mendelian!")[51]

Bateson did not attend the meeting of the Sociological Society in May 1904, where Galton read his eugenics paper, but instead submitted a written response that, after expressing general sympathy with the eugenic project, dwelt exclusively on a throwaway remark of Galton's about how much "the knowledge of what may be termed the *actuarial* side of heredity has advanced in recent years."[52] No, wrote Bateson: recent advances owe far more to the clear and definite results of experimental breeding than to anything learned from actuarial biology.[53] In defense of the latter, Weldon raised two objections. First, as he had shown with Mendel's own data, statistical methods are indispensable for assessing whether or not a given set of experimental results are to be believed. Second, even if they are to be believed, there remains the question of what those results reveal about phenomena outside of the vastly simplified conditions of the laboratory—a question whose answer lies with the analysis of data collected outside the lab. Weldon addressed the Mendelian:

> Your laboratory experiment is purposely simplified: you deal with one set of phenomena at a time; and by that very fact, you establish a degree of unlikeness between your laboratory experiment and the infinitely more complex experiment which is being conducted all round you from generation to generation. Before you can be sure that in simplifying your laboratory conditions you have not neglected some important factor which affects the result under the complex conditions of Nature's experiments, you must view your own result in its proper relation to that which occurs under more complex conditions; you must compare the conclusions drawn from your laboratory experiment with those drawn from an actuarial study of the more complex natural experiment. If the two agree, you have realised at least as much of the truth as will suffice for a working generalisation; if they do not agree (and at present the results of Mendelian experiment have not led to a single conclusion which holds for masses of human populations), then in this case there can be no doubt whatever that for the student of human Eugenics or of organic evolution generally, the conclusions drawn from the larger mass of complex material are far more valuable than those drawn from the simpler, smaller laboratory experiment.[54]

Backstage, however, the period was one of creative ferment scientifically. In the same month, August 1903, when he was finishing the Johannsen critique, Weldon used his correspondence with Pearson to pursue a discussion about how, as Bateson had hand-wavingly suggested, a blended character such as animal stature could be Mendelized. The trick, Weldon explained, was to suppose that, at the gametic level, the number of allelomorph pairs contributing to the character is indefinitely large, and that when opposite gametic elements or "units" meet in a heterozygote, the result is intermediate. To illustrate, Weldon set *AA* equal to

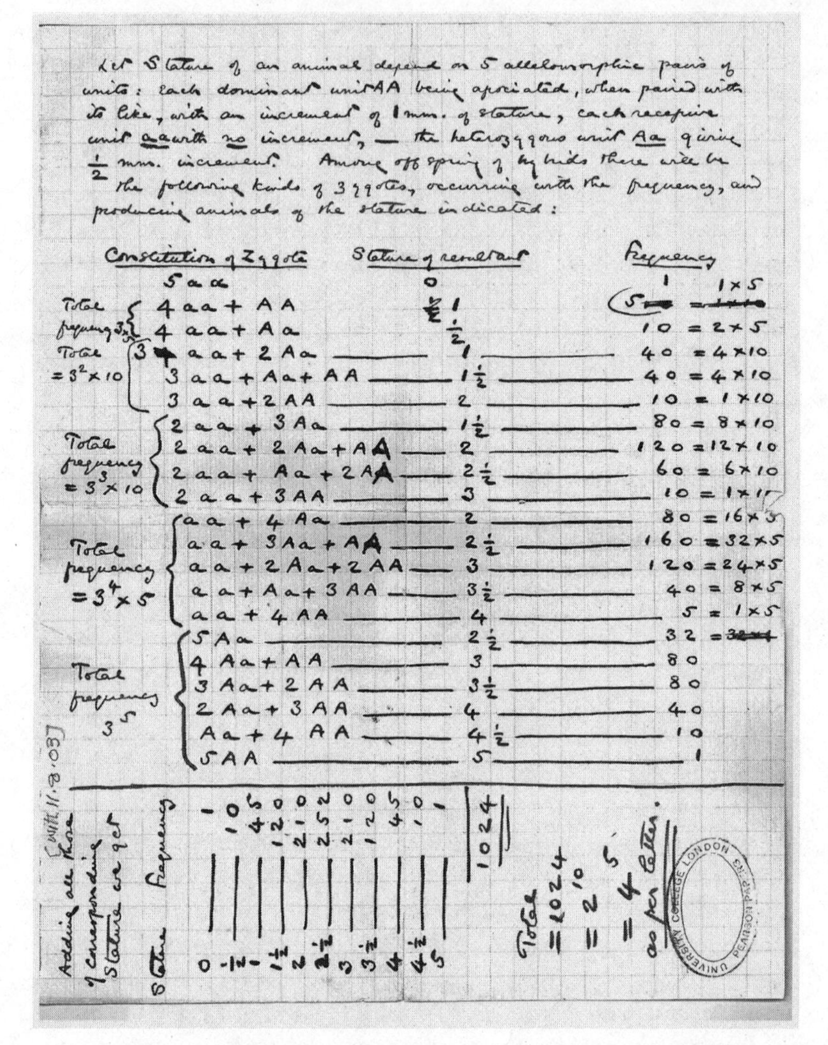
Let Stature of an animal depend on 5 allelomorphic pairs of units: each dominant unit AA being associated, when paired with its like, with an increment of 1 mm. of stature, each recessive unit aa with no increment, — the heterozygous unit Aa giving ½ mm. increment. Among offspring of hybrids there will be the following kinds of zygotes, occurring with the frequency, and producing animals of the stature indicated:

	Constitution of Zygote	Stature of resultant	Frequency
	5 aa	0	1 = 1×5
Total frequency 3×5	4 aa + AA	1	5
	4 aa + Aa	½	10 = 2×5
Total = 3²×10	3 aa + 2 Aa	1	40 = 4×10
	3 aa + Aa + AA	1½	40 = 4×10
	3 aa + 2 AA	2	10 = 1×10
Total frequency = 3³×10	2 aa + 3 Aa	1½	80 = 8×10
	2 aa + 2 Aa + AA	2	120 = 12×10
	2 aa + Aa + 2 AA	2½	60 = 6×10
	2 aa + 3 AA	3	10 = 1×10
Total frequency = 3⁴×5	aa + 4 Aa	2	80 = 16×5
	aa + 3 Aa + AA	2½	160 = 32×5
	aa + 2 Aa + 2 AA	3	120 = 24×5
	aa + Aa + 3 AA	3½	40 = 8×5
	aa + 4 AA	4	5 = 1×5
Total frequency 3⁵	5 Aa	2½	32
	4 Aa + AA	3	80
	3 Aa + 2 AA	3½	80
	2 Aa + 3 AA	4	40
	Aa + 4 AA	4½	10
	5 AA	5	1

Adding all those of corresponding Stature we get

Stature	0	½	1	1½	2	2½	3	3½	4	4½	5	
Frequency	1	10	45	120	210	252	210	120	45	10	1	1024

Total = 1024 = 2^{10} = 4^5 as per Table.

[with 11.8.03]

FIG. 8.2. "Let Stature of an animal depend on 5 allelomorphic pairs of units: each dominant unit AA being associated, when paired with its like, with an increment of 1 mm. of stature, each recessive unit aa with no increment, [and] the heterozygous unit Aa giving ½ mm. increment. Among offspring of hybrids there will be the following kinds of zygotes, occurring with the frequency, and producing animals of the stature indicated." The symmetrical distribution resulting is represented at the bottom of the page.

an increment of 1 mm in stature, *aa* equal to no increment, and *Aa* equal to an increment of ½ mm, letting the number of allelomorph pairs be the just-about-manageable number of 5. Cross a pure 5 *AA* parent (with maximal stature of 5 mm above the minimum) with a pure 5 *aa* parent (with minimal stature), and we get a generation of 5 *Aa* hybrids, with an exactly intermediate stature of 2½ mm above the minimum. Cross two hybrids, and we get, by Weldon's calculations (fig. 8.2), one 5 *aa* individual, one 5 *AA* individual, and a beautifully symmetrical bell-curve distribution in between.[55]

Where, exactly, was the error here? Why not accept that, for all the shortcomings of elementary Mendelism, a sophisticated Mendelian theory of inheritance was not only possible but, by these calculations, plausible—indeed, virtually indistinguishable from a sophisticated Galtonian theory? By summer 1903, Weldon was far from alone in bringing a new seriousness to such questions. In the monograph Weldon and Pearson were just then criticizing, Johannsen had sought not only to break up bell-curve distributions into discontinuous units, but to do so in the name of Galton, whose 1870s-vintage stirp theory Johannsen saw himself as reviving and vindicating.[56] Nearer to their immediate circle, Udny Yule in late 1902 had published, as a review of *Report 1* and the *Defence*, a closely argued manifesto for a new integration. Time was up, he concluded, on treating Mendel's laws and the law of ancestral heredity as if they were "necessarily contradictory statements, one or other of which must be mythical in character," rather than "parts of one homogeneous theory of heredity."[57] The two laws could not contradict each other, in Yule's view, because they concerned altogether different domains: in the case of Mendel's laws, hybridization *between* races; in the case of the law of ancestral heredity, heredity *within* a race. (Hence Yule's title: "Mendel's Laws and Their Probable Relations to Intra-racial Heredity.") Furthermore, in their own, duly delimited domains, both laws appeared to be true. In Yule's view, Bateson's inflated rhetoric and occasionally confused ideas should not obscure the genuine potential of Mendelian phenomena and the linked gametic explanation to throw light on the fundamental nature of heredity and related problems.[58] As for ancestral heredity, Yule drew a distinction between the basic form of the law and, as he saw it, the unconvincing attempts by Galton and Pearson to formulate it with numbers that hold across the board. According to Yule, the law of ancestral heredity, pared to its essence, was an equation stating that the average value of a given character in the offspring is equal to the average value of that character in the race plus—in line with regression—some fraction of the value of that character in the parent (since, as ever, abnormality, on average, decreases). In other words, one could predict the value of the character in the offspring more accurately if one knew not just about the race but about the particular lineage.[59]

Empirically, that approach seemed to work in every instance examined so far, Yule reported, both for continuously varying characters such as stature and for either/or characters such as normality/insanity (Yule's examples). And it generalized: knowing the values of the character in the grandparent, great-grandparent, and so on increased the accuracy of prediction, although the increase diminished the further back one went. Indeed, if it were otherwise, Yule noted, it would be impossible to make sense of the fact that breeders take such care to know the pedigrees of the animals they breed from. As a final point of clarification on ancestral heredity, Yule urged wariness of what he considered unhelpfully loose talk of ancestors "contributing to" (Galton) or "affecting" (Weldon) the character of descendants, as if there were some kind of transgenerational physical influence. To understand the biological underpinnings of the law of ancestral heredity, it was enough, Yule counseled, to allow, in the already conventional way, that the germ line is separate from the rest of the body (the "soma"), and that how a character appears somatically is the upshot not only of what is in the germ line, but of lots of other variables impinging during growth.[60]

With these preliminaries out of the way, Yule took up the challenge he had set himself: how to bring together the two domains and their respective laws. A good way in, he reckoned, was to ask what would happen if two either/or Mendelian races, dominant *A* and recessive *a*, freely interbred "*as if they were one race*." Tracking the fate of the dominant attribute first, he showed that it behaved just as the law of ancestral heredity predicted, in that the more *A*s in a lineage, the greater the chance of *A* in the next generation, with diminishing increases in the accuracy of the prediction the further back in the lineage one looked. Yule was triumphant: "Mendel's Laws, so far from being in any way inconsistent with the Law of Ancestral Heredity, lead then directly to a special case of that law, for the *dominant* attribute at least."[61] The recessive attribute, alas, was more complicated to deal with, but instructively so. To get from Mendel to Galton for the recessive, it was necessary, Yule judged, to relax one or the other, or perhaps both, of two assumptions of elementary Mendelism. The first was that dominance always holds: something Bateson was of course already prepared to jettison. The second was that fully developed attributes are, as Yule put it, so "rigidly *predetermined*" that they never deviate from their gametically assigned natures, and certainly not so much that a pure *A* individual might actually end up, due to the influence of circumstances, classified as an *a* individual and vice versa.[62]

Here Yule was picking up on an aspect of Mendelism about which

Bateson was more or less silent. But though it may not have been Mendelism, it was hardly news, in Yule's view, that sometimes circumstances can make a huge difference to whether and how an inherited tendency becomes manifest. The example he gave was of hereditary illness: "The chances of an individual dying of phthisis [consumption/tuberculosis] depends not only on the phthisical character of his ancestry, but also very largely on his habits, nurture, and occupation."[63]

So, for either/or characters, restricted versions of the law of ancestral heredity could be shown to follow from Mendel's laws, at least when charitably glossed. Yule guessed that the same could be shown for continuously varying characters, either because the predetermining gametes sometimes vary continuously or—in line with Bateson's preferred view—because such characters are the combined result of large numbers of discontinuous germ-cell elements.[64]

Lucid, insightful, and open-minded, Yule's two-part article, published in a new journal, *The New Phytologist*, was sure to antagonize everyone, and so it did. Robert Lock wrote to Bateson from Ceylon, "I had read Yule in New Phy. & thought it a villainous case of special pleading."[65] Castle judged it a merely puzzling case of flawed reasoning.[66] Yule sent a copy to Weldon, who "hurriedly looked through it," he wrote to Pearson, only to discover that, under the Mendelians' spell, Yule seemed to have lapsed into "an un-quantitative frame of mind," hence his uncritical embrace of Mendelian categories as if they were genuinely descriptive. Up came an old Weldonian gripe about mathematicians dealing with biology: "I think he always is inclined to neglect the complexity of the actual data, and to accept anything which can be easily turned into symbols."[67]

Pearson, though likewise inclined when it came to data and symbols, was also not impressed. He hastily added a note on Yule's paper at the end of a long article in press for *Biometrika* on the law of ancestral heredity, taking into account the latest data as well as responses to Bateson's recent attacks. In the note, Pearson signaled broad agreement about there being no incompatibility with Mendel's laws, though he doubted they held as securely as Yule thought—and, unsurprisingly, he judged Yule's reports of difficulties with the Pearsonian formulation of the ancestral law to be much exaggerated.[68] Over the next six months, Pearson basically rewrote Yule's paper, putting a distinctive spin on more or less the same derivations and conclusions—yet, spitefully, with no mention of Yule. "On a Generalised Theory of Alternative Inheritance, with Special Reference to Mendel's Laws" was submitted to the Royal Society in September 1903 and read out at a meeting in November. Instead of *A* and *a* elements, Pearson introduced new terminology: "protogenes" and "allogenes," respectively. But the irenic upshot could have come from Yule: "It is of interest to find 'Mendelian Principles' when given a wide analytical expression leading up to the very laws of linear regression, of distribution of frequency, and of ancestral inheritance in populations, which have been called into question as exhibiting only a blurred and confused picture of what actually takes place." For Pearson, although the theory of the pure gamete could not be the whole story about heredity, it could well be a part—and the biometrician stood at the ready to help with further testing.[69]

5

Ignore environmental conditions as a source of variability and the problem of analyzing inheritance becomes much simpler. That was Pearson's view, in autumn 1903 as much as ever. It was implicit in his approach to reconciling Mendelism and Galton's law, in which he treated the mathematized play of protogenes and allogenes as the exclusive item of business. And it was explicit in his Huxley Lecture, delivered to the Anthropological Institute that October on the subject of physical and mental inheritance in humans. To be sure, Pearson allowed, "size of head and size of body are influenced by nurture, food, and exercise." Furthermore, nothing could be less surprising than to find—as he had—that brothers and sisters, sharing the same home environment, resemble each other in these measures. But they had also turned out to resemble each other to almost exactly the same degree in cephalic index (the ratio of maximum head breadth to width), hair color, and eye color: physical characters far less plausibly under the influence of environmental shaping, as Pearson saw it. That same degree of resemblance also held, he went on, for a range of mental characters, from "vivacity" and "assertiveness" to "ability" and "handwriting." For Pearson, it followed that if the influence of environment could be discounted for some measures in this uniform set, then it could be discounted for all of them. In conclusion, he spelled out the predictable larger lessons, cosmic (humans do not stand apart from the rest of nature) and political (education cannot compensate for inferior brains, so Britain needed to alter the relative fertility of its best and worst stocks to succeed in the Darwinian struggle of nations). But there was, in addition, a methodological lesson for students of the science of heredity. When determining fraternal resemblance, Pearson declared, "environmental influence . . . is not to the first approximation a great disturbing factor."[70]

Neither Weldon (who attended) nor Galton saw it that way, as letters they exchanged within weeks of Pearson's lecture make vivid.[71] The correspondence dealt not with the lecture but with Weldon's latest research.

He was back to his snails, seeking to understand how individual development, environmental change, and natural selection go together: his enduring preoccupation as a biologist, and the oldest themes in his discussions with Galton. Even more fully than before, Weldon now saw snails as eminently suited to the problem of discriminating the adaptive effects of selective elimination from the adaptive effects of developmental convergence. Not only did their shells record their ages unambiguously but, as he had come to appreciate, their morphologies overall were strikingly, anomalously unresponsive to changed environments. That meant that if, as he hoped, he found that in his recently collected Italian *Clausilia*, as in his earlier German and British samples, variation narrowed as juvenile populations matured into adulthood, he could be even more confident in identifying natural selection as the cause.[72] By way of illustrating the promise of his material, he sent a box of snail shells to Galton, who wrote back to say that he had gone through them three times with "immense interest," and was now full of questions about selection and adaptation. Among these questions was whether existing data threw any light on, as Galton put it, "the possibility of acquired faculties having become hereditary after many generations." He went on, "We all agree that it is insensible in one generation, but do not possess evidence of its possible cumulative power after many generations."[73]

Weldon's answer was as refreshingly undogmatic as Galton's query. Weldon had two recommendations. The first was to get hold of some rabbits from the Portuguese island of Porto Santo, as Weldon himself had tried and failed to do when he was on Madeira at the height of the Boer War. Readers of Darwin's *Variation* encounter these rabbits as a prime example of the sometimes remarkable (and utterly unsnailish) plasticity of animal organization under changed conditions. Sprung from common rabbits set loose on the island in the fifteenth century, the Porto Santo rabbits had come to look and behave so differently from common rabbits that a naturalist innocent of the backstory would, Darwin was sure, rank them as separate species. And the Porto Santo rabbits retained their potential for rapid transformation. In just four years, as Darwin himself had observed, a pair of Porto Santo rabbits that had arrived in England with all the distinctively red fur of their kind had reverted to the black-tipped grayness of the common rabbit.[74]

The second recommendation was to read a 1901 book that Weldon had cited in his critique of de Vries: *Formative Reize in der tierischen Ontogenese* (*Formative Stimuli in Animal Ontogenesis*) by Curt Herbst, one of the German founders of the new experimental embryology. (About Herbst, Weldon wrote to Galton, "He and another heterodox friend of mine Hans

Driesch run a journal as wicked in its way as *Biometrika*. They call it *Archiv für Entwicklungsmechanik*, which is not so pretty as *Biometrika*?")[75] Galton struggled with German, as Weldon knew. But he reckoned it was worth Galton's effort since the book was, Weldon wrote, "a summary of a lot of the evidence that seems to me very important," above all for underscoring the inadequacy of the idea of an "acquired character":

> It seems to me that every character is at once inherited and acquired. The distinction between the two seems experimentally hopeless: and it seems that at every stage in its development an organism requires not only a certain organic structure, derived from its parents, but a certain definite environmental stimulus, which shall quicken its structural "mechanism" into activity.

Hence, for Weldon, the trouble with trusting to experiment alone as a guide to understanding the causal underpinnings of living form: it was too easy to contrive conditions in the laboratory that accidentally removed the most illuminating real-world dependencies. As an example, he mentioned the symbiotic union in lichens between algae and fungi, and the peculiar but persistent modifications wrought in the algae when, and only when, they were so closely associated with fungi. "Here is an environmental effect which has existed for generations," wrote Weldon, "and disappears within the limits of a cultural experiment." Another example was one he discussed in his de Vries critique: the way that progression through the normal embryonic stages can be suppressed by removal of a necessary environmental stimulus. "If that view of embryology should prove true,—and it looks truer every day," Weldon continued, "a prophecy I once made in lecture at Cambridge might be verified, and by removal of all the appropriate stimuli a hen's egg might be made to develop into a mass of undifferentiated cells."[76]

Weldon raised these same points the next month with Pearson, by way of challenging the anti-environment stance of his Huxley Lecture. The immediate prompt was a little note Pearson wanted to publish in *Biometrika* on homotypic similarity of leaves from the same beech tree. The note began with assurances about how any environmentally induced variability could be safely ignored. That claim, wrote Weldon, "will have to be looked at rather carefully":

> Take cephalic index. You say scornfully that environment does not affect it. Read with prayerful attention the remarks which the late Mr. Shandy senior made to Dr. Slop, and remember that all your brothers in your

> tables have passed, when still plastic, through the same pelvis. . . . I call the maternal pelvis environment, just [as] I call eggshell a part of the environment of a hen's embryo, just as I call the cell in the comb and the food supplied by workers a part of the environment of a bee's embryo,—just as I call the scarf which ties the baby's head to a board a part of the environment of a Southern baby.[77]

This was not a message calculated to appeal to Pearson, and Weldon found himself a few days later explaining that he "did not intend to be captious," nor was he suggesting that homotyposis was a bust. But he nevertheless insisted that the effects of environment had to be taken far more seriously than was Pearson's wont. For the most part, Weldon went on, an embryo was "not like a chronometer, which can either go at a certain rate or stop altogether:—it is like a watch with a really bad balance-wheel, so that it can go under a wide range of conditions, at a rate which is a very direct function of those conditions."[78]

If any of that made the slightest impression on Pearson, he kept it to himself. Certainly no one reading through to the "Miscellanea" at the end of the January 1904 *Biometrika*, where Pearson's beech-leaves note was published alongside Weldon's on albinism in Sicily, together with a further note from Pearson defending the law of ancestral heredity against the latest Mendelian criticisms (from Castle), could have known that a major disagreement had opened up between the *Biometrika* editorial chiefs.[79]

In the months that followed, Weldon's usual progress reports to Pearson came with occasional barbs. "I am at present rather more mixed than usual about your Mendelian regression surface," Weldon wrote in March 1904, referring to a diagram in Pearson's now-published Royal Society memoir (fig. 8.3).[80] Weldon was then in the thick of efforts to account mathematically for the reversion patterns in his primula data and in the ever-expanding mouse data. He was sure that the answer lay not merely in the particular gametic elements on which particular characters depended—the Mendelian and, now, Pearsonian assumption—but in the full range of gametic elements that came along with a race's ancestry and constituted a set of internal conditions. The missing formula eluded him, however.[81] In the meantime, as Weldon dispiritedly wrote to Pearson in April, his memoir's gift to the Mendelians of "refuge in your demonstration of the values which Mendelism gives to ancestral correlation" was one they would certainly take shelter in.[82]

Lest Pearson miss the links between ancestry, variability, and conditioning environments, Weldon spelled them out the next month. The

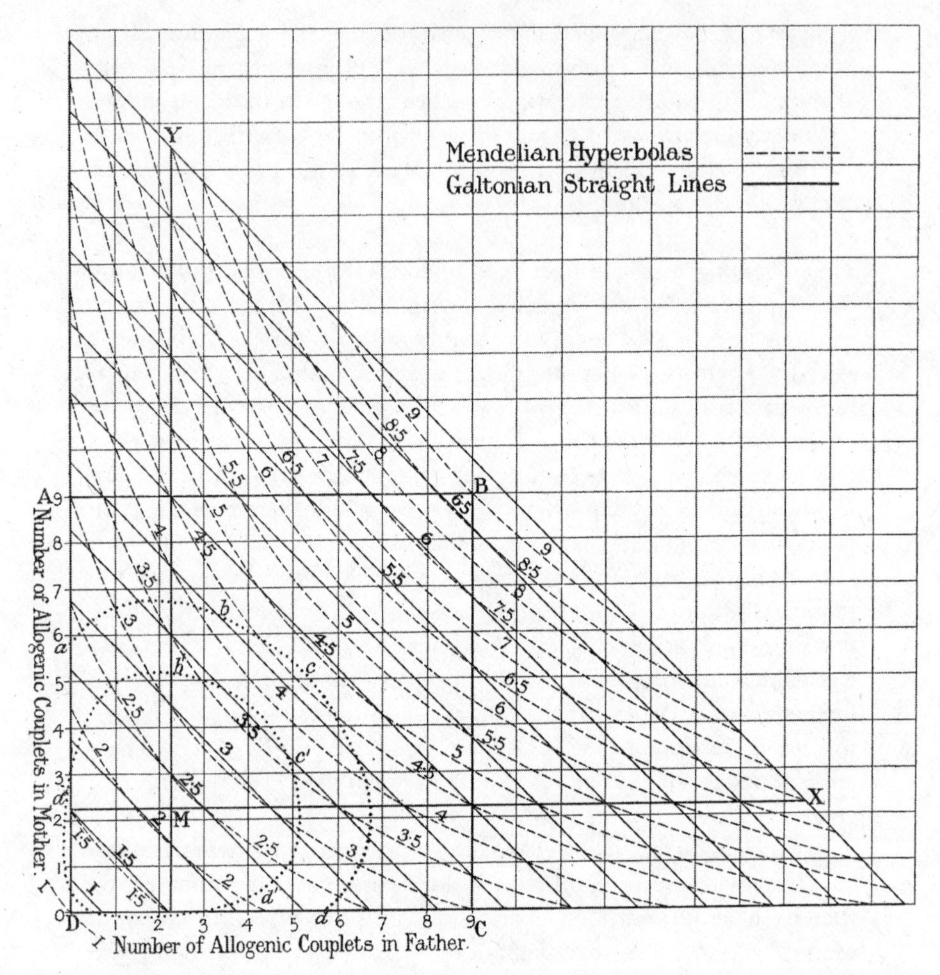

Midparental contours.

FIG. 8.3. Pearson's diagram showing how closely Mendelian and Galtonian expectations match in large populations in which mating is random and the character in question is under the influence of multiple gametic "couplets." (Here he considers nine, but in the paper he discusses up to one hundred.) Expectations diverge markedly only for the tiny fraction of cases in which nearly pure or, at the rarest extreme, pure recessive ("allogenic") parents mate.

prompt was a new paper from Leonard Doncaster. While based in Naples, conducting Mendelian crossing experiments in the home of Herbst-style *Entwicklungsmechanik*, Doncaster had found that the form of his hybrid sea urchins—the "dominant" form—depended on the temperature at which the larvae developed.[83] "I am sure you have got to pay attention to these things, sooner or later!" Weldon urged in recommending the paper, going on to explain how, in dealing with primula colors, he was attending to ancestry in just the way he attended to the influence of conditions generally:

> I am trying to see if the pedigrees give any sort of indication about the conditions which determine whether a colour shall be in patches on a white ground or scattered.—*By conditions I mean ancestry as well, of course.*—It seems to me of the first importance to get some knowledge of that. Some flowers on a hybrid plant may be red, others pink, others red and white,—[it is] as if quite local conditions, inside the organism, quite apart from things which happen at fertilisation, could make a "blend" or a patchwork. Half the whole thing seems to me in the knowledge of that, and I can't see any way to test or attack it, unless something "turns up" by looking at the pedigrees.[84]

Weldon stood his ground after the inevitable Pearsonian pushback. "I know you do not want to be worried," begins Weldon's next letter, "but I think my point about environment is worth considering." He then discussed recent work from his own lab, by Edgar Schuster, on bone sizes in mouse lineages, and an emerging suspicion about why, as Schuster was finding, mice from bigger families tend to have smaller bones. "This seems," wrote Weldon, "almost certainly an effect of nutrition,

> and I do <u>not</u> suppose it to be inherited. What I do suppose is, that the length of a mouse, or of its femur, is a function of two sets of things,—(a) something in the ovum & spermatozoon, and (b) a complex set of conditions of nutrition, etc. I suppose that the (a) things are inherited, and the (b) things not. But in such a case (b) must be a considerable factor in determining the amount of variability inside a fraternity and must therefore have a considerable effect on the fraternal homotyposis? . . . You will say, how does environment affect hair-colour, eye-colour, etc.,—and I do not know: but no character has yet been shown to be independent of nutrition; there is no character of which you can definitely say that a given embryo must either develop into exactly so and so, or die.[85]

Again, in Weldon's view, developing organisms were not tightly regulated machines, with characters that either appeared in a predetermined way or did not appear at all, independent of conditions. When conditions changed, characters could change, sometimes dramatically. Plainly, Mendel's principles could be extended to cover such condition-dependent plasticity. Mendelians could even study plasticity, as Doncaster showed.

But his research stood out precisely because it ran against the grain in Bateson's Cambridge. What would a science of heredity otherwise disposed—in which variability in inherited characters due to conditions, internal and external, was the norm rather than the exception—look like?

9

An Unfinished Manuscript

In no single case has it ever been asserted [by medical experts consulted]... that hereditary diseases affect the new born of the rich and the poor unequally.... The interpretation would seem to follow that Nature gives every generation a fresh start.

DR. ALFRED EICHOLZ, TESTIFYING BEFORE THE
INTER-DEPARTMENTAL COMMITTEE ON PHYSICAL DETERIORATION,
18 DECEMBER 1903

If, as is usual, the philanthropist is seeking for some external application by which to ameliorate the course of descent, knowledge of heredity cannot help him.... We have no experience of any means by which transmission may be made to deviate from its course; nor from the moment of fertilisation can teaching, or hygiene, or exhortation pick out the particles of evil in that zygote, or put in one particle of good.... Education, sanitation, and the rest, are but the giving or withholding of opportunity. Though in the matter of heredity every other conclusion has been questioned, I rejoice that in this we are all agreed.

WILLIAM BATESON, PRESIDENTIAL ADDRESS, SECTION D (ZOOLOGY),
BAAS, CAMBRIDGE, 18 AUGUST 1904

Some there are doubtless already who question whether the general policy pursued with regard to the lowest classes of the nation is a sound policy, who are troubled with the suspicion that hygiene and education are fleeting palliatives at best, which, in postponing, but augment the difficulties they profess to solve.... Permanent progress is a question of breeding rather than of pedagogics; a matter of gametes, not of training. As our knowledge of heredity clears and the mists of superstition are dispelled, there grows upon us with an ever increasing and relentless force the conviction that the creature is not made but born.

REGINALD PUNNETT, *MENDELISM*, 1905

Since the character of any organ depends, not only upon a specific something transmitted to it through the germ-cells out of which it was developed, that is to say upon something inherited, but also on two sets of conditions external to the organ itself, namely its relation to the parts of the body to which it belongs and its relation to the environment in which that body exists, we may say that every character of every animal is both "inherited" and "acquired."

W. F. R. WELDON, "THEORY OF INHERITANCE," MS, 1904–5

If Bateson had tried, he could hardly have invented a better—or more perfectly timed—foil for dramatizing the radical social meaning of emerging Mendelism than the report of the Inter-departmental Committee on Physical Deterioration. A vast document, published in installments between late July and mid-August 1904, it was the last great expression of Victorian meliorism. On the basis of months of evidence gathering, the committee members—most of them educationalists—reached the reassuring and widely publicized conclusion that worries about the nation's biological decline were groundless, and furthermore, that improved living conditions, kept up for a generation or two, would suffice to raise up those segments of the population most in need of it. They allowed that "hereditary taint" had to be reckoned with. But they were optimistic about the effects of more salubrious surroundings—education, sanitation, and the rest—on the growth and development of even the most hereditarily disadvantaged individuals.[1]

The report's final installment came out as Bateson took to the podium to deliver his address as president of the Zoological Section of the BAAS, meeting that August in Cambridge (fig. 9.1). After a bravura survey of the Mendelians' recent progress on heredity, he touched on the utility of their science in the making. "We are asked sometimes, Is this new knowledge any use?" Truth, whether useful or not, was of course the only fit subject for a BAAS address, Bateson insisted. But, he went on, so astonishingly useful was the new knowledge already proving, even at this early stage, that a few remarks from him would not go amiss. Above all, he said, Mendelism's usefulness was being felt in the domain of plant and animal breeding, "the greatest industry to which science has never yet been applied." At last, myth and guesswork were being expunged from the fixing of new types. As for human welfare more generally, his message was more mixed. The good news was that, should Mendel-minded breeders ever be allowed to operate on humankind, they could rapidly breed out some of our worst diseases and our most troublesome vices. Potentially, these breeders could even guide into being a higher form of humanity, nobler in intellect and morality. One piece of bad news, however, was that our species would never allow itself to be thus improved anytime

FIG. 9.1. A sketch in the *Daily Graphic* showing the Cambridge meeting of the BAAS in mid-August 1904. A portrait of Bateson as zoology section president appears in the center left of the image.

soon, due to our sentimental attachment to old ways of arranging these matters. Another was that, in light of the new understanding of heredity, time-honored social interventions now stood revealed, for the most worrisome cases, as wasted effort. To the Mendelian, the zygote was father to the man. Fate was sealed at fertilization.[2]

The next day's proceedings were what Punnett later recalled, the crowded section room becoming, in the words of the *Times*, "the arena of a sharp encounter between the Mendelians and the biometricians on the question of heredity." The Mendelians were on the program: a full day of papers from Saunders, Darbishire (now convinced that some of his mouse-crossing data corroborated Mendelian principles), Hurst, Punnett, Biffen, Lock, Doncaster, and Bateson. Two biometricians, Weldon and Pearson, were in the audience. But when Pearson spoke up, it was mainly to identify himself as the originator of the most mathematically elaborate Mendelian scheme yet devised, and to point out that Mendelian heredity so conceived looked indistinguishable from Galtonian heredity. The reported sharpness of the occasion arose from the exchanges between just one Mendelian, Bateson, and one biometrician, Weldon. At the start of the discussion period after the morning session, Weldon gave what amounted to a paper-sized dissent from Bateson's address of the day before, stressing Bateson's departures from the "Versuche," his ill-motivated indifference to the possibility of reversion to more distant ancestral characters, and the fuzzy categories that hid all the variability that might support a conclusion in favor of ancestral heredity. Hitting back in the afternoon at what were now labeled "the Ancestrians," Bateson dismissed them as comparable to the Flat Earthers who, long after the discovery of the Earth's true shape, continued to regard their theory as admirably in accord with observation. Whether the ancestrians liked it or not, gametic purity was now an incontrovertible fact, thanks to which the science of heredity was no longer an incoherent morass.[3]

Who won? Bateson thought he did (and Punnett, unsurprisingly, agreed).[4] Weldon, writing subsequently to Galton, showed no sign of having found the Mendelians any more impressive than usual.[5] At the event itself, the chair more or less declared a draw, after another participant voiced the hope that the combatants would ignore Pearson's conciliatory example and, for the good of biology, keep fighting.[6]

Only months later was Bateson announced as the victor, and in a way that made the victory decisive—or so it seemed to Weldon. The Royal Society, he learned in early November, was awarding the 1904 Darwin Medal to Bateson. "I suppose Bateson's Darwin medal will help the heresy on a bit further for a little while," was Weldon's brave-face initial remark on it to Pearson. A few days later, Weldon was more open about finding the now-public news bothersome:

> The only reason why this medal business annoys me is that it will make it still harder to interest men here [in Oxford]. Those who were at Cambridge will regard it as a sort of official answer to our little row, which they may accept, without taking any further trouble to make up their own minds.

> Some, who have seen this morning's *Times*, have been good enough to express that view to me already. Otherwise, Bateson is trying according to his lights, and he deserves the encouragement he wants as well as any one else, I suppose.[7]

The award citation, published in *Nature* a few weeks later, conveyed no sense at all that debate over Mendelism was ongoing. On the contrary, Mendelian Cambridge under Bateson's pioneering and energetic leadership was hailed as the place other biologists now looked to for advance notice of where the rest of the field was heading on heredity, variation, and evolution.[8]

Mendelism's upward glide indeed continued smoothly on. A second report to the Evolution Committee, with contributions from Punnett, Hurst, and others alongside Bateson and Saunders, came out in 1905.[9] So did a pocket-sized work of skillful popularization from Punnett, titled *Mendelism*.[10] Privately that year, and publicly the next year, Bateson proposed a new, product-differentiating name for the Mendelian science of heredity: "genetics."[11] When, in summer 1906, a third international hybridization meeting was held in London, once again organized by the Royal Horticultural Society, it was, at Bateson's suggestion, retrospectively canonized as "the Third International Conference 1906 on Genetics." The published proceedings began with a portrait of the abbot of Brünn.[12]

Those are some of the markers along the familiar path of the Mendelian success story circa mid-1904 to mid-1906. This chapter follows a parallel path, less well known. After the 1904 BAAS meeting, Weldon went on to develop the most potent alternative to Mendelism that there would ever be. He never lived to complete the manuscript, "Theory of Inheritance," in which he set out his ideas most fully. But the surviving drafts, letters, lectures, and other materials give a thrilling sense of what was in the offing.

1

No one who knew Weldon's publications up to summer 1904 would have guessed, on hearing him in Cambridge, that he was about to reframe his criticisms of Mendelism in a fundamental way. At Cambridge, he once again seemed every inch the exasperated naysayer, complaining, in the name of the law of ancestral heredity, about the scientific equivalent of corner cutting. After Cambridge, however, he began work on a much grander critical project.

Central to the new project was an original reinterpretation of Galton's writings every bit as far-reaching as Bateson's reinterpretation of the "Versuche." Recall that, in the *Defence* and in *Report 1*, Bateson had reversed Mendel's own priorities, promoting Mendel's explanatory hypothesis about gamete formation in hybrids to the "essence" of the paper while demoting his "law valid for *Pisum*" to secondary status. Weldon now began to do likewise for Galton. No longer would the law of ancestral heredity be represented as the definitive Galtonian achievement. It remained important, but as a downstream consequence of something much more basic: Galton's conception of how, in the developing organism, inherited capacities become variably manifest or not depending on surrounding conditions, physiological and physicochemical.[13]

Galton had last emphasized that point back in the early 1870s. But he never repudiated it, and, like a latent hereditary trait, it returned to visibility in his writings, public and private, as occasions called it forth, decade after decade. At the end of the 1880s, in *Natural Inheritance*, in a chapter titled "Processes in Heredity," his discussion of "particulate inheritance" included that passage quoted earlier about how, in the developing embryo, any subcellular trait-determining particle that escapes from the latent mass of them "must owe its success partly to accident of position and partly to being better qualified than any equally well placed competitor to gain a lodgment," with the ultimately patent trait going on, like the developing embryo overall, "to be influenced by an incalculable number of small and mostly unknown circumstances."[14] In the mid-1890s, in correspondence with Pearson, Galton made the case for thinking that, even within a single family, the effects of "nurture" could vary so greatly from child to child that to treat those effects as constant and therefore ignorable was to introduce room for error. (Pearson was unmoved.)[15] In 1903, in the newspaper commentary welcoming the Interdepartmental Committee on Physical Deterioration, Galton drew attention to the salutary effects of outdoor living on Britain's Boer War recruits as showing the good that can come from improvements on the "nurture" side of the nature-nurture dyad.[16] At the Sociological Society the next year, he opened with a definition of eugenics that, as ever with Galton, touched on both sides of the dyad: "Eugenics is the science which deals with all influences that improve the inborn qualities of a race; also with those that develop them to the utmost advantage."[17]

So there was material enough for an interaction-with-conditions reading of Galton on heredity, right up to summer 1904. A first glimmer of it as Weldon's reading shows up in a letter Weldon sent to Galton in late August, just after the Cambridge BAAS meeting. "I am more and more certain," wrote Weldon,

> that a theory of particulate reversion, such as you indicated in *Natural Inheritance*, will do the whole thing. I think you want to work out such a theory for units of various sizes,—so that the limit on one side would be reversion by particles so small that the result is a blend, on the other by units so big that large pieces of the individual are affected, as in your Bassett Hounds. Is not this the theory you have had in your mind always?[18]

Until now, Weldon had never shown much interest in this aspect of *Natural Inheritance* or Galton's wider oeuvre.[19] Weldon's own, career-long concern to give the interactions of heredity, development, and environmental conditions their biological due was, as we have seen, a legacy of his days as the Darwinian Balfour's student—a legacy recently reinforced thanks to the new experimental embryologists and their interest in normal development as utterly dependent on the presence of enabling environmental stimuli. As late as November 1903, when Weldon and Galton were corresponding about acquired characters, Weldon was in no way treating Galton as the originator of a perspective now being advanced so admirably by Herbst and others. Post-Cambridge, however, that was exactly what Galton was becoming for Weldon. Maybe, between all those convinced Mendelians he saw in action at the BAAS meeting, and the ever-proliferating breeding experiments recounted in the second report of the Evolution Committee (which Weldon was reading as a referee for the Royal Society), he appreciated just how powerful a theory Bateson had built around Mendel—and how useful it would be, in articulating an alternative, to build it around an individual whose work could likewise be represented as foundational.[20]

However the idea came about, Weldon was now inviting Galton aboard just such an enterprise. Had he not, after all, always stressed that reversion—as tracked in the law of ancestral heredity—happens when the particles or blocks of particles underlying a visible character reencounter the conditions, internal and external, required for that character's expression? If the relevant units are small—or, more precisely, when they are numerous (Galton annotated Weldon's letter with a query to that effect)—the character is a blending one, like stature. And if the units are big, and so necessarily few in number, the character is all or nothing, like coat color in basset hounds. Thus construed by Weldon, Galton's theory gave all the flexibility of the neo-Mendelian theory, but with no marginalizing of the role of conditions in the development of the characters, no indifference to all the variability that can arise as conditions change, and no incuriosity about the way that characters of different vintages can return as particles interact in new/old combinations with each other and new/old conditions. Here was a theory that really could "do the whole thing," whether Mendelian or otherwise.

By early October, Weldon's ambitions had grown. He now saw that, pleasing though it was in its basics, the Galtonian particles-in-context picture needed updating for the age of chromosome biology, and Weldon was accordingly attempting to master a burgeoning new literature. (He reported to Galton that the evidence on whether individual chromosomes remain constant over time was so contradictory that it made the evidence for the 3:1 ratio look good by comparison.)[21] The new theorizing was now taking the form of chapters for a book—Weldon's first since his abandoned manuscript for a thinking biologist's guide to Pearson. "I began the revised version of the section of Chap. II on some relations between the organism and its environment," he wrote to Pearson on 9 October, by way of explaining a delay in correspondence. That chapter was proving difficult; but the upcoming one, on "hereditary units," was a good bet "to tax my poor powers of moderation to breaking point."[22]

Even so, Weldon continued in the letter, the broad outlines of what he wanted to say on heredity, development, and environment were becoming clear. For one thing, he understood the pieces in a dividing egg to be like particles in "Galton's stirp," in that "each is so affected by its position in the embryo that some of its qualities become latent." For another, "the phenomena of regeneration"—as when a crab regrows a claw, the tissue at the base of the stump suddenly manifesting previously unguessed-at claw-making potential due to the removal of its usual surroundings—could be marshaled as evidence for how latent properties become active when conditions become congenial. The upshot would be "a general theory of the latency of characters . . . with Mendelism as a special case,—or what there is of truth in Mendelism." As for the main tenets of that general theory: "Take as the three fundamental things (1) equipotentiality of pieces into which an egg divides; then fate of cell determined by (2) position in organism and (3) environment and you have a big part of the game."[23]

Pearson proposed that Weldon present his book-in-progress in lectures at University College, and Weldon immediately agreed. "Your suggestion is extremely kind, any how, and I should be very proud to talk once more in U.C.L. where there are live people," he replied.[24] As the two planned what became an eight-lecture series titled "Current Theories of the Hereditary Process" (the title echoing that of Galton's *Natural Inheritance* chapter), to begin in late November, Weldon returned again and again to how excited he was for the chance to get back to London. "I am thoroughly enjoying the preparation for talking at U.C.L. again, but it makes me wish more than ever that we could work again within reach of one another, and that I were back among the live people!" Oxford, by contrast, "is the greatest danger to England I know. I hate it, and I hate myself because I have sold myself to it for money, instead of sticking to good old Gower Street, where there are live people who can be made keen."[25] Not even the news of Bateson's Darwin Medal—and then the news that Bateson himself would be attending Weldon's lectures—dampened his spirits.[26]

A first sketch of the lecture contents had Weldon starting with the motley of phenomena embraced by "inheritance"—from the persistent mean that distinguishes a species or race to the mosaic-like way that, within a single body, characters form independently of one another—followed by a look at the two groups studying inheritance, referred to as "the Breeders" and "the Statistical Workers." Then came the following:

> Complexity of relation between an organism and (a) its ancestry
> (b) its environment.
> Relation between "environmental stimuli" and "hereditary stimuli."
> Function of embryonic envelopes, etc., as "environmental stimuli"
> The blood as an "environment"—Born. etc.
>
> Complexity here indicated shows need for statistical treatment.[27]

That last line summarized the original meaning of the quincunx more pithily than anything Galton ever wrote. Galton's stirp indeed came up next, along with Darwin's gemmules, Mendel's units, and other submicroscopic hereditary "determinants" postulated in the intervening years. At the end of Weldon's outline, after a nod to the need for reflections on the difficulties in making physical sense of all such theories, Weldon wrote, "Fundamental difference between Galton and Mendel—Exclusion or latency of determinants." Not "the noticing or neglect of ancestry," as per Weldon in 1902, but the noticing or neglect of something concretely physiological and investigable in the here and now: the possibility that a dominant character might be absent even when its determinants are present. According to Mendel, a dominant character could be absent only when its determinants were excluded, as—on the Mendelian analysis—in green- or wrinkled-seeded F_2 peas. According to Galton, however, those peas could harbor determinants for yellowness or roundness, but just be missing the surroundings in which those characters are manifest. With a change of conditions, internal and/or external, a muted determinant could become active again. A further change of conditions could return it to latency.[28]

In between Weldon's list of theories and theorists of hereditary substance and his exclusion-or-latency statement, we read: "Conception of every character as 'acquired' and 'inherited' at the same time renders much of the discussion as to relation between the 'hereditary substance' and the soma futile." Here he was going well beyond Galton, and knew it. In the letter to Pearson accompanying the lecture outline, Weldon wrote that Galton "is rather annoyed with me, I fear, because I cannot see the difference between 'inherited' and 'acquired' characters.—I don't believe there is any: and although he is very kind about it, of course, he thinks me a muddle-headed ass, all the same."[29]

Weldon's enthusiasm for his new way of counterposing Galton to Mendel remained undimmed. With the lectures drawing closer, Weldon wrote to Pearson of having "got into my head the outlines of a little essay on Galton & Mendel, which I hope can see daylight after this report of Bateson's is published." As the latest from Bateson and Cuénot showed, the Mendelians' breeding experiments increasingly turned up patterns that, while anomalous according to their own theory, were anything but according to Galton's. "Good old Galton's stirp, in which some of the ancestral characters are latent, is still the only 'machine' which will work: and the proper line of research is an enquiry into those embryonic stimuli which make a given character evident or latent. That is my fixed belief."[30]

2

Gower Street delivered: around 180 people signed up for Weldon's first lecture, so many that Pearson had to arrange a larger room.[31] As expected, Bateson was there. He returned most weeks, taking notes.[32] But there were no confrontations. Hurst, out at his nursery in Burbage, in rural Leicestershire, was in regular correspondence with Bateson, but did not even know the series was happening. In December, when Weldon was three lectures in, Hurst wrote to Bateson about "the opposition": "After the smashing you gave them at Cambridge they have lost heart! W. appears to be silenced & K. P. seems to be giving off a few dying sighs before adapting Mendel to his purpose!"[33] Only in February did Bateson respond with a correction. We got it wrong, he explained to Hurst, in "supposing the opposition crushed. Weldon is giving a course of Lectures at University College on Heredity, of which most are now over, and last time (the 6th time) he reached the promised land of Mendelism. It was exceedingly amusing. Apparently they have taken new hope."[34]

Before addressing Mendelism, Weldon used his run-up lectures to hammer home the message that, at all points in the scale of life, a cell takes on a particular character not because it lacks the hereditary potential for anything else, but because its relations to its surroundings, from neighboring cells to ambient salt and temperature levels, end up making a selection from the full range of possibilities. Here, pulled together for the first time, were the basic biological facts adding up to the "general theory of latency" that Weldon had promised Pearson. The basic statistical facts of heredity—that the average value of a character tends to remain constant down the generations, with average parents having average offspring and exceptional parents having exceptional offspring—featured only briefly, at the very start of the initial lecture. From there, Weldon's subject, illustrated with slides and diagrams, was development, with a special emphasis on regeneration.[35]

He began his survey of the animal kingdom at the bottom, with the single-celled freshwater protozoan *Stentor*. As he explained, the body of a full-grown *Stentor* is no homogeneous blob, but a highly differentiated, pear-shaped structure, with a mouth at the wider end and a long, sausage-links-shaped nucleus running throughout. Yet if one slices a *Stentor* into three parts, then, provided each part has at least a bit of nucleus, each part will grow into a complete new *Stentor*. Slices as thin as a half millimeter will suffice as material for a new whole. Weldon spelled out the lesson for a general theory of latency (to quote from the lecture summary subsequently published in the medical press):

> A stentor is thus composed of "units" each of which, removed from the rest, is capable of rebuilding the whole body system of units. These units as they exist bonded together in the body must have many properties latent, but when, for example, the head region is cut off the neck region is roused from its latent or dormant state and regenerates a fresh head region.

The same goes for sliced *Hydra*, another microscopic pond dweller that, although multicellular, is able to reproduce not just asexually (like *Stentor*) but sexually.[36]

Next Weldon moved to the sea urchin, a far more complex invertebrate whose minute study, in the laboratories of Driesch and others, had made it the exemplary organism for understanding development in sexually reproducing animals, from fertilization onward. It was well understood that a sea urchin zygote divides first into two, then four, then eight, then sixteen cells. These cells (or "blastomeres") go on to form a hollowed-out ball before proceeding through the buckling, contractions, and differentiations that turn the ball into a functioning larva. Up to that tightly choreographed transformation—called "gastrulation"—sliced-out parts of the ball, from anywhere on the ball, remain capable of growing into functioning larvae. What determines the role any particular cell plays in gastrulation depends on the position it happens to find itself in once the mass of cells is sufficiently large and ball-shaped. All the cells to that point are at what Weldon called the "equipotential stage." After gastrulation, however, equipotentiality is gone.[37]

The further we travel up the animal scale of complexity, Weldon stressed, and the further a complex animal travels from zygotic beginnings, the lower the power of regeneration. With worms, crustaceans, and frogs, we are dealing not with the power of parts to regenerate whole organisms, but with the power of nearly whole organisms to regenerate lost parts. With higher vertebrates, not even that much regenerative power is retained. Nevertheless, for Weldon, the larger point about cell fates remained the same: they have potentialities that go unrealized because conditions render them latent.[38]

In Weldon's fourth lecture—his last before the Christmas break—he turned to consider "the environment and its special action on developing organisms," noting that "the investigation of the special action of elements of the environment on development is a recent achievement in biology." For normal gastrulation to take place in a sea urchin embryo, Weldon taught, it needs to grow in water where the right salts (sodium chloride, potassium chloride, magnesium sulfate, and so on) are present in the right concentrations, and where the temperature is not much warmer or colder than the normal environment in which sea urchins live. Deviations from these environmental requirements produce deviant morphologies. The same general lesson held for butterflies, for amphibians, for everything living. And if environment was always a factor in development, then the outcomes of development were never just hereditarily fixed, and—the great Weldonian theme—there were no grounds for dividing those outcomes into two kinds, the inherited and the not-inherited (i.e., the acquired). "Every phase of development of an embryo was influenced by thermal, physical, chemical or toxic, and nutritive agencies," reads the published summary. "No character could be shown in an animal to be completely germinal in origin and independent of environment, and the environment (and its elements) affected even closely allied animals very differently."[39]

Weldon used the 1904–5 vacation to figure out what he wanted to say about chromosomes, in themselves and in relation to Mendelian patterns. Letters and manuscripts accumulated, all of them impressively reasoned, none of them conclusive. "It would be rather fun if an arithmetical theory of Mendelism led to an opinion about the 'permanent individuality of the chromosomes'!" he wrote to Pearson, four days before the lectures were to resume. "I shall hope to bring a lot of stuff for you to look at on Tuesday.—Only, how in the world to lecture, when I am in such a state of flux myself, I don't know."[40]

Bateson, who had recently returned from a trip with his wife to Vienna and Brünn ("the Mendelian Mecca," in Hurst's phrase), could not attend the lecture, so he sent Saunders instead. She reported back that, after fifty minutes of outlining the consensus on nuclear division, Weldon offered a ten-minute "not over clear statement of the chances that one half or all the chromosomes will consist entirely of either maternal or paternal units on the supposition (1) that the chromosomes persist in the resting nucleus [between divisions], & (2) that chromosomes are not permanent structures."[41] Although the impermanence of chromosomes was an idea Weldon took seriously—on the view that, if chromosomes sometimes acquire a mixture of maternal and paternal units, that would account for otherwise unintelligible blending and mosaicism (and blast a hole in "gametic purity")—his main concern in this lecture was rather different, and more in keeping with the overall "conditions matter" thrust of the series up to then. What, he asked, should we think about Weismann's theory of cell-lineage differentiation? According to Weismann, as development proceeded, there was, in Weldon's phrase, a "sorting of determinants," with the ones needed for making, say, heart cells ending up only in the heart cells, the ones needed for making liver cells ending up only in the liver cells, and so on. As this distribution took place, the chromosomes—bearers of the invisible determinants—in different cell lines would become qualitatively different. But that conclusion looked doubtful, Weldon argued, given what had already been established about potentiality depending on relative positions and environmental conditions. A better bet, he suggested, was that, in a developing embryo, every cell got the same determinants, but positions and conditions brought out certain potentialities while the rest remained latent.[42]

By the time Weldon reached Mendelism, he was a full lecture behind schedule. In the time remaining, he nevertheless alighted on a fair amount of what he had set out to cover: the need to separate Mendel's data from Mendel's attempt to explain those data; the failure of any "purity of the gametes" explanation to account for the data collected by Galton; the probability that "both views, the Galtonian and the Mendelian, will be reconciled in time in a wider generalisation of the facts of inheritance and descent—a larger theory of heredity"; the role in that endeavor of a perspective likening the dominance of one character version over another to the absence of regenerative growth when an animal is uninjured (dominance as conditional even when it seems it is not); the power of such a perspective in making sense of what looks anomalous from a Mendelian view and atavistic from a Galtonian view; the extent to which Galton himself articulated a conditional theory of dominance in his writings on patent and latent elements; and the extent to which

Galton's law of ancestral heredity is the most straightforward statistical upshot of that theory.[43]

Bateson, of course, shrugged it all off, scoffing to Hurst about Weldon's continued reliance on "reversal of dominance" cases—wrinkledness can be dominant, what a shock!—as evidence for the prosecution. Nevertheless, Weldon seems to have made an impression. Immediately after bringing Hurst up to date on Weldon's lectures, Bateson commented on some new work of Hurst's on human eye color as obeying Mendel's principles, with brown eyes dominant and blue eyes recessive. Bateson was uncharacteristically wary about the capacious categories being used to make the case:

> About the eye-colour. Your figures are extraordinarily interesting. If you have really cleared up eye colour that far, it will be a great help. The question I feel most concerned about is the classification. In your sense blue, if it is not to contain any yellow or brown at all, must be very blue. So far as I have ever noticed (not much) such blues as that seem to be excessively rare. . . . Personally I have never tried classifying eye-colour, but I have always had the notion that it was a very indeterminate affair. Perhaps you have a lot of true-blue Scandinavian blood in Leicestershire—though that seems unlikely. . . . I notice you have no case of blue × blue. Surely the biometricians profess to have cases of blue × blue giving various colours.[44]

As for Galton, originator of the existing body of data on eye-color inheritance, he remained as even-handed as ever, expressing supportive curiosity about Weldon on dominance while also seeing in the Mendelian view new opportunities, not least for eugenics.[45] On the day Weldon gave his final lecture, 14 February 1905, a new paper of Galton's, "National Eugenics," was read before the Sociological Society, meeting a short walk away at the School of Economics and Political Science. Inspired by the appointment of Edgar Schuster to the recently established (and Galton-funded) Francis Galton Research Fellowship in National Eugenics at the University of London, Galton had now come up with a seven-point "to do" list for eugenics workers. Under point 4, "Heredity," he set out the general desirability for more work on both "actuarial" and "physiological heredity." But he also suggested a particular study. For several generations now, intermarriage had been taking place between English people and Hindus, on a scale large enough to generate "trustworthy results." Here, Galton advised, was a subject meriting study in its own right, given its "national importance," but also one that could serve "as a test of the applicability of the Mendelian hypotheses to men."[46]

3

Now Weldon brought his customary energy to finalizing the "Theory of Inheritance" manuscript.[47] Throughout the spring, summer, and early autumn of 1905 he remained, as ever, busy on multiple fronts: lecturing; maintaining the mouse-breeding experiments; initiating new breeding experiments with silkworms (acquired during a month's holiday in Ferrara); meeting with Pearson for cycle rides and biometric seminars à deux (on the mathematics of determinantal heredity); and picking up new enthusiasms, from Gregorian chant and sunspot photography to the strange persistence of local legends about which famous person is buried where. Along the way he even found the time to take a new portrait photograph of Galton (fig. 9.2).[48] But the book was the priority. When the Weldons set off for Ferrara in late March, the manuscript came too. Weldon reported to Pearson from there that, with little else to do but sit in the university library and plug away at the book all day, he was making steady progress—and indeed, though he was only half done by the time they returned to England, he "never possessed such a stack of manuscript."[49] In early July, rumors began to circulate—they reached Bateson—that the book's publication was imminent.[50] In September, it seemed close enough at hand that Weldon asked for Galton's blessing in dedicating the book to him.[51]

The surviving manuscript departs most strikingly from Weldon's University College lectures at the beginning, in two ways. The first concerns his treatment of statistical methods. In the lectures, as we saw, he opened with a brisk overview of how statistical data matters for heredity before launching directly into his lengthy *Stentor*-and-up survey of growth and reproduction, noting in passing that these developmental facts would come in handy later on in evaluating rival hypotheses purporting to explain the statistical facts. In the manuscript, by contrast, the first chapter deals exclusively, and extensively, with statistics. Instead of asserting their importance to the scientific study of heredity, Weldon here makes the case, bringing in the most general considerations about descriptive statements in science, how they relate to what is actually observed, and why, even when we think we know what is going on in the world, we must never forget that there is an element of uncertainty.[52]

To illustrate, he gives an astronomical example: the latitude of the transit telescope at the observatory in Oxford. One might have thought, reasonably enough, that there is a stable fact of the matter as to what its latitude is, and therefore that the slightly discrepant values reported by careful observers down the decades can be safely ignored in favor of a

FIG. 9.2. Weldon's photograph of Galton, aged eighty-three, taken in July 1905.

single compromise value. And indeed, Weldon continues, in the physical sciences, for many purposes, much of the time, no harm is done by treating variable experience as if it were constant. Yet even in these sciences, sometimes fundamental discovery hinges on not forgetting that the constancy is a fiction. Such was the case with latitude. As astronomers came to appreciate only toward the end of the nineteenth century, latitude is not absolutely constant, but varies in line with periodic changes in the orientation of the Earth's axis of spin. (When the axis of spin changes, the Earth's equator gets relocated, with knock-on effects for locations reckoned relative to the equator.) A sensational revision in the understanding of the Earth as a physical system thus followed from a refusal to sweep variability under the carpet of make-believe constancy.[53]

Another example Weldon used to the same end shows up in a lecture he gave on inheritance in August 1905 in Oxford, as his contribution to an omnibus series there on "the method of science." After telling the latitude story—probably familiar to him from an Oxford friend and ally, the astronomer Herbert Hall Turner, who included a chapter on it in his 1904 popular book *Astronomical Discoveries*—Weldon underscored the methodological moral with a second discovery story, again from topical physical science.[54] The 1904 Nobel Prize in Physics had gone to Lord Rayleigh for his part, ten years before, in the discovery of argon. (Recall the lively Royal Society discussions of that discovery—and the consequences for Weldon and Bateson.) Argon got discovered, Weldon now told his audience, because when Rayleigh had attempted to determine the weight of nitrogen, he found that when he weighed a volume of it derived by extracting all known constituents except nitrogen from atmospheric air, he got a set of measurements fractionally higher—on the order of hundredths of a gram—than the ones he got when he weighed nitrogen derived from nitrogen-containing compounds. Rather than ignore the trifling discrepancies, he took seriously the possibility that the atmospheric nitrogen might contain hitherto unknown constituents. The periodic table of elements thus acquired a new column only because Rayleigh was unwilling to let himself off the hook when confronting contradictory data.[55]

So truly great science and scrupulous attention to variable experience go hand in hand. In the chapter as in the lecture, Weldon went on to say that what holds for the physical sciences holds with far more force in the biological sciences. Uncertainty about the location of a telescope is so marginal that, for most of the history of modern astronomy, not even experts' suspicions were aroused on hearing it described as being at such-and-such a latitude. Yet anyone told that "Englishmen are 5 feet 7½ inches high" immediately recognizes the statement as preposterously inexact.[56] Here is where Galton's statistics, introduced at length, come into their own. When we use them, our descriptions capture not merely the average value but the form and extent of deviation from the average, and in a way that, as Weldon put it, "will enable us to form a mental diagram of our whole experience," the better not merely to remember series of observations but to compare them.[57] As for why biological phenomena are so variable that they demand statistical descriptions, Weldon explained that especially clearly in his lecture: "All experience, which we are obliged to deal with statistically, is experience of results which depend upon a great number of complicated conditions, so many and so difficult to observe that we cannot tell in any given case what their effect will be." Hence, for Weldon, the instructive nature of dice rolls for the student of heredity. Anyone wanting to understand correlation and regression (which is obligatory: to study heredity the Weldonian way, it becomes clear, is to be concerned first with describing how far, for any given character, an individual's deviation from the generational average correlates with such deviations among ancestors and among descendants) should, Weldon advises, consider the results of 2 throws of 12 dice, where half the results from the first throw are carried over into the second. Whether dicey or biological, chance outcomes look the same.[58]

Running through the methodological and mathematical weave of the opening chapter is a subtle thread of Darwinism. When, at the outset, Weldon introduces character constancy and parental-filial correlation as the starting-point phenomena for anyone interested in heredity, he immediately stresses that these are not necessarily connected in a causal way. The average value of a character, and the proportions of individuals deviating from that average, can remain the same down the generations whether children inherit their parents' peculiarities with total fidelity or none whatsoever. Conversely, the actual, observed value of correlation between parents and children—roughly halfway between total fidelity and zero fidelity—is "perfectly compatible with a change in the race-character from generation to generation." In other words, when the average character remains constant, it is not heredity alone that keeps it so, but heredity operating in an environment; and if the environment changes in a way that favors a change in the average character, then, with no change in the laws governing heredity, natural selection will, other things being equal, gradually shift the average character.[59] Weldon reiterates the point from a different angle near the end of the chapter, in closing his discussion of regression to the mean. Notwithstanding the claims of some commentators on Galton's work, regression does not, Weldon explains, function to keep the average character constant, nor are there occasional, accidental offspring so divergent from this average value that they somehow escape its gravitational pull, like a rocket escaping the Earth's gravity in an H. G. Wells story. To make such claims is, again, to misunderstand the fundamentally probabilistic nature of regression. Yes, children regress on their parents; but equally, parents regress on their children. As you go forward or backward in time, given an initial value and the operation of chance, a correlated range of values will probably cluster more closely to the average, with dice as much as with people. Far from regression thus acting to resist the effects of natural selection, regression—having nothing to do with biology—can be expected even in a population like the shore crabs in Plymouth Bay, undergoing selective elimination so intense that its effects can be detected within a single year.[60]

4

The hero of Weldon's "Theory of Inheritance" is not Darwin, however, but Galton. According to Weldon, the study of heredity has two goals: first, to describe the relations that hold between the visible characters of parents and offspring; second, to explain those relations in terms of the invisible germinal process that brings them about. These goals, he continues, are so different from each other, requiring such different methods of inquiry, that most often they have been pursued by different people.[61] His second chapter introduces the two men who, up to then, had gone furthest in bringing statistical knowledge of character relations to bear on theorizing about inheritance at the germinal level: Galton and Mendel.

In Weldon's University College lectures, these two and their theories had come up only toward the end. Even some ways into the writing of his book, Weldon seems to have kept that overall organization.[62] Eventually, however, he decided—in the other major departure from his lectures—to move his comparative analysis forward. The effect of this change was to turn the extensive material on animal regeneration and the role of conditions in eliciting developmental potentialities into an answer to the question of how to choose between Galton and Mendel. That choice, it now became much clearer, is ultimately a choice between rival conceptions of dominance.

Weldon traces Galton's conditional conception to "On Blood-Relationship" (1872), Galton's first paper in that burst of creativity following his break from Darwin over pangenesis—the period that also brought the invention of the quincunx and the coining of "nature and nurture." As Weldon (with a little help from Galton himself, via comments on a draft) summarized it:

> In this statement [Galton] regards the germ, from which an individual is developed, as containing a number of elements, each of which is capable, under suitable circumstances, of determining the appearance of a particular character or group of characters in an adult body. He supposes that every such element is capable of existing either in an active, or as he says in a "dominant" condition, or in a condition which he calls "latent" and Mendel would have called "recessive." He considers that a particular element achieves dominance partly through being more vigorous than its

> immediate competitors, and partly through the unknown conditions under which the struggle takes place.[63]

Considered as a theory of inheritance, this statement, Weldon stresses, is incomplete. For one thing, it needs updating in the light of thirty-plus years of progress on the cellular basis of heredity—something Weldon promises to pick up again later. For another, it needs filling out with, as he puts it, "a knowledge of the law which expresses the dominance or latency of given elements in a given germ." Here, Weldon continues, the necessary work is already well advanced, in the form of the law of ancestral heredity. On Weldon's gloss, then, so long as mating is random, and whether the characters under study are blending (such as human stature) or non-blending (such as basset-hound coat colors), dominant characters—along with the elements determining them—get transmitted in proportions roughly conforming to a geometrical series.[64]

Turning to Mendel's theory, Weldon proceeds in a spirit of neutrality, even admiration. He points out, for example, that Mendel went far beyond his predecessors in establishing the equivalence of male and female germ cells; that his hawkweed paper made a start on extending his theory from non-blending to blending characters; and that his work overall is "in a sense complementary" to Galton's, in that where Mendel dealt "with the case in which selective mating is carried to its extreme limit among the ancestors of the stock observed, while the parents belong to distinct races," Galton dealt "with the stock produced by parents of a single race, in which selective mating is reduced to a minimum." In other words, although Mendel tells us only what heredity looks like in the special case of breeding between unusually homogenized lineages, what he tells us is nevertheless helpful for understanding that case, which Galton did not investigate. Moreover, in presenting Mendel's theory in detail, Weldon emphasizes how extraordinarily well it explains Mendel's data. That at least some of those data might be too good to be true is never even hinted at, let alone discussed.[65]

Weldon's quarrel here is not with Mendel's data, but with his non-Galtonian conception of dominance as a constant, permanent property:

> Mendel conceives dominance to be an invariable attribute of one determinant in each alternative pair; he believes that yellow-giving are always dominant over green-giving determinants, whenever the two meet in the same zygote, or fertilised germ. This is the first great difference between his conception of dominance and that held by Galton . . . [for whom] the same character of an alternative pair might be at one time dominant, at another recessive, so that he was led to regard dominance as depending

> not upon the character borne by a germinal determinant, but upon the condition of the determinant itself, and upon its relation to the other germinal determinants, at some moment during fertilisation or subsequent development.[66]

As Weldon weighs the two conceptions, he brings out, more vividly than before, how far the content of Mendel's theory hinged on his decision to cross parents from distinctive purebred sub-lineages: purebred yellow-seeded and green-seeded races, and so on. It was, Weldon reckoned, only because of the deliberately denuded internal contexts of the parental races that Mendel found, for every pair of character-version binaries that interested him, that one version of the character was uniformly dominant in the offspring. But that repeated observation was what then led Mendel to conclude that dominance must be a permanent property of that dominant version and its germinal determinants. This conclusion in favor of the permanence of dominance, in turn, meant that when, among the offspring of the hybrids, Mendel discovered the return of the recessive version, he had, for consistency's sake, no option but to suppose that those recessive plants must be free of dominant determinants (for if any such determinants were present, the plants would not show the recessive version). And such purity could have come about, given Mendel's conception of dominance, only if the gametes joined at fertilization had themselves been pure for the recessive version. Hence Mendel's framing of his hypothesis of gametic purity—a hypothesis that licensed Mendel (or at least Mendelians) to think that the recessive offspring of hybrids might as well have sprung directly from their recessive grandparent, so little did their dominant parents matter hereditarily.

So: Galton or Mendel? If comprehensiveness were the criterion, then, on Weldon's showing, the choice would have to be for Galton, since his theory, though in need of refurbishment for the age of chromosomes, could cope with Mendelian phenomena (though not designed for them) as well as a whole lot more. Where Mendel interpreted the uniformity of his hybrids as a sign that dominance is a permanent property of particular character versions and their determinants, the Galtonian instead sees their dominance as, in Weldon's words, "a property conferred upon them for a time by the peculiar conditions of the experiment"—namely, all that starting-point, breeder-engineered purging of unwanted variability until all that is left are character versions unmistakably contrasting with their partner character versions. And where Mendel interpreted the 1:2:1 ratio among the offspring of his hybrids as a sign that the gametes of the hybrids must not themselves have been hybrid, the Galtonian has no need for such a hypothesis, since the ratio is, notwithstanding all that experimental peculiarity, consistent with Galton's geometrical series: half the offspring are like their parents, a quarter like one set of grandparents, and the other quarter like the other grandparents. Beyond the Mendelian phenomena, Galton's theory reaches phenomena that Mendel's theory does not—and maybe cannot—touch, notably heritable variation within those Mendelian binaries (as witnessed over and over again in the post-*Telegraph* history of English pea varieties, reviewed by Weldon), blending characters (for stature is not, Weldon hints, ultimately due to the accumulation of little stature bricks, as the Mendelian requires), and the long-latent characters known as atavisms (whose existence, though outlawed by Bateson, was now, Weldon reports, increasingly well documented by other followers of Mendel).[67]

Yet Weldon pulls back from declaring a Galtonian victory and, in line with the chapter's fair-dealing tone, suggests at the end that, given how well both Mendel's permanent and Galton's conditional conceptions of dominance fare when tested against Mendel's data, what is needed is some "independent evidence" for choosing between those conceptions. And that evidence, which had been accumulating over the last twenty years of experimental studies of regeneration and development, is the subject of the succeeding four chapters—the final complete ones in the extant manuscript.[68]

In these chapters Weldon provides more detailed versions of the accounts he gave in his London lectures, punctuated now, however, with asides about how well the experimental results accord with the Galtonian conception. From sliced-up *Stentor* to variably spined water fleas to smallpox-scarred humans, artificially and naturally manipulated organisms reveal how important it is to disturb normal relations—of the parts of a body with one another, of the whole of an organism with its environment—to assay the range of structures to which a tissue can give rise. No structure becomes visible, and no underlying determinant dominant, just by intrinsic nature. Dominance is a conditional property, conferred when relations beyond a tissue permit it and removed otherwise.[69]

In closing, after explaining why, in this view, every character must be understood as both "inherited" and "acquired" (recall the final epigraph at the head of this chapter), Weldon comments on those quotation marks. We need, he urges, to distinguish between the case for classifying characters into the germinally determined and the environmentally determined, and the case for "the inheritance of acquired characters"—that is, Lamarckian inheritance. That form of inheritance could come about, he continues, only if the abnormal circumstances required to call forth an abnormal character from the relevant germinal determinants *change the determinants themselves*, in such a way that, in the next generation, the abnormal character will appear even under normal circumstances. The evidence against that happening, he insists, is overwhelming. But the classification is nevertheless, and fundamentally, "unsound,"

> for on the one hand there is no character which does not require some definite environmental conditions under which the embryonic "mechanism" which determines it may develop, so that the so-called blastogenic characters depend for their appearance on the reaction between an internal something of germinal origin and a set of external conditions, while the "somatogenic" characters, admittedly produced in part by the action of the surrounding conditions, can only be produced by the action of those conditions on a soma, which is itself a result of transmitted germinal constitution. The most that can justly be said is that the visible effect produced by a change in environment is greater in some cases than in others; but since neither this effect nor the effect of transmitted germinal constitution has ever been shown to be zero in the case of any character whatever, the establishment of two distinct categories appears to be unjustified, and for this reason, while we shall recognise the importance of environmental influence in every case, we shall neglect the distinction between "inherited" and "acquired" characters, treating every character as belonging not to one or to the other of these classes, but to both of them.[70]

5

How was the book supposed to end? Foreshadowings in the completed six chapters suggest that, after his tour of the evidence favoring the Galtonian theory, Weldon planned to do three things. First, he would modernize the 1872 conception of dominance, bringing it, as he wrote in the second chapter, "into harmony with the facts of germinal structure which have been discovered during the last thirty years." In practice, that would almost certainly mean reconstructing Galtonian dominance in chromosomal terms—showing, for example, that, according to ideas about chromosome behavior then current, one could explain the elementary Mendelian pattern while also explaining how a small proportion of the green-seeded grandchildren (say) could end up with unexpressed yellow-making determinants.[71] Second, he would collect the abundant evidence on reversals of dominance and other forms of atavism, in line with the predictions of Galton's theory, especially as documented by Tschermak with peas and by Weldon himself with his waltzing mice ("I am writing up the mice for the book," he told Pearson in October 1905).[72] Finally—by way of tying off that Darwinian thread stitched into the opening chapter—he would demonstrate how well this enriched version of Galton's theory meshed with the theory of gradual evolution by natural selection. Earlier in the book, Weldon promised to return to the question of the actual values to be plugged into Galton's law. Since selection made that question salient (unless individual variations are reliably passed on, there can be no cumulative improvement), Weldon would probably have dealt with it here, endorsing Pearson's selection-friendly raising of the parental contribution and eviscerating Johannsen's selection-unfriendly reduction of that contribution to nothing. In the latter connection, Weldon could also have made good on another promise: to expand on the mismatch between Mendelian theory and statistical reality when it came to blended characters such as stature. No one had ever detected the units into which the Mendelian was obliged to resolve continua, and no one ever would, because they did not exist.[73]

One measure of the book's potential to change the course of Mendelism's future is how unnerved Bateson became during the waiting period after Weldon's London lectures. These were, to be sure, in many ways months of solid Mendelian achievement, from the publication of Punnett's gospel-spreading *Mendelism*, to Darbishire's ever more complete "conversion to Mendelism" (as Weldon put it in a letter to Pearson), to the development at Cambridge of a clever grid system—eventually known as the "Punnett square"—that took the toil out of calculating probabilities for different kinds of offspring in a Mendelian cross.[74] There was even a minor-key public victory over Weldon, when a Mendelian exhibit at the Royal Society summer soiree provoked him sufficiently that Bateson heard about it afterward.[75] As ever, new projects proliferated, including one between Hurst and Bateson on coat colors in horses, aiming to show that chestnut—together with non-blackness in the tail, mane, and leg hairs—behaved as a straightforward Mendelian recessive, and that bay or brown—together with blackness in the hairs—behaved as the counterpart dominant. But Bateson was not at ease. "Are you ready for the outlying difficulties and the ever-lying difficulty makers?" he asked Hurst, as the work on horses began to pick up pace. "The question is, will a jury convict in view of the 1% exceptions. I am not sure." Any exceptions, Bateson stressed, would be seized on by the opposition and claimed as successful predictions of the ancestrian theory.[76]

The strain of anticipation became visible at a lecture Bateson gave at University College in mid-November 1905, shortly after submitting Hurst's horse paper to the Royal Society. Bateson was presenting his own, competing-with-Weldon evening course on heredity—much to Pearson's amusement. "B. gave 157 to 35 as a good 3 to 1 result the other night!" he reported to Weldon. "He also lost his temper when a member of his audience cited something he had heard in your lecture. 'Weldon knows nothing about this subject, he has never bred anything himself.'"[77] Yet as Bateson, too, began to deal with organisms he had not bred himself, the ratio-muddling exceptions became harder to ignore. The lecture syllabus had promised to address two male-only hereditary conditions in humans, hemophilia and color blindness. But as he admitted exasperatedly to Hurst afterward, much of the data just did not fit Mendelian patterns at all well.[78] It was in a state of growing alarm that he thus began getting ready for the upcoming reading of the horse paper, scheduled for 7 December, at what was shaping up to be another of the Royal Society's "discussion" sessions, with statistical work in heredity once again in the frame.[79] By way of trying to bulk up the Mendelian side, he hurriedly submitted a paper with Punnett and Saunders.[80] As the day approached, his communications with Hurst became increasingly frenetic. Then, on the day itself, traveling to the Piccadilly premises, Bateson noticed that one of the horses pulling their cab was chestnut, but with a black tail: a Batesonian nightmare come to life.[81]

Bateson's anxieties were not misplaced. After the horse paper was submitted, Weldon—who got sight of it early on as chair of the society's zoological committee—was initially bowled over. "Thank God I have not finished that futile book. There must be a chapter of Race Horses. Damn!!!" he wrote to Pearson.[82] To Galton, Weldon confessed that reading the paper made him want "to most publicly and quickly retract everything I had said against Mendel."[83] Just to be sure, however, Weldon went to check Hurst's source: the twenty-volume *General Stud Book of Race Horses*. And he could scarcely believe how little the evidence recorded there conformed to the Mendelian ratios reported by Hurst—or how willfully selective Hurst must have been with that evidence in order to make it appear otherwise. "I think I may feel free from any need to apologise for opposing Mendelism so far," Weldon wrote to Galton,

> but there is a difficult point to settle. This paper will be read before the R.S. directly, and I must go and discuss it. What on earth am I to say? Hurst has had the same marvelous luck that Bateson had with his Peas. He looks at the stud book, and finds Mendel. I look at it, and find what I have told you. I know I am sober, and I think I am about as sane as I ever am: what can I conclude? . . . It is horrible to think that a man has picked his evidence in a case of this kind. I feel a thorough blackguard for entertaining the idea: but I do not see any alternative, unless there really is a devil somewhere about Bletchley, who makes it possible to see the same thing in one way here and otherwise in Cambridge!

What was clear, Weldon continued, was that thorough study of the stud book presented an outstanding opportunity "to see how far the idea I have been trying to think out is true,—that your dominance of elements in the stirp is determined by the condition and relative numbers of elements in each germ cell at the moment of fertilisation."[84]

By the time of the Royal Society meeting, presided over by Lord Rayleigh—the very embodiment of the scientific virtue of careful scrutiny of the exceptional cases—Weldon had shed whatever inhibitions he may have had about landing a knockout blow. Bateson, full of self-doubt, struggled to respond. At one point, in desperation, he pulled out of his bag three taxidermied pigeons, two black and one white, to show the assembly that Mendelian inheritance was irrefutably, right-before-your-eyes real. But Rayleigh ruled them irrelevant to a discussion on horses. Defeated, Bateson retracted the paper.[85]

"There is no good trying to feel happy about today's battle," Bateson wrote to Hurst that night. "We leave with a bad wound, that won't heal for a long time." Nor was there any respite from battle in view, since, before the meeting was over, Weldon had agreed to return shortly with a printed version of his criticisms. Soon not just Bateson and Hurst, but Saunders, Punnett, Bateson's wife Beatrice, and Hurst's sister Alice, were hard at work in hopes of shoring up the original paper's case for explaining away the exceptions—non-chestnut horses from chestnut parents, and so on—as due to errors by breeders or by printers.[86] Unexpectedly, a bit of good news arrived: Bateson had not, it turned out, retracted Hurst's paper after all, since, technically, the rules of the society did not permit retractions, but only—a gloriously English workaround—indefinitely delayed postponement of publication. So, if desired, there was still a chance that Hurst's paper could appear in print, with a note appended answering Weldon's criticisms more convincingly than the Mendelians had managed to do at the meeting.[87]

Weldon, meanwhile, was deep in the stud book too. Before he left for a Christmas holiday with Florrie in Rome, he told Galton that, in a number of exceptional cases, the supposedly erroneously recorded coloring of a horse was subsequently *inherited*. Equally exciting was how well Galtonian theory made sense of what, on Mendelian theory, looked like the overproduction of bay or brown offspring when chestnut mares were crossed with bay or brown horses that had either a chestnut parent or a chestnut foal (so, in Mendelian terms, "*DR*" bays or browns). The solution lay, as it so often did, in expanding the unit of analysis to include older ancestors; for, with British racehorses, a taste for chestnut coloring was only very recent, which meant that only a few generations back, chestnut ancestors were very rare.[88]

In Rome, Weldon kept at it, filling card after card with painstakingly extracted pedigrees. He resented the drudgery, the more so because, with his book so close to completion, he had thought he was finally done with carping, controversial papers. But if the recent London triumph meant that his side now had the advantage, no effort could be spared. "I think the withdrawal [of Hurst's paper] will shake the credit of Mendelism in Cambridge, as well as at the R.S.," Weldon wrote to Pearson from Rome, "far more than anything else could have done." The short paper he was now preparing would consolidate those gains, for Weldon was sure, as he wrote to Galton, that the latter's concept of "changeable dominance . . . will fit all these results like a glove."[89]

Weldon submitted the paper in mid-January 1906. Offered as a preliminary statement backing up his remarks at the 7 December meeting, it also advertised the power of his forthcoming book's new interpretation of Galton's theory as pivoting not on the law of ancestral heredity (not mentioned in the paper) but on that concept of changeable, conditions-influenced dominance. As ever, Weldon warned of the dodginess of Mendelian categories, pointing out that the stud book recognized not three coat-color categories but six, each embracing a range of tints, themselves connected by intermediates. He noted that, far from non-chestnut offspring being born to chestnut parents in an irregular, patternless way (as would happen if they were all mis-recordings), on average, 1 in 60 offspring were non-chestnut, with the exact proportion going up or down depending on the ancestry of the lineage—"precisely what Galton's theory of dominance would lead us to expect," wrote Weldon, but "a difficulty in the way of any Mendelian theory involving a reasonably small number of elements." Weldon calculated the chance of chestnut being a Mendelian recessive and yet *DR* bay × chestnut crosses yielding the excess of bays recorded in the stud book at about one in a thousand. The excess of browns in *DR* brown × chestnut crosses made that recessiveness even less probable. Taken all in all, these were facts, he concluded, best "expressed in terms of the hypothesis outlined by Mr. Galton in 1872, and developed by him in his subsequent writings"—a hypothesis, Weldon added, that he hoped to discuss at greater length soon.[90]

At the Royal Society meeting three days later, Bateson was back to his usual, indomitable self. The momentum, however, stayed firmly on the other side. "I do not feel thoroughly satisfied with yesterday," Bateson wrote to Hurst afterward, "and I have an uneasy feeling that they saved their skins," adding a few days later that this latest fight was attracting so much attention that the little postscript to be appended to Hurst's paper was bound to be widely read.[91] Weldon was more upbeat, writing to Pearson that Bateson had been "certainly clever . . . and he produced the effect

FIG. 9.3. "He fell, the volume of life exhausted, fighting for the new learning": Pearson on Weldon, 1906. The bust that Pearson commissioned in memoriam is now at Oxford.

he wanted: but I think we have him." Weldon now threw himself with even more gusto into the horse pedigree work. Soon all his thoughts, and even his dreams, were full of "these infernal beasts," he told Pearson, joking to Galton, "I fully expect to find myself hissing when I brush my hair!"[92]

The letter to Galton was in response to one in which he had asked Weldon about progress on "Theory of Inheritance": "How does the book, the magnum opus, get on? Will any of it be in type this spring? You can hardly believe how much I thirst for its appearance, for your zoological facts are just those I am most deficient in."[93] Weldon explained that, with the horse work now pressing, the book would have to wait. "You are too kind to my poor little book," he replied. "It must lie in the drawer for the present. I must see the way through these horses, now. One could not write the chapter of the book without being a little clearer about this matter?"[94]

That was in early February 1906. Weldon carried on through March and into early April, analyzing ever more racehorse pedigrees, visiting the office where the records were kept (far too well-managed an operation for blunders of the scale Bateson required, Weldon judged), fulfilling his professional obligations, and making time as ever for arduous walks and learned pastimes. Then, during a joint Eastertime holiday with the Pearsons in rural Oxfordshire, he came down with what seemed like a bad cold. At his dentist's in London soon after, he collapsed. By the time he was taken to a nursing home, he was dying of pneumonia.[95]

Among the last things Weldon read was the new issue of the *Proceedings of the Royal Society*. It included his final, book-previewing paper. Printed just before it, however, was Hurst's paper, now—as Weldon discovered to his outraged dismay, expressed to Pearson (fig. 9.3)—ending with a postscript purporting to answer Weldon's criticisms while in fact ignoring them.[96] Once again, the Mendelians, with help from the Royal Society, had managed to neutralize those criticisms in advance; for who, reading the postscript, would bother even to read Weldon's paper, let alone take it seriously? He died believing that he had been outmaneuvered, now—with his unfinished manuscript doomed to remain in that drawer—for the last time.[97]

Part 3

BEYOND

10

The Success of a New Science

Working on Mendel's principles, Mr. Biffen has . . . made a most important contribution to our knowledge of the inheritance of disease, by proving that certain common diseases in wheat are transmitted to the offspring in strict accordance with Mendel's laws, so that they can be controlled, and in fact bred out . . . his work holds out the brightest hopes to farmers and landowners in these days of gloomy agricultural prospects.

CITATION ON THE AWARDING OF A MEDAL TO ROWLAND BIFFEN, LONDON, 2 AUGUST 1906

When I came to this Conference to hear of Mendelian theories I was rather doubtful; but now that I have been so much with you, and have heard all that has been said, especially (if I may be allowed to say so) this morning, by Mr. Biffen, some of whose characters at first sight seemed to be strangers to Mendel's laws, I am and will ever be an apostle of the theory.

FRENCH BREEDER PHILIPPE DE VILMORIN, LONDON, 2 AUGUST 1906

An award and an apostle were Rowland Biffen's souvenirs from the banquet at the third, and largest yet, in the series of international conferences on plant breeding and hybridization (fig. 10.1). Today, the London 1906 conference, held between 30 July and 3 August, is best remembered for the public debut of the term "genetics." Bateson introduced it in his opening speech as conference president, and it caught on fast. When Philippe de Vilmorin spoke at the banquet three days later, he concluded his remarks—a response to a toast to British and foreign delegates—with an invitation to the Royal Horticultural Society "to come and hold your next Conference on Genetics in Paris." But for Vilmorin too, plainly, the truly exciting innovation was Biffen's. Bateson underscored the excitement in a subsequent toast to the Board of Agriculture, Horticulture and Fisheries: "Those who listened to Mr. Biffen's paper this morning must have felt that here we had one of the first solid facts that has ever been discovered respecting the inheritance of disease." More such discoveries, Bateson continued, were virtually guaranteed if even a fraction of the

THE SOCIETY'S BANQUET IN THE GREAT HALL.

FIG. 10.1. The banquet at the Third International Conference 1906 on Genetics (as it was later known). The medal giving and multiple toasts took place between an eight-course dinner and violin-piano duets.

sums to be spent by the board on research into disease could be channeled in the Mendelians' direction.[1]

Publicizing Biffen's work as an emblem of the utility of Mendelism, and of the practical boons to come from greater investment in it, was already a staple of the Cambridge Mendelians' campaigning. In April, around the time Weldon died, the following passage had appeared in the *Quarterly Review*, in a fundraising puff piece titled "A Plea for Cambridge":

> Under Mr Bateson some twelve researchers are already at work following out Mendel's law in many varieties of plant and animal. The extreme importance of these studies, which, if they prove a key to heredity, will place in man's hands an instrument as powerful as Watt's application of steam, is shown by the fact that Mr Biffen has already discovered susceptibility to rust [a fungus] in wheat is Mendelian, and is thus a property which may be eliminated by breeding.[2]

With Mendel as biology's Watt, the new science was set to do for agriculture what the steam engine had done for industry. At the London conference, Biffen lost no opportunity to drive the message home. On the first evening, at the welcoming conversazione, he mounted a display illustrating what he had done. In one pot were two kinds of wheat: plants sickly unto death from infestation with rust; and plants that were the very picture of rust-free health. He reported that when the rust-susceptible variety was crossed with the rust-immune variety, the descendants all proved susceptible to rust. But among the descendants of these hybrids—as the besuited and begowned guests could see in another pot—the ability to resist infestation returned, in the ratio of three susceptible plants to one resistant one: the telltale Mendelian ratio.[3] In the paper to which Bateson referred, Biffen stressed the vast economic potential of resistance to rust being unveiled as a Mendelian recessive, noting that, within living memory, rust attack had cost German wheat breeders the equivalent of 20 million pounds sterling in a single year.[4]

In the published proceedings of the conference—with medal citations for Bateson, Wilhelm Johannsen, Becky Saunders, and C. C. Hurst as well as for Biffen, and with Mendelism-extending contributions not only from these five, but from Darbishire (on mice), Davenport (poultry), Doncaster (lepidoptera and rats), the Zürich naturalist Arnold Lang (snails), Lock (peas, maize, and jack beans), Punnett (sweet peas and poultry), recent Cambridge recruit Richard Staples-Browne (pigeons), Tschermak (peas but also wheat, rye, barley, beans, etc.), the St. Andrews breeder John Wilson (wheat, oats, peas, and foxglove), the Newnham-based Muriel Wheldale (snapdragons), Wood (sheep), and Yule (the mathematical meshing of Mendelism with the law of ancestral heredity)—we encounter a new science on a roll.[5] And so it went. Consider just a few of the achievements over the next few years. In 1907, Davenport and Hurst independently confirmed that eye color in humans obeys Mendel's law, with blue eye color inherited as a recessive. ("Two blue-eyed parents will have only blue-eyed children," as the Davenport paper's handy summary put it.)[6] In 1908, Bateson at last got the Cambridge professorship he had long craved, thus inaugurating the world's first professorship in genetics.[7] In 1909, Johannsen gave the Mendelian object par excellence the name by which it has been known ever since, "gene"—an etymological tribute, in his view, to the bit of "pangen" expressing the notion of tiny, independently acting, gametically located determiners of hereditary characters.[8] That same year, Bateson brought out a new edition of his *Mendel's Principles of Heredity*, turning what had been a defensive screed into the comprehensive textbook the fledgling field needed.[9] In 1910, a new statue of Mendel was installed in Brünn, with Bateson among the dignitaries present at the unveiling (fig. 10.2).[10] A different sort of Mendelian monument appearing that year was the single-volume *Reports to the*

FIG. 10.2. The unveiling of a statue of Mendel in the old town of Brünn, near the Abbey of St. Thomas, 2 October 1910. Bateson was not impressed with the statue, declaring it "banal and shocking."

Evolution Committee of the Royal Society, gathering what had grown to be five reports from the now-expired committee.[11] Its function was rapidly taken over by new specialist journals, including the *Journal of Genetics*, under the editorship of Bateson and Punnett, and a new magazine from the American Breeders' Association, bearing the subtitle *A Journal of Genetics and Eugenics*.[12]

If any story in science is a success story, it is Mendelism's. In chapter 12, we will track some of the twentieth- and even twenty-first-century repercussions of that success. This chapter is an attempt to get clearer on its nature and, especially, its ingredients—to step back from the details of the history reconstructed so far and ask what, exactly, accounts for the astonishing ascent of Mendelism even before its first decade was over. In other words, what explains Mendelism's success? The answer ventured below is that, in common with other successful sciences, Mendelism acquired a winning combination of teachable principles, tractable problems, and technological promise. Before setting out that answer more fully, however, I need to say more about what, as I see it, we are trying to explain when we try to explain Mendelism's success, and why some intriguing candidate explanations, for all their interest, should strike us as unsatisfactory.

1

A useful point of entry is provided by the three terms within our question: "explains," "Mendelism," and "success." Taking them in reverse order:

Success: Not just the term "genetics" but also, much more significantly, the science caught on, well beyond the research school that Bateson founded and energized. That is the kind of success whose explanation is this chapter's business.[13] Success in this sense has any number of hallmarks, institutional and intellectual. Above, I touched on some familiar institutional ones: university posts; an emerging canon of exemplary research publications; specialized journals; popularizing books for the curious (Punnett's) and the earnest (Bateson's); public memorials to the originator; a proliferating technical vocabulary. (The inventive Johannsen also gave Mendelism "genotype," for underlying genetic constitution, and "phenotype," for the associated visible character.)[14] Among the intellectual hallmarks is one that Bateson boasted about in that 1906 conference address: the new Mendelian science, he predicted, would soon "absorb and modify profoundly large tracts of the older sciences."[15]

Bateson's boast generalizes. A successful science secures bragging rights for absorbing its predecessors, including the practical successes linked to them. Intellectual ownership is claimed, and those claims become widely honored. The pace and scale of this honoring, like the pace and scale of institutional embedding, need not be uniform; and the absorption will almost always be less complete than advertised, with some elements of the older sciences only partially or superficially integrated, and others ignored entirely. The process of absorption, moreover, will leave the new science modified too, not just the older ones. Even so, there will be no mistaking whether a science gets absorbed or does the absorbing.[16] All of these points hold for Mendelism, as we will see. For now, savor the contrast between two editions of the *Encyclopaedia Britannica*. In the 1911 edition, there were separate articles on "heredity" and "Mendelism," with little overlap in their contents. In the 1974 edition, the entry on "genetics" defined it as "the study of heredity in general and genes in particular," going on to say that, although recognition of heredity's influence can be traced into prehistory, scientific understanding began with Mendel. That is what success looks like—and why it is now a job of work among historians and philosophers of genetics to expose smoothed-over heterogeneity.[17]

Mendelism: Again, Mendel was no Mendelian. He sought a new science of hybrids, not heredity; he evinced no sign of thinking that underlying each inherited character were paired somethings, countably discrete like balls in an urn; and his version of cell theory allowed for the possibility that character-making stuff sometimes mixed reversibly (in variable hybrids) and sometimes irreversibly (in constant hybrids). Yet, after 1900, a world-beating, absorption-claiming science of heredity grew from his "Versuche," with adherents who recognized in it vital pointers to the truths being discovered through experimental crossing with purified breeds. Accordingly, when Punnett, in his guise as bringer of Mendelism to the masses, tried to say what Mendelism was, he was inclined to describe it as that science of heredity developed by Mendelians.[18] Among other virtues, such a definition makes room for all of the heterogeneity rapidly contained *within* Mendelism. Even at the 1906 conference, there was contentious boundary pushing on the Mendelian side. Darbishire, for example, affirmed that he in no way saw a fatal challenge to Mendelism in the undoubted and un-Mendelian fact that, in his mouse crosses, the greater the number of albino ancestors in the pedigree of a hybrid, the greater the proportion of albino descendants. But, in his view, absorption of this hitherto anomalous pattern required modification of the Mendelian hypothesis about gametic purity.[19] Over the next twenty years, all sorts of intra-Mendelian debates would arise: between Darwinian (Ronald Fisher) and anti-Darwinian Mendelians (Bateson), Lamarckian (Paul Kammerer) and anti-Lamarckian Mendelians (Bateson again), and so on.[20]

But to leave the matter of definition there would be to suggest an amorphousness that is misleading. What gave Mendelism its distinctive character, across all its internal diversity of theory and practice, was the concept of the Mendelian gene, understood as that entity whose properties do the explanatory work in elementary genetics—for example, in explaining why two blue-eyed parents can have only blue-eyed children. This entity's chief properties are, first, that it comes in just two variants, dominant and recessive; and second, that it is completely determinative, in the sense that nothing other than gene variants or "alleles" (as Bateson's terminology got abbreviated) needs mentioning. If we want, in elementary genetics, to explain why two blue-eyed parents will have only blue-eyed children—or why some pairings of a blue-eyed parent and a brown-eyed parent will yield only brown-eyed children, and other pairings will yield a mix of blue-eyed as well as brown-eyed children—we need to consider only alleles for eye color. They explain without remainder.[21]

Of course, even before "Mendelism" was coined, the possibility that things were more complicated was openly acknowledged. That complication, however, was not central to Mendelism. The gene concept was. And that has made all the difference.

Explains: Consider again the explanation just gestured at, of why two blue-eyed parents have only blue-eyed children. "Because blue eye color is recessive" counts as an explanation within elementary genetics because it connects what we want to explain with what, within elementary genetics, counts as a fact, or maybe a whole family of facts, about causes. (To talk Mendelian for a moment: a recessive phenotype gets expressed only when there is a double dose of the recessive allele—that is, when an organism is a recessive homozygote. With both parents having nothing but recessive alleles to transmit, offspring can inherit nothing but recessive alleles, so offspring phenotypes must be recessive.) Also, notice that the question "Why do two blue-eyed parents have only blue-eyed children?" implicitly—via that "only"—incorporates a contrast case, a foil. If we were to make the implicit explicit, the question would read, "Why do two blue-eyed parents have children with blue eyes rather than some other eye color?" The presence of the foil, whether implicit or explicit, is crucial in directing us toward the facts that are *relevant*, given what we want to explain. After all, lots of causes bear on eyes, and there are lots of facts about those causes. The answer to the question "Why do two blue-eyed parents have children with blue eyes rather than children with no eyes?" will be entirely different from the answer to the question about blue-rather-than-brown eyes. (A witty complaint against genetics, from the 1930s, was that it failed to satisfy anyone "interested more in the back than in the bristles on the back and more in eyes than in eye color.")[22]

At least some of the time, then, to explain is to give an answer to a "why?" question, where the answer connects whatever is to be explained with relevant facts about causes, and the question specifies or implies a relevance-highlighting contrastive foil ("why X rather than Y?"). That definition seems to fit a good deal of scientific explanation, though not where mathematics, rather than causation, is doing the explanatory heavy lifting. And it seems to fit a good deal of historical explanation too, though we need to make allowances, in our connection-making, for facts about reasons as well as about causes.[23]

To summarize so far: when we seek to explain Mendelism's success, we are, I take it, in search of facts about causes and reasons bearing on the question of why Mendelism—the science of the Mendelian gene writ large—caught on in the extraordinary way it did, rather than remaining a local enthusiasm of Bateson's school.

2

With our question clarified, we can move on to consider some candidate answers. The following three ascribe Mendelism's success to causes so powerful that the doings and decisions of individual men and women—the eventful stuff of the previous chapters—look inconsequential: foamy nothings carried on the deep-running tide of history. In keeping with the boldness of these explanations, I will state them boldly.

1. Genes exist, and any scientific community worthy of the name will eventually enshrine a version of the Mendelian gene concept. Why that enshrining happened only after 1900 in the case of our scientific community is an interesting question. But its answer will illuminate only the *timing* of the concept's enshrinement, not why that concept, rather than some other concept, was so enshrined. The Mendelian gene concept reflects reality, and that is the reason the gene came to occupy the place it does in our science of biology.
2. Unequal societies such as ours depend for their stability on the widespread belief that inequality is foreordained. Back when the church was the main voice of authority in Western culture, people believed that their fates were in God's hands, and that He decided who would be rich and who poor. After science took over from the church, nature supplanted God as the great fate determiner. The Mendelian gene concept begat the latest and most advanced form of the notion that nature is destiny. (Recall Bateson's approval of the "scientific Calvinism" tag.)
3. Empire and capitalism have left their imprint on the sciences, biology very much included. The idea that parents pass on some kind of defining, constancy-preserving essence grew conspicuous only thanks to the post-Renaissance transplantation of plant species, animal species, and human races from environment to environment. And the genic form of that idea owes more than a little to circa-1900 enthusiasm for rationalized production and data management. The Mendelian gene concept is modernity made biological.

These answers are inspired by my reading of the secondary literature on the rise of Mendelism. What matters is not that anyone would endorse them as stated, but that, when stated in this way, they reveal themselves as belonging to a common genre. For all their outward diversity, they depict the Mendelian gene concept as summoned forth: by reality (1); by the functional needs of an unequal society (2); or by modernity (3). They can even be run together, on the view that (3) adds welcome detail to (2), and this combination in turn neatly explains, apropos of (1), why the concept emerged when it did rather than at some other point in history.[24]

The sweep of these summoning-forth explanations makes them exciting. But it also makes them obscure, since we cannot easily understand how, exactly, the summoning forth could have happened. To start with (1): it posits that, when the time was ripe, reality delivered the Mendelian gene concept into the scientific inventory of real entities. The trouble is that this inventory formerly included crystalline spheres, phlogiston, and caloric. On the face of it, then, an entity's being real or not is irrelevant to its catching on. Did reality's delivery mechanism fail to function in the cases of crystalline spheres, phlogiston, and caloric, but function properly in the case of the gene? What could such a mechanism be?[25] Comparable mysteries about mechanism arise in relation to (2), which asks us to accept that, somehow or other, a well-functioning society in need of a pacifying idea can communicate this need to the individual members best placed to satisfy it; that—whatever they think they are doing—these members respond in the needful way; and that everyone else, on encountering the idea, is duly pacified.[26] The obscurity of (3) lies less in its mechanism than in its seemingly arbitrary promotion of one strain of modernity above others in its potency. If we take modernity to mean standardization, purification, universalism—the modernity of, say, Ford and Le Corbusier—then yes, the Mendelian gene looks eminently modern. But if we replace Ford and Le Corbusier with Einstein and Picasso as exemplary moderns (and why not?), then modernity means doing away with the one, privileged standard. In that case, Weldon's relative conception of dominance looks like a better match than the Mendelian absolute conception.[27]

Even if (1), (2), and (3) could be made fully intelligible, there would remain the question of whether we should believe them. To start again with (1): take, as a surrogate for the wider class of explanations within elementary genetics, the Mendelian explanation of eye color—a success story within the larger Mendelian success story. "Many people who know nothing else about genetics," writes the evolutionary biologist John H. McDonald, "think that two blue-eyed parents cannot have a brown-eyed child." Indeed, the spread of this idea ensures that, when two blue-eyed parents do have a brown-eyed child, insinuations of infidelity reliably follow. So not only historians have a stake in knowing whether, as per (1), the Mendelian gene for eye color is real. The short answer is no. The evidence against it began accumulating almost immediately after Davenport and Hurst published. A 1909 study of Wisconsin university students and their families found that "eye colors cannot be divided into sharply defined classes" because "all sorts of intermediate shades occur as well as irregularities in the distribution of pigment over the surface of the iris," and furthermore, that although blue-eyed parents generally have only blue-eyed children, there are exceptions. A 1918 study on families in New York City's immigrant Italian, Czech, and eastern European Jewish communities reported that, within the total set of families in which both parents had blue eyes, 12% of the children had brown eyes. These were the most recent data that McDonald could find when he went looking in the early 2010s for published data on parent-offspring eye color. (Another mark of the success of a science is that its best-known results cease to be the subject of critical investigation.) On the underlying molecular biology, however, a great deal of research has been done, and it is wholly undermining of the idea that there is a Mendelian gene for eye color. As McDonald summarizes, "Eye color is not an example of a simple genetic trait, and blue eyes are not determined by a recessive allele at one gene. Instead, eye color is determined by variation at several different genes and the interactions between them, and this makes it possible for two blue-eyed parents to have brown-eyed children."[28]

Let us allow, in charitable spirit, that the loss of one Mendelian gene, however prominent, from the scientifically approved list of real genes does not in itself spoil (1). But then, how many genes would need to be delisted before we would be entitled to wonder whether the general concept indeed latched onto reality sufficiently to explain its success? My quotations from McDonald come from his pamphlet *Myths of Human Genetics* (2011), where seventeen other supposedly binary, single-gene characters, from arm folding and "asparagus urine smell" to tongue rolling and widow's peak, come in for similarly rough treatment.[29] In chapter 12, I will draw a comparable lesson from the molecularization of seed shape in the garden pea. The point is not that the Mendelian gene concept's latching onto reality will turn out to have zero evidence in its favor, but that even the briefest inspection of the evidence we might use to assess (1) plunges us back into those puzzles over its obscurity.

The same holds for the evaluation of (2) and (3), albeit the relevant evidence looks nothing like the evidence relevant to (1). For (2), consider that not everyone, on encountering Mendelism, lost the desire to shake up the status quo. In 1909, Gaylord Wilshire, who at that time ran a small socialist press in New York, published the American edition of Punnett's *Mendelism*. At the back of the book Wilshire advertised two socialist-scientific classics in English translation: Marx's *Contribution to the Critique of Political Economy*; and *Socialism and Modern Science (Darwin, Spencer, Marx)* by the Italian socialist Enrico Ferri. At the front was a preface from Wilshire extolling Mendelism's socialist virtues. Here, in Wilshire's view, was the biological theory for socialists like him, convinced that the longed-for shift from a competitive to a cooperative system would never come about through half-measures but only through revolutionary transformation. He took courage from the theory's mathematical precision:

> If then all evolution proceeds by mutation I think the case is still strong for my declaring the change in society must proceed by mutation. The most interesting part of the Mendelian theory is that it is a mathemati-

> cal one, and this is what charms me regarding the theory of mutation in society. It, too, is a mathematical one. You can count up the number of machines and count up the number of men, and can prophesy the time almost exactly when Socialism must come in order to make a balance between production by machines and consumption by men.

Nor was Wilshire in any way put off by Punnett's insistence that men are born, not made. If the environment counts for little against our inherited qualities, then, for Wilshire, critics' fears that socialism would "reduce us all to a dead level" were proved groundless.[30]

The above is, to put it mildly, not what (2) leads us to expect. Should that be held against it? If not, why not? And if so, how many more such instances must we collect before we give up on (2)? It is hard to escape the conclusion that (2) may be correct, but we are in no position to find out, or even to know how to find out. That holds for (3) too. Recall its claim that, if not for early modern travels and transportings, the conviction would never have arisen that a similarity-preserving something gets transmitted from parents to offspring. Now consider the evidence of two passages from the writings of a well-known early modern author. The first comes from the latter part of *Julius Caesar* (1599), when, in the aftermath of Caesar's assassination, Brutus and Cassius are in fraught discussion. Afterward, Cassius apologizes for losing his temper, blaming "that rash humour which my mother gave me." Brutus, accepting the apology, replies, "from henceforth / When you are over-earnest with your Brutus, / He'll think your mother chides, and leave you so." According to (3), this exchange, in which the idea of a transmitted quick temper is played for friendship-restoring laughs, must in some way be a response to Elizabethan imperialism. But is it? What reason could there be for thinking so? Would it not be much more straightforward to suppose that, in the vernacular culture of the time, transmission of certain traits from parents to offspring was just taken for granted? Such questions arise with even more force in relation to the second passage I want to mention, from a slightly earlier play, *The Merchant of Venice*. Having eloped with her boyfriend, Lorenzo, who is a Christian, Jessica, the daughter of the Jew Shylock, is en route to her new life when she encounters Lancelot the clown. Lancelot tells her that she should hope that someone other than Shylock fathered her—a "bastard hope," he wittily calls it—or else she is "damned." Her conversion to Christianity, in other words, will do nothing to free her from that destiny of damnation that her father, as a Jew, gave her (if he is her father): an idea that emerges much more obviously out of ancient and medieval religious teachings than from early modern imperial traffic.[31]

3

On balance, then, we create at least as many problems as we solve if we try to explain Mendelism's success as a grand-scale summoning forth. If we try instead to explain it by reference to human-scale causes and reasons, we have a shot, at least, at producing an account that is both intelligible and well supported empirically. When it comes to Mendelism, a remarkable example of an explanation at this more promising scale lies to hand. "Biometrician versus Mendelian: A Controversy and Its Explanation" (1974), by Donald MacKenzie and Barry Barnes, was an early case study in the sociology of scientific knowledge, associated with the Science Studies Unit at the University of Edinburgh—the "Edinburgh school"—where Barnes was MacKenzie's PhD supervisor. A look back at this important and impressive paper affords opportunities to consider the biographies of Bateson, Weldon, and Pearson afresh, and also to ask whether the really interesting question is not about why Mendelism succeeded, but why these men and their allies could not bridge their differences, make common cause, and put the battle over Mendel behind them.[32]

"Among the many controversies littering the history of science," MacKenzie and Barnes began,

> that between the Mendelians and the biometricians in the early Twentieth Century is as notable as any. Its study is central to the understanding of the rise of modern genetics, and is also important in the history of statistics. Moreover, the controversy is well suited to sociological study in that its history has been thoroughly explored and described, and among the workers who have undertaken this task there appears to be little disagreement about the course of events. There is, in fact, a consensus upon all the main points in the story. The biometric school was a small and tightly knit group. Its leaders were the mathematician, Karl Pearson, and the zoologist, W. F. R. Weldon. The Mendelian side of the debate was more amorphous, and we will concentrate our attention on the leading British Mendelian, William Bateson, and his immediate co-workers.

They followed with a superb overview of the existing historical scholarship up to that point. On their reading, the main issue in contention was whether change in nature is continuous (biometric) or discontinuous (Mendelian). Turning from description to explanation, they asked not just what explained this entrenched divergence but what kind of explanation could satisfactorily do the explaining. One by one, they examined and dismissed various possibilities: the psychologies of the individuals (deemed irrelevant because the behavior of groups, made up of psychologically diverse individuals, was what mattered here); the real structure of nature (irrelevant because the question of nature's real structure was precisely what was being disagreed about, and not even with hindsight can the Mendelians be held to have believed what they did because of how nature is, while the biometricians believed something else because of intellectual defect or ideological distortion); and different kinds of training (irrelevant because Weldon and Bateson were both products of Balfourian Cambridge in the late 1870s and early 1880s). With alternatives thus eliminated, the only kind of explanation remaining, according to MacKenzie and Barnes, was the kind of sociological explanation that their Edinburgh colleague David Bloor was just then defending, "invoking the whole range of socialising experiences of the actors involved, and the general milieu of their time."[33]

MacKenzie and Barnes then proceeded to give such an explanation. On the Mendelian side, they noted, we have the socially elite Bateson, who in class terms—as the son of the master of St. John's, Cambridge—was an aristocrat. Typically we expect aristocrats to be conservative upholders of the social and political status quo, and Bateson, they claim, fitted the mold. In society, he liked things to stay just the way they were; in nature, he saw not ceaseless change but long periods of stasis. Over on the biometrician side, they concentrated on Pearson. Pearson was middle class and, in line with class interests, favored gradual, progressive change as both a sociopolitical ideal and a scientific reality. Much of the rest of the paper fleshed out these alignments, the better to help us see how each man's class-bound "social experiences" and "general milieu" would have biased him to one side or the other on the matter of what change is like in nature, with the result that their respective positions on natural change matched the social and political interests of their respective classes. That, for MacKenzie and Barnes, is why the debate had the charged character it did, and why it was in practice irresolvable.[34]

With the benefit of nearly another half century of further historical study of the debate, what should we make of this explanation? Consider the Mendelian side first. In characterizing Bateson as a conservative, MacKenzie and Barnes drew heavily on a brilliant but flawed paper by the historian of biology William Coleman, "Bateson and Chromosomes: Conservative Thought in Science," published in 1970. How, Coleman asked, could Bateson have got it so right about Mendelism but so wrong about chromosomes? How could he have been so radical and innovative for the first decade after 1900 and then so crustily conservative thereafter, resisting the identification of Mendelian genes with bits of chromosome? Coleman's ingenious answer was that, in both phases of Bateson's scientific career, he was in the grip of an all-embracing, Enlightenment-rejecting "conservative style of thought," as identified in the 1920s by the Marxian sociologist Karl Mannheim. For Coleman, the most important feature of this style of thought was its anti-materialism. Bateson's anti-materialism led him to admire Mendel's work, with its emphasis on discontinuous characters and whole-number ratios, as a biological counterpart to the vibrational physics then in vogue in Cambridge. Yet that very same stance later made Bateson resistant to the idea that ultimately Mendelian factors would turn out to be bits of chromosomal matter.

Coleman's basic insight is sound: Bateson's quest for a mathematically framed vortex biology, grounded not in the chemical properties of particular, static substances but in the physical properties of general, dynamic systems, was indeed, as we have seen, a big part of his initial attraction to Mendel, and a big part as well of his arm's-length attitude toward chromosomes. But Coleman, in the spirit of Mannheim's precedent, represented this preference of Bateson's as an expression of an anti-materialist conservatism pervading the whole of his life and thought. Other supposed expressions included Bateson's passion for surrounding himself with beautiful paintings; his doubts about democratic faith in the equal distribution of intellectual capacity (and his related belief that significant intellectual advance was the gift of rare, intuitive genius); and his disdain for commerce and its sullying bourgeois concern for the utility of scientific knowledge.[35]

Here is where the difficulties start. Weldon collected art too (fig. 10.3), and Pearson took as determinist a view as Bateson did about potential at birth. So there was nothing distinctively Mendelian, let alone Mannheim-style conservative, about Bateson on those fronts. As for Bateson's supposed anti-utilitarianism, he was indeed, into his later twenties, a typical product of Rugby School, high-mindedly dismissive about inquiry for anything but inquiry's sake. But his time at the Plymouth station tempered his arrogance, showing him how much he stood to learn from practical men. Thereafter no one was more energetic in courting those men, in exhorting his scientific colleagues to work with them, and—in the sixteen years he spent as director of the John Innes Horticultural Institution, after leaving Cambridge in 1910—in sponsoring as well as conducting research aiming to throw light on their problems. In 1911, when the BAAS again met in Cambridge, Bateson used his position as president of its new agricultural section to bemoan the snobbery that still regarded applied science as suitable only for second-class minds. In 1921–22, his standing as a specialist in the science of garden peas was so high that he served as an expert witness in a lawsuit concerning the identity of peas marketed under one varietal name but, it was suspected, really belonging to another, inferior variety. And from first to last, he stressed the utility of Mendelism as a boon to breeders and a badge of its truth.[36]

FIG. 10.3. A 1928 oil sketch for a portrait of Florence Weldon. The sketch is now in Oxford's Ashmolean Museum, along with the paintings in the Weldons' collection, some of which hang on the wall behind her.

So the image of Bateson as an all-around conservative does not survive contact with a more generous sampling of the evidence than Coleman furnished. As for MacKenzie and Barnes's characterization of the other, biometrician side, the trouble, of course, is their focus on Pearson. To be sure, Pearson backed Weldon in his attacks on Mendelism, going on, after Weldon's death, to build on his critical efforts by making the Galton Eugenics Laboratory at UCL a home to scrutiny of the human data on which Mendelians tried to anchor a number of unit-character, single-factor claims. But Pearson never shared anything like Weldon's animus toward Mendelism, or suggested that the choice between Mendelism and a Galtonian science of inheritance was somehow fundamental. On the contrary, as early as summer 1904, Pearson publicly self-identified as someone whose own work showed how to transcend that divide—something that Pearson, at ease with Mendelism's men-are-born-not-made message and its indifference to the effects of environment, found straightforward to do. Weldon could hardly have been more different. From the start, the fight against Mendelism was Weldon's fight. And for him, it was never about continuity versus discontinuity—a contrast he hoped (in vain) that his dice rolls would make an irrelevancy. For Weldon, the problem with Mendelism was something he never got Pearson to care about at all: its enthronement of a conception of dominance that licensed incuriosity about all the variability that comes when context changes. Put the same determinant in a different ancestral-germinal, and/or physiological, and/or physicochemical context, and its effect on a body can be different. Weldon wanted a science of heredity that made this insight the starting point. That was what he was fighting for.[37]

In sum, the answer that MacKenzie and Barnes gave to their question has become, in light of fuller evidence, untenable. When we restore Weldon to his rightful place as the central oppositional figure (and so restore dominance, not discontinuity, to its rightful place as the central issue for the opposition), and restore the practical-utility dimension to the image of Bateson (thus undermining his status as a counter-Enlightenment, stasis-besotted conservative), it becomes impossible to hold that Mendelism became as controversial as it did because of divergent alignments over natural and social change.[38] The loss of MacKenzie and Barnes's answer should also, however, lead us to question their question. Once we know what drove Weldon, we need no longer puzzle over why he was so bothered about Mendelism's increasing success with his peers, and why he did not simply give up or give in. We see more clearly too that, then as now, talk of "synthesis," "reconciliation," and so on between his perspective and the Mendelian perspective masks what was actually a takeover. A science of inheritance can start from changeable, relative dominance (Weldon's peas) or from unchangeable, absolute dominance (Mendel's peas), but not from both. Lying behind the puzzlement about the protracted debate over Mendel is a misapprehension about what happened after Weldon died: there was conceptual merging, but it was done on Mendelian terms.[39]

We will return to the fate of the Weldonian perspective in genetics. For now, as a case in point, consider the fate of the law of ancestral heredity. One might have guessed that, with the triumph of Mendelism, ancestral heredity must have gone the way of crystalline spheres, phlogiston, and caloric. Not at all. Recall that Yule in 1902, and then Pearson the next year, demonstrated how to derive versions of Galton's law from versions of Mendelian presuppositions. Their aim was to dispel a sense of conflict between Mendelian and Galtonian approaches. At the third international hybridization-cum-genetics conference in 1906, Yule picked up where Pearson had left off, but now to vindicate Mendelism. According to Yule, for a character like stature, the law should be interpreted as showing that, thanks to myriad Mendelian factors interacting in chancy ways within diverse environments but without dominance effects, the correlation coefficient could be expected to halve with each succeeding generation. Today, Yule's paper is honored as the opening statement of "quantitative genetics": the branch of genetics concerned with quantitative (i.e., continuously varying) characters. In its methods, quantitative genetics is biometric through and through. But in its name and identity, it belongs to Mendelism.[40]

4

"There is a tide in the affairs of men / Which, taken at the flood, leads on to fortune: / Omitted, all the voyage of their life / Is bound in shallows and in miseries." Thus spoke Brutus to Cassius, shortly after their exchange on hereditary rashness.[41] In summer 1906, with Mendelism at the flood, Yule repackaged his old results as an unambiguously Mendelism-extending contribution. Darbishire did likewise. No theory of success in science is needed to make sense of their actions. Other things being equal, moderately ambitious people, in and out of science, tend to go with the flow.[42] What cries out for explanation about Mendelism is its success among people who previously had no idea that there was a science of heredity, let alone which way the tide in it was running. If we grant that, for the reasons I have given, invocations of the veracity, functionality, and/or modernity of the Mendelian gene cannot do the job of explanation, what can?

Thomas Kuhn's 1962 classic *The Structure of Scientific Revolutions*—source of that ubiquitous jargon phrase, "paradigm shift"—seems, on the face of it, an unlikely ally. Kuhn's famous schema of scientific change through paradigm-shifting revolutions is a manifestly poor fit for the rise of Mendelism. Only the grossest Mendelian partisan could argue that, as per Kuhn, the science of heredity matured into Mendelism around 1900, or that, as the anomalies piled up around a hitherto all-conquering Galtonian paradigm, there ensued a crisis resolved only with the installation of a new, all-conquering Mendelian paradigm. Again, Bateson was a *Galtonian* partisan when he first welcomed Mendel's law, seeing in it a promising extension of Galton's law. Moreover, Bateson's subsequent transition, from someone who took Galton's law to be the great generalization and Mendel's law to be the special case into someone who took the reverse view, involved no Gestalt-switching leap from one "incommensurable" paradigm to the other. Weldon and Bateson had no trouble understanding each other. They just disagreed.[43]

Even so, Kuhn's book remains rich with insights into how successful sciences become entrenched. On my reckoning, these insights will take us two-thirds of the way toward an explanation of success that deals in human-scale causes and reasons (cf. summoning-forth explanations) while taking seriously what those involved took seriously (cf. MacKenzie and Barnes). Although I will dwell exclusively in what follows on Mendelism's success, the same combination of teachable principles, tractable problems, and—going beyond Kuhn—technological promise can, I suspect, explain other, comparable success stories in the history of modern science.[44]

Teachable principles. The Mendel of our textbooks is largely the creation of Bateson and his fellow Mendelians. The following précis is from the introductory textbook in genetics used until recently at my own university:

> The first person to obtain some understanding of the principles of heredity—that is, the inheritance of certain traits—was Gregor Mendel. From his breeding experiments with garden peas, Mendel proposed two basic principles of genetics. In modern terms, the principle of segregation states that the two members of a single gene pair (the alleles) segregate from each other in the formation of gametes. For each gene with two alleles, half the gametes carry one allele, and the other half carry the other allele. The principle of independent assortment, proposed on the basis of experiments involving more than one gene, states that genes for different traits behave independently in the production of gametes. Both principles are recognized by characteristic phenotypic ratios—called Mendelian ratios—in particular crosses. For the principle of segregation, in a monohybrid cross between two true-breeding parents, one exhibiting a dominant phenotype and the other a recessive phenotype, the F_2 phenotypic ratio is 3:1 for the dominant : recessive phenotypes. For the principle of independent assortment, in a dihybrid cross, the F_2 phenotypic ratio is 9:3:3:1 for the four phenotypic classes.[45]

Compare the basic choreography of ideas here with that in the "Versuche" itself, as recounted in chapter 1. In the textbook, the motion is from a simple experiment in heredity to its explanation (segregation), followed by a more complicated version of the experiment, then by its explanation (independent assortment). In Mendel's original paper, by contrast, the description of his initial, single-character crosses gives way not to an explanation but, in keeping with the topic announced in his title, an inquiry into the long-run fate of the hybrids—the 2s in the 1:2:1 version of the 3:1 ratio. Next we get a description of his two-character crosses, but though we learn that plants grown from the seeds of his hybrids yielded 315 round-and-yellow seeds, 101 wrinkled-and-yellow seeds, 108 round-and-green seeds, and 32 wrinkled-and-green seeds, the possibility of smoothing and joining those numbers into a 9:3:3:1 ratio is never mentioned. Instead, Mendel proceeds to report what happened when he planted those seeds, and to show that the resulting plants fell into the categories derived when he multiplied the "law valid for *Pisum*," $A + 2Aa + a$, by its *Bb* version. He presents this fitting of his union-of-characters crossing data with his union-of-combination-series mathematics as the main principle established, though he points out too—as something established in addition, rather as an explanation—that the united characters behave independently of each other. Only at this point does he consider what, at the level of the gametes, might explain all of these results. In summarizing, he writes not of a principle of segregation, but of egg and pollen cells from the hybrids representing, in equal numbers, the constant forms of their true-breeding parents.

Historians of science like to moan about textbook distortion. Kuhn helped them see why, in successful sciences, bad history and good pedagogy often go together. A science becomes, and then stays, successful only if its students become invested in it, cognitively and emotionally. Imparting a sense that they are joining a proud tradition, with an illustrious founder, plays a role. But what holds and keeps them is their stepwise induction into that tradition. You might be intrigued to learn about Mendel's discovery of the 3:1 ratio, and impressed with the experimental design that brought that ratio to light. But when, at the next step, you see how beautifully the segregation of alleles explains the pattern, you are (textbook) Mendel's. That second step is not difficult—recall Weldon's remark about booming Mendelism as "so simple that everyone can understand it"—but you are starting to have to work to keep up. The effort required increases considerably when, at the next step, you have to wrap your mind around the much more complex pattern you meet when Mendel crosses plants that differ in two characters. But the Punnett square aids you in tracking the sixteen possible combinations, and then in spotting the groupings that form the 9:3:3:1 ratio. And so on, step by step, extension by incremental extension, as you venture ever further into what

FIG. 10.4. Punnett and Bateson in 1907, in a photograph captioned "The Partners" (*above*); and the "Punnett square" included in the 1907 edition of Punnett's *Mendelism* to show how the 9:3:3:1 ratio of a two-character or "dihybrid" cross comes about (*right*).

geneticists know and how they know it. At no point, through all that you will learn about incomplete dominance, codominance, effects of the environment, and so on, do you query the soundness of that Mendelian foundation. It will never occur to you to ask whether, say, garden peas might be anything other than round or wrinkled, yellow or green. After a short while, the only important question is, can you see how to extend the principles you have learned to the next problem?[46]

The enterprise of making Mendelism was not separate from the enterprise of making Mendelism teachable. The science-defining interpretive choices that Bateson made in reading Mendel's paper were exactly what turned the paper into something twentieth-century biologists interested in heredity, along with everyone else, could learn from: the elevation of gametic purity to the essence of Mendel's discovery; the privileging of his ratios, explicit and implicit, over his combination-series algebra as bearers of that discovery's mathematical credentials; the simultaneous promotion of dominance (which went with the ratios) and demotion of

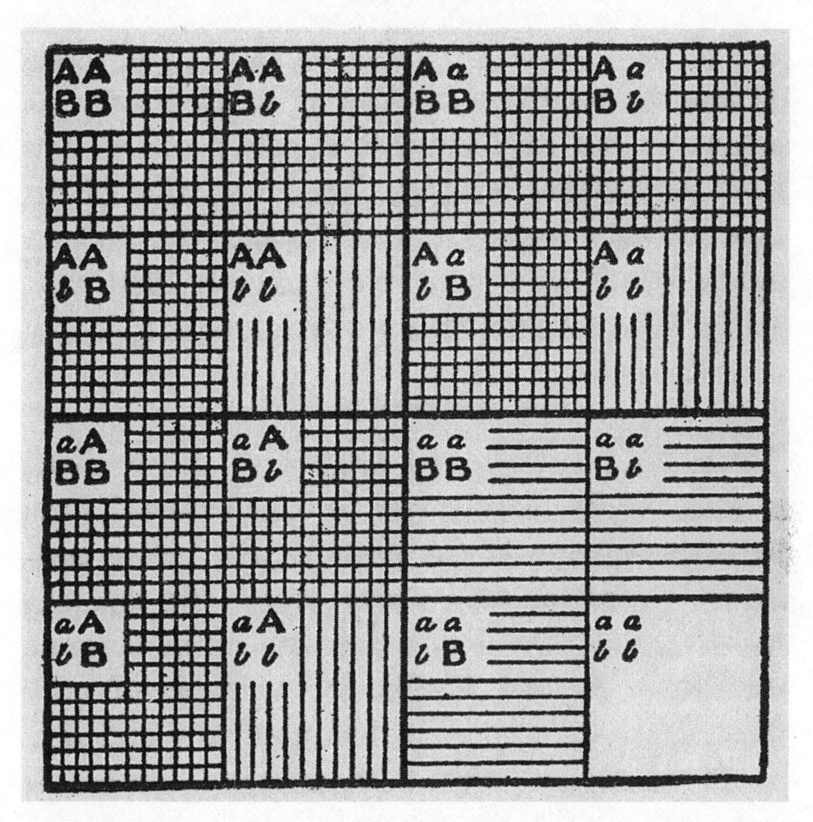

dominance (as no part of the discovery's essence); the insistence that his method of experimental crosses between organisms cleanly differentiated by this-or-that characters be the basis for all future knowledge of heredity.[47] And of course, Bateson wanted people to learn, and had the good teacher's instincts for theater. We saw those instincts at work in that 1902 lecture to the breeders in New York, with his enticing talk of hereditary qualities now becoming as swappable as elements are for the chemist. Or consider the massed Mendelian exhibits at the conversazione at the start of the 1906 conference, from Biffen but also from Saunders, Darbishire, and others. Hurst took Mendelian visual pedagogy to the next level the following summer, when he invited guests from that year's BAAS conference, held in Leicester, to his nearby home village, Burbage, where schoolchildren obediently arrayed themselves to illustrate the Mendelian pattern for eye color.[48]

It is Punnett, however, who was Mendelism's outstanding pedagogue in its first decade (fig. 10.4). His *Mendelism*, especially in its 1907 second edition, is a model of clarity and challenge. After introducing his square to help readers consolidate their understanding of two-character, 9:3:3:1

crosses, he next features it in an explanation of how the seemingly anomalous case of a one-character, 9:3:4 cross can easily be brought into the Mendelian fold by supposing that there is "interaction," as he called it, with another, impinging character pair.[49] A sign of the book's success, along with the number of copies sold (seven thousand by 1910 in Britain), is its effect on one of the readers of the American edition quoted from above. In autumn 1909, Alfred Sturtevant was a seventeen-year-old freshman biology student at Columbia University in New York, where *Mendelism* was not on the curriculum. But Sturtevant read it, then continued reading around the subject, eventually finding his way to Hurst's paper on the coat colors of horses. Sturtevant reckoned that data on American horses were better suited to the purpose. When, in spring 1910, he showed his five-factor Mendelian analysis to his teacher, the experimental embryologist T. H. Morgan, Morgan—who had omitted mention of Mendelism from the introductory lectures he was giving—enthusiastically encouraged publication.[50]

Tractable problems. Some students, like Sturtevant, go on to become researchers. According to Kuhn's analysis, that transition is one of degree. For researchers too, the aim remains the extension of earlier, exemplary investigations. The difference is that whereas students are given problems to solve, researchers must find them. The successful researchers—the ones whose work goes on to enhance the success of the science—acquire the knack of identifying interesting problems to which the approved methods and concepts can ultimately be made to extend, ideally in ways that in turn suggest new interesting-yet-soluble problems, whose solutions in turn suggest new interesting-yet-soluble problems, and so on.[51] From Mendelism's beginning, there was always a new character, or a new variety, or a new species to be examined via experimental crossbreeding to see whether or not it conformed to baseline expectations. Mendelian problems were thus triply tractable: because they arose within a matrix of extendable methods and concepts (including exculpatory explanations, insuring against the possibility that a new result might scupper the whole science); because the materials required—plants and animals—could be found anywhere, in endless and often inexpensive supply; and because, to discover something of potential value, one needed only to ensure that breeding took place, à la Mendel, under controlled conditions, starting with suitably purified parents. Consider again the range of Mendelian investigations that participants at the 1906 conference had under way. Had Mendelism's problem-solving tool kit enlarged no further, Mendelian research was set to run and run.[52]

But it did enlarge, and decisively, thanks most of all to Bateson's least-favorite development: Mendelism's integration with the chromosome theory of heredity. That took place in a full-blown way in Morgan's "fly room" at Columbia between 1910 and 1915, starting around the time Sturtevant shared his Mendelian paper on horse coats, and scaling up as Sturtevant and then other Columbia students—Calvin Bridges, Edgar Altenburg, and Hermann Muller—joined the lab. Throughout Mendelism's first decade Morgan had not been a fan, dismissing talk of "factors for unit characters" as objectionably outdated, given all that had been learned about how different environments draw out different potentialities from embryos. For similar reasons, he was doubtful that the sex of an animal was determined exclusively by the chromosomes it inherits. (Morgan would have loved Weldon's "Theory of Inheritance," which cited Morgan's work on regeneration approvingly.) At the beginning of 1910, Morgan was keeping true-breeding stocks of fruit flies, not because he wanted to inaugurate chromosomal Mendelism but because he hoped to induce de Vriesian mutations. But when, that spring, he found a white-eyed male among the standardly red-eyed *Drosophila* in one of his pedigreed cultures, he proceeded to cross that male with a red-eyed female, then to cross their red-eyed offspring with each other. He was not surprised that white eyes turned up again, and in the expected ratio of 3 red-eyed flies to 1 white-eyed fly. But he was surprised to find that the white-eyed flies were all *male*. The most straightforward explanation, he eventually persuaded himself, was that the factor for eye color was carried on the X chromosome.[53]

The historical standing of Morgan's experimental study would be hard to overstate. The modern-day textbook I quoted from above gives a detailed exposition, adding at the end: "Most significantly, Morgan's results strongly supported the hypothesis that genes were located on chromosomes." That is certainly true; other results from the fly room would strengthen it even more. But at least as significant were the new frontiers opened up for Mendelian problem solving.[54]

It was already well known, for example, that independent assortment of characters did not always hold. In *Mendelism*, Punnett noted that in sweet peas, longness in the pollen grains and purpleness in the flowers often (but not always) went together, as did roundness and redness. He called that phenomenon "coupling" and, without suggesting an explanation, predicted that "it will be found of peculiar importance in the elucidation of the gamete." Morgan and company called it "linkage" and explained it easily: characters are linked when the factors for them are grouped on the same chromosome. With *Drosophila*, the linked characters fell into four groups of different sizes, correlating pleasingly with the fly's four differently sized chromosome pairs. Furthermore, linkage had its own exceptions—character versions that were usually joined sometimes came apart—and these, too, could be explained, by the view that during the production of gametes, chunks of paired paternal and maternal chromosomes sometimes swap places, so that, for example, in some sweet peas, the factors for longness and redness end up together as part of the same chromosome. Sturtevant realized that if, as Morgan suspected, the probability of a linkage break increases the farther apart two factors are, then data on the frequency of breaks (or, in their vocabulary, of "crossing-over") could be used to calculate the relative positions of factors on a chromosome. Thus began the chromosomal mapping of Mendelian genes.[55]

The virtuoso Mendelian problem solver in the fly room was Muller. He had come to Mendelism earlier than the others, via Robert Lock's 1906 book *Recent Progress in the Study of Variation, Heredity, and Evolution*. Where Punnett's primer ignored chromosomes, Lock's emphasized the "extraordinary similarity" between the behavior of chromosomes and of Mendelian allelomorphs, coupling included. And where Punnett stuck with Mendelian binary characters, Lock gave generous coverage to Pearson's and Yule's theoretical work on how more complex characters might be explained on the assumption of multiple Mendelian factors. Muller's most far-reaching achievement was to make that old idea newly investigable. The trick was to find a simple character associated with a chromosome also associated, in some way, with the complex character in question. Then the appearance of that simple character in a fly could serve as an unambiguous signal of the presence of that chromosome. Through carefully designed experiments, individual flies could then be constructed so that each component of the complex character could be examined in isolation and in various combinations.[56]

The book summarizing the Morgan lab's work, *The Mechanism of Mendelian Heredity* (1915), was a masterpiece, in part because the research reported was dazzlingly executed and lucidly recounted, and in part because the whole was shot through with an understanding of heredity and evolution that, though presented as a vindication of Mendel, could have come directly from Weldon. Everything he had stood for, selection included, was here not merely admitted but embraced as what every thoughtful Mendelian should have believed all along. Multiple factors, distributed across multiple chromosomes, were now upheld as the norm, even for characters as apparently simple as eye color in *Drosophila*. Complex dependencies on developmental and environmental backgrounds were also the norm. That meant that a Mendelian claim to have found the "factor for" a character version in no way implied that the factor is somehow single-handedly and directly responsible for the production of that character version. Rather, the claim is always the much more modest one that, given two organisms bred to be identical in all other respects, nature-wise and nurture-wise, *that* version of the character is visible in that organism because of a difference in *that* factor. Characters are not units, they are "the realized result of the reaction of hereditary factors with each other and with their environment." When their environment stays stable, selection becomes possible, since selection boosts the frequency of whatever combinations of factors make for the most advantageous forms of a character. When environments change, or development changes, or the factorial background changes, then it is anybody's guess whether a Mendelian "factor-difference"—the more accurate shorthand they suggested—will result in the same character difference. The Morgan group adduced case after case to suggest otherwise.[57]

Technological promise. Mendel's garden and Morgan's fly room are two of the most iconic experimental sites in the history of biology, maybe even of science. A traditional appreciation of such places goes something like this: Laboratories and their like are important because they are where skillful investigators assemble the apparatus and other materials needed to isolate causes and effects that can be studied outside the laboratory only in obscuring interaction with other causes and effects. If that job of investigation is done well, then the conclusions drawn will be true, and their application in the wider world can begin in earnest. Getting nature right boosts our chances of developing new techniques and technologies that will actually work. Conversely, when those techniques and technologies work, it is because we have got nature right.[58]

Anyone unfamiliar with the above picture could infer it from Bateson's promotion of Mendelism. Over and over again, as we have seen, he talked up the scope for applying the truths revealed through Mendelian experimental breeding to agricultural and social problems, representing the success of those applications as deriving from the truth of Mendelian principles. Members of his school followed suit. In Lock's book, the chapters on Mendelism end with admiring coverage of Biffen's studies of wheat, and the book's concluding pages, addressing "the science of genetics applied to human affairs," echo Bateson's heredity-is-all-so-breeding-is-all remarks at the end of his 1904 BAAS address. ("The demand for a higher birth-rate," wrote Lock, "ought to apply strictly to desirables. Instead of this the cry is for education and physical training, processes which can have no permanent beneficial effect upon the race.")[59] Punnett's book concluded similarly, with a note added in the second edition of 1907 on Biffen's achievements in identifying unit characters in wheat.[60] The next year, in an article in *Harper's Monthly Magazine* titled "Applied Heredity" and included as an appendix in *Mendelism*'s 1909 American edition, Punnett spelled out the implications in more detail:

> The fact that resistance to yellow rust is a unit character exhibiting Mendelian inheritance makes it a simple matter to transfer it to wheats which are in every way desirable except for their susceptibility to rust. From the knowledge gained through his experiments Professor Biffen has been able to build up wheats combining the large yield and excellent straw of the best English varieties with the strength of the foreign grain, and at the same time quite immune to yellow rust. During the present year several acres of such wheat coming true to type were grown on the Cambridge University Experimental Farm, and when the quantity is sufficient to be put upon the market there is no reason to doubt its exerting a considerable influence on the agricultural outlook.[61]

Except, perhaps, to the reader of Coleman or MacKenzie and Barnes, the constant publicizing of Mendelism's utility is not remotely surprising. New sciences have been winning supporters through the promise of practical boons since the seventeenth century. Of course a science on a roll publicizes the applications to come. It stays on a roll, however, only if those applications succeed, or seem to succeed, at least enough to keep hopes alive. And Mendelism acquired a reputation for delivering.

How? "By getting nature right" will not do, for the reasons I gave earlier in this chapter. Nor will "by extending the exemplar," not because it is wrong, but because it leaves unanalyzed what needs explaining: namely, how, exactly, exemplar-based problem solving is made to work beyond the controlled confines of the experimental garden or farm or laboratory.[62] (In *Structure*, Kuhn deliberately ignored the social and technological dimensions of scientific change, judging them to be irrelevant to his purposes.)[63]

Here are three observations collectively pointing toward the kind of answer that I think we need. First, well before practical Mendelism had generated anything genuinely new—Biffen did not release his first Mendelian wheat variety, the rust-resistant *Little Joss*, until 1910—Mendelism's power to illuminate past failures and successes, and to suggest new efficiencies in experimental crossbreeding, struck some breeders as impressive enough for them to give the new science the benefit of the doubt. The scale and speed of Mendelism's uptake among breeders can easily be exaggerated, and has been. Not even the gushing Vilmorin proved, in practice, an especially eager convert. Nevertheless, by 1910, exposure to Mendelian principles was becoming a standard part of breeders' training.[64] Second, with predicted technologies, we should be alert to the mechanics of self-

FIG. 10.5. A spoof image of Sir Rowland Biffen from a 1926 Cambridge student magazine. By then, Biffen's reputation was so secure that the relentless hyping of Mendelian crosses as the key to breeding success could be gently satirized.

fulfilling prophecy. Sometimes the belief that something will happen helps to make it happen by inspiring people who want it, are willing to persist in pursuit of it, and are happy to credit the initiating idea for any eventual success.[65] In Biffen, practical Mendelism found not just a tireless advocate but one who melted his own considerable ambition into his ambition for Mendelism. "The undoubted success of plant breeding work at Cambridge," announced A. B. Bruce, superintending inspector of the Board of Agriculture and Fisheries, in late 1918, "is due primarily to the fact that in recent years an entirely new science has been built up as the result of the discoveries made by the monk Mendel in the early sixties." Bruce went on to praise Biffen's Mendelian wheats, citing *Little Joss* as well as two newer Biffenian varieties, *Yeoman* and *Fenman*.[66] Such was Biffen's reputation as the Mendelian wizard of wonder wheats that when, a few years later, his students at Cambridge published an homage, they depicted him holding an outlandishly large successor to *Yeoman* (fig. 10.5).[67]

My third observation relates to the traditional account of the relationship between experimental truths and their worldly application. The term "application" invites us to think that something in nature is truly discovered, that this discovery is successfully put to use, and that the success derives from the truth. But there is another way to look at how experimental knowledge becomes practically useful. Consider that in a successful experiment, investigators produce effects that, in the general course of nature, rarely or never arise by imposing special conditions on a protected corner of the world—the experimental site—along lines suggested by background theory. In the fly room, for example, Morgan and his students bred into being fruit flies with the conspicuously Mendelizing characters that Mendelian theory predicted but which wild fruit flies on the whole do not exhibit. The Morgan group could not have been clearer in stating that their "factor for" claims held for flies outside the fly room only to the extent that those flies' internal and external environments were sufficiently like those of the Columbia flies, with like impacts on embryological development.[68]

Biffen was less careful in stating that the rust resistance he introduced from a Russian wheat variety worked as advertised in *Little Joss* in part thanks to whatever complementary context came along for the ride, and in part thanks to *Little Joss* being taken up on farms where conditions were similar to those on Biffen's experimental farm. But he understood it well enough. When *Yeoman*, shortly after its release, began to show worrying signs of throwing off rogue forms, he looked for a difference-making difference on the farms. He reckoned he had found it in the form of the threshing machines that traveled from farm to farm. The machines, he concluded, were responsible for mixing seeds from rogue plants with his seeds. His solution was to start over, but now to exercise more control over the fate of his seeds, courtesy of a new Cambridge-based National Institute of Agricultural Botany, partnered with Biffen's own Plant Breeding Institute. Seeds of the new variety, *Yeoman II*, were for sale only from NIAB, in sealed sacks bearing NIAB's logo. Farmers were to acquire the seed from no other source, and to keep threshing machines away from the fields where *Yeoman II* seeds were planted.[69]

What looks, then, like successful "application" from a distance can look, closer up, like the extension beyond the experimental site of the conditions making for success within it. As Ian Hacking, philosophical expositor of this alternative account, has noted, "few things that work in the laboratory work very well in a thoroughly unmodified world—in a world which has not been bent toward the laboratory." That bending typically takes at least as much creative problem solving as the initial, exemplar-extending research. In the end, worldly success gets credited to the theory behind the research, even though the ties binding, for example, gametic purity and threshing-machine traffic may be weak at best.[70]

Let us turn now to that other great domain of Mendelism's potential application, human society. In "Applied Heredity," Punnett cautioned that, though the laws of heredity governing plants and animals surely govern humans too, human heredity is immensely difficult to study, given our species' complexity, our slow breeding, and the need to draw conclusions about ourselves from "the haphazard and inextricable experiments which are the outcome of civilized marriage customs." Even so, some progress had been made, with eye-color inheritance shown to be Mendelian ("the biologist would regard with some surprise the brown-eyed child sprung from parents whose eyes were blue"), and the list of Mendelian conditions now including brachydactyly (shortened fingers and toes) and congenital cataracts as well as alkaptonuria. Further progress on hereditary disease would come, he predicted, through study of pedigrees interpreted in the light of breeding experiments like Biffen's on immunity. "With increase in knowledge will come powers of prevention far greater than those we have to-day," Punnett concluded. "How far we may use these powers must rest with the future to decide."[71]

Bateson, in his 1909 *Mendel's Principles of Heredity*, which ended with a section titled "Sociological Application," was more directive. There was no need, in his view, to postpone decision-making on preventive measures, since "the elimination of the hopelessly unfit is a reasonable and prudent policy," and evidence was mounting that "in the extreme cases, unfitness is comparatively definite in its genetic causation, and can, not unfrequently, be recognized as due to the presence of a simple genetic factor." Concerns that society was not ready for the obvious next steps, that sentiment would surely triumph over science, were exaggerated, Bateson counseled:

> Society has never shown itself averse to adopt measures of the most stringent and even brutal kind for the control of those whom it regards as its enemies. Genetic knowledge must certainly lead to new conceptions of justice, and it is by no means impossible that in the light of such knowledge public opinion will welcome measures likely to do more for the extinction of the criminal and degenerate than has been accomplished by ages of penal enactment.[72]

11

What Might Have Been

It may seem surprising that a work of such importance [as Mendel's 1866 paper] should so long have failed to find recognition and to become current in the world of science. It is true that the journal in which it appeared is scarce, but this circumstance has seldom long delayed general recognition. The cause is unquestionably to be found in that neglect of the experimental study of the problem of Species which supervened on the general acceptance of the Darwinian doctrines. . . . Had Mendel's work come into the hands of Darwin, it is not too much to say that the history of the development of evolutionary philosophy would have been very different from that which we have witnessed.

WILLIAM BATESON, *MENDEL'S PRINCIPLES OF HEREDITY: A DEFENCE*, 1902

The discovery of argon was, therefore, directly due to a refusal to replace the variable and discordant experience of the weight of nitrogen by an ideal uniformity based on the mean of the actual experiences. If Lord Rayleigh had replaced all the results in his table by a single compromise between them, and had then been content to stop there, we should not know the existence of argon to-day. On the other hand, if the chemists of the nineteenth century had paid more attention to the discrepancies in their own experience of nitrogen, if they had not been content to replace their experience and that of Cavendish by a compromise which neglected his residual bubble, argon must have been discovered long ago.

W. F. R. WELDON, "INHERITANCE IN ANIMALS AND PLANTS," 1906

Thus did two well-trained scientists make use of the scientific past in trying to shape the present and future of the post-1900 science of heredity. Nothing in Kuhn's *Structure of Scientific Revolutions* prepares one for the existence of such passages. On a Kuhnian view, curiosity about the history of science is typically a casualty of scientific training, along with independence of mind generally. Only once debate has subsided, and the time has come to write an inspiring backstory into the new textbooks, does the history of science, briefly, matter. Among the many ways in which historians of science have moved beyond Kuhn is in learning to appreciate, on the contrary, how pervasively and variously the scientific past has featured as a part of scientific debate.

Another departure bears more directly on what our two debaters had to say about that past. Strikingly, each addressed himself not only to why something happened as and when it did—its cause—but to what might have happened had that cause been absent. Moreover, the effect in both instances was to emphasize the role of accident, happenstance, contingency in the advance of scientific knowledge. If not for Darwin failing to read Mendel, Bateson suggested, the Gärtnerian tradition of experimental hybridization would have continued to progress after the mid-1860s, with Mendel's work as its invigorating new model, and without the deadening authority of Darwinism slowing progress. If not for Rayleigh fussing over discrepant weight-of-nitrogen data instead of averaging the differences away, Weldon suggested, the list of known chemical elements would have continued to seem satisfyingly complete, and no one would have guessed that anything (indeed, a whole class of gases) was missing. What held for the physical sciences, Weldon warned tacitly, held even more forcefully for biology, just then in danger of following Bateson's lead into compromising complacency.

For historians of science, Kuhn put contingency on the map. In his vision of scientific change, the path of the past was studded with forking points, where the coalescing of communities around one science instead of its competitors—Copernican astronomy instead of Ptolemaic astronomy, oxygen chemistry instead of phlogiston chemistry, Einsteinian dynamics instead of Newtonian dynamics—sent the development of scientific knowledge in one direction rather than in other possible directions (fig. 11.1). For Kuhn, those coalescings were both indispensable and indefensible. They were indispensable because, as he saw it, wide agreement on the starting point for inquiry in an area is a precondition for progress. But they were indefensible because the commitments that go along with a given starting point—the components, intellectual and material, of the grid that each science lays down on the world, making it investigable—are agreed to before they can be fully vetted and vouched for. As a result, "an element of arbitrariness," as he put it, is always built in from the beginning, ensuring that, as a science grows, so will its deficiencies, until eventually the community faces its next major fork in the path.[1]

But would our current sciences really be so different now, other than in trivial ways, if in the past, the coalescings had gone differently? On what basis can we form judgments one way or the other? Kuhn's book raised these questions but left it to readers to work out the answers. When, in the 1970s and '80s, some of those readers took him to be licensing a no-

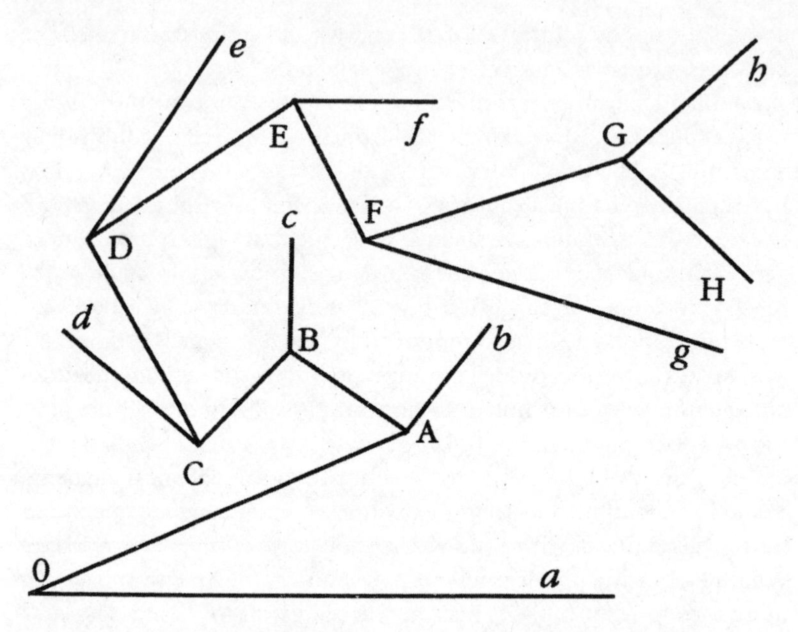

FIG. 11.1. The past represented as paths taken and not taken: from O to A rather than to *a*; from A to B rather than to *b*; and so on. From the 1876 edition of the novel *Uchronie* by the French philosopher Charles Renouvier.

tion of scientific truth as radically contingent on social arrangements that might have been otherwise—science as a "social construction," in a phrase the Edinburgh school helped popularize—Kuhn demurred, but without quite managing to say on what grounds. In the 1990s, controversy over whether or not science was a social construction became bad-tempered enough for the period to be called the "science wars." My own graduate studies in history and philosophy of science took place in the middle of it all. Out of that time, for some of us, came an eagerness to treat the contingency of scientific knowledge not as an article of faith, to be accepted or rejected, but as an object of inquiry, to be investigated by any means necessary.[2]

This chapter is an attempt to bring this recent work to bear on the science of heredity in the first decade of the twentieth century. I chose the epigraphs for this chapter in part to show that an interest in the might-have-beens of science is no alien import from present trends. By their nature, all historical explanations, of whatever vintage, whether concerning science or anything else, bring in tow claims about what did not happen but might have happened; about what was not, but might have been—the counterfactual past, as it is now known. Historians of science should not (to be counterfactual) have needed the stimulus of Kuhn in order to take that side of the job seriously.[3]

1

A closer look at the counterfactuals proposed by Bateson and Weldon will help get us under way. Bateson's, from his 1902 *Defence*, begat a line of conjectural longing that, as we have seen, has run and run. If only Darwin had read Mendel . . . Notice how, in its original setting, this counterfactual shores up Bateson's explanation for why Mendel's "Versuche" was so little recognized in Mendel's day. In Bateson's view, work as outstanding as Mendel's can be expected to make its mark, however obscure its venue of publication. Recognition, in other words, is inevitable, and ordinarily, it will be rapid. So why the long delay in Mendel's case? According to Bateson, responsibility lies with something extraordinary: the all-pervasive influence of Darwinism, which, notwithstanding the superb experiments Darwin himself reported in his *Variation*, directed those interested in the problem of species away from experimental studies toward speculative reconstructions of the evolutionary past, laced with further speculations about how gradually evolving structures might have kept species adapted in a changing world. By asserting that the actual scientific past would have unfolded very differently had an unrealized possibility within it—Darwin reading Mendel's paper—been realized, Bateson underscores what he sees as the accidental nature of the initial neglect of that paper. In so doing, moreover, Bateson helps us understand what he takes to be important about his explanation. The point for him is not just that Darwinism caused Mendelism's delay, but that, if not for Mendel's paper failing to fall into Darwin's hands, *nothing* would have delayed Mendelism.

Weldon, too, stresses the accidents of history, but to opposite effect—and, of course, to opposite purpose. His double-barreled counterfactual is from his lecture to Oxford students in summer 1905, on the methods of scientific research into heredity. He has just told the students that anyone who measures anything will, upon repeated attempts, get different values, and so face the question of whether to represent variable experience with a single compromise value. Can the measurer be certain that the differences are all due to imperfections on the measurer's part? If so, then little harm will be done in averaging. But if not, and the differences instead reflect differences in the world, then averaging will disguise what one is trying to discover. On Weldon's telling, argon was discovered only because Rayleigh refused to average away what anybody else would have regarded as negligible discrepancies in his data on the weight of compound-derived versus atmospheric nitrogen. No Rayleighan refusal, then no argon. Conversely, had earlier investigators been similarly, superhumanly scrupulous, argon could have been discovered at any time over the previous century. Here the counterfactuals drive home not the robustness of a scientific advance, but its fragility. For Weldon, the point is that the most momentous discovery he had lived through—Florrie called 1894 "Argon year"—came close to never being made at all, remaining an unrealized possibility, because it was so dependent on the rarest circumspection. That showed, in Weldon's view, how much can be at stake when fictional uniformities are imposed on variable facts. And if the stakes are so high for physicists and chemists, then, he counsels his students, just imagine how much biologists—whose materials are so conspicuously variable—stand to lose by treating variability as an ignorable nuisance.[4]

Explanations and counterfactuals go together because the former imply the latter. To explain, with Bateson, the delayed recognition of Mendel as due to Darwinism is to suggest that, if not for Darwin's innocence of Mendel, Mendelism would have arrived much sooner. To explain, with Weldon, the discovery of argon as due to Rayleigh's rare circumspection is to suggest that, if not for Rayleigh involving himself in the weighing of nitrogen, argon's discovery might never have arrived at all. The counterfactuals implied by explanations are not always made explicit, in the manner of Bateson and Weldon. But when they are made explicit, they help us to interpret the explanations in a way that can deepen our understanding, as we have seen. Explicit counterfactuals can also help us in evaluating explanations, since we can ask, for any implied counterfactual, how plausible we find it. The more plausible the counterfactual, the greater our confidence in the explanation. The less plausible the counterfactual, the lower our confidence.[5]

2

The term "counterfactual" entered the English language from mid-twentieth-century studies in logic, as a name for contrary-to-fact conditionals roughly of the following form: if some actual *X* were not, or had not been, the case, then our world would be different from the actual world in such-and-such a way (if at all). The philosopher David Lewis, at the start of his 1973 classic *Counterfactuals*, gave, as an example of what his book was about, the sentence "If kangaroos had no tails, they would topple over." The term "counterfactual" serves well enough to bring into focus questions about the logical interest of such sentences, and the troubles they raise for standard accounts of necessity, possibility, meaning, truth, and knowledge. But when "counterfactual" is construed, as it sometimes is, to suggest that all such sentences, and the reasoning supporting them, inhabit a fact-free zone, where evidence from the actual world, with its actual past, can have no bearing, it is profoundly misleading.[6]

By way of preliminary illustration, let me turn to the evidential testing not of Lewis's example (too grisly), but of Bateson's. Earlier I marshaled evidence from several sources to argue that, had Darwin read Mendel, the impact would probably have been minimal. There was, first of all, the way Darwin dealt with his snapdragon crosses, in which one parental form was absent from the hybrids, but then came back in the next generation in something close to a quarter of the flowers. In *Variation*, Darwin reported the pattern as an instance of form-over-form "preponderance" in action, and also made a distinction between the characters that offspring visibly inherit and the characters that they are capable of transmitting. When plucked from their context, as Bateson did in his 1902 *Report 1*, Darwin's snapdragons can be represented as showing just how tantalizingly close Darwin got to Mendelian heredity. In context, however, they show only that Darwin regarded such patterns as one of the great many that confront the student of inheritance—and inheritance itself, of course, as just one of the nine categories of biological fact that he was trying to explain in a connected way with his pangenesis hypothesis. Then there was the evidence of Darwin's response to Thomas Laxton's work on hybrid peas. To Darwin, no one was more authoritative than the pangenesis-vindicating Laxton on that topic. And, as Weldon was keen to emphasize, Laxton's 1866 paper on experiments with hybrid peas (which Darwin had on his shelves) reported results far more variable than those reported in Mendel's paper from the same year. Finally, there was the evidence of Mendel's snarkily phrased remarks in his paper casting doubt on whether the domestication of plants induces greater variability—remarks unmistakably targeting Darwin, and so hardly calculated to curry favor with him, let alone encourage him to ditch pangenesis for pure-gamete theory.

Quite a lot of evidence, then, goes against Bateson's counterfactual.[7] Not every counterfactual, of course, will lend itself to testing against evidence from the actual past. But that just makes some counterfactuals like untestable explanations, such as "Mendelism succeeded because reality/an unequal society/modernity summoned the Mendelian gene concept into being." The better class of counterfactual, like the better class of explanation, invites us not only to look again at old evidence but to seek out new evidence. By investigating what might have been, we end up understanding more about what actually happened and why it happened that way. Imaginatively fictionalizing the facts takes us not farther away from them, but closer to them—by an angle of approach otherwise unavailable.[8]

In the previous chapter, I argued that Mendelism succeeded on the outsized scale that it did because, initially through the efforts of Bateson and members of his research school, Mendelism became embodied in teachable principles, tractable problems, and technological promise. My explanation implies that, absent this winning combination, Mendelism would not have become the success that it did. A related implication is that, had some other post-1900 science of heredity acquired this winning combination, then that science could have become our science of heredity, not Mendelism.

Is that right? By way of answering, I shall consider first whether, against the counterfactual (and so against my explanation), the actual past in fact indicates that the Mendelian gene concept was irresistible around 1900, and furthermore that the success of a science of heredity enthroning that concept was inescapable. Evidence for the concept's inevitability as the basis for some research school or other lies, it seems, with the famous triple rediscovery of 1900; while evidence for the inevitable success of gene-centered biology lies with the miserable failures of midcentury Soviet agriculture, whose architects rejected the Mendelian gene concept. After, as I hope, defusing these skeptical challenges, I will start building the case for thinking that at least one alternative to Mendelism—Weldon's—could have grown into a successful science of heredity. Much of this evidence will come from Weldon's time. But some of it comes from our time, and in particular, from an attempt to see what happens when students are introduced to genetics not with a Mendelian starting point, but a Weldonian one.

3

Contemplating Mendel's long neglect and then the sudden, multiple rediscoveries of 1900 inspired the anthropologist Alfred Kroeber, in a famous 1917 essay, to stirring eloquence:

> There may be those who see in these pulsing events only a meaningless play of capricious fortuitousness; but there will be others to whom they reveal a glimpse of a great and inspiring inevitability which rises as far above the accidents of personality as the march of the heavens transcends the wavering contacts of random footprints on clods of earth. Wipe out the perception of De Vries, Correns, and Tschermak, and it is yet certain that before another year had rolled around, the principles of Mendelian

> heredity would have been proclaimed to an according world, and by six rather than three discerning minds. That Mendel lived in the nineteenth century instead of the twentieth, and published in 1865 [*sic*], is a fact that proved of the greatest and perhaps regrettable influence on his personal fortunes. As a matter of history, his life and discovery are of no more moment, except as a foreshadowing anticipation, than the billions of woes and gratifications, of peaceful citizen lives or bloody deaths, that have been the fate of men. Mendelian heredity does not date from 1865. It was discovered in 1900 because it could have been discovered only then, and because it infallibly must have been discovered then.

To the extent that de Vries, Correns, and Tschermak arrived at the Mendelian gene concept in isolation from each other, and from Mendel, that concept can indeed, as Kroeber suggested, be judged as independent of the precise historical circumstances that carried it into scientific consensus. Note that such convergence thus offers, at least in principle, a straightforward empirical test of inevitability. The greater the number of past trajectories that converged on the same conclusion, and the greater the independence of those trajectories, the more plausible will be the idea that the conclusion was inevitable. Kroeber indeed laid great store by convergence, piling up famous examples of multiple, simultaneous discoveries alongside the Mendelian one: Leibniz and Newton converging on the calculus; Darwin and Alfred Russel Wallace on evolution by natural selection; and so on. He took it that the best explanation for so much convergence was the inevitability of a discovery or invention, given the accumulation of human knowledge to a certain threshold.[9]

An inference from convergence to inevitability was not new with Kroeber. Nearly a century earlier, it featured in the English historian Macaulay's reflections on the phenomenon of belonging to one's times.[10] In our day, it is most familiar less as a historical argument about the nature of biology than as a biological argument about the nature of history. I refer to the well-known debate between the paleontologists Simon Conway Morris and Stephen Jay Gould centered on the interpretation of the fossils in the Burgess Shale in Canada. In the "contingentism" camp was Gould, arguing that modern animals might well have evolved radically different anatomies, had contingent events at the end of the Cambrian period resulted in a different selection of survivors. In the "inevitabilism" camp is Conway Morris, arguing that, whatever happened in the Cambrian period, modern animals would look broadly similar to the animals that now inhabit the Earth, since natural selection optimizes anatomical design, and the set of optimal designs is narrow. The ace up Conway Morris's sleeve is convergence: over and over again, it seems, independent lineages have evolved the same designs—eyes, wings, fishiness, intelligence, social organization, and so on.[11]

For the inevitabilist, much depends on establishing that two or more convergent pathways were in fact independent. Otherwise, the specter of convergence due to the contingent sharing of a local, constraining inheritance arises. That specter often proves harder to exorcise than the inevitabilist might hope. Certainly the convergence argument for gene inevitabilism now looks decidedly shaky. For one thing, as we have seen, the monk in the garden did not regard his work as revolutionizing the science of heredity. For another, there is now little doubt that de Vries, Correns, and Tschermak were all aware of Mendel's paper at earlier stages in their research than they later said. The contribution of de Vries, who published first, was arguably the most dependent on a reading of Mendel, for he seems to have hit on neither the ratios nor their explanation independently. Correns does appear to have completed the bulk of his hybridization research before searching the literature and finding the 1866 paper. But he was a former student of Mendel's botanical correspondent Karl von Nägeli, so a prior acquaintance cannot be ruled out. Moreover, the slow gestation of Correns's own paper may well be a sign that the late reading of Mendel brought new clarity. Be that as it may, Correns's trumpeting of Mendel's priority in discovery was probably intended less to bestow honor on Mendel than to take it away from de Vries, an envied rival. (As Correns pointed out, de Vries in his first-published Mendelian paper had quietly abandoned his own vocabulary for Mendel's, though not mentioning Mendel at all.) Tschermak, it seems, did discern the ratios independently of Mendel, but made little of them until he noticed the Mendelian bandwagon starting to roll, then clambered on.[12]

Some mention should be made as well of the small to nonexistent roles that de Vries, Correns, and Tschermak went on to play in establishing Mendelism as the basis for a new science of heredity. A reader of *The Mechanism of Mendelian Heredity* would never guess that these three ever converged on anything, let alone the Mendelian gene concept. Praise goes to de Vries for his theory of intracellular pangenesis because it correctly represented every cell in a developing organism as inheriting the same nuclear materials, and so correctly predicted that, whatever accounted for tissue differentiation during development, it would not, in any direct way, be those materials. Correns comes up as a leading student of cytoplasmic inheritance—a topic about which Morgan and company were supremely relaxed. They were keen to show that Mendelian inheritance is chromosomal, not that all inheritance is chromosomal, and they devoted several pages to the compelling evidence, from Correns and others, that some diseases and other characters that are transmitted only from the mother might be due to self-perpetuating, chemically influential bodies in the egg's cytoplasm. And Tschermak appears, without ceremony, as someone who had shown for hybrid peas what many other people had shown for many other hybrid plants and animals: that variability in the F_2 sometimes takes the form not of a 3:1 ratio but of a continuously variable spectrum. Only Bateson, with his distinctive interests, attitudes, and ambitions—as someone seeking to geometrize biological patterns at all scales, to drive historical reasoning out of biological science, and to bring the rest of the world with him—saw in Mendel's paper the great generalization upon which to build a new science of heredity.[13]

For our purposes, the important point is that, like independence, convergence is not straightforwardly assessed. We quickly find ourselves asking what criteria have to be met for different historical trajectories to count as independently convergent. How similar do the pathways need to be, in what respects, and why? Other putative cases of independent convergence present similar difficulties. Consider those "co-discoverers" of evolution by natural selection, Darwin and Wallace. They were different from each other in all sorts of interesting ways. But they shared much as well, including ideas generative of their evolutionary theories, notably the Malthusian population principle and the Lyellian view that history explains biogeography. Even if one overlooks their common inheritances, intellectual and cultural, there remains the fact that Wallace's theory was not strictly identical to Darwin's.[14] On what grounds shall they be declared the "same" theory? Moving from the history of biology to biology itself, consider the emergence of eyes in animal lineages constituting distant branches of the evolutionary tree. Eyes can be found among the vertebrates, arthropods, and mollusks, but not among their common ancestors. Yet, in each phylum, the development of eyes is regulated—so our genic biology teaches—by genes similar enough to work when swapped into organisms belonging to the other phyla. Those eyeless ancestors seem thus to have bequeathed not eyes, but the biochemical apparatus making eyes possible.[15]

In sum, the triple "rediscovery" of the Mendelian gene concept in spring 1900 is not at all, on closer inspection, compelling evidence for its inevitability. In order to break the spell of the Mendelian triple—as much a textbook staple as Mendel's peas—we need to go further, however. For just as awareness of the Mendelian triple leads to a sense that around 1900, biology was bound to embrace something like the Mendelian gene concept, our awareness of the centrality of that concept in our biology leads us to invest special significance in the supposed triple rediscovery. Yet the Mendelian triple was not the only triple in biology around that time. In 1896, the American psychologist James Mark Baldwin, the American paleontologist Henry Fairfield Osborn, and the British comparative psychologist Conwy Lloyd Morgan published on what subsequently became known as the "Baldwin effect." Widely discussed over the next decade, it reconciled the appearance of Lamarckian inheritance with—as Weismann had shown—the reality of its nonoccurrence. The basic idea is that whenever the environment favors individuals that, over their lifetimes, learn to adapt themselves in certain ways, natural selection will favor variations that eliminate the need for learning by making those adaptive characters hereditary. "That three workers independently thought of the Baldwin effect at the same time demonstrates that the idea was in the air, that it was an inevitable outgrowth of the intellectual atmosphere of the time," wrote the American paleontologist G. G. Simpson, in a 1953 essay that gave the effect its name. He went on: "That time was at the height of the neo-Darwinian *versus* neo-Lamarckian controversy and shortly before the rediscovery of Mendelism gave a radically different turn to biological thought." As for the neglect of the Baldwin effect from the mid-1900s until Thomas Huxley's grandson Julian revived awareness in his 1942 book *Evolution: The Modern Synthesis*, Simpson put that down to the spread of Mendelism, which made theoretical accommodation with Lamarckian inheritance "seem unnecessary."[16]

As with modernity generally (recall the previous chapter's discussion), so, then, with biological modernity: what we see as future-setting trends depends on where we look.

4

The twentieth century contains at least one famous and amply documented example of a Mendelism-rejecting biology: Lysenkoism. It ran in parallel with Mendelian biology through the middle decades of the century. It enjoyed enviable institutional clout. Its practitioners took a robustly skeptical stance toward the existence of Mendelian genes. And it was, by common consent, an utter fiasco. On the face of it, the failure of Lysenkoism seems to show that no scientific community deserving of the name can ignore the Mendelian gene concept forever.

The movement took its name from the Soviet agrobiologist Trofim D. Lysenko, who in 1936 put his doubts about the gene plainly. What "we deny," he said, was "that the geneticists and the cytologists will see genes under the microscope." Microscopic studies, Lysenko continued, will no doubt reveal structures of interest, but these "will be particles of the cells, nuclei or chromosomes, and not what the geneticists mean by genes." In Lysenko's view, the fundamental mistake of "Mendelism-Morganism" was the notion that heredity could be identified with a mere part of the organism—a bit of chromosome, if Morgan's school of fruit fly geneticists was to be believed—rather than the whole set of an organism's ever-changing relations with its environment. More perniciously, as Lysenko saw it, the geneticists' false theory of heredity drew support from, and lent support to, a related doctrine: that acquired characters could not be inherited. A major Lysenkoist ambition was to show, on the contrary, that permanent improvements to crop varieties could be engineered through directed transformations of the environment. These views were much debated in the Soviet Union from the 1930s to the mid-1940s. But by the late 1940s, Lysenkoism had won monopoly rights as the Soviet science of heredity. Mendelian genetics, previously thriving, was shut down, its leaders marginalized, even imprisoned.[17]

Lysenkoism had a well-known ideological dimension. Although the Bolsheviks had supported genetics from early days, Lysenko gradually tarred it with the brush of elitist, mystifying reaction. Whereas, according to Lysenko, the revolution needed energetic guidance on rational breeding, the academic geneticists taught that breeders could do nothing but await fortuitous mutations. "Mendelism-Morganism is built entirely on chance," declared Lysenko, in a notorious 1948 address. "There is no effectiveness in such science. With such a science it is impossible to plan, to work toward a definite goal; it rules out scientific prediction." The political credentials of Mendelism-Morganism were little helped by its class affiliation. Unlike Lysenko, the son of peasants, the geneticists came overwhelmingly from bourgeois backgrounds. A further liability was its association with lab-bound fruit flies. As symbols of the otherworldly sterility of academic genetics, flies were second to none. A cartoon from the Soviet periodical press of the era shows a geneticist, pockets full of test tubes of fruit flies, marching arm in arm with a Klansman and a policeman. Another shows a fat-cat businessman admiring the flies that, the Lysenkoists alleged, had distracted Soviet scientists from the task of improving agriculture (fig. 11.2). The geneticists had wanted to learn about flies when the people had needed them to learn about wheat.[18]

Yet, for all its ideological correctness, Lysenkoism at last fell from favor. Why? According to one of the most distinguished Western historians of Soviet science, Loren Graham, the Soviets had a belated reality check. He suggests that, as the theoretical and practical failures of the Lysenkoists mounted, and as the excuses for failure became less and less convincing, what became obvious was that, to put it bluntly, Lysenkoism was wrong. The abandonment of Lysenkoism thus shows that science is a social construction only within certain bounds. Even totalitarian societies cannot pursue congenial but false sciences forever. Graham compares the Soviets' denial of the realities of the marketplace to their denial of the

FIG. 11.2. "Mendelism-Morganism" as capitalist ideology: cartoons illustrating a 1949 article titled "Fly-Lovers and Man-Haters," published in the Soviet equivalent of *Life* magazine.

realities of the gene. "Despite all the social constructivist support" for Lysenkoist biology and Marxist economics, he argues, "both have fallen into eclipse. Both the gene and the market have reemerged, and, one is tempted to add, 'with a vengeance.' Natural and economic realities have obtruded." Graham lingers especially over the famous image of Nikita Khrushchev visiting an Iowa farm in the late 1950s, amazed to learn of its productivity, and of the Mendelian principles that, as Khrushchev was told, guided the development of that American agricultural wonder, hybrid corn (fig. 11.3).[19]

On Graham's reading, then, the failure of Lysenkoism is a kind of divergence argument for gene inevitabilism. Eventually, he suggests, a path of inquiry angled away from the gene becomes intolerably steep. Under Lysenkoism, the ignored genes wreaked their vengeance throughout So-

viet agriculture, leaving it enfeebled compared with gene-savvy American agriculture. The message borne home to Khrushchev—that, in Graham's gloss, Western genetics had developed "much more effective agricultural practices than Lysenko's genetics"—amounts, for Graham, "to the intrusion of reality into Lysenko's socially constructed worldview." No player of favorites, reality, it seems, would have intruded just as rudely into the Western market economies if Mendelian genetics had somehow failed to find its footing there in the early twentieth century.[20]

Graham, we should note, was echoing interpretations that circulated widely as part of the controversy over Lysenkoism, when a number of Western commentators on what they saw as the scandal and tragedy of midcentury Soviet genetics stressed, contra Lysenko, how useful Mendelian truths had been to agriculture. "Practically all the crops now grown in the United States consist of varieties which have been synthesized by the Mendelians," reported the botanist-turned-historian-of-science Conway Zirkle in *Death of a Science in Russia* (1949), at the end of a discussion highlighting the greatest Mendelian success up to that point, hybrid corn. The incredible yields associated with that variety, and the Mendelian reasoning that led to its invention, also featured in a book by Julian Huxley on Lysenkoism, published in the same year. Ever since, the contrast between American fields full of healthy Mendelian corn and the underperforming Russian fields under Mendelism-rejecting Lysenko has stood in for a set of ideologically weightier contrasts: the market versus the com-

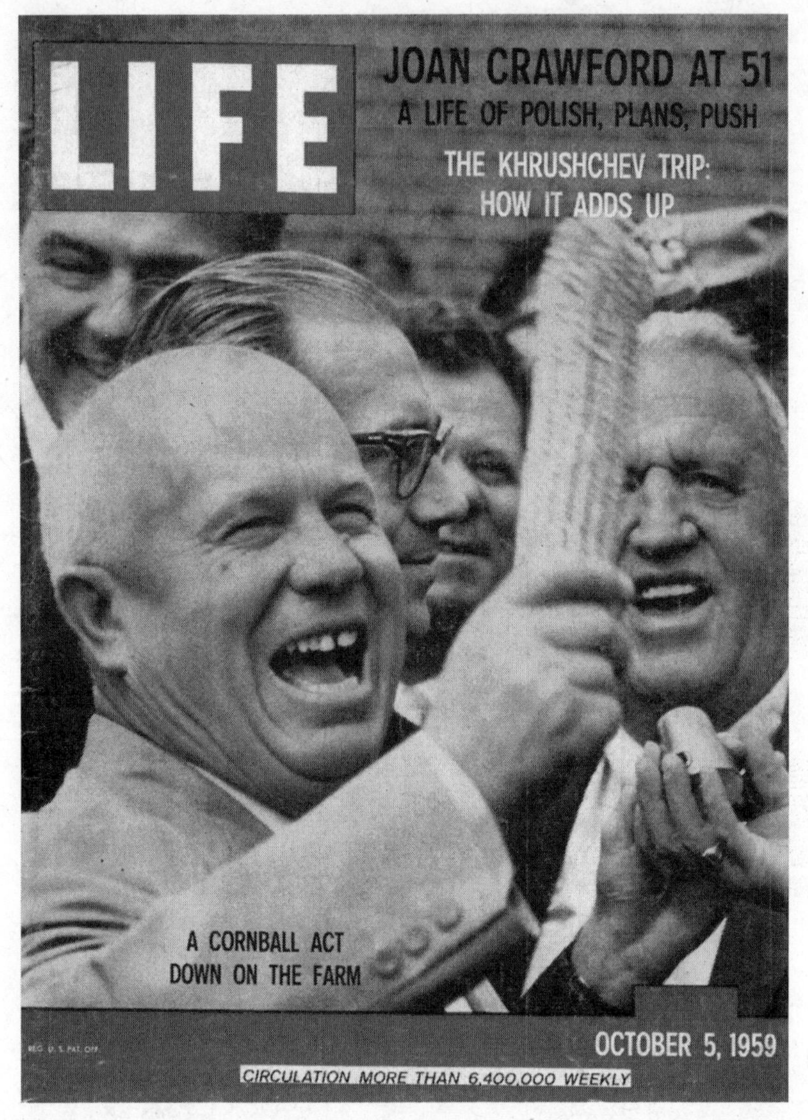

FIG. 11.3. "A Cornball Act Down on the Farm": *Life* magazine's take on a jolly Nikita Khrushchev visiting the Iowa farm of Roswell Garst, a champion of hybrid corn, in September 1959.

mand economy; freedom versus utilitarianism; reality versus its social construction.[21]

But is the testimony of agriculture indeed so unequivocal? The legacies of the Cold War are such that even now, one needs to consult the Marxists for a contrary view. The case for doubt has been put most effectively by the self-described dialectical biologist Richard Lewontin and his colleagues. Their arguments have not, to my knowledge, been answered. Graham does not even acknowledge them. Here are three of the most impressive in the Lewontonian brief. First, comparative data on annual crop yields do not show the Soviet Union to have lagged behind the United States—perhaps because pesticides, fertilizers, and farm machinery, common to both nations, have mattered more than divergent theories of inheritance. Second, the rejection of Lysenkoism should be seen as part of a wider generational shift in the Soviet Union, as idealist revolutionaries, willing to bear the costs of establishing a socialist alternative, gave way to pragmatic bureaucrats unwilling to bear those costs. Third, there is room to doubt that only Mendelism can guide the breeding of corn as productive, other things being equal, as hybrid corn. Horticulturalists might well have achieved as much using pre-Mendelian methods, selectively crossing the highest-yield individuals from the highest-yield variety, generation after generation. The attraction of Mendelism for American seedsmen was less its promise of otherwise unattainable yields—the phenomenon of "hybrid vigor" was familiar enough—than the prospect of forcing farmers to return to the seedsmen year after year (since plants born of seeds produced through Mendel-style, intervarietal hybridization would not themselves breed true).[22] Here are historical and scientific grounds on which to hesitate before concluding, with Graham and others, that agricultural reality crushed a would-be rival to Mendelian genetics.

A Lamarckian biology with a bad attitude toward laboratories and chromosomes is about as different from our science of heredity as could be. If not even such an extreme alternative can be shown to be fatally incompatible with reality, then we should be all the more wary of supposing, counterfactually, that other, less extreme alternatives must inevitably have gone the way of Lysenkoism. When the alternative in view is the Weldonian science that might have been, we consider a science of heredity about as much like our own as could be. It is non-Lamarckian, lab-based, and chromosomal. It even acknowledges a limited role for Mendelian patterns. But there is no Mendelian gene concept. Let us see what happens, counterfactually but also actually, when we give this science a try.[23]

5

We come now to what, in methodological terms, seems to mark a difference, and for the better, between counterfactual history of science and counterfactual history of pretty much everything else. If I am curious about what might have happened had, say, the losing side in a siege or election been the winner, the methods at my disposal for satisfying that curiosity are all, in one form or another, representational. I can think about it, write about it, maybe even program a computer to run a simulation. I cannot actually go back and meddle. But if I am interested in what might have happened had the losing side in a scientific debate been the winner, or even just not as big a loser as it actually was, there is nothing in principle preventing me from undertaking an intervention: the bringing back into being, now, of that losing-side science.

I take the intellectual essence of the science that lost out in the battle over Mendel to be well expressed in the following credo, quoted earlier but bearing repetition, from Weldon's "Theory of Inheritance" manuscript:

> Since the character of any organ depends, not only upon a specific something transmitted to it through the germ-cells out of which it was developed, that is to say upon something inherited, but also on two sets of conditions external to the organ itself, namely its relation to the parts of the body to which it belongs and its relation to the environment in which that body exists, we may say that every character of every animal is both "inherited" and "acquired."

I cannot, anytime soon, usher into existence the Weldonian world of my counterfactualist dreams, full of laboratories, fields, and clinics that are themselves full of biologically and biomedically trained people who approach inheritance in such a thoroughly interactionist, inextricably nature-and-nurture manner. But the materials that might educate such people are another matter. Why not try to develop a Weldonian curriculum in introductory genetics and then—passing from representation to intervention—teach it to students? To get them pointed in a Weldonian direction from the start, the first meeting would dwell not on Mendel's pea experiments—as per the traditional "genes for" curriculum—but on an example in which the effects of genes have to be understood alongside the interacting effects of many other causes—as with, say, the healthiness or otherwise of a human heart. From that beginning, the teacher, the readings, and the associated activities would seek to press home, at every possible opportunity, the message that genes have the effects that they do on bodies only in developmental and, more broadly, environmental contexts. Where the traditional curriculum typically treats the Weldonian emphases on variation, ancestry, and environments as largely ignorable luxury items, the Weldonian curriculum would treat them as primary. It would likewise dispense, so far as possible, with "dominance." And when the time came, eventually, to introduce Mendelian patterns, the students would learn to regard these not as showing inheritance at its most basic—as the great generalizations—but as special cases, exceptions to the rule, interesting precisely because, for explicable reasons, they can be understood as if genes acted independently of their contexts.[24]

Suppose one could actually teach this curriculum, and get it, as pedagogy, working well enough that the students would not simply rush for the exits in dismay and confusion after the first meeting, but would instead stay the course, right through to the end. What would these students be like? In particular, how different would they be when it comes to holding that exaggerated view of the power of genes that our academic culture calls "genetic determinism"? The most exciting possibility, of course, is that these students would be interestingly different, and in a particular way. They would be less prone than students coming out of a traditional Mendelian course to accept unquestioningly the latest news reports about "genes for" autism or schizophrenia and so on. They would be more sensitive to the ways in which individual differences matter in whether a *BRCA1* mutation or other "disease gene" mutation will lead to illness in any particular individual. They would be more willing to demand more information from the experts, not just about genes, but about the genetic backgrounds and the range of environments in play in the research warranting claims to know what genes are "for." They would, in other words, be less determinist.[25]

In collaboration with Jenny Lewis, a genetics education specialist, and Annie Jamieson, who trained in developmental biology before becoming a historian of science, I got a chance, at pilot scale, to find out. Our Genetics Pedagogies Project, as we came to call it, started at the University of Leeds in autumn 2012 and ended two years later.[26] The basic design was to work with two groups of students, one taking Weldonian genetics (the experimental group), the other taking Mendelian genetics (the control group). Both groups were assessed beforehand and afterward on their "belief in genetic determinism," to use the phrase of the pioneer questionnaire-makers in this area. Under ideal conditions, the project would have recruited from a single homogeneous pool of students; randomly selected which individuals got put into which group; used the same teacher for both groups; offered those groups courses having the same number of lectures of the same length delivered in the same style; and so on. Alas, our project did not take place under these conditions. Our Weldonian group consisted of about thirty humanities (mostly philosophy) students at the start of their second year, while our Mendelian group consisted of about the same number of first-year biology students. Where the students in the Weldonian group met for nine extracurricular one-hour lectures taught by a member of the project team, and had associated reading and blog tasks to do, the students in the Mendelian group met for more than thirty credit-bearing lectures, in a large lecture hall with hundreds of other students, were taught by a member of the School of Biology, and had associated reading plus labs and problem classes. The incomparabilities introduced by all these differences are considerable. Even so, our students entered their courses with more or less the same attitudes toward genes and their determinative power.

Annie Jamieson wrote the Weldonian course, taught it, and collected all of the relevant data. She knew Weldon's work intimately, having assisted me with my earliest studies of Weldon's "Theory of Inheritance." With her dual training, she was extraordinarily well-positioned to "get" Weldon. And she is a truly wonderful teacher.[27] For present purposes, a brief and imperfect overview of her first lecture, on the question "What is genetics?" can serve to give at least a feeling for what it was like to be introduced in Annie's classroom to Weldonian genetics. She starts with inheritance as we observe it in the relations of parents and their offspring. Humans have human babies, horses have foals, tigers have tiger cubs, and so on. There is always a sameness that gets preserved, and yet there is difference as well, and different degrees of difference. The children in a human family more closely resemble their parents than they resemble the parents of other children, but are nevertheless different from their parents, and different from one another. To be interested in inheritance scientifically is to want to understand all of that: the sameness that gets preserved over time and the differences too.

From there, we move on to consider inheritance and development. In typical biological educations, genetics is handled in one course and development in another, separate course. In our Weldonian classroom, by contrast, students encounter genetics from the first as fundamentally tied up with development. Mature characters are not what are inherited. What are inherited are genes plus cytoplasm and, more generally, elements of a developmental context. To think about how genes affect bodies is to think all the time about the developmental context in which genes function.[28] In the next step, that point is affirmed and enlarged with a look at organism and environment, with an invitation to the students to think expansively about environments. These are not, on reflection, just one thing, but many different kinds of things. Your family is an environment of a sort. So is the climate. So is your diet. What are your environments, and how do they affect you? While the question lingers, we move on to two examples of genes and environment in interaction. In one, the same genome in different environments produces very different results (queen and worker bees are genetically identical, but the queen develops by being fed exclusively on royal jelly). In the other, different genomes

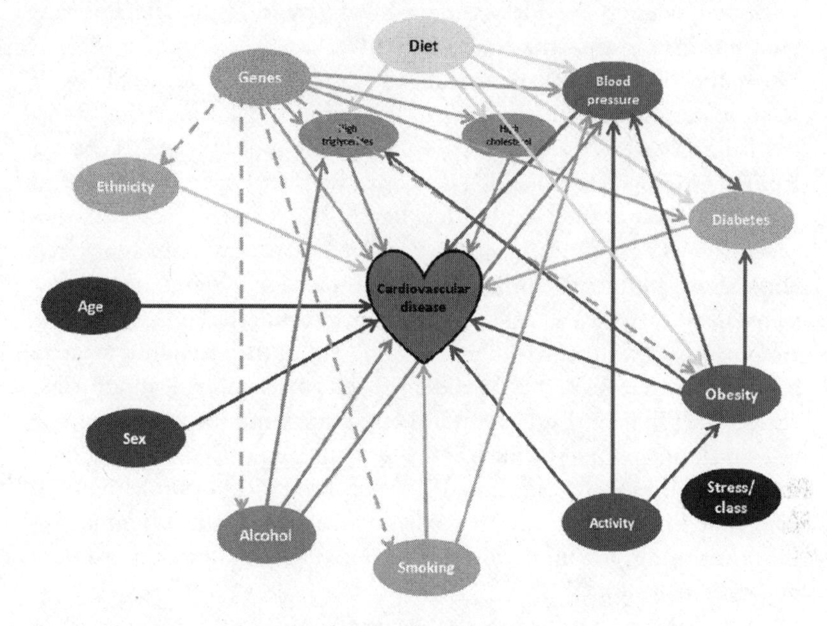

FIG. 11.4. Introducing a Weldonian perspective on genes to students via the web of causes linked to cardiovascular disease.

in the same environment give different results (mice given access to the same amount of food get fat or not depending on whether they have a mutation in the *OB* gene, which, in its functional state, encodes a hormone called leptin, which regulates appetite and energy expenditure).

I admire, and would like to think Weldon would admire, that pair of examples as a vehicle for impressing upon students a sense of genes and environment not just as interactive but as symmetrical. Genes are not presented here as super causes, as what inheritance is "really about," with everything else involved playing a supplementary role. Genes are there in the mix, making things happen, but along with lots of other things. And so that first lecture carries on. If we skip ahead to the very end of the lecture, we come—as in our anticipatory sketch—to the heart, specifically to cardiovascular disease and the question of the contribution that genes make. One by one, causal lines of influence are traced. Yes, single-gene mutations can affect cardiovascular condition directly. But they are among at least a dozen other factors that need to be considered, including diet, activity levels, blood pressure (itself affected by genes, diet, and activity levels), and social class. The students then get an image representing all of those causal lines put together. It is as tangled a web of influence as could be (fig. 11.4). And where in there, the students are asked, is the "gene for" cardiovascular disease? A hopelessly inapt question, as, we hope, they now see too.[29]

To our delight, almost all the students who began the course stuck with it. So, by the end, did these students' determinist beliefs change or stay the same? And how did they compare with the control group? Of course, from the perspective of counterfactualist methodology, any data are interesting, whether they support inevitability (no change in determinist-belief levels) or contingency (the levels go down). Our data supported contingency. Students coming out of the traditional Mendelian course were, on average, just as determinist about genes at the end as they were from the start, whereas students coming out of the Weldonian course showed, on average, reduced belief in genetic determinism, and to a statistically significant degree. Mindful of the small numbers involved, the previously noted imperfections in the experimental design, and Mendel's own difficulties with exaggerated statistics (and Weldon's role in publicizing them), I am reluctant to make too much of statistical significance here. But within the limitations of a proof-of-concept study, these were encouraging results.[30]

6

If Weldonian genetics can be taught, then its principles are teachable, and—since nothing in the overall framing of the Leeds course derived from anything later than Weldon's "Theory of Inheritance" manuscript—could have become teachable much sooner. But what of those other success-making features behind Mendelian genetics, tractable problems and technological promise? Let me take them in turn.

Tractable problems. When Weldon wanted a research study that compactly exemplified his views on heredity and how to study it, he reached first of all for his former student Ernest Warren's investigation of the water flea *Daphnia*. Warren had shown (confirming Weldon's earlier observations) that the spine visible on normal water fleas gradually diminishes in size over generations as the water they grow up in becomes more crowded and so more polluted with a poison carried in their excreted waste. Raise *Daphnia* in overcrowded water, and you will produce spineless *Daphnia*, generation after generation. But remove one of the spineless individuals to pure water, and its descendants—*Daphnia* is parthenogenetic—will grow up to have a long spine. Here were results, as Warren put it in his paper, "illustrat[ing] how closely the organism is knit to its external conditions of life."[31] Weldon, in what appears to be a draft for a public lecture on inheritance—possibly his evening discourse at the 1902 BAAS meeting in Belfast—went further in spelling out their significance for the student of heredity:

> Now clearly the condition of the spine . . . is not exclusively acquired; and it is not exclusively inherited. It belongs to both categories. A *Daphnia* of certain ancestry, growing in a solution of certain chemical composition, has a spine of a certain length. You cannot assert that either factor is more essential than the other; for if the *Daphnia* remain the same you can change the result by changing the solution; while if the solution remain the same you can (within considerable limits) change the result by changing the *Daphnia*.

Alter the solution, and the developed character will alter, because you have changed the environment. Keep the environment constant but alter the *Daphnia*—by substituting in, say, a race of the same species but from another locality, adapted to different conditions—and again the developed character will alter, but now because you have changed the germinal constitution. As Weldon summarized, we see here "the complex way in which the characters of an organism are determined by the interaction of the two sets of factors, the environmental and the ancestral."[32]

We know that *Daphnia* could have become the basis for a flourishing, exemplar-extending research program in heredity-environment interaction because, barely three years after Weldon's death, that is what happened. In June 1909, at a meeting in Germany marking the centenary of Darwin's birth, Richard Woltereck, a Weismann-trained zoologist in charge of a hydrobiological research station—a kind of freshwater, inland Naples—in central Austria, presented the results of extensive experimental studies on how changes in both environment and ancestry affect the development of characters in *Daphnia*. Woltereck's paper is best remembered for the introduction of what are known as norm-of-reaction graphs. Consider, for example, one he drew representing the relationship between nutrient level and head height in *Daphnia* (fig. 11.5). The *x*-axis represents quantity of nutrient in the solution, from poor to average to rich; the *y*-axis represents average head height, as a percentage of shell length; and the points on the curves A, B, and C represent the average developed head height, at each nutrient level, for three different races of a species called *Hyalodaphnia cucullata*, with each race from a different locality. We see that when the solution was nutrient-poor, individuals of all three races grew short heads, with A's average head height a little higher than B's and B's a little higher than C's. As Woltereck increased the nutrient level, however, the races reacted very differently. In A, developed head height spiked up rapidly and then plateaued. B experienced

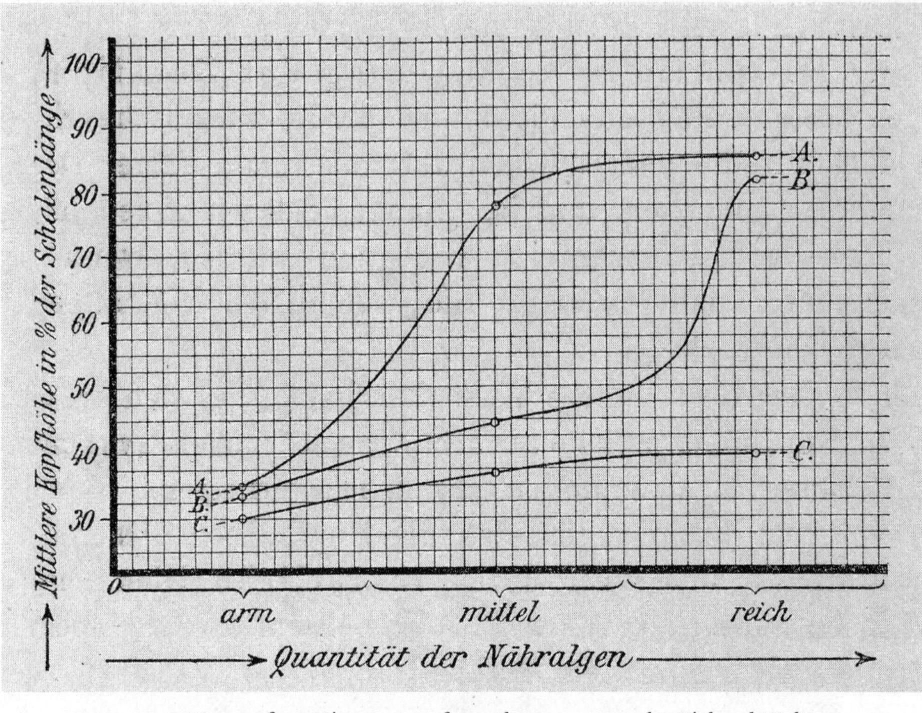

FIG. 11.5. Norm-of-reaction curves from the 1909 paper by Richard Woltereck, using head height to show how differently three different races of *Daphnia* react to different quantities of nutrients.

the reverse: a near-plateau until the solution became quite rich, then a sudden spike. In contrast to both A and B, C, with the shortest average head height to start with, barely budged upward, ending, in the nutrient-rich solution, with an average height still below those of the other two.[33]

Until the interruptions of the Great War, the paper set the research direction for Woltereck and his students, along the way securing him impressive promotions and an international reputation. In *The Mechanism of Mendelian Heredity*, the Morgan group cited a 1911 Daphnian paper by Woltereck as evidence that, whatever the germinal makeup, sex could sometimes be determined environmentally.[34] So intrigued was Morgan that in 1913 he suggested that Altenburg do his PhD thesis on *Daphnia*, readily available in the aquaria across from the fly room. Altenburg declined, dismissing the proposal as a fobbing-off insult. (He was in such high dudgeon that he also turned down Morgan's invitation to become a coauthor on the *Mechanism* book: a costly fit of pique.)[35] Norm-of-reaction graphs nevertheless soon found their way into the drosophilists' world, thanks to research begun in 1916 in a new midwestern outpost, in the lab of Morgan's friend Charles Zeleny. A student of Zeleny's, Joseph Krafka, found that the average number of eye facets in the flies he raised depended on which stock they came from as well as on the temperature during development. His plotting of the facet-number-to-temperature curves for three of his stocks looked like a mirror image of Woltereck's curves, and led to a similar conclusion: "The number of facets is determined by a specific germinal constitution plus a specific environment."[36]

Krafka's three-part paper on his research was for specialists. But it got picked up on by the left-wing, London School of Economics–based physiologist-turned-science-popularizer Lancelot Hogben, who pressed it into service as part of his anti-eugenics campaigning in the early 1930s. For Hogben, to understand the eye facet–temperature graph was to understand why, in his words, "in so far as a balance sheet of nature and nurture has any intelligible significance, it does not entitle us to set limits to changes which might be brought about by regulating the environment."[37] Ever since, the norm of reaction has figured in reflective and polemical writing aiming, along Hogbenian lines, to strike a blow against limit-setting genetic determinism.[38] At the same time, it has become an increasingly familiar part of breeding research because it shows at a glance, in graphical form, how breeds compare in their sensitivity to environmental differences. In deciding which of two plant varieties to release, a seed company will want to know not just which variety produces the highest average yield in the best possible environment, but which produces the most uniform yield over the widest range of possible environments.[39]

Technological promise. Toward the end of the draft quoted from above, after mini-lessons on correlation and regression, Weldon turned to selection. In his view, nothing could testify more powerfully to the efficacy of selection than the history of human breeding of domesticated plants and animals. Indeed, as we have seen, Weldon had enormous respect for practical breeders such as Laxton, and would not have presumed to tell them their business. If we ask, then, what agricultural advice might have issued from the applied wing of the Weldonism that wasn't, the short version is this: Select for the individuals most closely approximating to your ideal, while being maximally creative in identifying the nurturing conditions that, for the character of greatest interest to you, will enable its fullest development.

But there would have been more. On a Weldonian view, in making your selections, you must attend not only to the character of interest, and to developmental responses to different environments, but to the role of ancestry—the hereditary background that a particular breed carries with it—in mediating those responses and, generally, in making a breed more or less variable. Weldon touched on this theme later on in his mini-lesson on selection. He was concerned, he said, that the testimony of selection past was in danger of being disregarded under the influence of the obscurantist likes of de Vries, who took persistent variation in a character to be a sign of selection's inefficacy. But that was a mistake through and through, Weldon insisted. As Pearson had shown theoretically, and Weldon empirically (in his study of selection on the garden pea under domestication, and in his *Clausilia* study of selection in nature), variability will persist even where selection long ago fixed the mean value. And just as different breeds, with different ancestries, respond differently to environmental variation, different breeds, for the same reason, differ in the range of variation they regularly generate. Often the commercially successful breed will be drawn from stock whose peculiar germinal constitution makes for minimal variability. Here is Weldon:

> It is of little use to a breeder to fix the mean character of a race, if the generations he produces are very variable. If I want to make butter, I buy cows of a race which yields a great deal of milk. But by this I mean, cows of a race in which each individual cow yields well. I shall not care what the average yield per cow may be, unless I can feel sure that each of my own cows will yield about that quantity; for if half my cows run dry, the average yield of their race will not help me to make butter. *What a breeder must do, if he is to succeed commercially, is to obtain a race not only of high average character, but so little variable that he can predict with fair certainty the character of every individual.*[40]

Where Mendelism led to a focus on the purification of originating stocks, Weldonism would thus have stressed the recognition that different originating stocks (i.e., different genetic backgrounds) make for different degrees and kinds of variability, within and across different environments. Where Mendelism upheld hybridization as the primary means for improving characters of interest, consigning selection to a minor, supporting role, a Weldonian perspective would have reversed these priorities. And where, for the Mendelian, persistent variability was a sign that the job of purification had been badly done, for the Weldonian it would have been simply a fact of life, minimizable but not eliminable.[41]

Beyond agriculture, the other major domain of applied Mendelism was, alas, eugenics. Weldon never had much to say about it, either publicly or privately. But had he lived to complete, and publish, his "Theory of Inheritance," and on its basis begun to attract adherents eager to press its conclusions into useful service, the more eugenically inclined among them would probably have adopted views similar to those put forward by J. Arthur Thomson. Recall that in a popular lecture given in spring 1900, on the eve of the Mendelian surge, Thomson had presented a summary of the state of hereditary science whose major themes included its chromosomal basis, the importance of the Galtonian-Pearsonian ancestral law, and the close (albeit impermanent) interaction of the germinal with the environmental. Independently of Weldon, then, Thomson had assembled a picture with many of the same elements. Over the years of debate that followed, both Weldon, in his "Theory of Inheritance," and Thomson, in *Heredity*, a lengthy survey completed in 1907 and published in 1908, made room for Mendelism but without it forcing any larger changes of mind. In *Heredity*, Thomson devoted a chapter to Mendelism, but along the way drew on Weldon's UCL lectures for the view that dominance was dependent on changeable relations, so that caution was needed before accepting apparent gametic purity as genuine.[42]

Unsurprisingly, Punnett was unimpressed, writing a dismissive review incorporated into the American edition of *Mendelism*.[43] Equally unsurprisingly, Thomson's message on the implications of hereditary science for human society was far more upbeat than Punnett's (and Bateson's) men-are-born-not-made message. Yes, acquired characters do not seem to be transmitted in the germ line. But when we remember that "every inheritance requires an appropriate nurture if it is to realise itself in development"; that the interaction of heredity and environment begins before birth, in the womb; that, for better or worse, the same environments can induce the same modifications, generation after generation; that, in situations where a repeated modification is adaptive, it can become hereditary through natural selection, thanks to the Baldwin effect; that in humans, what gets transmitted is not just gametes and physiologies but a rich cultural and social heritage, which affects the capacities we develop no less than our physical environments do; then "we are led to direct our energies even more strenuously to the business of re-impressing desirable modifications, and therefore to developing our functions and environments in the direction of progress."[44] And yes, of course, nature is not infinitely plastic under the operation of nurture, so any scheme of progress must take germinal quality seriously. But all eugenic selection must proceed with extreme caution, for we so easily overestimate the extent to which a hereditary illness is uniformly devastating, under every circumstance, and underestimate the extent to which a change of circumstance could bring about a change of outcome (e.g., "one can hardly doubt that the high rate of criminals among illegitimate children . . . is artificially created").[45]

7

Of course, the advent in the early twentieth century of a Weldonian science of inheritance, trailing teachable principles, tractable problems, and technological promise, depended on more than Weldon surviving his cold-cum-pneumonia in April 1906. He would have needed, first of all, to go on and actually finish his "Theory of Inheritance," publishing it in time to capitalize on the momentum growing from his UCL lectures and his work on racehorses. (Recall again how unsettled Bateson had become in the year before Weldon's death.) Then he would have needed, on the basis of the book's success, to attract researchers eager to extend its reach to new audiences, new problems, and new applications.

On the one hand, there is room for doubt about how Weldon would have managed. His track record as a finisher of projects was, at best, spotty, while the example of Darbishire suggests a less-than-faultless ability to recruit and retain brilliant, loyal disciples. On the other hand, the "Theory of Inheritance" book was no ordinary project, which is why he got as far with it as he did (much further than he got with his textbook on Pearsonian evolution for biologists).[46] And, counterfactually, a counterexample to Darbishire lies to hand, in the form of the Lankester-trained evolutionary anatomist Edwin Goodrich. Eight years Weldon's junior, Goodrich became a fellow at Merton College in 1900, shortly after Weldon arrived there. He served as Weldon's demonstrator—Weldon described him to Pearson in 1902 as "the only decently trustworthy assistant I have"—and then, after Weldon's death, in effect as his successor. Goodrich was an outstanding biologist, elected an FRS in 1905.[47] And in *The Evolution of Living Organisms*, a popular introduction published in 1913, he communicated the core Weldonian teaching:

> An organism is moulded as the result of two sets of factors: the factors or stimuli which make up its environment, the conditions under which it grows up; and the factors of inheritance, the germinal constitution, transmitted through its parent by means of the germ-cells. No single part or character is completely "acquired," or due to inheritance alone. Every character is the product of these two sets of factors, and can only be reproduced when both are present. Only those characters reappear regularly in successive generations which depend for their development on stimuli always present in the normal environment. Others, depending on a new or occasional stimulus, do not reappear in the next generation unless the stimulus is present. In popular language the former are said to be inherited, and the latter are said not to be inherited. But both are equally due to factors of inheritance and to factors of environment; in

> this respect the popular distinction between acquired and not acquired characters is illusory.[48]

Like Thomson, then, Goodrich responded positively to the main argument of Weldon's book even in the less developed version in which it circulated during his lifetime. And the Merton-based Goodrich was in a position to become just the sort of ally a fledgling Weldonian research school in heredity needed—a Punnett to Weldon's Bateson.[49]

Taken all in all, then, the evidence of diverse kinds surveyed in this chapter suggests that, had Weldon survived beyond spring 1906, his "Theory of Inheritance" might well have become the nucleus of a science of heredity rivaling Mendelism in the teachability of its principles, the tractability of its problems, and the technological dimensions of its promise. Twice over, that is a conclusion worth reaching. As noted earlier, a counterfactual's plausibility serves as a test of the associated explanation: in our case, the explanation ventured in the previous chapter for why sciences like Mendelism catch on, rather than remaining essentially local enthusiasms. But the plausibility of the Weldonian counterfactual is also of interest in its own right, for reasons we shall consider in some detail in this book's final chapter. Before we do so, however, let us stay a while longer in the actual past, in order to look more closely at what, exactly, we owe to Mendelism's success.

12

Mendelian Legacies

Each step in the progress of this branch of science has rather compelled the recognition of genetic determinism; and the hope that by change in the conditions of life or by any external influences significant alteration can be induced in succeeding generations, whether of organisms amenable to experiment or of the human population, must be abandoned.

WILLIAM BATESON, "MENDELISM," *ENCYCLOPAEDIA BRITANNICA*, 1926

Go to some homes in the village,
Look at the garden beds,
The cabbage, the lettuce and turnips,
Even the beets are thoroughbreds;
Then look at the many children
With hands like the monkey's paw,
Bowlegged, flat headed, and foolish—
Bred true to Mendel's law.
This is the law of Mendel,
And often he makes it plain,
Defectives will breed defectives
And the insane breed insane.
Oh, why do we allow these people
To breed back to the monkey's nest,
To increase our country's burdens
When we should only breed the best?

JOSEPH DEJARNETTE, A VIRGINIA PHYSICIAN WHOSE EXPERT TESTIMONY WAS CITED IN THE 1927 *BUCK V. BELL* US SUPREME COURT DECISION UPHOLDING THE LEGALITY OF STATE INSTITUTIONS FORCIBLY STERILIZING PEOPLE IN THEIR CARE

Mendelism has been established once and for all in biological thinking and nothing will oust it, for it expresses a considerable part of the real basis of life itself. It colours willy-nilly our manner of observing[,] experimenting, reasoning and thinking.

JEAN ROSTAND, "JOHANN GREGOR MENDEL," *UNESCO COURIER*, APRIL 1965

Jean Rostand had a flair for dramatic statement to rival that of his playwright father Edmond, author of *Cyrano de Bergerac*. (In the *Oxford Dictionary of Scientific Quotations*, Jean—a prolific biologist-popularizer in the J. Arthur Thomson mode—is featured for the following *bon mot*, from 1939: "Kill a man, and you are an assassin. Kill millions of men and you are a conqueror. Kill everyone, and you are a god.") But effusive tributes to Mendel and his legacies were ubiquitous during the centennial of Mendel's 1865 lectures. At the start of a four-day memorial symposium in Brno in August 1965, the city's mayor welcomed guests who had come from around the world to celebrate "a great scientist who was born in this country and who by the foresight of a genius la[id] the foundations of a science whose practical application has found its way into all branches of modern science, into life and of course into man himself." In Fort Collins, Colorado, the next month, a five-day symposium sponsored by the Genetics Society of America resulted in a collection of papers titled *Heritage from Mendel*, whose front cover showed a pea plant's tendrils spiraling into a double helix (fig. 12.1). Lest anyone fail to register the message that the molecular genetics of the 1960s was a direct outgrowth of the pea-crossing experiments that Mendel had concluded in the 1860s, the back cover spelled it out: "Scientific advancement since 1865 has displaced neither Mendel's main concept of inheritance nor his methodology. Rather, his contributions have been steadily built upon, expanded, and refined, supplemented in major ways from other sources, and used in developing the elaborate theoretical structure of genetics that is now central to all biology."[1]

This emphasis on the debt owed to Mendel by molecular genetics and everything that has come in its wake, from genetic engineering to the various "omics" (genomics, post-genomics, epigenomics, transcriptomics, proteomics, and so on) to gene editing, is one of the legacies that we have inherited from the 1965 centenary.[2] Another is the downplaying of connections between Mendelism and eugenics. Jean Rostand had been an enthusiast for both, but his biographical article on Mendel omitted all mention of eugenics. At the Brno meeting, the Berkeley-based Curt Stern touched briefly on "class and race prejudice within the eugenics movement" as one among several causes of the disappointingly slow advance of human genetics, but otherwise participants gave eugenics a wide berth. Eugenics as a progress retardant also came up, also in passing, in the Columbia geneticist L. C. Dunn's centennial tie-in book, *A Short History of Genetics*. Alfred Sturtevant, in his tie-in, *A History of Genetics*, gave the topic greater coverage, stressing that Mendelism's main contribution was to dampen the hopes of those who dreamed of eliminating hereditary defects. If a defect is dominant, then, if afflicted individuals survive to reproductive age at all, the condition, like Huntington's cho-

FIG. 12.1. An emblem of continuous growth in the science of heredity, from Mendel's peas to the DNA double helix, as illustrated on the cover of one of several volumes published to commemorate the centenary of Mendel's "Versuche."

rea, may not become manifest until long after they have reproduced. And if the defect is recessive—as so many had turned out to be—then eliminating the culprit genes would depend on identifying heterozygous carriers: an impossibility at that time. What was more, Sturtevant went on, there were signs that heterozygosity for certain recessive conditions could confer a survival advantage, which explained how the conditions came to be prevalent in the first place, and why they would probably be maintained into the future.[3]

At the Fort Collins symposium, one of the few speakers who brought up eugenics was the Chicago geneticist-turned-biochemist George Beadle, in a talk titled "Mendelism, 1965." For the rising generation of molecular geneticists, the name "Beadle" immediately called to mind the words "and Tatum," since all biologists learned—as they continue to learn—about Beadle and Edward Tatum's classic experiments with the bread mold *Neurospora* in the early 1940s, showing that genes function by directing the formation of enzymes ("one gene, one enzyme") and, more generally, of proteins. Most of Beadle's talk at the symposium was a recital of the continuous growth of genetic knowledge from Mendel to the latest developments, from the cracking of the genetic code, whereby cells translate DNA nucleotide sequences into the amino-acid sequences of proteins, to the regulation of gene expression by other genes, as elucidated by two other symposium speakers, the Paris-based François Jacob and Jacques Monod. But toward the end, Beadle turned to what 1965-vintage Mendelism had to say about the advance of civilization. In Beadle's view, it taught two lessons. First, any inequality in the cultural achievements of different peoples was due to the accidents of history, not to what DNA-determined nervous systems were or were not capable of. Second, cultural heritage was so much more malleable than genetic heritage that our efforts at self-improvement as a species should concentrate on the former, under the guise not of eugenics but of what had recently been called "euphenics."[4]

It is not difficult to understand why, amid the scientific and social tumult of the mid-1960s, the old guard in genetics were keen to underscore the continuing relevance of Mendelism while minimizing its role in the discredited enterprise of eugenics. But their polemical reading of the historical record can no longer stand. Thanks to superb scholarship over recent decades, we can now see more clearly than ever that molecular genetics and its successors, pure and applied, owe little to Mendelism, whereas Mendelism played an outsized role in eugenics in the two countries—the United States and Germany—where, before the Second World War, eugenics was translated most quickly and consequentially into action. This chapter will review the new understanding of just what

Mendelism's legacies have been in these areas before turning to a third, more durable legacy: the determinism about genes that is built into our "start with Mendel" way of teaching genetics.

1

Like Newtonian mechanics, Mendelian genetics is, in the phrase of the physicist Werner Heisenberg, a "closed theory." No one thinks of Newtonian mechanics as an accurate description of what nature is like. If we want an accurate description, we turn now to relativity theory: a science whose fundamental concepts are at odds with those of its predecessor. But for other purposes, Newtonian mechanics is not just serviceable but, within its limits, unimprovable. We use it, and teach it, because in its domain, it works.[5] So too with Mendelian genetics. As a research enterprise, it proved, within a few decades, to be a dead end. Molecular genetics, and the more accurate description of hereditary processes that came with it, arose out of other sciences, first of all microbiology, biochemistry, and X-ray crystallography, and later the "-omics" alliance of sequencing technologies and computing power.[6] Whether biology students in general benefit from learning Mendelian or "classical" genetics in the way they typically do is a question to which I will return. But classical genetics continues to have its uses in biology, pure and applied.

At the 1965 Fort Collins symposium, Beadle touched briefly on the early twentieth-century debate over Mendel, quoting from an address that T. H. Morgan had presented before the American Breeders' Association in 1909. In Morgan's jaundiced view at that time, Mendelism was a license to postulate factors in whatever numbers and variety Mendelians needed in order to fit their results into their explanatory scheme. Do not be taken in, Morgan counseled:

> The superior jugglery sometimes necessary to account for the results may blind us, if taken too naively, to the common-place that the results are often so excellently "explained" because the explanation was invented to explain them. We work backwards from the facts to the factors, and then, presto! explain the facts by the very factors that we invented to account for them.[7]

That was unfair, ignoring all the effort that could, and sometimes did, go in to checking, by supplementary crosses, the soundness of a conjectured explanation. Even so, Morgan here captured something of the ultimately circumscribed creativity of Mendelian reasoning. On its own patch, Mendelism was, and is, endlessly, invincibly, fertile, for there will always be another character or organism to be studied by crossbreeding, another anomalous result to be resolved, another change to ring on the theme of factorial explanation—the more so after the innovations of the fly room. But already by the mid-1930s, when Beadle was a postdoctoral student in Morgan's lab (which had by then relocated to California), ambitious Mendelians were itching to strike out, to transcend the limits to which their methods tied them. And that meant, in effect, abandoning Mendelism for other sciences.[8]

Beadle made this point himself, in relation to the new molecular knowledge of how crossing-over works. "For years," he recalled, "many of us were convinced that, if we could learn enough about crossing over in higher organisms by classical genetic methods, we would understand the details of the mechanism by which it occurs. We were discouragingly unsuccessful." What finally opened up the problem, he continued, was recent work with viruses infecting the gut bacterium *E. coli*: organisms seen, into the early 1940s, as beyond the pale for geneticists, for whom research meant manipulating sexual reproduction (which these tiniest of microbes had seemed not to do) in order to map easily detected characters onto genes and then genes onto chromosomes (none of which bacteria and viruses seemed to have).[9] The experiments for which Beadle was famous involved an organism whose genetics had been explored via the classical methods, along with a mutation-inducing technique—X-ray irradiation—introduced by Muller. But "Genetic Control of Biochemical Reactions in *Neurospora*," Beadle and Tatum's epochal 1941 paper, made minimal use of such methods. To establish what genes do physiologically, Beadle and Tatum had looked at what mutant strains of *Neurospora* need nutritionally in order to survive, on the assumption that if a strain's survival depends on adding an extra ingredient to its growth medium, then that strain must have suffered mutation in a gene "controlling" the synthesis of that ingredient—in the first instance, by controlling the synthesis of whatever enzyme catalyzes the synthesis reaction. It was brilliant research, but its brilliance began where its links with the Mendelian past ended.[10] And in what Beadle declared to be "the most significant single advance in biology of this century, comparable to Mendel's and Darwin's nineteenth-century contributions"—the publication by James Watson and Francis Crick in 1953 of the double-helical structure of DNA, by then recognized as the molecule that genes are made of—the links were broken entirely. Watson and Crick drew not on experiments with crossbred organisms but on a combination of X-ray crystallographic data, the mathematical theory of helices, nucleic-acid biochemistry, and model building: resources about as alien to Mendelism as could be.[11]

No wonder that by the mid-1960s, Max Delbrück, the ex-physicist who had pioneered genetic research with bacteria-infesting viruses (called "bacteriophage," or "phage" for short), could be found pounding tables and shouting, "Genetics is dead! Genetics is dead! Genetics is dead!"[12] Genetic knowledge remained, of course—and remains still. But often, in the course of becoming "molecularized," it has undergone a kind of mutation.[13] Consider, once again, eye color in humans. From the elementary Mendelian account, we expect that when a brown-eye allele and a blue-eye allele meet in the same zygote, the brown-eye allele somehow or other overmasters the blue-eye allele, hence the child has brown eyes. (Here is how a study guide for American "pre-med" students in the mid-1990s put it: "How does [a] woman become brown-eyed when one of her alleles 'wants' her eyes to be blue and the other allele 'wants' her eyes to be brown? . . . For an organism that is heterozygous for a trait, both alleles can be thought of as competing for expression of that trait. The allele which 'wins out' is called the dominant allele.")[14] In the biochemical description of the case, the image of rival allelic expressions disappears utterly. There is no brown-eye pigment, pumped out by the brown-eye allele, and no blue-eye pigment, pumped out by the blue-eye allele (unless thwarted by the brown-eye allele). Biochemically, there is, in the most basic situation, just one kind of pigment, melanin, whose quantity in a human iris determines whether, on the spectrum from darkest brown to lightest blue, the iris looks brownish (lots of melanin) or hazel-greenish (an intermediate amount) or bluish (little or none). The quantity of melanin, in turn, depends on the quantity, as well as on the quality, or "strength," of a precursor protein encoded by DNA at a locus on chromosome 15. The gene at that locus, called *OCA2*, comes in a large number of variant forms, which result in more or less precursor protein, of greater or lesser quality. And the level of expression of that gene, in turn, is under the influence of genes at other loci, notably one called *HERC2*, nearby on the same chromosome. Also in the mix are other genes, elsewhere in the genome, some encoding the precursor protein, some encoding other proteins needed to turn the precursor into melanin, and some influencing the level of expression of those genes.[15]

Or consider one of the supposedly dichotomous characters that Mendel himself studied: roundness and wrinkledness in the seeds of garden peas. Again, we expect, on the elementary account, that in the heterozygote, wrinkling will be suppressed under the thumb of roundness. But again, biochemically, these two qualities are not in contention. There is simply more or less of one thing: an enzyme, called SBEI, that is responsible for the synthesis of a particular component of starch (branched amylopectin). If a pea has two copies of functional, SBEI-encoding DNA, at a locus on chromosome 3, then a high proportion of sugar molecules will

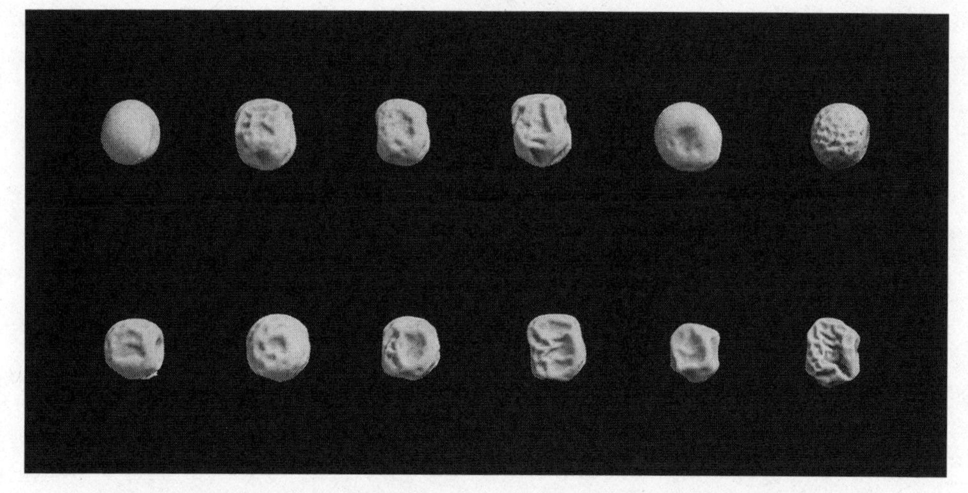

FIG. 12.2. Diversity in wrinkledness in the garden pea, not from Weldon's 1902 paper, but from a 1991 paper from John Innes Centre researchers studying the range of mutations that can affect this character.

get converted into starch molecules, less water will be lost from the developing seed, and when the seed dries, the seed surface will look smoothly rounded. If a pea has just one copy of SBEI-encoding DNA (i.e., the pea is a heterozygote), the pea will still produce enough functional enzyme for a similar level of sugar-to-starch conversion to make the mature seed look round. Only if both DNA copies are nonfunctional, and so no SBEI enzyme is produced, is the outcome notably different. Thanks to a high proportion of small sugar molecules in the seed, it absorbs lots more water. As the seed dries, it shrinks, and the seed coat now wrinkles.[16]

The working out of the molecular basis of seed shape in the garden pea was a triumph of late-1980s molecular genetics. "Mendel—Now Down to the Molecular Level" was the title of a celebratory commentary in *Nature*.[17] Again, as with human eye color, a qualitative, this-or-that character has been replaced with a quantitative, more-or-less one.[18] Indeed, the same era saw researchers discover new alleles and new loci affecting seed shape, yielding a roundness-to-wrinkledness spectrum similar to that documented by Weldon (fig. 12.2). (Of the photograph above, we read, "Mendel was extremely fortunate that no other genes that affect the wrinkling of the seed . . . were segregating in his experiments. If this had been the case, his analysis would have been rather more complicated.")[19] And just as the polygenic nature of eye color explains why two blue-eyed parents can have a brown-eyed child, the polygenic nature of seed shape means that a breeder so inclined could generate a round-seeded pea variety from two wrinkled-seeded ones.[20]

By the end of the twentieth century, the molecular dissolution of the simple Mendelian image of how heredity works had become so complete that, in the judgment of the historian and philosopher of science Evelyn Fox Keller, the time for pretending otherwise was over. Her goodbye-to-all-that, *The Century of the Gene* (2000), argued that whether one asked about gene structure, gene function, genes in development, or genes in evolution, nothing remained of the old perspective on heredity as centered on the action of stably persisting, character-determining genes. As for stable persistence, it had turned out that such stability as stretches of DNA have is the upshot of active maintenance by cells, and indeed, that evolutionary success for a lineage may depend on maintenance regimes becoming less stringent in times of environmental stress. As for character determination, it had become clear, first of all, that the number and diversity of proteins vastly exceeds the number and diversity of genes, thanks to the ability of cells to conduct a kind of protein alchemy, drawing on coding sequences from physically distant DNA stretches and then further modifying the products according to the moment-to-moment needs of the organism; and second, that the complex choreography of gene expression that makes for a developed and then a surviving organism depends on interplay not just with other genes but with a series of environments, from the epigenetic and cytoplasmic on up (and out).[21] As Keller summarized in a 2014 paper updating her argument in light of more recent affirming advances, especially in understanding how noncoding DNA (the "dark matter" of the genome), too, functions in the regulation of gene expression:

> Today's genome, the post-genomic genome, looks more like an exquisitely sensitive reaction (or response) mechanism—a device for regulating the production of specific proteins in response to the constantly changing signals it receives from its environment—than it does the pre-genomic picture of the genome as a collection of genes initiating causal chains leading to the formation of traits. . . . We have long understood that organisms interact with their environments, that interactions between genetics and environment, between biology and culture, are crucial to making us what we are. What research in genomics shows is that, at every level, biology is itself constituted by those interactions—even at the level of genetics.[22]

2

Turning next to the question of Mendelism's relationship to eugenics, we face a tradition not of overstatement but of understatement. As we have seen, in the centennial outpourings of 1965, eugenics barely came up at all, and the few remarks addressing it tended to depict Mendelism's contribution as critical and undermining. Since then, the historian Pauline Mazumdar has gone furthest in making the case for lack of connection between Mendelian theory and eugenic practice.[23]

For Mazumdar, the key point is that compulsory sterilization of the eugenically undesirable got under way on a large scale only from the 1930s, first in the United States and then in Germany, at a time when Mendelism as a research program, in human genetics as elsewhere, was in decline. Among German researchers into heredity, Mendelism was never more than a minority enthusiasm: those who worked with plants and animals typically preferred to examine the interplay of inheritance and development, while those who studied human heredity, although excited by early reports that schizophrenia and certain racial traits were Mendelian recessives, found further success along these lines elusive, and gradually abandoned Mendelian data crunching for other methods. In the United States, where Mendelism was at the research forefront throughout the 1910s and '20s, Charles Davenport and his allies—notably Harry Laughlin, who collaborated with Davenport in running the Eugenics Record Office (founded in 1910, adjacent to the Cold Spring Harbor station), and Henry H. Goddard, director of research at a training school for "feeble-minded" boys and girls in Vineland, New Jersey—became well known for publishing pedigrees apparently showing that feeble-mindedness was a Mendelian recessive. But their research came under attack, initially by Pearson and his students, then more widely, to the point where, in 1928, Goddard recanted, while the Eugenics Record Office's Carnegie Institution funders started investigating what increasingly looked like a scientific scandal.[24]

The lesson Mazumdar draws from this history is that, since many research experts in heredity did not endorse Mendelism when the eugenic dream of compulsory sterilization became a reality, "eugenic sterilization was a legal and political matter, in which the science of genetics had a mainly rhetorical role." A conflation and a distinction, both untenable, are being made here. Against the conflation, consider that a science is not equivalent to the views held by its research elite. Indeed, part of the challenge for anyone trying to understand a science as successful as Mendelism became is to do justice analytically to the semi-autonomous life of older, outmoded claims circulating beyond elite publications: in textbooks and manuals, in popular-science books and news articles, in courtrooms and living rooms. I shall return in the next section to consider whether something about the structure of Mendelian knowledge might make it especially prone to keeping defunct, determinist simplifications current alongside their more sophisticated debunkings.[25]

As for the distinction Mazumdar tacitly draws: a science can play "a mainly rhetorical role" in something's happening and yet still be causally consequential. Rhetoric, after all, is the art of persuasion. In a culture respectful of scientific knowledge, a standard way of persuading others to adopt a course of action is to represent it as based on scientifically attested facts or, better still, a law of nature. "The science of Eugenics owes its basis to the law of the Augustinian monk, Gregor Mendel" was a line that American newspapermen began copying from each other throughout spring 1921. On the page, in the *Great Falls Tribune* (Montana) or the *Chickasha Daily Express* (Oklahoma), it appeared as just one of an endless—and zanily heterogeneous—supply of did-ya-know curiosities with which to fill otherwise blank space.[26] But 1921 proved a major year for American eugenics and its Mendelian pedigree. In early May, in the lead editorial for the *Saturday Evening Post*, with 2 million subscribers, editor George Horace Lorimer thundered:

> If we should ever erect a monument to the Abbé Mendel it should be not because he taught us to breed the black sheep out of our flocks and to produce hens with unrivalled laying records but because he supplied the data that have enabled scientists to study intelligently the beginnings of our racial degeneration and to utter those authoritative warnings which must be presently heeded if we are not, as a people, to forfeit our high estate and join the lowly ranks of the mongrel races.

Lorimer explained that the likes of Madison Grant, New York lawyer and author of *The Passing of the Great Race* (1916), had pressed Mendel's insights into service to explode what Lorimer called the "great American myth" of the melting pot. According to the myth, the world's racial flotsam and jetsam washes up at Ellis Island and American conditions somehow transform it all into good American stock. In Mendelian reality, however, the many races do not become one, but retain their distinctive qualities. The result was that, with the arrival of ever more Italians, Jews, and other southern and eastern Europeans, the country's "Nordic" stock, from which Washington and Lincoln were sprung, was fast disappearing—and with it, the country's high standing.[27]

Later that month, the US Congress passed a law inaugurating a new era of eugenic restrictions on immigrants, which limited their numbers by national quotas.[28] A few months later, the new public partnership between eugenics, Mendelism, and America was sealed at the Second International Congress of Eugenics, held over six days at the American Museum of Natural History in New York City, virtually at the doorstep of the Morgan school, and featuring a mix of eugenics and genetics papers, all overseen by Grant's ally, the Baldwin effect co-discoverer H. F. Osborn. (Morgan himself was never a keen supporter of eugenics; Muller, however, was even trying to convince Stalin of its virtues.)[29]

For the eugenically vigilant, the threat from the enemy within was at least as alarming as the threat from the enemy without. What made eugenic sterilization seem legally and politically justified in 1920s America was, again, the authority of the Mendelian science in which campaigners for eugenics wrapped their proposals. In 1926, when the *Encyclopaedia Britannica* printed Bateson's article on Mendelism, with his declaration that the advance of genetics had continually confirmed what he termed "genetic determinism" (recall the first epigraph of this chapter), the American Eugenics Society mounted an exhibition in Philadelphia aiming to help visitors not only to acquire the Mendelian basics, but to conclude from them that "some people are born to be a burden on the rest"—a burden that those armed with Mendelian science could take steps to reduce (fig. 12.3). The next year, the US Supreme Court confirmed the legality of compulsory sterilization nationwide in its decision in *Buck v. Bell*, a case that eugenicists had brought in hopes of extracting exactly this outcome, and that Mendelian expertise had carried up through the Virginia judicial system.[30]

"Buck" was Carrie Buck, the supposedly feeble-minded daughter of a supposedly feeble-minded mother; "Bell" was John Bell, superintendent at the colony housing Carrie and eager to sterilize her. The superintendent at another Virginia institution, Dr. Joseph DeJarnette, testified that, when it came to the inheritance of incurable feeble-mindedness, "Mendel's law covers it very well," hence his opinion that, as the published summary of his testimony put it, Carrie "was the probable potential parent of socially inadequate offspring by the laws of heredity, that her offspring would probably be so affected and that her welfare and the welfare of society would be promoted by her sterilization." (DeJarnette's paean to Mendel's law furnishes this chapter's second epigraph.) An expert from the Eugenics Record Office, Arthur Estabrook, agreed, explaining that, because feeble-mindedness was a recessive condition, someone like Carrie was expected to produce feeble-minded offspring in predictable Mendelian proportions, depending on whether her mate was himself feeble-minded (in which case all of their children would be feeble-minded), or normal-seeming but from "bad stock" (in which case half the children would be feeble-minded), or normal-seeming and from "good stock" (in which case all the children would be normal-seeming but carry the "taint of feeble-mindedness.") Estabrook's boss, Harry Laughlin, who had served as "expert eugenics agent" for the House committee behind the 1921 immigration law, and was the author of the model sterilization

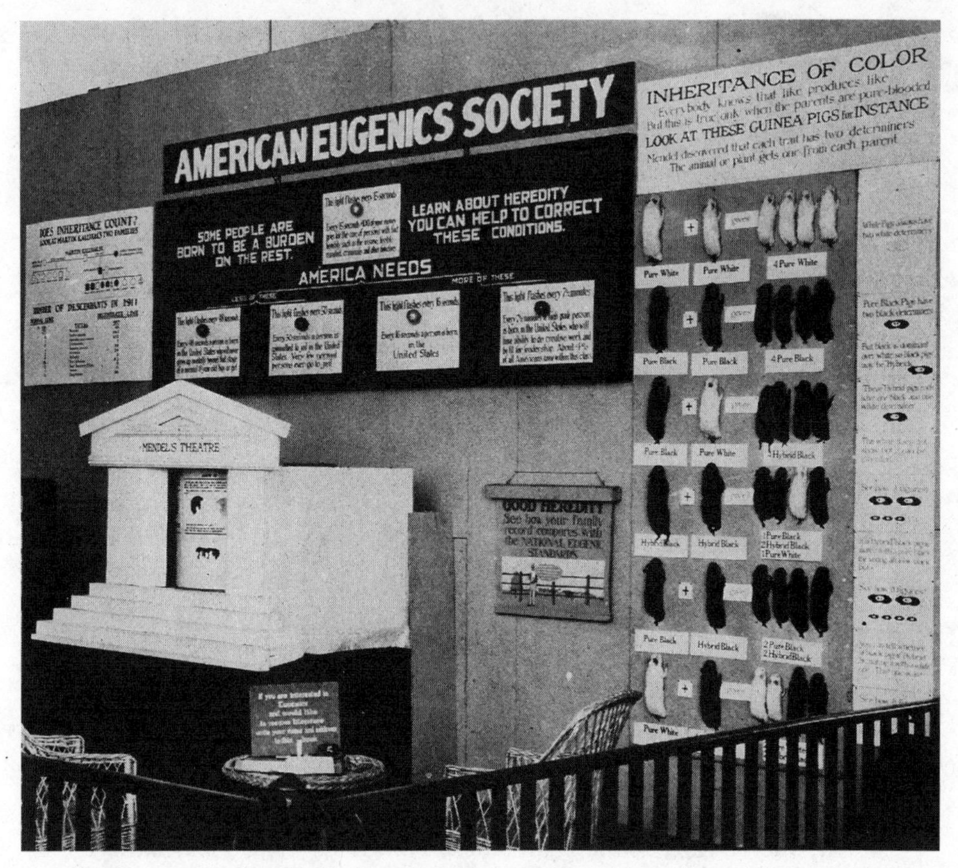

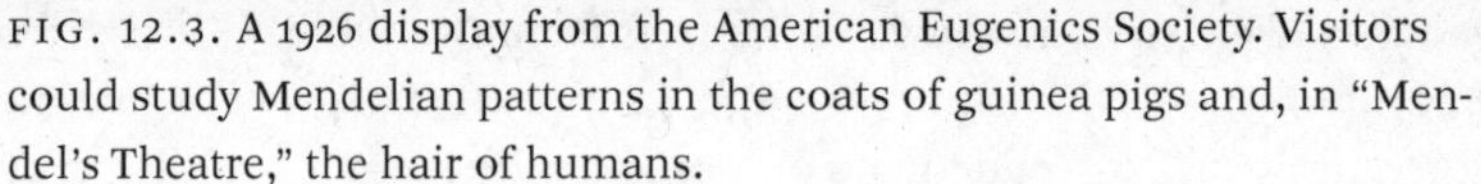

FIG. 12.3. A 1926 display from the American Eugenics Society. Visitors could study Mendelian patterns in the coats of guinea pigs and, in "Mendel's Theatre," the hair of humans.

law on which the Virginia statute was based, provided a deposition vouching for the hereditary nature of Carrie's feeble-mindedness.[31]

From first to last, every court affirmed the state's right to sterilize. "Three generations of imbeciles are enough," wrote Supreme Court Justice Oliver Wendell Holmes, citing DeJarnette's testimony as establishing the factual basis for the 8–1 decision, and referencing not just Carrie and her mother but also Carrie's infant daughter.[32] Carrie was duly, legally sterilized. Eventually, over sixty thousand other Americans were too.[33]

In Germany, after Hitler came to power in early 1933, Mendelism and eugenics became even more closely, and perniciously, bound together.[34] A theatrical hit of the early Nazi state, *Erbstrom* (*Hereditary Stream*), which played before tens of thousands as it toured throughout Germany from 1933, featured among its plotlines the quest of a young German couple to follow the science in making up their minds about whether to marry. True, many who saw it were obliged to do so, including schoolchildren taken by their teachers and factory workers sent by their bosses. No matter. For officials in the Health Department of the Reich Ministry of the Interior, the play's didactic magic was irresistible, ensuring that spellbound viewers became emotionally as well as intellectually caught up in "the exceptional meaning of hereditary health for the race and for the future of the nation."[35]

The core message was that hereditary defects can lurk invisibly even in seemingly healthy people, for reasons illuminated by Mendelian genetics. Anyone unfamiliar with that science got the needed lesson during Act 1, when one character helpfully explained the essentials to another. The already standard Mendelian example of eye color was used, but with some distinctively German vocabulary added in for the case of a match between two heterozygotes. Among the offspring of these "hybridized" parents (*Mischlinge*—literally, "mixture"), one-fourth would be expected to inherit the combination of the recessive gene variant from the mother and the recessive gene variant from the father. The recessive character, blue eyes, would thus be "mendeled-out" (*herausgemendelt*) in these children. For the sake of audience members who might have struggled to follow along, the character receiving the lesson on stage professed not to hear that last word properly. "What did you say?" "Mendeled-out—m-e-n-d-e-l-e-d—after the famous Augustinian pastor Gregor Mendel." The point bore stressing because, as the betrothed couple went on to learn, so many of the conditions imperiling the *Volk* were recessive.[36]

We talk still of voltage and wattage, of pasteurizing milk, of being galvanized into action by a mesmerizing speaker. The transforming of an honored surname from the history of science into technical jargon and then into everyday language is familiar enough. By the 1930s, Mendel had become esteemed throughout the world. In Nazi Germany, furthermore, the esteem was tinged with race pride. As Amir Teicher has shown in *Social Mendelism* (2020), his revelatory history of Mendelism under the Nazis, children were taught not merely that Mendel was the father of genetics and so, by extension, the father of race hygiene, but that he was the *German* father. (A poem concluding a booklet on Mendelism for children hailed Mendel as "*ein echter, ein deutscher, ein ganzer Mann!*"—"a genuine, a German, a real man!")[37] Even so, in Nazi usage, "mendeling-out" typically referred to something shameful. When apparently healthy, apparently German parents mendeled-out offspring who were feeble-minded, or schizophrenic, or Jewish-looking (to cite conditions widely accepted in the Nazi period as Mendelian recessives), the tragedy was not merely the birth of these unfortunates and all the trouble they brought. It was also evidence of continued contamination of the hereditary stream, despite the Nazis' best and increasingly strenuous efforts, first through Hereditary Courts and the sterilizations they enforced, then the outlawing of sexual relations between the "German-blooded" and Jews (with the Nuremberg laws), then the "euthanizing" of the inmates of asylums and other institutions, and eventually the Final Solution. The specter of recessive genes circulating imperturbably, awaiting their moment to join together and wreak biological-social havoc, haunted Nazi racial thinkers, Hitler included. In May 1942, he expressed his exasperation at how "descendants of Jewish [*Mischlinge*] even after four, five, six generations continue to mendel-out pure Jews. These mendeled-out Jews constitute a great danger!"[38]

Maybe all this Mendelian talk, unappealing though we find it, was just talk, of no real consequence? Teicher argues convincingly to the contrary. To be sure, the Nazis helped themselves opportunistically to whatever science lay to hand in dressing up their hatreds as legitimate. But that did not make the legitimations irrelevant or, once in place, inert. In a classic article on social Darwinism in Germany, Richard Evans suggested that widely discussed ideas about the German nation as engaged in a Darwinian struggle for existence, where only the fittest would survive, "helped remove all restraint from those who directed [Nazi state terrorism], carried it out and drove it on, by persuading them that what they were doing was justified by history, science and nature."[39] Teicher mounts a parallel case for Mendelism, as signaled in his provocative book title and in his attention throughout not only to Mendelian propaganda like *Hereditary Stream* and the children's booklet, but to occasional resistance from the targets of that propaganda. We see, for example, that for some trainee science teachers, Mendelism was a harder sell than anticipated, with their students casting doubt on whether something that worked for peas and fruit flies could really apply so straightforwardly to humans. Persuasion was not redundant.[40]

Teicher shows, too, that in at least three ways, Mendelian reasoning made a difference in determining not only who cooperated with the Nazis, but who suffered. The first concerns the Nazi sterilization law, officially named the Law for the Prevention of Hereditarily Diseased Offspring, passed in July 1933 and put in practice from 1934. It mandated sterilization for individuals expected to transmit a range of listed conditions. At the top of the list were the usual suspects, feeble-mindedness and schizophrenia. Mental infirmity was the standard target of compulsory sterilization laws in an ever larger number of US states (a precedent that German eugenicists were eagerly emulating). But the German law additionally, and uniquely, covered Huntington's chorea, hereditary blindness, and hereditary deafness, which in eugenics circles were regarded as marginal, in the number of people afflicted as well as in the economic burdens imposed. By contrast, criminality was seen to be a scourge that sterilization would deal with handily—yet criminality did not appear in the German list. Why the inclusions, and why the exclusion? Because, Teicher argues, the evidence for Huntington's, blindness, and deafness following a Mendelian pattern was excellent, whereas the evidence for criminality doing so was nil. By curating their list in the way they did, the law's framers maximized the appearance of bowing to the dictates of Mendelian science—an appearance further enhanced in the Mendelism-emphasizing official commentary and in the materials used to train the judges in the Hereditary Courts. (Among the law's international admirers were DeJarnette and Jean Rostand.)[41]

Mendelian reasoning also left its imprint on the workings of the Hereditary Courts themselves. They processed the cases of nearly half a million people before 1945, more often than not deciding in favor of sterilization. The sole question before them was whether an illness listed in the 1933 law had been correctly diagnosed as running in a family. To challenge a diagnosis, you could try to show that, say, your blindness was caused by a well-attested accident and so was not hereditary; or that your son, labeled as feeble-minded, was manifestly capable. What you could not do was adduce your family's history of healthiness as evidence that, although one of the culprit recessives was probably getting transmitted, it was not, in your family's case, affecting anything. Here, Teicher suggests, was a distinctively Mendelian indifference. He notes that under an alternative, Galtonian understanding of heredity, ancestry was seen not just as where genes come from, but as one of the contexts conditioning the effects of genes. The Galtonian wanted to know not just about genes but about genetic background. But for the Mendelian, genes were what mattered. "According to their own self-understanding, Hereditary Courts cared less about the actual health of individuals, which they saw as merely external or phenotypic," writes Teicher; "their aim was to annihilate malignant genes, not to cure diseases."[42]

There was, finally, a Mendelian imprint on something the Nuremberg laws made especially salient: the definition of a Jew. By the Nazi period, anthropologists had long acknowledged that, racial typologies notwithstanding, all the European races were about as impure as could be. There were, however, two notable exceptions, united by histories of comparative isolation. One was the village-bound German peasant. The other was the (until recently) ghetto-bound Jew. Mendelism, with its vision of ineradicably persistent genes and of traits depending only on those genes, held out the hope that the German race of old, preserved in that pure rural remnant, could be bred back to its former glory. Yet that same vision put this reconstruction fantasy under threat, since the Jewish recessives already mixed in were bound to cling stealthily, silently to life. When it came to safeguarding the German race from Jewish genes, then, any Jewish ancestry, however far back in someone's genealogy, was too much. In practice, however, there was concern that if the criteria for counting as a Jew were overly inclusive, there would be a backlash. Experts argued for different proposals, all of them couched in Mendelian terms, all of them equally bogus. A remark Teicher quotes catches the absurdity beautifully. On learning in 1935 that, with just one Jewish grandparent, he was not deemed Jewish, the nuclear physicist Fritz Houtermans decided he would deem himself so anyway. As he explained to his wife, "I wanted to mendel-out myself."[43]

Mendelism's thorough Nazification comes as a surprise nowadays because, as Teicher documents near the end of *Social Mendelism*, so much effort was expended after the war to dissociate Mendelism from its political past. That project became urgent less because of embarrassment over Nazi enthusiasms than because of the rise of Lysenko in the Soviet Union and the brutal suppression there of Mendelian genetics, condemned as a reactionary science serving bourgeois capitalist (and indeed racist) interests. By the 1960s, Lysenkoism was a spent force, but Mendelism's depoliticalization became permanent.[44] When, in 1965, geneticists around the world marked the centennial of Mendel's lectures, UNESCO devoted a special issue of its magazine, the *Courier*, to the theme of race and prejudice. Included was an update of the organization's famous statement on race, declaring that, according to modern biology, there is no such thing. Humans belong not to races but to genetically variable populations, none biologically superior to any other. Further on in the issue was Rostand's profile of Mendel, representing him as the great begetter.

3

If Mendelism were a person, its ego would be badly bruised by now. We have seen that subsequent advances in the science and technology of heredity have depended little on Mendelism, whereas what is most shameful and distressing about the eugenicist past often bears a Mendelian stamp. My aim in dispatching supposed legacies in these areas is not to trash a reputation, however, but to draw out a genuine yet often overlooked legacy: Mendelism's extraordinary capacity to simultaneously disavow and promote elementary Mendelian explanations, in which characters are binary units—brown or blue, round or wrinkled, normal or feeble-minded, German or Jewish—completely determined by "genes for" those characters. To be sure, thoughtful Mendelians have more or less always regarded the "genes for" idea as an oversimplification, allowing for all kinds of complexity, in characters and in the interactions behind them. Yet the power of Mendelian explanations to cut through complexity and expose underlying simplicity has nevertheless been relentlessly showcased, from Bateson's day to our own.

What accounts for this robust coexistence? Of course, simplicity has a perennial appeal, in heredity as elsewhere. By contrast, to quote Richard Lewontin, a tireless critic of genetic determinism, "measured claims about the complexity of life and our ignorance of its determinants are not show biz."[45] Even so, some attractively simple scientific theories do eventually die, which is why students now need help in interpreting the exchange in *Julius Caesar* about inheriting a mother's choleric humor. To come to grips with the peculiar, continuing life of "genes for" Mendelism in our culture, we need to dig deeper.

A good place to start is with that ur-statement of sophisticated, Weldonized Mendelism, *The Mechanism of Mendelian Heredity*. Although the term "gene" is not used in the book, the Morgan group's discussion there of how Mendelian talk should and should not be construed—and how much scope there is for misconstrual, even among experts—remains peerless. The book's retention throughout of the older term "factor" was certainly idiosyncratic. As the bibliography testifies, "gene" had spread rapidly among geneticists after its introduction in 1909 by Johannsen. Morgan's own coauthors—Sturtevant, Bridges, and Muller—had used "gene" in the titles of cited papers. But *The Mechanism of Mendelian Heredity*, aiming at a more general biological readership, carried a message for which "factor" was indeed a better fit than "gene," and not just because the latter was still for specialists. Over and over again, the reader is reminded that whenever chromosomally borne hereditary factors are identified as "for" a visible character (e.g., eye color), or for a version of a character (e.g., red eye), that identification is limited to situations in which the many other factors—hereditary, but also developmental and environmental—influencing that character or character version are held constant. Absent such constancy of background, all bets are off.[46]

After a first chapter introducing the pairing and segregation of chromosomes as underpinning the pairing and segregation of Mendelian factors, the second chapter, misleadingly titled "Types of Mendelian Heredity," drives home how phenomenally complex the relationship between a factor and an associated character can be. It is not merely, as Bateson had stressed, that simple "dominance"/"recessiveness" relationships cannot be counted on, or that a single factor can affect multiple characters, or that a single character can be affected by multiple factors. A change in temperature, for example, can change the character version associated with a given genetic constitution, as Erwin Baur had found in work with primroses. Raised at around 20°C, the white primrose, *P. alba*, produces white flowers, and the red primrose, *P. rubra*, produces red flowers. But when the temperature rises to 30°C, *P. rubra* produces not red flowers, but white ones. The lesson drawn is as much linguistic as biological:

> Strictly speaking, we should say, not as we generally do for brevity's sake, that the difference between the two races is that one has white, the other red flowers, but we should say rather that P. rubra reacts at 20° by producing red, at 30° by forming white flowers; P. alba, on the other hand, reacts both at 20° and at 30° by producing white flowers. The constant difference between these races is not their color, but in the possibility of producing specific colors at certain temperatures.[47]

More examples follow, of environmental influences but also developmental influences, glossed expansively:

> "Age," too, is in a sense an environmental condition, which influences the development of characters. Thus a white flower may change to purple as the plant gets older, or the flaxen hair of a child may turn to brown when he becomes a man. But, as in the case of other "environmental" conditions, age may not have the same effect on individuals with different factors; in this way it comes about that animals or plants which differ by certain factors may show a difference in character only at certain ages, or may not show the same difference at all ages.[48]

The Morgan group has acquired a reputation for relegating development to the sidelines as no business of geneticists, whose job is to understand transmission. But that is, at best, an unfair caricature. They stressed that if, as seemed clear, each cell in a developing organism receives the same set of hereditary factors, then embryological differentiation can take place only if, as regional peculiarities establish themselves, factors are differentially affected. No less than temperature differences or age differences, then, the chemical differences distinguishing one region from other regions could be necessary for a factor's expression. ("Thus when we speak of factors for eyes or for legs, we really mean factor-differences which can produce effects only in the eye, the leg, or other regions of the body.")[49] Yet another context that could matter was the complex of other factors present in an organism. Without the factors required to make an eye, say, an eye-color factor can have no effect; and such effects as it does have can be different depending on interactions with other eye-color factors present.

These diverse complications come across not as problems for chromosomal Mendelism but as taken-for-granted presuppositions. It is precisely because everything is interacting with everything else, all the time, with the most complex consequences, that we need to study heredity in the Mendelian manner, using organisms purged of variability and raised in controlled environments. Anyone who understands that, as Morgan and company put it at the end of the chapter (in a line quoted from earlier), "every character is the realized result of the reaction of hereditary factors with each other and with their environment" will grasp the wisdom of a method of inquiry that imposes constancy on the factorial and environmental background in order to investigate how differences at a single locus of a single chromosome can affect a body.[50]

Of course, there is a catch: the need not to forget all the contrived homogeneity behind Mendelian "factor for" and "unit character" talk and, conversely, the need to remember the status of that talk as abbreviating much more cumbersome talk about what was observed when, against a backdrop of factorial, developmental, and environmental constancy in a particular set of organisms, a particular change took place at one locus. Concern that new Mendelians, like some old ones, might need extra coaching in order to keep the limitations in mind surfaces throughout the rest of the book. A discussion of the "sex factor" known to be carried on the X chromosome, for example, disabuses the reader of the notion that the factor is, in any simple way, for sex:

> As in the case of sex limited characters, so in the case of sex itself there must be many factors in the fertilized egg that are as essential to the development of sex as are the sex factors themselves, but as they are distributed to all individuals alike, they are not thought of as differentiators, but as forming the chemical background on which the single or the double amount of the sex factor gives its result. It is quite conceivable that one or more of these other factors might so change that the sex differentiators would become inoperative or even change so that those other factors themselves become the differentiators that determine sex.
>
> The environment—the outer world—is also one of the components that enters into the development of every individual. A specific environment is one of the conditions of development. Why then, it may be asked, may not the environment turn the scale and determine sex? As a general proposition this must be acceded to at once—it is entirely a matter of proof. If there is an internal mechanism to determine sex in a normal environment it is quite conceivable that it might be supplanted in a new world. It is a question of evidence as to how often, if ever, this occurs. It is furthermore quite conceivable that some animals have no internal mechanism to regulate sex but depend on a difference in their medium.

> If such an environment can be discovered it would be sex determining in the same sense in which the term is here employed when the sex differentiators are hereditary factors.[51]

Morgan and company even declare at one point that it would be no challenge to their "factorial hypothesis of sex determination" should, in a given case, the sex-determining factor prove to be environmental rather than genetic.[52] For them, to be a student of inheritance is to be interested in differences that make a difference, whatever the source—though, for the student of Mendelian inheritance, chromosomal differences (e.g., XY = male, XX = female) take pride of place.

The book's concluding chapter, titled "The Factorial Hypothesis," reinforces all these lessons. It begins by setting out two ways of expressing the relationship between factor and character, using the factor for red eyes in fruit flies as an example. In the first way, a factor is a segregating, intra-gametic something whose effects, upon combination and recombination, explain the 3:1 ratio of red-eyed flies to white-eyed flies in the second hybrid generation. In the second way, there are gametes that produce red eyes and gametes that produce white eyes, and these gametes differ by just one factor-difference. Like pretty much every character, eye color in flies turns out to be under the influence of multiple loci (twenty-five were known by 1915). Can we nevertheless talk of a mutation associated with, say, pink eyes as "the cause of pink"? Yes, they say, because "we use cause here in the sense in which science has always used the expression, namely, to mean that a particular system differs from another system only in one special factor."[53] And again, they stress, the ramifications of a change at a single locus may affect multiple characters, not just the character on which the effect is so conspicuous as to tempt ill-advised talk of a "unit character." ("So much misunderstanding has arisen among geneticists themselves through the careless use of the term 'unit character' that the term deserves the disrepute into which it is falling.")[54]

So the two ways of understanding "factor" are not equal, because the first, "factors for unit characters" way needs the second, "differences that make a difference" way in order not to be misleading unto falseness. As the Morgan group put it, by way of sharpening the contrast between their Mendelian picture of development and inheritance and the Weismannian picture, with its vestigial conception of characters as existing preformed in associated determinants:

> The factorial hypothesis does not assume that any one factor produces a particular character directly and by itself, but only that a character in one organism may differ from a character in another because the sets of

factors in the two organisms have one difference. This point is not likely to be misunderstood by any one who grasps the meaning of the factorial hypothesis.[55]

As to how, exactly, factor differences manifest themselves developmentally as character differences, they acknowledged not merely the interest of that problem for experimental embryology, but the possibility that whatever would be learned would help improve our understanding of heredity. Even so, there was, in their view, no need to wait for that future science of development in order to proceed with the current science of heredity, since, understood as they understood it, Mendel's law "stands as a scientific explanation of heredity, because it fulfills all the requirements of any causal explanation."[56]

4

By 1915, then, we find, among the foremost practitioners of Mendelian genetics, exasperated acknowledgment that the Mendelian gene had all too often been mistaken, even by geneticists, for a simple character-maker rather than a complex difference-maker. To talk of a "gene for a character" was, for the Morgan group, to use shorthand for a causal relationship manifest only when, against a particular genetic, developmental, and environmental background, differences at a chromosomal locus produce differences in a character. Change something in the background, and you may well change, or even eliminate, the effect on the character. When the chromosomal differences produce character differences so cleanly differentiated as to be binary—red or white eyes in flies, blue or brown eyes in humans—that is an artifact, contrived or accidental, of the system, not an intrinsic property of the allele pair.

It was, and remains, hard to make this more complex gloss on the Mendelian gene stick.[57] Why? An important clue lies, I think, with the nature of the relationship between "gene for" talk and the "gene as difference-maker" notion. That relationship is not, on inspection, straightforwardly that of an abbreviation to what it abbreviates. Abbreviations do not, after all, usually cut loose semantically. No one who understands English is in danger of mistaking, say, "USA" for anything other than the United States of America, or "USB" for anything other than the sort of computer port so labeled (though only experts—along with nonexperts who read instruction manuals—will know that the label is an acronym of Universal Serial Bus). Furthermore, abbreviations typically arrive on the scene *after* what they abbreviate. In the case of the Mendelian gene, however—and as Morgan and company well knew—it was the early critics of Mendelism, among them Morgan, who had insisted that, contrary to Mendelian emphases on the factorial and the binary, the same hereditary factor could have different effects under different internal and external conditions. In *The Mechanism of Mendelian Heredity*, the Morgan group was fighting a rearguard action in declaring entrenched "genes as character-makers" talk to be shorthand for "genes as difference-makers" talk.

The "gene for" locution is, I submit, straightforwardly shorthand for "genes as character-makers" talk and only un-straightforwardly shorthand for "genes as difference-makers" talk. In principle, becoming educated in genetics involves the displacing of the character-maker conception by the difference-maker conception, even as "gene for" talk remains constant. In practice, however, the situation is much trickier, and not just because some people go further in their genetics education than others. The very structure of a typical genetics education endows the character-maker conception with independent life by investing it with heuristic power. Begin your education in genetics with Mendel's peas, and you will learn not merely about a case in which, you are told, binary characters are determined by genes for those characters and by nothing else. You will learn too that many apparently more complicated cases can be made tractable by treating them in the first instance like Mendel's peas. (And if you don't learn that, you won't pass.) Of course you will go on to learn about all sorts of exceptions to your rule of thumb, and the reasons why those exceptions are the way they are: the effects of other genes, epigenetic modifications, the interplay of development and environment, chance. By the end of your education, you will know, of course, that "it's not all in the genes," and become annoyed with anyone who suggests that you think otherwise. But the Mendelian "treat 'em like the peas" rule of thumb will remain in place. It will guide your reasoning and even—in the way of heuristics—perhaps your unreasoned reflections and reactions too, with much reinforcement from the wider culture in the form of gene-personifying 23andMe ads, "gene for" discovery stories, jargon talk of what is in an organization's DNA, and so on. You will affirm genes-for-characters determinism in your actions and attitudes while rejecting it if asked about it because you know that it's false.[58]

It is striking, in this light, to look with renewed attention at how determinism leaks into genetics textbooks even as their authors attempt to help students avoid it. Consider two examples, one from the 1930s, the other from the present day. I take the first from the second edition of *Principles of Genetics: A Textbook, with Problems*, published in 1932 by the Columbia University geneticists Edmund Sinnott and L. C. Dunn (who would go on to write one of the centennial histories). "Sinnott and Dunn"

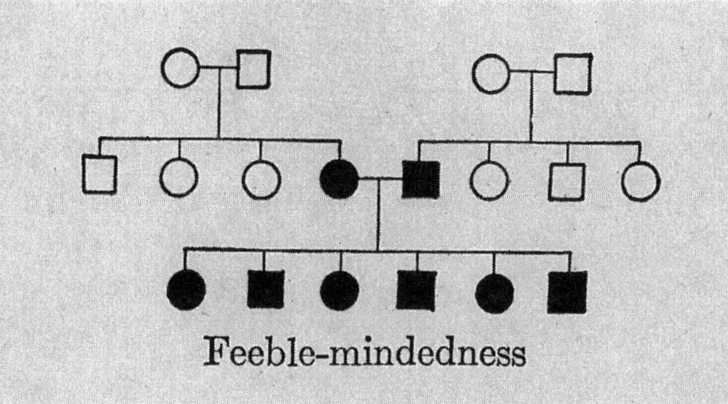

FIG. 12.4. "Determine . . . the *method of inheritance* of the trait in question (whether dominant or recessive); and, as far as possible, determine for that trait the *genotype of each individual.*" Problem from a 1932 genetics textbook.

was the leading textbook in genetics through the second quarter of the twentieth century. Just as Mazumdar would lead us to expect, we find, between the first edition of 1925 and the second edition, a backing-off from eugenics and associated claims. In the first edition, an upbeat treatment of eugenics gets the final chapter all to itself. In the second edition, that is gone entirely, replaced with a chapter on genetics and development, and a new question is appended to the chapter introducing Mendel's laws: "Criticize evidence for the inheritance of feeble-mindedness in man as due to a recessive factor." Turn, however, to the problem set at the back of the book, and we find, retained from the first edition, a feeble-mindedness pedigree, presented as one of six trait-exhibiting pedigrees for which the student needs to work out whether the trait is due to a dominant factor or a recessive factor (fig. 12.4). The feeble-mindedness pedigree is, truly, a textbook case of recessiveness, with two heterozygote couples each giving birth to three normal children and one feeble-minded child, and the coupling of the feeble-minded individuals resulting in exclusively feeble-minded offspring.[59]

The second example is a witty problem from a textbook already quoted from, Peter J. Russell's *iGenetics: A Mendelian Approach*, likewise from the chapter introducing Mendelian genetics.[60] Before the story of Mendel and his peas, there is first a section titled "Genotype and Phenotype," where we read: "The genetic constitution of an organism is called its *genotype*, and the *phenotype* is an observable characteristic or set of characteristics . . . produced by the interaction between its genotype and the environment." Another couple of paragraphs elaborate the theme of inter-

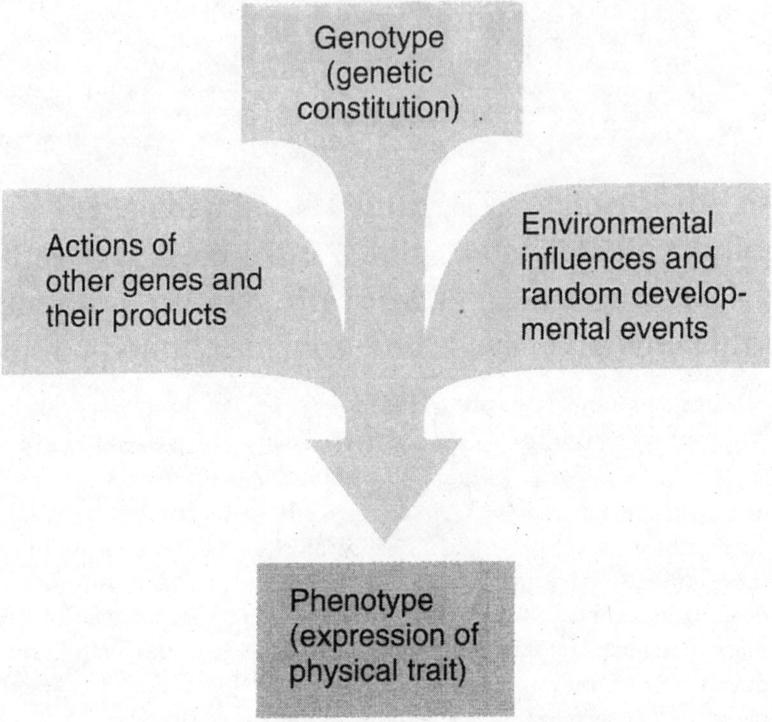

Genotype (genetic constitution)

Actions of other genes and their products

Environmental influences and random developmental events

Phenotype (expression of physical trait)

FIG. 12.5. Genotype, phenotype, and the complex interactions that can condition the effect of the former on the latter: the last thing the student reader of *iGenetics* sees before Mendel and his peas.

action as central to genetics, since genes "provide only the potential for developing a particular phenotypic characteristic," and a combination of other genes, the environment, and chance determines how far that potential is realized. An example given is height: many genes influence it, but there are also "internal and external environmental influences such as the effects of hormones during puberty (an internal environmental influence) and nutrition (an external environmental influence)." From genotype to mature phenotype, then, a great deal of complexity interposes as biochemical pathway interacts with biochemical pathway—a notion visualized in an image of merging causal arrows (fig. 12.5).[61]

But the student who gets to the end of the chapter, where the questions and problems are, finds this:

> After a few years of marriage, a woman comes to believe that, among all of the reasonable relatives in her and her husband's families, her husband, her mother-in-law, and her father have so many similarities in their unreasonableness that they must share a mutation. A friend taking a course in genetics assures her that it is unlikely that this trait has a genetic basis and that, even if it did, all of her children would be reasonable. Diagram and analyze the relevant pedigree to evaluate whether the friend's advice is accurate.[62]

The genetics-educated friend tries to allay the woman's fears in two steps: first, by dismissing as "unlikely" the hypothesis that unreasonableness "has a genetic basis"; second, by arguing that even if her hypothesis were true, the woman's children would not fall victim to hereditary unreasonableness. Notice what has happened. From expressing doubts about the trait's having a genetic basis, the friend—and now the student aspiring to evaluate the friend's advice—proceeds to accept the posit of a mutation for unreasonableness. That posit is what makes unreasonableness analyzable by Mendelian principles. And what makes the posit possible is the quiet replacement of "genetic basis" by "mutation." The terms are not, on the face of it, equivalent. "Genetic basis," after all, suggests something messily genes-plus. If unreasonableness has a genetic basis, then presumably one could inherit the mutation (the "gene for unreasonableness") yet not be unreasonable because whatever extra causal ingredients are needed to make for unreasonableness are absent. Unreasonableness, so understood, would be better described as the upshot of genes and environment, where "environment" might encompass everything from the other genes in the genome to upbringing, nutrition, and stress levels. Indeed, it is easy to imagine that the causal package making for unreasonableness could be so complicated and heterogeneous that, even if genes are part of that package, there would be no warrant for talking about "a gene for unreasonableness," even as shorthand.[63]

The beauty of this problem, for purposes of Mendelian pedagogy, is that the Mendelian friend seemingly acknowledges all of that complexity, with the initial scoffing, but then, in getting serious and analyzing the lineage, puts it all to one side. That, the successful student learns, is what one must do in order to generate the percentages that Mendelian analysis so attractively delivers. In this case, the student who understands the principles and the task should go on to conclude that the friend's advice is inaccurate, for, whether the mutation for unreasonableness is dominant or recessive, the woman's children will, alas, have a 50% chance of being unreasonable.[64]

Beyond the power of the treat-'em-like-the-peas heuristic lie other ways of thinking, talking, and doing that will tend in the same insulating direction. Since the 1920s, for example, having variable effects in different contexts has come to be discussed in terms of a genotype's "expressivity" (that is, the extent to which its full effect is realized) and "penetrance" (that is, the proportion of individuals bearing the genotype that show the associated phenotype, at any level of expression). In principle, such talk sensitizes the talker to the importance of context for gene expression. In practice, however, "expressivity" and "penetrance" have become like "dominance": just another context-independent property of genes (indeed, a kind of modifier of "dominance" when it manifests only to a certain extent).[65] And since around the same time, students who have gone on to become researchers themselves have tended to work on problems that look likely to give clean results, using model organisms whose well-behaved genomes are maintained for exactly that purpose. To do otherwise, if it occurred to someone at all, was to risk career suicide when a career had barely begun.[66] As for the students who have gone on to lead non-geneticist lives, if they have remembered anything at all from their time in the genetics classroom, they have remembered Mendel and his peas, dominance and recessiveness, round and wrinkled, brown and blue, no brown-eyed babies from blue-eyed parents . . .

Could it have been otherwise? Could it *be* otherwise?

13

Weldonian Legacies

It is evident that personality clashes were as important as scientific arguments in sustaining the conflict [between the Mendelians and the biometricians]. If Weldon had adopted Mendelian inheritance, instead of opposing it, Pearson's whole attitude toward Mendelism might have been different. If Pearson, Weldon, and Bateson had worked together, population genetics might have begun in earnest fifteen years sooner than it did.

WILLIAM B. PROVINE, *THE ORIGINS OF THEORETICAL POPULATION GENETICS*, 1971

The words "penetrance" and "expressivity" . . . have led to a great deal of confusion in the field, and this sort of categorical thinking tends to miss complexity. . . . As such, perhaps we should get rid of the two terms altogether and just discuss the expression of each trait in the context of a phenotypic spectrum, which is of course what led Walter Frank Raphael Weldon to establish the field of biometry.

GHOLSON J. LYON AND JASON O'RAWE, "HUMAN GENETICS AND CLINICAL ASPECTS OF NEURODEVELOPMENTAL DISORDERS," 2015

When, in the late 1960s, William Provine spun his counterfactual lament, in a PhD dissertation that, in published form, went on to stand for decades as what to read (and assign) on the battle over Mendel, his aim was to chart the beginnings and early career of a science that he admired unreservedly. His *Origins of Theoretical Population Genetics* is an extended, splendidly scholarly version of the sort of history of science commonly found in science textbooks. On the whole, it is history in the service of the continued progress of the science chronicled—history as a means of strengthening allegiance to that science, chiefly by deepening appreciation for the founders, their achievements, and their fortitude in facing down difficulties. History of genetics in this celebratory mode looks at Weldon and sees squandered opportunity for faster progress.

But history of science can serve other ends. Notably, it can rescue figures and ideas on the losing side of a debate from what the social historian E. P. Thompson called "the enormous condescension of posterity."[1] Pursued in this more critical mode, history of science can, if not dispel, at least diminish the aura of givenness that settles around an established science, to the benefit not only of historians but also of scientists who find themselves in the uncomfortable position of questioning what they have been trained to do and to want. In a small way, I got to see up close how such scientific bolstering can work in autumn 2012, when Gholson Lyon, a psychiatrist researching the genetics of mental disorders, emailed me after coming across a video of a lecture I gave setting out the themes of this book. "As an active clinician (child, adolescent and adult psychiatrist), I have spent the past 8 years (at least) carrying on with detailed phenotyping and immersion in human variation," he wrote. "I can see the complexity and major expressivity differences, even in the same families and even in monozygotic twins." He was grateful to have made Weldon's acquaintance, and eager to read all the Weldoniana I could send him.[2] A couple of months later, after he read a manifesto for the Weldonian curriculum experiment then in the planning, he replied:

> I agree with this completely. I just attended the CSHL [Cold Spring Harbor Laboratory] Personal Genomes and Medical Genomics meeting for the past 3 days, and this element of genetic determinism was in full force, particularly among certain people pushing for prenatal genetic diagnosis for anyone. I objected to this pervasive genetic determinism. . . . My own thinking has evolved on this topic in the past year. Therefore, even some of my own earlier papers have some wording in them that I would change in retrospect, although I do discuss this issue of "causation" in my papers much more than some other human genetics folks have done.[3]

It was a boon for Lyon, as he began pushing back against a research world that paid no more than lip service to taking variability, ancestry, and environment seriously, to discover that he had an ally—indeed, a brilliant one—in the scientific past. So resonant did Lyon find Weldon's writings that the extraordinary, Weldon-citing review article from Lyon quoted in this chapter's second epigraph, on the genetics of neurodevelopmental disorders, begins by quoting Weldon in 1902 on how the "fundamental mistake which vitiates all work based upon Mendel's method" is "the neglect of ancestry."[4]

Perversely, in the twentieth century, the main legacy of Weldon's paper was the aspect that Weldon cared least about: his calculation of the improbably good fit between Mendel's observations and Mendel's theory. A chapter on Weldonian legacies has to make room for the "too good to be true" data problem; as we will see, the story of its public life turns out to throw new light on how the midcentury Mendel anniversaries we met in the last chapter intersected with the larger cultural politics of science. But in the twenty-first century there are other, more constructive, more distinctively *Weldonian* legacies in the making, as new history—with its attendant counterfactuals—gives new life to old science. In the final parts of this chapter, I want to explore some of these incipient legacies for the way we understand where genetics has been and for the choices we make about where it goes from here. As a teaser for the latter, let me first expand a little on the intellectual evolution that made Lyon so welcoming of Weldon's point of view on Mendelism.

1

Anyone wondering about the benign legacies of Mendelism need look no further than the discovery of Ogden syndrome. The 2011 paper announcing the discovery, in the *American Journal of Human Genetics*, has twenty-eight, internationally distributed authors. But the team leader was Lyon, then based in the Department of Psychiatry at the University of Utah, where he had arrived in 2009 after completing clinical training in psychiatry in New York. Between the relative isolation of the Mormons, their large families, their tending to stay put, and their outstanding genealogical record keeping, Utah had long been a promised land for human geneticists. At the center of the Lyon team's investigations was the Black family, from the town of Ogden. Across two generations, the family had repeatedly endured the tragedy of a baby boy dying in infancy, after first developing a strangely wrinkled appearance, large eyes, heart arrhythmia, and a range of other problems (fig. 13.1). None of these boys had managed to sit up, let alone talk. The symptoms did not fit any known disease profile. Nevertheless, for someone with Lyon's Mendelian training, the pattern was immediately suggestive of X-linked inheritance, and he seized on the opportunity to work with the family, telling an interviewer that "a rare, lethal, highly penetrable mutation that was passed down from the grandmother to two of her sons . . . [and then] from two of her daughters to two of their sons" seemed ideal for purposes of gene discovery.[5]

At that time, medical geneticists had begun collecting and sequencing "exomes" (protein-expressing genomic DNA) from people diagnosed with rare Mendelian disorders in order to identify—far more cheaply and rapidly than could be done with full-genome sequencing—the hitherto elusive mutations behind the disorders. But no one before had used exome capture and sequencing to identify a rare Mendelian disorder new to medical science. By concentrating on the X chromosomes of members of the Black family and then—using a new algorithm developed by Utah colleagues—systematically comparing the sequence data with data from reference exomes with no known disease associations, Lyon and his team

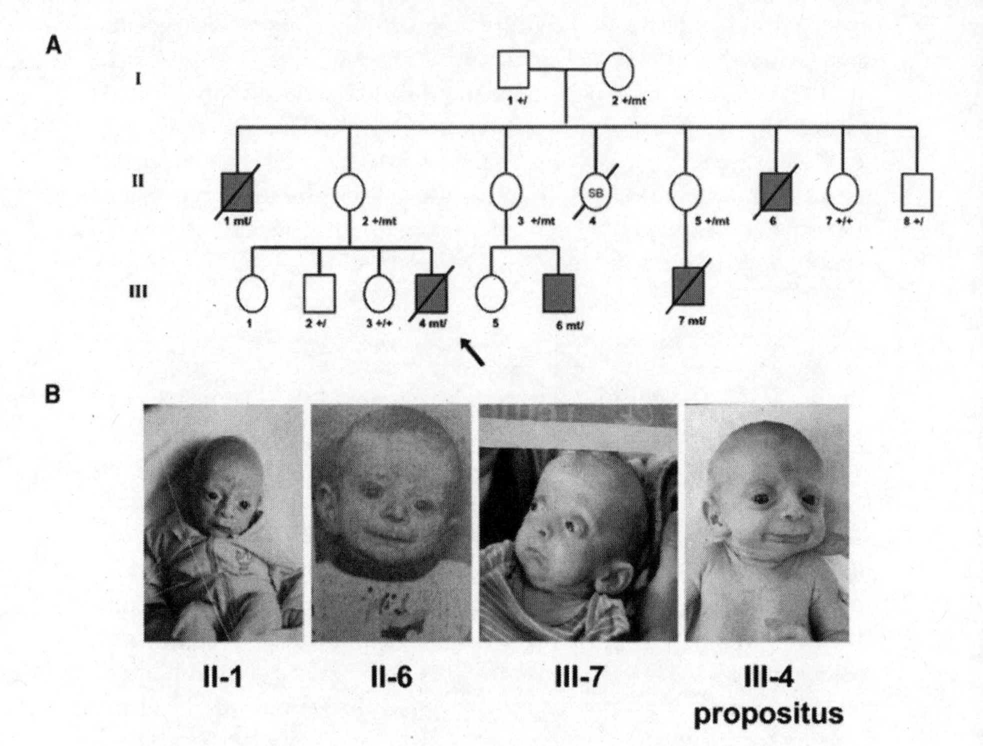

FIG. 13.1. The Black family's pedigree, with photos of four boys with Ogden syndrome. A fifth, labeled "III-6," died a few days before the publication of the paper describing the syndrome.

eventually closed in on a variant of the gene known as *NAA10*. Further studies of the variant and its encoded protein, along with the serendipitous discovery of an unrelated California family whose X chromosomes carried the same variant, and in which boys were occasionally born who suffered the same fates as the Ogden boys, corroborated the identification of the *NAA10* mutation as the culprit.[6]

Here was a project showing how new-generation molecular technologies and bioinformatics could yield valuable new knowledge and with it, through genetic testing, new power to avoid human suffering. For Halena Black, matriarch of the Ogden clan, understanding what caused her sons and grandsons to die was, she was reported as saying, "a relief." "It's so nice because now other grandchildren can be born into our family, and we can help other people."[7] And for Lyon—as well as for James Watson, who subsequently recruited Lyon to Cold Spring Harbor—the big next step was supposed to be from easily studied but rare conditions like Ogden syndrome to conditions like schizophrenia and autism, affecting far more people, but far messier in their symptom profiles and causal mix.[8] A report about the discovery on *Nature*'s website quoted a medical geneticist on the promise of the approach taken by the Lyon team: "This exemplifies an exceptionally rare disease, but the same type of strategy is now going to be applied to more common diseases to get the root cause. . . . We're going to take the term 'idiopathic,' which basically means 'we don't know,' and eliminate it."[9]

But Lyon's interests took a different turn. Rather than trying to assimilate the likes of schizophrenia and autism to Ogden syndrome, he became fascinated by the reverse approach: examining all the ways in which, in its own right, Ogden syndrome—in common with other apparently simple, cleanly Mendelian disorders—is a universe of complication, particularity, and messiness due to the complexity of interactions within a genome as well as between a genome and its environments. He was, in retrospect, well prepared for the possibility of such a reorientation by his time as an undergraduate student at Dartmouth, working in a lab studying the developmental disruptions wrought when a pregnant woman lacks sufficient iodine in her diet; by the huge variability in the children with Down's syndrome he got to know during his clinical training; by his experience with the trying out of a new form of therapy for velocardiofacial syndrome that brought about improvements in a five-year-old child seemingly doomed to a severe form of autism. Only now, though, as Lyon set about puncturing the pretenses of "gene for" genetics, did those aspects of his CV start to seem preparatory for something.[10]

Whereas in the 2011 discovery paper he stressed the track record of exomics in identifying the mutations behind an ever-expanding list of Mendelian disorders, in a review article published the next year, he and a Utah colleague, Kai Wang, noted that, as public databases had filled with ever more sequence data from ever larger human populations, an increasing number of what had been classed as rare disease-causing variants had needed to be reclassified, since—it was now clear—they were not in fact all that rare, and occurred in healthy people too. Plainly, in these cases, the mutations, wrote Lyon and Wang, "are thus unlikely to be truly causal or at least are modified by genetic background effects."[11] Those background effects got attention in their own right in a discussion of the importance of families not just as donors of DNA for sequencing and as beneficiaries of the results, but because family members share the genomic contexts that can determine what a mutation determines. That old idea was now catching on again under a new term, "clan genomics." Lyon and Wang's summary posited "that there are unique combinations of rare variants in recent family lineages, playing a causative role in disease," and hence, "for a specific patient, it is more important to consider the recent 'genetic history' of the patient's extended pedigree or clan, rather than their overall ethnic background."[12]

These discontented murmurings gave way in short order to a full-on assault on Mendelian categories, and on complexity-missing categorizing generally. The Weldon-citing passage above comes, as noted, from a review article published in 2015 by Lyon and a graduate student working with him at the time, Jason O'Rawe. Here is a longer sample, taken almost at random:

> Classifications can sometimes lead people to try to force round pegs into square holes, and so we are reluctant to further promulgate these classifications. Such classifications include terms such as: "Mendelian," "complex disease," "penetrance," "expressivity," "oligogenic," and "polygenic." For example, some have used the word "Mendelian" to refer to a disease that appears to be "caused" by mutations in a single gene. As such, cystic fibrosis, Huntington's disease, and Fragile X are all diseases that some people refer to as being "caused" by mutations in single genes. However, the expression of the phenotype within these diseases is extremely variable, depending in part on the exact mutations in each gene, and it is not at all clear that any mutation really and truly "causes" any phenotype, at least not according to thoughtful definitions of causation that we are aware of. For example, some children with certain mutations in *CFTR* may only have pancreatitis as a manifestation of cystic fibrosis, without any lung involvement, and there is evidence that mutations in others genes in the genomes can have a modifying effect on the phenotype. In the case of Huntington's, there is extreme variability in the expression of the phenotype, in time, period, and scope of illness, and all of this is certainly modified substantially by the number of trinucleotide repeats, genetic background and environmental influences. Even in the case of whole chromosome disorders, such as Down Syndrome, there is ample evidence of substantial phenotypic expression differences, modified again by genetic background, somatic mosaicism [when there are genomic differences across different parts of an individual's body], and environmental influences, including synaptic and brain plasticity. The same is true for genomic deletion and duplication syndromes, such as velocardiofacial syndrome and other deletions. And, of course, there is constant interaction of the environment with a person, both prenatally and postnatally. As just one example, cretinism is related to a lack of iodine in the mother's diet, and there is incredibly variable expression of this illness based in part on the amount of iodine deficiency and how this interacts with fetal development.[13]

And Ogden syndrome? It, too, turned out, like more or less every other condition and character, to be spectral. Throughout the 2010s, as the cost of sequencing fell and doctors used it ever more frequently to help with diagnosis, more and more children were found to have a variant form of the *NAA10* sequence, but showed many permutations in their symptom profiles. This phenotypic variability was due not to some *NAA10* mutations being more powerfully "penetrant" or "expressive" than others, but to their variable protein products having the diverse effects they did within diverse internal and external environments. Yes, sometimes, for ill-understood reasons, all that diversity could seemingly cancel out; in the original study, within both the Black family and the unrelated California family, the same mutation had led to the same recognizable symptom profile, despite different genomic and environmental settings. Yet even when such "canalization," as the geneticist C. H. Waddington called it, was robust, phenotypic heterogeneity was noticeable for anyone prepared to take notice. Whereas, in the discovery-announcing paper, *the* phenotype associated with the mutation in the California family was declared "indistinguishable" from *the* Ogden phenotype, and a summary table of syndrome features accordingly downplayed variability, follow-up papers in 2018 and 2019 instead dwelt in detail on differences across those individual boys and the other children—girls as well as boys—studied subsequently (fig. 13.2). Hence a new name was proposed for the overall syndrome, "*NAA10*-related syndrome," with "Ogden syndrome" reserved for those extreme cases whose severity was what attracted Mendelism-minded medical attention in the first place.[14]

2

What is now known as the "Mendel-Fisher controversy" traces to a paper that Ronald Fisher published in 1936, when he was Galton Professor of Eugenics at University College London.[15] Fisher was already internationally famous not only for his comprehensive Mendelizing of Darwinian biometry but for his innovations in statistical methods, especially in agricultural research.[16] Bringing to bear that side of his expertise, Fisher presented a detailed reconstruction of Mendel's program of experimental work year by year. Reanalyzing Mendel's data statistically, Fisher found, as Weldon had, that they are improbably good.[17] But what that showed, Fisher suggested, was what a great thinker Mendel was. For relatively soon after his crossing experiments were under way, Mendel must have worked out his theory in the abstract; and from that moment, he knew how his data ought to look. His program of experiments thus became, in Fish-

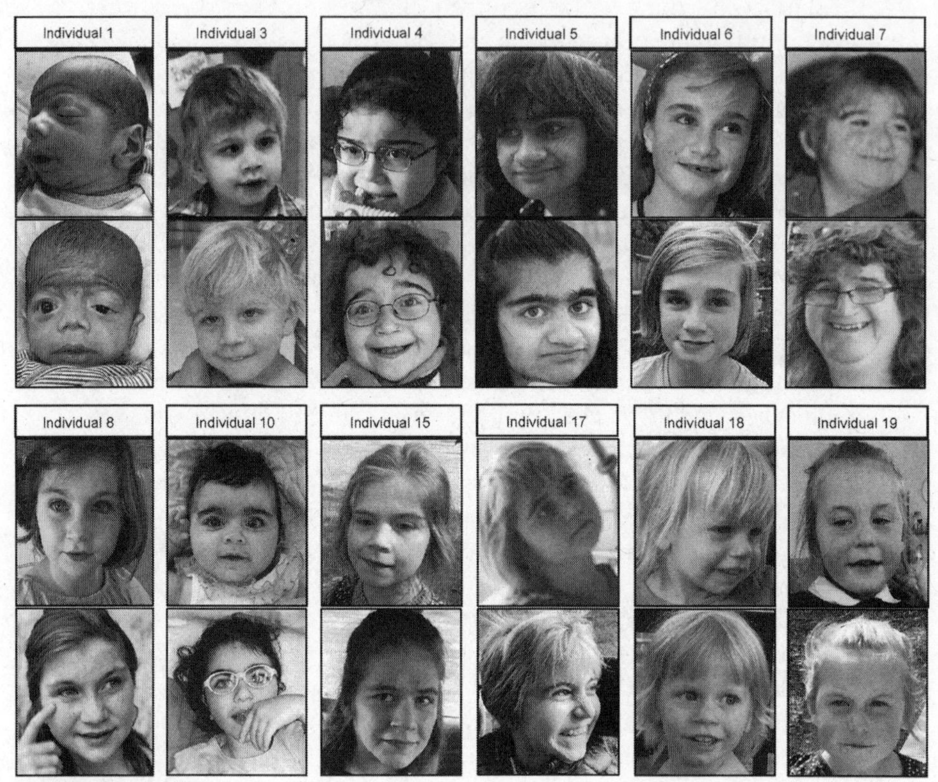

FIG. 13.2. Thirteen children (mostly girls) shown at two different ages. All were born with a mutant variant of *NAA10*, sometimes inherited, sometimes arising de novo. Individual 1 died at six months of age, and his facial features and other characteristics resembled those of the boys with Ogden syndrome from the original study. Although the other children were all classified as having what is now called "intellectual disability," or ID for short, the degree of impairment ranged from severe to moderate to mild. As for their faces, Lyon and colleagues concluded, "A recognizable, regular pattern of dysmorphologic facial features was not discerned among the cases, other than perhaps thicker eyebrows and broad philtra."

er's words, "a carefully planned demonstration of his conclusions."[18] In Fisher's view, the data's limitations were thus largely to Mendel's credit, and such blame as Fisher was willing to consider, he meted out to a well-meaning but misguided underling who, Fisher surmised, must have quietly gotten rid of whatever plants threatened to mess up the master's ratios: “Mendel was deceived by some assistant who knew too well what was expected.”[19]

Again like Weldon, Fisher in private correspondence was less reserved. He wrote to the Oxford ecological geneticist E. B. Ford in January 1936: “I have had the shocking experience lately of coming to the conclusion that the data given in Mendel’s paper must be practically all faked.” Calling this an “abominable discovery,” Fisher nevertheless affirmed his faith in Mendel’s honesty: “I don’t believe that this touches Mendel’s own *bona-fides* or the reality of the experiments he carried out.”[20] In keeping with Fisher’s intentions, over the next quarter century, his analysis of the improbably close match between Mendel’s expected and observed results became familiar, in and out of genetics, not as a shameful instance of scientific fraud but as an illustration of how to run a chi-square test and why doing so is important.[21] Whatever disquiet geneticists may have felt in making sense of Fisher’s work they kept in-house.

From 1948, Lysenkoism turned public silence into a political imperative. There was no mention at all of the “too good to be true” data problem in Huxley’s *Soviet Genetics and World Science* or Zirkle’s *Death of a Science in Russia*, two 1949 books whose boosterism about Mendelian breeding I touched on earlier.[22] Accentuating the positive became official policy for the Genetics Society of America, which used its 1950 conference on the “Golden Jubilee” of the triple rediscovery to celebrate Mendel and his legacies, above all in agriculture, recording the whole of the upbeat proceedings for broadcast beyond the Iron Curtain by the propagandizing Voice of America. Again, there was no hint of a problem with Mendel’s data.[23] When Zirkle, a speaker at the conference, published a short piece in *Science* four years later titled “Citation of Fraudulent Data,” he dealt exclusively with the Lysenkoists and their British fellow travelers, whose fondness for citing the work of Paul Kammerer—an Austrian physiologist notorious in the 1920s for supposedly showing Lamarckian inheritance in breeding experiments with midwife toads, and whose suicide was widely interpreted as an admission of the fraud he was accused of by Bateson and others—Zirkle found exasperating.[24] Zirkle remained in full Cold Warrior mode through the rest of the decade. The most overtly political contribution to the 1959 *Origin of Species* centennial was his *Evolution, Marxian Biology and the Social Scene*, a lengthy treatise combining a history of the rise of Malthusian Darwinism, Mendelian genetics, Galtonian eugenics, and their synthesis with a history of the emergence of the Marxian negation of all that, together with a study of Marxian influence among humanities scholars and social scientists in the United States. Again, fraud comes across as exclusively, and endemically, Lamarckian.[25]

1959 was also, of course, the year that Khrushchev let himself be photographed amid that emblem of Mendelian bounty, an Iowa cornfield. As de-Stalinization loosened Lysenkoism's grip in the Soviet sphere, the sense of emergency among geneticists outside that sphere subsided. Soon geneticists began to participate in a wider trend within Western, and especially American, culture in that phase of the Cold War: a new emphasis on the capacity of the human mind in a state of freedom to think and create without limits. From government departments and think tanks to the cognitive science laboratories being set up at prestigious universities, there was a growing interest in exploring and, crucially, promoting the "open mind" as what defined humans at their rational and democratic best. An open mind was one not captive to dogmatic certitudes but free to probe even cherished beliefs as potentially mistaken. Habits of criticism, including self-criticism, were held up as vital, along with the freedom to speak and write honestly about one's conclusions. Freedom meant the freedom to think about anything, to question anything, to be transparent about flaws and difficulties. For those eager to inhabit or, for propagandistic purposes, exhibit the distinctive strengths of what was known as "the free world," self-critical science was thus of a piece with civil-rights marches and abstract expressionism.[26]

It was at just this moment that the "Versuche" centennial approached. Now silence over the Mendel data problem gave way to volubility. An early entrant into the new conversation was Zirkle, in a 1963 lecture (published the next year) on "the delayed discovery of Mendelism."[27] Other public commentators over the anniversary period included Beadle (in his "Mendelism, 1965" talk), Sturtevant, Dunn, the Darwinian population geneticists Sewall Wright and Theodosius Dobzhansky, and two British biologists, the evolutionary embryologist Gavin de Beer and the marine biologist Sir Alister Hardy.[28] None thought that an accusation of fraud was justified; all thought there was nevertheless something to be explained or, preferably, explained away.[29]

To understand why this inconclusive discussion inaugurated a small but steadily active—indeed, ongoing—academic industry, we need to turn to Cold War culture on the other side of the Iron Curtain. Recall that among the centennial meetings in 1965 was one in Brno itself. Yes, Khrushchev in Iowa in 1959 had signaled to the world the beginning of the end of Lysenkoism.[30] Even so, when Lysenko visited Prague the next year, he was received as an honored guest. Around that time, Mendel's statue in Brno was removed from public view.[31] Lysenkoist deprecations of him aside, Mendel—as a Catholic prelate in a fabulously wealthy monastery, and as a cultural German beloved of the Nazis, whose brutal occupation of Czechoslovakia was still a fresh memory, and who had planned

FIG. 13.3. A Czechoslovakian stamp commemorating the Mendel Memorial Symposium in Brno. Note the UNESCO branding and, again, the iconography uniting a pea plant (plus a fuchsia) with a double helix.

to set up a Mendel Research Institute in Brno to advance genetics "in the spirit of German National Socialism"—was not at all an easy figure for Czechoslovak Communist officialdom to celebrate.[32] So the politics were delicate when, in 1963, the Czechoslovak Academy of Sciences got behind the project to hold an international, UNESCO-funded Mendel symposium (fig. 13.3). The man who eventually made it happen, Vítězslav Orel, a former poultry researcher turned historian of science, understood his brief, and used his symposium address to stress Mendel's credentials as someone who, as befitting a socialist hero of science, belonged to the world of practical agricultural and horticultural improvement.[33] If ever there was a time *not* to mention the data problem, the Brno meeting was it. And no one did mention it, not even Zirkle, who courteously left it out of a rerun of his delayed-discovery lecture.[34]

Only a few years later, when the Czechoslovak rehabilitation of Mendel was well in hand, taking institutional form in a new Mendel museum in Brno and a new research journal (both run by Orel), did Orel face down that less-than-heroic theme of the centennial. Unsurprisingly, Orel's view was that much ado had been made of nothing, and in relation to someone whose meticulous record keeping, as the documents preserved in the Brno collection showed, was beyond reproach. Yet Orel's 1968 article "Will the Story on 'Too Good' Results of Mendel's Data Continue?" backfired completely. Far from killing the story, Orel gave it permanent vigor, partly by publicizing it for readers who would never have bothered reading the anniversary papers and volumes en masse, and partly by summarizing so ably the abundance of surmises already in play: about what Mendel had or had not done as well as about which statistical tests should be applied, how to apply them, and how to interpret the results.[35]

So forbiddingly technical were the issues raised that the "Mendel-Fisher controversy" might well have remained a matter strictly for specialists. But word began to spread. In the pro-Kammerer bestseller *The Case of the Midwife Toad* (1971), Arthur Koestler turned the tables on the likes of Zirkle, suggesting that it was not the Lamarckians, but the Mendelians, who had a foundational fraud problem. "It is rare," wrote Koestler, after introducing Fisher's analysis, "to find this historical scandal mentioned in the literature. It was not so much hushed up as shrugged off. Since Mendel's Laws had been shown to be correct, what does it matter if he cheated a little?" Quoting at length from Hardy, who had speculated in Fisherian fashion about well-meaning assistants, Koestler added, "Tolerance and broadmindedness towards the dead are no doubt laudable; but what if that obscure monk in Brünn had been caught red-handed doctoring his statistics—or even neglecting to check the gardeners' statements?"[36]

What transformed this minor theme of modern Mendeliana into a major one was the congruence with a larger image problem that overtook Western science in the 1970s. Somewhere between, let us say, Rachel Carson's revelations about the chemical industry's poisoning of the environment, the ceaseless images of technoscience-delivered devastation of the people and landscapes of Vietnam, and the struggle for justice for children born deformed because their mothers had taken the pregnancy drug thalidomide (to cite just three of the best-remembered lows), the moral high ground taken for granted in the polemics of the Cold War at its frostiest was lost. Along with that détente-accelerated shift came an enlarged public appetite for learning about scientific fraud, present-day but also historical.[37] The American science and mathematics popularizer Martin Gardner caught the mood in a 1977 piece for *Esquire*, "Great Fakes of Science," written in the wake of the scandal surrounding the British psychologist Sir Cyril Burt, whose twin studies purporting to show that intelligence is genetically determined had been exposed as fraudulent. "Politicians, real-estate agents, used-car salesmen, and advertising copywriters are expected to stretch facts in self-serving directions," Gardner wrote, "but scientists who falsify their results are regarded by their peers as committing an inexcusable crime." Alas, he continued, "the sad fact is that the history of science swarms with cases of outright fakery and instances of scientists who unconsciously distorted their work by seeing it

FIG. 13.4. From Martin Gardner's scientific rogues' gallery, "Great Fakes of Science," in *Esquire* (1977), subheaded "*Yes, even Brother Mendel lied.*"

through lenses of passionately held belief." High up on Gardner's list was Mendel (fig.13.4)—"such a hero of modern science that scientists in the thirties were shocked to learn that this pious monk probably doctored his data. . . . They are too good to be true."[38] Not long after, a book on the fraud problem in science placed Mendel in a concluding chronology of scientific frauds. (He was in excellent company: before him were Hipparchus, Ptolemy, Galileo, Newton, and Dalton.)[39]

And there Mendel remains. Anyone who knows anything about Mendel typically knows, along with the dominance of brown eyes over blue eyes, that, hero-of-heredity status notwithstanding, there is something suspicious about Mendel's data. He may be the father of genetics, but he is also "the father of scientific misconduct."[40] Orel achieved a great deal—to the point where, as we saw in chapter 1, his corrective emphasis on Mendel's immersion in the world of Brünn breeding now needs correction—but he failed utterly in scotching the data story. It has rumbled on and on, for the most part to little effect.[41] The consensus view is more or less where it was at the start: the data are indeed improbably good, but that in itself is not evidence of fraud, nor is there any other evidence to suggest fraud. As Fisher said in a talk he gave on heredity to the Cambridge Eugenics Society in 1911, "It may have been just luck; or it may be that the worthy German abbot, in his ignorance of probable error, unconsciously placed doubtful plants on the side which favoured his hypothesis."[42]

Luck or unconscious bias (or worse): that framing, set by the eugenicist Fisher and solidified by the peculiar geopolitics of the Cold War, has ensured an ultimately sterile debate. What has been consistently overlooked is the possibility that, as Weldon thought, Mendel's "too good to be true" data show the shortcomings of working with only two categories. Assume that pea seeds must be either "yellow" or "green," "round" or "wrinkled," and you will indeed be at risk of assigning seeds that are, in Fisher's terminology, "doubtful" to whatever category will vindicate your prediction. You can try to figure out a way of deciding which of the two categories each pea *really* goes in. Or—with Weldon—you can instead give up on the categorizing and take the variability you actually see as what it is you are trying to explain: variability brought about because genes have the effects that they do in complex contexts.[43]

3

I shall come below to the question of how an awareness of what Weldon wanted for biology can change our sense of the present and its possibilities. But I want first to explore some repercussions for our sense of the past and its possibilities, looking in turn at three topics that, in various forms, have mattered a great deal in this book: the debate over Mendel and what it was actually about; Mendelism's success and how to explain it; and the prospects for a technologically impressive but nongenic biology, had the debate over Mendel gone differently or Mendelism never arisen.

The debate over Mendel. For decades, those who teach and write about the history of biology have had a fairly standard story to tell, lucidly set out in Provine's *Origins of Theoretical Population Genetics*. The 1900 rediscovery of Mendel's experiments with hybrid peas, they have said or written, got swept up into in an older, wider, already bad-tempered debate about whether evolutionary change is or is not Darwinianly continuous. Both sides took Mendel's emphasis on all-or-nothing character versions (seeds as yellow or green, never anything in between) to support a discontinuous picture of evolutionary change, in which nature leaps in a single bound from one evolutionarily stable species to the next one. Although the scope for reconciling Darwinism and Mendelism was pointed out early on, personal enmities and professional ones—notably over whose methodology was to be preferred, the biometricians' ancestrian statistics or the Mendelians' breeding experiments—ensured that positions remained polarized. The debate ended only when, with Weldon's removal by premature death and Pearson's by preoccupation with eugenics, others got on with Mendelizing biometry and then Darwinism. Their efforts culminated in the synthesis achieved around 1930 by Fisher along with Sewall Wright and the British geneticist J. B. S. Haldane: the "theoretical population genetics" whose origins Provine traced.

The trouble with the history above, as we have seen, is that it is winners' history. Yes, for Bateson, discontinuity in evolution was *the* great issue. But that was never true for Weldon, who, with a sigh, responded to such talk with a tutorial on dice rolling and what it showed about the operations of chance. To Weldon, the debate over Mendel was never just a proxy for a debate over Darwinism. What vexed him above all about Mendel-style experiments and explanations coming to be at the center of an emerging new science of heredity was not their compatibility or otherwise with Darwinism, nor the statistical dodginess of Mendel's data, but the way that Mendelian conceptions—and especially dominance—misled about what was already well understood about how heredity works and how variable organisms are. In Weldon's view, the outstanding lesson of the biological science in his day, from the evolutionary embryology of Balfour to the experimental embryology of the Naples station, was the ever-greater appreciation for development as dependent on contexts, internal and external. Two outwardly similar organisms raised in the same environment might nevertheless develop differently thanks to differences in their germinal backgrounds or, in Weldon's Galtonian shorthand, ancestries. Likewise—and the symmetry was important for Weldon—two organisms with identical hereditary constitutions could develop very differently if raised in different environments. As Weldon saw it, Mendelism treated such context-dependent variability not as the central fact about inherited characters, but as an inconvenience to be avoided or, where that was not possible, ignored.

Were not Bateson and Weldon agreed on the Mendelian doctrine of "purity of the gametes"—the production by crossbred organisms of gametes that are not themselves hybrid but "pure for" one or the other of the parental-character versions—as a major point of contention? Yes and no. Bateson promoted gametic purity as the "essence" of Mendelism and also as the negation of the Galton-Pearson law of ancestral heredity (which posited the steady diminution of parental influence, while gametic purity wiped out that influence in one generation). And Weldon over and over again drew attention to cases where observed variability in the descendants of hybrids was unintelligible under the Mendelian doctrine but easily made sense of, and even predictable, by anyone who knew their Galton. Yet for Weldon, again, Mendel's erroneous hypothesis about what happens to parental characters in the gametes of hybrids flowed from the same, more fundamental error that led to his dodgy data: his reliance on overly simple, unstatistical descriptive categories. According to Weldon's reconstruction, once Mendel had decided to work with "yellow"-or-"green," "round"-or-"wrinkled," and so on, then, when he discovered that just one character version in the pair was visible in the hybrids, he had no choice but to declare that character version "dominant"—that is, visible even in the presence of the germinal ingredients for the other character version—and the other one "recessive"—that is, visible only when the germinal ingredients of the dominant-character version are excluded. And having gone that far, he had no choice, on discovering the return of the recessive character version in a quarter of the hybrid offspring (the extracted recessives), but to suppose that a process of segregation had taken place in the hybrids.

A potential stumbling block here has the shape of a chromosome. We are accustomed to thinking of the relationship between Mendelism and the chromosomal theory of heredity as an exclusive one, thanks to winners' history turned into textbook tradition. (In the chapter on Mendelian genetics in *iGenetics*, Mendel's crosses with round and wrinkled peas are illustrated twice: the first time just with peas; the second time with peas plus the relevant allele-bearing chromosomes, paired up in the individual plant and segregated in its gametes. The student following along is advised, "Keep in mind that the segregation of genes from generation to generation follows the behavior of chromosomes.")[44] Although Bateson's doubts about chromosomes are well known and are even, as we have seen, the subject of much interesting speculation, those doubts are usually held against Bateson, not against the irresistible charms of chromosomal Mendelism.

Even specialist historians of biology can be surprised to learn that Weldon, repudiator of Mendelism, took nuclear chromosomes for granted as the cellular basis of heredity. If we ask how, had Weldon lived to complete his "Theory of Inheritance," he would have rendered these commitments compatible, we can do worse than return to Provine. He was too good a historian—and too good a biologist—not to acknowledge that, on inheritance, Weldon had scored some points. Here, for example, is the gloss that Provine gave on Bateson on gametic purity in his 1902 *Defence*:

> In defense of Mendel's law of segregation [against Weldon's "plates of peas" attack in *Biometrika*], Bateson expounded the "purity of the germ cells." He cited experiments which showed that extracted recessives were "identical" to their recessive grandparents, a phenomenon which could not be explained by Pearson's law of ancestral heredity. Bateson's choice of words here was unfortunate because segregation of other factors might make the organism with the extracted recessive distinctly different from either grandparent. Weldon later insisted that Bateson believed any organism with an extracted recessive factor must be exactly like one of its ancestors. Bateson's exposition of the "purity of the germ cells" also ignored interaction effects, which he discovered only later.[45]

Bateson here was both gently admonished and generously excused for overdoing it in declaring gametes pure and extracted recessives identical to grandparental ones. Yes, it would have been better had he been more forthright in noting that, as Weldon emphasized and the Morgan group went on to emphasize, character versions get transmitted as identical units mainly as an experimental artifact, when all of the identity-wrecking interference has been purified away. (Recall again the present-day investigators of seed shape in peas remarking on Mendel's good fortune "that no other genes that affect the wrinkling of the seed . . . were segregating in his experiments.") And it was not quite true to say that Bateson "discovered" interaction effects only later, since, plainly, Weldon regarded them as exactly what anyone familiar with the era's biological science would have expected, given everything already known about inheritance and development, going back at least to Galton's quincunx and Balfour's Darwinian embryology.

Of course, by the time of the Morgan group's *The Mechanism of Mendelian Heredity*, the classic work on chromosomal, difference-maker Mendelism, a thoughtful Mendelian was indistinguishable from a thoughtful Weldonian—until, that is, it was time to teach students, or give a public lecture, or write a popularizing book or article, or provide expert testimonials to judges or legislators seeking to put genetic determinism into eugenic practice. Then, often though not always, the character-maker gene concept, in all its cut-to-the-chase simplicity, came to the fore.

Mendelism's success and its explanation. Successful sciences come to seem inevitable, as though some unstoppable historical pressure brought them into being. As we have seen, Mendelism's success has been put down to a range of supposedly bulldozing agencies, from the reality of the Mendelian gene to the functional needs of unequal societies to the totalizing tendencies of modernity. I find the opacity of these explanations, and the opportunistic use of evidence in support of them, off-putting, but each has its value. Descending from the grand scale to the human scale, we saw that, notwithstanding the 1900 triple rediscovery, there was no army of biologists ready to take Bateson's place as champion of Mendelism had Bateson been otherwise engaged. On the contrary, no one besides Bateson took the "Versuche" to be the foundation for a new ism. Yet once he made the securing of that ism's future his mission, he brought to it a rare blend of brilliance, tenacity, and entrepreneurial energy—and changed everything. The later catastrophe of Lysenkoism has been interpreted as a sign that, had Bateson and his research school never existed, our biological science would have gone Mendelian eventually, on pain of death through failed agriculture. But that reading of the record turns out to depend on an exaggerated notion of what successful agriculture owes to Mendelism—a notion put into circulation by Bateson and his school, then spread via the same Cold War dynamics that, in a roundabout way, led to so many people knowing that Mendel's data may have been too good to be true.

So Mendelism's success was not inevitable. How, then, to explain it? What, exactly, did Bateson and his school do with all of that brilliance and tenacity and entrepreneurial energy? The most important things, I argued, were the drawing out from the "Versuche" of teachable principles, tractable problems, and technological promise. It was by way of testing the plausibility of that explanation that I pursued an implied counterfactual: that had some other science of inheritance proved capable of being taught, researched, and technologized in comparable ways at around the same time, then it, too, might have become as successful as Mendelism became. Putting Weldonian flesh on the bones of that counterfactual—in the form of, respectively, the Leeds curriculum experiment, the norm-of-reaction research tradition, and the translation of that tradition to agricultural and social improvement—gave us, I suggested, a plausible alternative to Mendelism. As often happens with counterfactuals, however, this one does more than merely strengthen confidence in the begetting explanation. When the successful Weldonian alternative that never was looms out from the space of possibilities around the success of Mendelism, our sense of the explanatory challenge before us changes. Earlier, we asked, "Why did Mendelism succeed rather than remaining a local Cambridge enthusiasm?" Now we ask instead, "Why did Mendelism succeed rather than the Weldonian alternative?"

It is not enough to answer "Because Weldon died before finishing his book," since that would leave untouched what makes the question worth asking in the first place: the prospect that a successful alternative to Mendelism was not merely a possibility but a close-run thing. A tantalizing cultural clue to how close biology came to going Weldonian rather than Mendelian lies with that other circa 1900 triple, the discovery of the Baldwin effect—in G. G. Simpson's judgment, an event more reflective of wider trends in the biological science of the day than the rediscovery of Mendel's ratios. A biology in which connection-making across learning, development, adaptation to changing environments, and Darwinian evolution was "in the air" was a biology apt to welcome a vision of inheritance that, unlike Mendelism, invited those connections.

Another cultural clue is the manner in which Weldonian emphases did not so much disappear from genetics as take up permanent residence on the conceptual sidelines. We have already seen how that marginalizing works pedagogically. Historically, there is no shortage of important biologists who, down the decades, have made a fuss of the determinism-dissolving power of context as pivotal to understanding heredity. A full inventory of these biologists would be a chapter unto itself, but a few instances will do. I was both chagrined and gratified to come across my hard-won interpretation of Galton's quincunx as a "context matters" machine set out in a 1952 popular introduction to genetics, called *Understanding Heredity*, by the German émigré geneticist Richard Goldschmidt.[46] (As one would expect from that date, Goldschmidt's book was explicitly aimed at countering Lysenkoism. Publicity surrounding the Golden Jubilee conference likewise played up the scope for genetically based traits to be environmentally modified.)[47] In Provine's history, the Castle-trained Sewall Wright—in many respects the anti-Fisher—emerges as a champion of interaction, with a theory of evolution profoundly shaped by a deep appreciation for how, in Provine's summary, "in one genetic background allele A_1 might be dominant to allele A_2, but in another genetic background A_2 might be dominant to A_1," so that, whether on the farm or in nature, what was selected were less individual genes than genic systems.[48] The Nobel-winning immunologist Sir Peter Medawar, in a 1977 essay on the Burt affair in the *New York Review of Books* (in which—as one would expect from *that* date—the Mendel data problem was dealt with in passing), explained why the whole notion that a character can be partitioned into the inherited and the acquired is nonsense because, as the likes of Lancelot Hogben and J. B. S. Haldane had long ago made plain, "the contribution of nature is a function of nurture and of nurture a function of nature, the one varying in dependence on the other, so that a statement that might be true in one context of environment and upbringing would not necessarily be true in another." Superb teacher that he was, Medawar went on to give compelling examples. In humans, the mutation behind an inability to digest the amino acid phenylalanine behaves like a Mendelian recessive. But the mental retardation associated with a double dose of the mutation—and so the condition, phenylketonuria, that the mutation is "for"—can be avoided by bringing up the affected child on a diet free from phenylalanine. In a world where normal diets exclude phenylalanine, wrote Medawar, the disease will shift from being a genetic disease to being a disease "wholly environmental in origin. It will manifest itself in the presence of phenylalanine but not in its absence, and will thus present itself as a disease caused by phenylalanine." Or, he urged, consider eye color in the brackish-water shrimp *Gammarus chevreuxi*. Some genetic combinations make eye color appear to be wholly under environmental control; some ambient temperatures allow for patterns in color inheritance conforming to Mendelian rules.[49]

The more we permit this not-all-that-hidden history of contextualism to seep into our overall picture of the history of biology, the more that Mendelism's success, while explicable, nevertheless looks strange, surprising, fragile. Let us follow that path to its conclusion. Suppose, counterfactually, that the ties binding Mendelism to history were not merely weak, but broken. In this revisionist rerun of the history of biology, there is no course-deflecting irruption of Mendelism, either because there is no Mendelian triple in 1900, or because Bateson takes no notice. Without that deflection, biology would, I venture, have developed in a Weldonian direction (whatever Weldon's own degree of involvement), centering on Darwinian natural selection, chromosomes and their modifying contexts, and the statistical study of variation. The study of chromosomes—inward to DNA, outward to protein synthesis—would have precipitated out something close to our molecular understanding of heredity, development, and evolution. To the extent that researchers needed the method of crossbreeding, their tool kit would have provided it, since its manifest usefulness, so plain to de Vries and other readers of Darwin, would have ensured its continual elaboration. Eugenics, with its multiple roots, would have emerged but, deprived of fertilizing Mendelism, might have been less prone to the genes-are-all excesses that blighted our actual history. Agriculture would have continued on its successful way, with breeders continuing to refine their time-honored blending of hybridizing and selection, eventually with help from the new molecular tools, down to the CRISPR-Cas9 genome editing system whose developers were awarded the 2020 Nobel Prize in Chemistry.[50] And no one would talk or think of heredity in terms of invisible string-pullers.

A technologically impressive nongenic biology? The molecular knowledge underpinning our age of biotech wonders may, as we have seen, owe little to Mendelism. But to render that lack of dependence not just plausible but *believable*, we need new examples to think with. I offer the spider goats in that spirit.

Recall that in 1902, Bateson predicted that Mendelism would endow breeders with the power to pull out and plug in hereditary qualities at will. A century later, that prophecy seemingly found dramatic fulfillment. Amazingly enough, spider DNA can be spliced into a goat genome and then *work*. "What we're doing here is ingeniously simple," Jeffrey D. Turner, CEO of Nexia Biotechnologies, told an interviewer from the *New York Times*. "We take a single gene from a golden orb-weaving spider and put it into a goat egg. The idea is to make the goat secrete spider silk into its milk." The milk of these transgenic goats contained, not spider silk exactly, but a component protein of the silk that spiders use to hold themselves, and their webs, aloft. Stronger than steel, yet biodegradable, this "dragline silk" is an enormously promising biomaterial. Historically, the major barrier to commercializing it was the aggressiveness of spiders, which renders them impossible to farm. Nexia's solution was to turn docile goats into spider surrogates.[51]

On closer inspection, however, the spider goats are less than perfect as vindications of Mendelian simplicity. For one thing, nothing in the discovery of the transferred gene depends on Mendelian genetics. When, in the late 1980s, Randy Lewis, a protein chemist then based at the University of Wyoming, set out to find the gene for a dragline-silk protein (in order to work out the protein's amino-acid sequence—something frustratingly hard to do by standard means), he did not first perform breeding experiments with spiders to check whether the gene's inheritance conforms to Mendelian patterns, then use that information to pinpoint the chromosomal location of the gene, then sequence the DNA at that stretch of chromosome. For his purposes, a far more efficient strategy was to zoom in on the coding DNA directly by collecting messenger RNA (mRNA) from the silk-production glands of stimulated spiders, convert those mRNA transcripts into their DNA counterparts (using an off-the-shelf kit), then winnow the candidate pool until the most promising candidates could be identified and sequenced. The techniques used came from biochemistry, and assumed no Mendelism-derived skill or even knowledge.[52]

When, a decade later, Nexia's staff wanted to introduce this gene (as well as another gene that Lewis and his colleagues discovered—dragline silk is a liquid-crystal mixture of the two proteins) into goat eggs, the gene was not, in fact, just plugged in, and for good reason. In the context of a spider's body, the silk glands regulate the expression of the gene. When a silk gland empties of silk, there is a knock-on physical change to glandular cells, which, as a result, express a protein that in turn, by binding to the silk-protein gene, serves as a molecular signal: time to make more silk protein. As the silk-protein gene gets expressed, and the gland replenishes with silk protein, the glandular cells gradually return to their resting state. When their signaling stops, so does expression of the silk-protein gene. But in the context of a goat, with no dedicated silk glands, nothing like those physical and physiological arrangements would hold. A silk-protein gene that ended up in a goat's genome could thus potentially be expressed anywhere in the goat's body, or nowhere, or everywhere. A goat that produced no silk at all would be a bad result; a goat whose brain, lungs, or joints filled up with spider silk would be a terrible one. So, before Nexia's technicians introduced the silk-protein DNA sequence, they modified the start of the sequence to include an additional, "promoter" DNA sequence, which would serve to ensure (they hoped) that the subsequent code got expressed only in the goats' mammary glands, and only during lactation. The expectation was that, whenever a female spider goat suckled, the oxytocin released would stimulate the production of both milk and the silk protein.[53]

What they inserted into each goat egg, then, was not "a single gene . . . from a spider," but an artificial amalgam, furnishing the spider gene with the new micro-context needed to function in the new macro-context of a goat. That macro-context was, in turn—and as ever—a source of variability, the more so because multiple copies of the amalgam were pipetted into the egg shotgun-style, with no way of controlling where in the genome, if anywhere, a copy got incorporated, or how many copies got incorporated. The female goats that grew from the shotgunned eggs were, predictably enough, variable in how much milk they produced and in how much silk protein that milk carried. In an interview with me, Lewis reckoned that, for silk-protein production, there is probably a "sweet spot" between having too few copies of the amalgam and too many copies.[54]

Had Nexia been a blue-skies research institute rather than a biotech start-up, its employees might have busied themselves investigating all sorts of questions about how the diversity in the amalgam's uptake affects expression of the silk protein. But with investors to satisfy, a reliable old technology now came into its own: selective breeding. In the *Origin*, Darwin quoted William Youatt's statement that selection is "the magician's wand" because of the power it gives to skillful breeders to conjure up any form they please. In the case of the spider goats, selection helped to generate the illusion that to make goats that secrete spider silk in their milk, all you need to do is pull a gene out of a spider and plug it into a goat.[55]

Parting questions from Kuhn and Provine. If twenty-first-century biology at its best has in effect done away with the Mendelian gene concept, while twentieth-century biology at its worst was often proudly Mendelian, should we conclude that biology—and the wider world that biology has had a hand in shaping—would have been better off had the Mendelians not won the argument against Weldon in the early twentieth century? For a final time, I would like to return us to Kuhn's writings, to a passage not in *The Structure of Scientific Revolutions*, but in a paper published the next year, "The Function of Dogma in Scientific Research." For Kuhn, the belated embrace of the point of view of the losing side at an earlier branching point in a science's history did not necessarily imply that a wrong turn had been taken in the past:

> The historian can often recognize that in declaring an older paradigm out of date or in rejecting the approach of some one of the pre-paradigm schools a scientific community has rejected the embryo of an important scientific perception to which it would later be forced to return. But it is very far from clear that the profession delayed scientific development by doing so. Would quantum mechanics have been born sooner if nineteenth-century scientists had been more willing to admit that Newton's corpuscular view of light might still have something to teach them about nature? I think not. . . . Or would astronomy and dynamics have advanced more rapidly if scientists had recognized that Ptolemy and Copernicus had chosen equally legitimate means to describe the earth's position? That view was, in fact, suggested during the seventeenth century, and it has since been confirmed by relativity theory. But in the interim it was firmly rejected together with Ptolemaic astronomy, emerging again only in the very late nineteenth century when, for the first time, it had concrete relevance to unsolved problems generated by the continuing practice of non-relativistic physics.[56]

Consider Ogden syndrome again. Maybe the lesson to learn from its discovery is that, when it comes to heredity, beginning a search where the light is good, in the manner that Mendelian training promotes, is sound. Yes, predictably, refinement came after the initial discovery. But there was something to be refined in the first place only because the discovery was made—and it was made only because Gholson Lyon was looking for a hereditary pattern suggestive of a disease caused by a single mutation. Then again, a counterfactual history in which biology took its bearings from Weldon's "Theory of Inheritance" would not, after all, have been a history of science devoid of Mendelian reasoning. As we saw, by the time he wrote the book, Weldon put Mendel alongside Galton as one of the great pioneers. For Weldon, Mendelian patterns and even, appropriately glossed, explanations had their place. But that place was a limited one. In a Weldonian past, Mendelian reasoning would have become a resource, but without the hypertrophy of Mendel*ism*. Had Lyon trained within Weldonian biology, he would have been just as prepared to seek for Mendelian patterns, but with greater sensitivity from the start to how much phenotypic variability he would probably find once he expanded his search window.

In a remarkable afterword that Provine appended to a 2001 edition of his book on the population genetics that emerged within our biology in the first half of the twentieth century, he expressed dismay at how durable the simplisms of that foundational era had proved to be. At pains to distance himself from the University of Chicago zoology graduate student that he had been when he wrote the book, and who, he saw in retrospect, had uncritically embraced the Darwinian Mendelism then celebrated—literally, during the 1959 *Origin* centennial (headquartered in Chicago) and the 1965 "Versuche" centennial (when the university's president was George Beadle)—as all-conquering, the older Provine was no longer prepared to indulge what Weldon had found so problematic in Mendelism: the treatment of binary inheritance patterns as exemplary, with all the indifference that treatment licensed toward actual variability in characters, as well as toward the causal complexity underpinning that variability.[57] "My skepticism regarding the usefulness of [such] one-locus, two-allele models has increased steadily since this book was published," wrote Provine. He now viewed them as "an invitation to misunderstandings," tasking historians of the future with inquiring into the "amazing persistence" of these unhelpful oversimplifications in biology teaching. "Surely," he added, "the misleading limitations of these models are understood by evolutionists and geneticists. Do teachers think that students must first learn what they did as students, and later correct these beliefs?"[58]

4

Awareness of the pasts that might have been alters our perception, and enlarges our understanding, of the past that actually was as well as the imprint of that past on our present. When, as here, the upshot is to loosen the grip of past and present upon us, we stand to gain even more. The aim in trialing a Weldonian curriculum in introductory genetics in the classroom was historical: to put to empirical test the seeming inevitability of the traditional, determinist-Mendelian organization of scientific knowledge of heredity. But in showing that genetics could, in fact, be taught differently—that teachers do not have to start with Mendel and his peas, or some other genes-cause-characters simplism, but could instead start with the variability that students can see all around them, and the multiple, interacting causes that bring it about—the study opened up new options for teachers unsatisfied with the curriculum they have inherited but resigned to teaching it regardless.

In summer 2020 I hosted a small workshop for British secondary-school biology teachers interested in refreshing their genetics teaching. In Britain as elsewhere, such teachers have relatively little room to maneuver in what they teach and how they teach it. The end-of-year exams that their students take at ages sixteen and eighteen are set nationally, and the textbooks and associated materials used are designed to be preparatory. On the genetics sections of these exams, it is a safe bet that, in some form or other, there will be Punnett squares. Elementary Punnett-square reasoning is, of course, as unit-character determinist as it gets: a student can reach the approved answers to the set questions only by taking for granted the binary nature of characters—the example in the recent exam the teachers and I looked at was dimples (contrasted with no dimples)—and explaining their distribution exclusively by reference to combinations of dominant and recessive alleles. Manifestly, teachers who fail to help students solve Punnett-square problems correctly do those students a disservice. The time-honored way to counteract the possibility that students dutifully served will end up with a profoundly misleading picture of how genes work is to tack on dispiriting caveats, after the learning is done, about how really it is all so much more complicated than that . . . An alternative strategy, I suggested, is what I called "the Weldonian reversal." Start with the complexity of a real-world character, in all its variable, interactive, context-mattering messiness, then introduce Punnett-square reasoning as applying when, by whatever route, that complexity gets dialed down to the point where everything appears cleanly Mendelian. In effect, students learn to imagine a sign hanging above any Punnett square, reading "If all else is equal." It is important for them to know what will happen when all else has been made equal, in order to get the approved answers to the set question. But it is no less important for students to know—and to remember, long after they are through with the exam-taking stage of life—that all else being equal is, in heredity, not the rule but the exception.[59]

The responses from the teachers were gratifyingly encouraging. "For me the instant change I could make is to the teaching sequence," one remarked during the discussions. "Rather than teaching these simple examples, and going straight in with Punnett squares, and calculating percentages, what about instead starting with that bigger picture and then talking about removing variables to make it simple? So students can still get those marks in the exam but actually they've got a broader view of how genetics works in the real world." After the workshop she wrote:

> It gave me an option to think about in order to combat the confused faces when teaching genetics rather than giving up and saying "well really it's more complicated than that but we'll just do it this way for now." One of the vision statements for our department is that we prepare students for science in the real world and our current method of genetics teaching doesn't do that. Being able to have discussions with our students about the complexity and variability of genetic inheritance isn't something I have considered before but I will [be] moving forward as a result of the session. I believe that this will help our students have a clearer more accurate understanding of genetics and give them the tools to be better prepared to understand the topic in the "real world." . . . I think [the Weldonian approach] provides a clear basis for teaching genetics to students. I'm not sure how many teachers delivering genetics lessons will even have heard of Weldon and his work, but my guess would be that nearly all of them have struggled explaining inheritance using examples like eye colour when students can clearly see that what we are telling them about genes doesn't work. Being aware of [this approach] will help teachers facilitate a discussion around context and the environment.

Another participant wrote afterward that "at the moment the curriculum encourages us to teach allele inheritance as absolute, despite us teaching (in different parts of the topic) that variation is a result of both genes and environment," and that, as a result of the workshop, she was now planning to use more complex "real-life examples," and to "spend more time teaching about the effect of both genes and the environment."[60]

Even the most traditional curriculum, when inspected with Weldonian emphases in mind, will offer up opportunities for giving variability and complexity their due. Consider a hands-on classic of Mendelian pedagogy: the provision of maize kernels or pea seeds or fruit flies to students who then count their way up to the 3:1 or the 9:3:3:1 ratio. The rationale, of course, is to give students a chance to see for themselves the truth of what they have been taught, and to experience something of the thrill of empirical discovery as Mendel himself must have known it. But those goods come at the cost of discouraging critical thinking about what the exercise does and does not reveal about heredity. A few, small, Weldon-inspired changes could help put that right. Before students begin counting, they could, for example, be invited to reflect a little on the artificially bred nature of the organisms to be studied, how different their wild counterparts would look, and what those differences suggest about genes, genetic backgrounds, wider environments, and Mendelian crossing as a means for learning about their interactions and effects on variability. After the students have totted up their ratios, they could swap samples with other students and compare the results, in the expectation that—as the biometrician Raymond Pearl (an admirer of Weldon's) long ago reported—no two observers will derive exactly the same ratio for any one sample.[61] The interest of the questions thus raised about how much phenotypic variability persists even in purified lineages could then be underscored by asking the students to tot up once again, but now availing themselves, for each character, of three categories: the dominant phenotype; the recessive phenotype; and an intermediate phenotype. In a fascinating study of the "too good to be true" data problem in the 1980s, Robert Scott Root-Bernstein found that students given the third category used it.[62] The lesson could conclude with some discussion of the data problem, how it eventually led to unfair accusations of fraud being leveled against Mendel, and why a better take-home is that nature does not come in binary categories (and Mendel never said otherwise).

Elsewhere, too, a little novelty—especially in drawing attention to the limitations of research methods—can go a long way in neutralizing lurking determinism. "Look to your left; look to your right: one of those people will end up dead from a genetic disease" is, as I have witnessed, an attention-grabbing way of launching an undergraduate genetics course. But I wager that undergraduates would be at least as intrigued to learn that when researchers hunting for the genetic secret to longevity sequenced the genomes of forty-four centenarians, the sequence data revealed not "the longevity gene" but the opposite: loads of gene variants previously classified by medical science as lowering life expectancy by causing disease.[63] And where the former starting point dramatizes the authority of the genetic knowledge that students are about to acquire, the latter instead puts them on notice about the complexity of heredity and the provisional nature of knowledge based, so often, on the study of correlations found where the light is good; that is, in populations conspicuously exhibiting a condition. When, later in the course, students encounter more sophisticated versions of that associational theme—the polygenic risk score is, at the time of this writing, the latest turn of the DNA-predicts-all wheel—they will be better able to keep their skeptical wits about them.[64]

In even the most suffocatingly constrained curriculum, then, there will be air-pockets of opportunity for giving Weldonian teaching a go. Spotting and using those opportunities is creative work for the present. So is imagining what a fully Weldonized biology would look like.[65] If Weldonian curriculum reformers could operate with a free hand in their introductory courses, what would they do? Among other things, I expect they would

- tell the story of genetics as starting not with a discovery, or a rediscovery, but with a *debate*—a debate about questions on how best to organize our knowledge of heredity, at a time when there was wide and growing interest in applying that knowledge to improving human welfare.[66]
- dwell less on domesticated organisms in invariant ecologies (e.g., garden peas and lab-bred fruit flies) and more on wild organisms in potentially variable ecologies, the better to bring out the role of the external environment in selecting from the range of potential phenotypes that an organism may develop.[67]
- stress the development of characters as tracing the trajectories they do because of interactions between and across, in the first instance, individual genes, genetic backgrounds (a.k.a. the internal environment), developmental noise, and external environments, using norm-of-reaction graphs to help students grasp the relationships visually.[68]
- go beyond standard presentations of internal interactions as "epistatic" (when one character is affected by multiple genes) or "pleiotropic" (when one gene affects multiple characters) by looking at, for example, noncoding DNA, the epigenome, the proteome, and the microbiome (in organisms with microbial symbionts) as additional sources of causal complexity and so phenotypic plasticity.[69]
- introduce Darwinian natural selection, along with genetic drift, niche construction, and other evolutionary processes, as shaping populations characterized less by genotypes (à la Fisher) than by developmental potentialities, which are by no means the work of genes alone.[70]
- introduce the body of predictive statistics that goes under the name "quantitative genetics" (and which, we have seen, is only honorifically Mendelian) not as a specialized branch of genetics, but as what the study of heredity in general looks like when all the causal com-

plexity involved in development has not been simplified away, naturally or artificially.[71]

- embed a survey of genetics applied to agriculture within ever-widening perspectives on relevant contexts, from the genomic (e.g., as taken advantage of via selection on the transgenic spider goats) to the ecological (by considering, e.g., the effects on biodiversity and climate of conventional chemical farming using genetically purified seeds).[72]
- embed a survey of genetics applied to medicine within ever-widening perspectives on relevant contexts, from the genomic (e.g., a drug for sickle cell anemia that works by removing the molecular brake from a gene that normally produces hemoglobin in the fetus and then, after birth, gets switched off) to the ecological (by considering, e.g., how public health measures such as reducing air and water pollution reliably do more than biotech wizardry to improve quality and length of life for the largest number of people).[73]

Race is such a sensitive topic that sometimes it is avoided altogether in the biology classroom. If addressed at all, it tends to be in connection with Punnett-square reasoning about human diseases that—all else being equal—arise from inheritance of a double dose of a recessive mutation that is generally rare but notably more frequent in certain populations. Thus students learn that Tay-Sachs disease is not just a Mendelian condition but a "Jewish disease," that sickle cell anemia is a "Black disease," and that cystic fibrosis is a "white disease." The trouble here is not merely that, in Gholson Lyon's phrase, category thinking misses complexity. (In Angela Saini's *Superior: The Return of Race Science*, she writes about a Black child whose cystic fibrosis went undiagnosed for years until an X-ray of her chest happened to be seen by a radiologist who had no idea what color her skin was.)[74] In an experimental study with middle-school and high-school students in the United States, the educational psychologist Brian Donovan reported that students taught to associate particular Mendelian diseases with particular races come away from the classroom with exaggerated notions of the genetic differences between the races. Worse, compared with students not so taught, these students also become more likely to cite genetic differences in explaining why academic attainment is higher for some races than for others, less likely to support policies aiming to reduce attainment gaps between races, and less likely even to want to socialize across racial lines.[75] Once again, a better way seems to lie with introducing real-life variability and complexity sooner rather than later. In a subsequent study, Donovan and colleagues found that presenting students and adults with scientifically up-to-date data

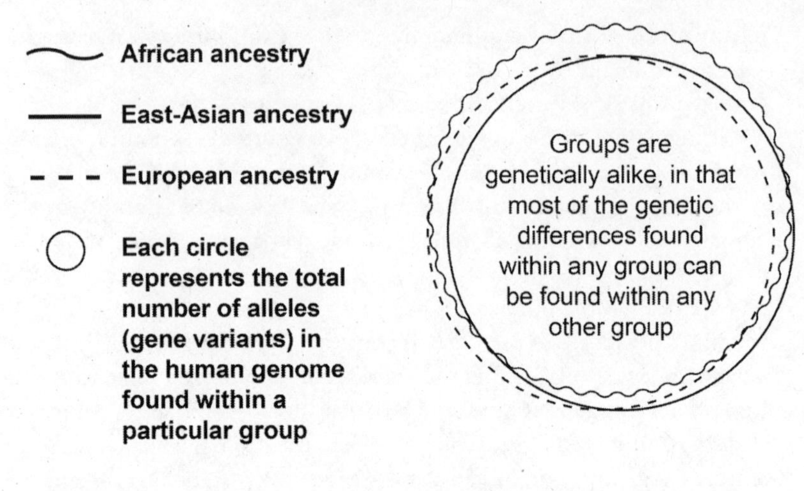

FIG. 13.5. These nearly overlapping circles visualize an important "big-picture" point from recent studies of human populations worldwide: although particular genotype combinations can be used to infer, say, East Asian ancestry, on average, two individuals from different groups will be about as similar genetically as two individuals from the same group will be.

about human genetic differences (fig. 13.5) reduced not only perceptions of racial discreteness and uniformity but, as a consequence, racial bias.[76]

If Weldonian emphases on variability, internal and external contexts, and the malleability of the traits that genes influence can lessen the scope for confusion about what is meant by "genes for" (since, in a Weldonian curriculum, that terminology really does function as shorthand for the difference-maker conception), while also helping students avoid or overcome the "essentialism" about kinds of people that is a short trip from determinism about genes, then those emphases will have earned their keep. Wherever life takes those students, the decisions that they go on to make in situations when genes come up, from the personal (what to do in the light of a genetic test, sequenced genome, or polygenic risk score?) to the political (what sort of seeds should be preferred for use in an ecologically sound agriculture?), will be made with a clearer-eyed view of how complex the issues really are, and how important it is not to underplay that complexity.[77]

The social benefits of genetics education going more Weldonian could potentially be even greater. Suppose that some form of Weldonian teaching, with its "context matters" message about genes, lives up to its promise of imparting knowledge of genetics without inadvertently inculcating genetic determinism. Suppose further that on the whole, the less determinist one is about genes, the more optimistic one will be about whether people can change. And suppose as well that on the whole, increased optimism about people's capacity for change goes along with an increased willingness to confront racism, sexism, and other forms of prejudice—and conversely (and as Nazi propagandists intuited), that Mendelism-tinctured pessimism about prospects for change goes along with acquiescence to prejudice and its injustices. The testing of these suppositions and their causal links, through a combination of computer-based randomized controlled trials and cognitive "think alouds" (when students explain the thinking behind their answers to questions), is the mission of a large-scale project begun in 2020 and led by Donovan, the biology education researcher Michelle Smith, and me, with the collaboration of over fifty teachers of genetics at colleges and universities across the United States. What concerns these teachers is the possibility that, in spite of their best efforts, their students end up with an exaggerated impression of the controlling power of single genes. Whatever the outcomes of the research, we are bound to learn a great deal about student learning of genetics, how to improve it, and how to use those improvements to help make genetics—so persistently part of the problem when it comes to social justice—into part of the solution.[78]

Conclusion

Why, in the early years of the twentieth century, did biology take a Mendelian turn? How did that turn matter for what followed? And what might biology look like now had it not taken that turn? In bringing this book to a close, I want to review what I take to be my answers to these questions. I am mindful, of course, that some historians will look askance at the third, counterfactual question. But historians who say they loathe counterfactual history will often be found to have perpetrated it.[1] That is because, typically, they wish their answers to their own questions about why what happened went as it did—their explanations of the past—to be not merely accepted but understood. And there is no better way to clarify what accepting a historical explanation amounts to than to spell out how different the past would have been had the explanatorily picked-out actions or accidents or whatever been otherwise. "I am tempted," the philosopher Robert Nozick once wrote, "to say that explanation locates something in actuality, showing its actual connections with other actual things, while *understanding* locates it in a network of possibility, showing the connections it would have to other nonactual things or processes."[2] That view tempts me too. Accordingly, I think that if we want not merely to explain why biology went Mendelian but to understand the significance of its doing so, we have to consider—as explored in this book—other, nonactual histories, along with the other, nonactual biologies that might have emerged within those histories.

We saw in part 1 that a science of heredity of some kind was in the cards around 1900, in a way that it was not in 1865, when Mendel lectured on his experiments, Darwin drafted his pangenesis hypothesis, and Galton published on the potential for eminence in humans as inherited. For the growing corps of professional biological researchers plying their trade in the marine biological stations popping up on both sides of the Atlantic, heredity became an attractively investigable element in Darwin's theory of evolution by natural selection. Meanwhile, for the reform-minded likes of Zola, Galton, and Pearson, human heredity emerged—partly thanks to work in medicine that Darwin himself drew upon—as something that no one who wanted to improve society could afford to ignore. By 1900, heredity's time had come, biologically and socially. But there was nothing inevitable about the next century's becoming "the century of the gene." In that century's early years, as Mendelism took shape and drew recruits under Bateson's leadership, the new science's limitations provoked Weldon to develop an alternative science of heredity, over the course of the events examined in detail in part 2. It was Bateson who transformed Mendel's experiments, which (for Mendel's own, very different purposes) deliberately minimized the influence of internal and external contexts on blocky, binary characters, into a conceptual and methodological foundation. By contrast, Weldon aimed to give maximal prominence to the role of development and environmental contexts, and to the spectrum-spanning, statistically describable variability that changes in those contexts can cause. Both Bateson's Mendelism and Weldon's alternative enjoyed prestigious institutional bases. And, as stressed in part 3, both fitted well with elements of cultural modernism and were well supplied with the potential for the teachable principles, tractable problems, and technological promise that, I argued, make for field-sweeping success in modern science. If not for Weldon's death in 1906, with his "Theory of Inheritance" uncompleted and unpublished, biology might have gone Weldonian rather than Mendelian.

With this "network of possibility" around the actual, factual success of Mendelism in view, we can now ask, What difference did it make to our biological science that Mendelism, rather than Weldon's nonactual, counterfactual alternative, did the sweeping? A good place to start is with another point emphasized in part 3, having to do with advances since the 1930s and 1940s in the study and manipulation of heredity at the molecular level. Although Mendelism has acquired a reputation for making those advances possible, that reputation looks, on closer inspection, largely undeserved. Without Mendelism, we would, I venture, have learned everything we now know about—and know how to do with—chromosomes. The centrality of chromosomes as the cellular bearers of heredity was, as we saw, well established in 1900, indeed, was already featured in classy popular science in London. Nothing known about segregation and independent assortment depends on Mendelism because in cells, what segregate and assort independently are chromosomes, and their doing so was bound to become textbook science one way or the other. So too was the molecular structure and functioning of chromosomes, down to DNA and up to the proteome and beyond. If anything, Mendelism slightly retarded progress on this front, between Bateson's own cold-shouldering of chromosomal work and the inaptness of the Mendelian technique par excellence, crossbreeding, for elucidating the DNA-to-protein causal chain. And so far as crossbreeding has turned out to be valuable in biological and medical research, pure and applied, it would have become so anyway had the textbooks gone Weldonian instead of Mendelian, since, for Weldon, Mendel's achievement was considerable. In Weldon's view, Mendel had shown how much one can learn, here and there, by experimentally stripping out all the differences except a difference of interest. The trouble comes only when the results of such homogenizing experiments are treated as opening a window into how heredity works in general. Biologists trained in Weldonian biology would have been able to avail themselves of crossbreeding theory and practice when needed, but with heightened sensitivities to the limited nature of the conclusions to be drawn.

Another reputation that does not survive scrutiny is Mendelism's status as the science of inheritance that the theory of natural selection needed. For one thing, that theory was in a much healthier state around 1900 than tends to be remembered. The fame of the triple convergence of de Vries, Correns, and Tschermak has made it easy to neglect the earlier triple convergence on the Baldwin effect, and so to overlook how creative and wide-ranging was the discussion from which the latter discovery precipitated—a discussion as much about heredity, environment, and development as about Darwinian evolution. As we saw, Weldon's own way of connecting those topics took shape through his studies with the Darwinian morphologist Balfour, his contacts with peers among the new generation of experimental embryologists, and his admiration for the work of Galton: to Weldon, not the arch-hereditarian remembered nowadays, but the originator of a conception of dominance as context dependent. In "Theory of Inheritance," Weldon contrasted that conception favorably with the absolute conception of Mendel. The integrations that, I suggested, Weldon would have made toward the end of his book with the chromosome theory and with Darwinian theory were made in the next decade by Morgan and his students in *The Mechanism of Mendelian Heredity*. Again, what mattered was not Mendelism per se but knowledge about chromosomes and their behavior—and that knowledge would have emerged at least as rapidly under a Weldonian as a Mendelian dispensation.

So, if, in the early twentieth century, incipient Mendelism had been absorbed within the Weldonian science of heredity, rather than vice versa, biologists would have gone on to find out everything important that they subsequently found out about life anyway, from the molecular level to the microevolutionary level. And yet, in a deep sense, the accumulated knowledge would have been utterly different, because it would have been organized not around the Mendelian gene—the entity that Punnett-squareably determines the outcome in a Mendelian cross—but around the norm of reaction, in which the possibility of variable effects depending on context is made unmissable. Had the shaping role of context been made central rather than peripheral, the growing body of facts about heredity would have been thought about differently, and so acted upon differently, with consequences both for biological knowledge itself—with different questions being asked, maybe by different kinds of people—and for its applications.

At this point, we are teetering on the edge of what we can plausibly know about what might have been. Even so, throughout part 3, I touched on work that seems to me to throw some light on the possibilities. As with biological knowledge generally, so with its applications: much of what took place would have taken place whatever science of heredity had won out. In agriculture, the changes that Mendelism brought about—notwithstanding the claims of Mendelian publicity and then Cold War propaganda—amounted to modest improvements in pre-existing breeding methods. The resources behind those improvements would have been no less available to breeders educated within Weldonian biology. Indeed, in line with Weldon's remarks on farmers wanting cattle whose milk yields are as advertised across variable environments, the norm-of-reaction concept is most familiar nowadays in agriculture. Nevertheless, plant breeders educated in Weldonian science, with its bullish attitude toward selection and its attentiveness to the evolutionary impact of industrially polluted environments (recall Weldon's Plymouth crab study), might have been more interested than Mendelian breeders turned out to be in trying to develop high-yielding varieties through selection for existing environments rather than through crossbreeding for fertilizer-enhanced ones. And farmers educated in Weldonian science might have demanded wider choice from the breeders, the better to ensure that the varieties they purchased would be as high-yielding as possible in the particular environments those farmers worked in. (In the right environment, environmental sensitivity can be a good thing.)

Would that the breeders of men had been as excited about norms of reaction as the breeders of maize . . . Alas, as we saw, nowhere was the distinctively Mendelian concept of the either/or unit character, unvaryingly determined by alleles "for" that character, more prominent or more problematic than in materials promoting eugenics. Of course, had Mendel never existed, or Bateson never developed an enthusiasm for those hybrid pea experiments, there would have been forms of eugenics in the twentieth century, including racist forms, associated with policies supposedly rooted in textbook biology. (Indeed, that would have been so even had Galton never existed. Recall that Pearson's eugenic epiphany came not from reading Galton but from learning about Weismann's demonstration that acquired characters are not inherited. And that was in the late 1880s. According to the historian Theodore Porter, among asylum doctors in several countries, "by 1859, eugenics, in a broad sense, was old hat.")[3] So Mendelism was in no way *responsible* for the eugenics bandwagon. But Mendelism greased the wheels. Again, Nazi propagandists knew what they were doing. They grasped that to learn that human hereditary traits are like yellowness and greenness in the garden pea is to become the readier to accept the legitimacy of homogenizing categories and the conclusions that seem to follow from reasoning with and from them. Far from decrying such uses of Mendelism, the early Mendelians led the way. Had there been early Weldonians, it is hard to believe they would have followed suit. Probably the closest anyone came to hammering out a Weldonian eugenics was J. Arthur Thomson, for whom the accent fell much more strongly on the search for environments where hereditary constitutions might develop most flourishingly. To the extent that, despite Weldonians' best efforts, Mendelian simplisms made their way into the public domain, Weldonian pedagogy might have functioned—here and there and from time to time, if not across the board—as a counter.

It still can. Like our biological inheritances on a Weldonian view, our scientific inheritances are ours to mold: not in any way we please, but in more ways than we might have guessed before taking a closer look at the scope for alternative possibilities. Although they are only beginning, classroom experiments in "Weldonizing" introductory genetics suggest that our biology teachers are not doomed to produce students whose first thoughts about genes are of character-makers, dominance, recessiveness, and Punnett squares rather than difference-makers, contexts, variability, and norm-of-reaction graphs; or who come away from lessons on the genetics of disease convinced that people of different races are, in essence, different kinds of animals.[4] Students educated to be less liable to deterministic thinking about genes will be better equipped, in later life, to ask probing questions when confronted not only with genetics-invoking social prejudice but—as they increasingly will be—with complex genetic information and powerful genetic technologies.[5] And if, as part of classroom instruction in genetics, these students learn a little of its history, I hope its basics are presented not as emerging fully formed from Mendel's forehead, but as forged in the crucible of an early twentieth-century debate about his experiments—a debate so fierce that its proponents felt themselves to be in battle. A biology education along those lines belongs to a future worth fighting for.

Postscript 1

On "Genetic Determinism" and "Interaction"

[The editor remarked] that I could "get my own back" on opponents of "genetic determinism." I don't know which annoyed me more: the suggestion that I favoured genetic "determinism" (it is one of those words like sin and reductionism: if you use it at all you are against it) or the suggestion that I might review a book for motives of revenge.

RICHARD DAWKINS, ON BEING INVITED TO REVIEW S. J. GOULD'S *EVER SINCE DARWIN* (1977)

"But wait," the exasperated reader cries, "everyone nowadays knows that development is a matter of interaction. You're beating a dead horse."

SUSAN OYAMA, *THE ONTOGENY OF INFORMATION*, 1985

When helping students to develop their understanding of basic genetic concepts, it can be useful to reduce complexity by adopting a traditional, linear view of gene expression (one gene, one protein, one characteristic) but there is a risk that this will result in a deterministic view of genetics in which every characteristic is determined by a single gene. The reality, unexpectedly confirmed by the Human Genome Project, is that there are very few single gene characteristics or disorders in humans. Rather, the relationship between the genome (the entire DNA sequence), gene expression and the environment was shown to be considerably more complex than anticipated. The result is a move away from a focus on single genes (genetics, understood narrowly) and towards a consideration of the whole genome and its interactions with the environment, both internal and external (genomics).

JENNY LEWIS, "GENETICS AND GENOMICS," IN *TEACHING SECONDARY BIOLOGY*, 2011

One of my concerns in this book has been with the organization of knowledge: how a given body of knowledge comes to be organized in one way rather than another, and what the consequences are, for individuals and for collectives, of different schemes of organization being adopted, especially in teaching. I have dwelt in particular on a contrast between the Mendelian organization of knowledge of heredity, from William Bateson's pioneering pedagogy to twenty-first-century genetics textbooks, with its single-gene, binary-character starting point, and the Weldonian organization that might have been, and still might be, with its multifactorial, spectral-character starting point. If, in delineating the downstream effects of the Mendelian starting point, I could avoid the term "genetic determinism," I would—as I would likewise avoid the term conventionally used to express the rejection of genetic determinism, "interaction." But whatever my preferences, these terms belong to the historical record I have been examining as well as to the history of reflections on that record. As I want to be true to the former while making contact with the latter, avoidance is not an option. Still, some comments are in order.[1]

Both terms bring heaps of baggage with them, cluttering up my otherwise conceptually tidy presentation of the issues, threatening to draw attention away from what matters to me. For some readers, "genetic determinism" will be the more problematic. If I had to define genetic determinism as though it were a doctrine, I would go with the definition of the historian, philosopher, and biological educator Kostas Kampourakis: "Genes invariably determine characters, so that the outcomes are just a little, or not at all, affected by changes in the environment or by the different environments in which individuals live."[2] But no thoughtful person would endorse such a doctrine, so even that definition can end up making genetic determinism look like a straw-man position.[3]

Those who grant that "genetic determinism" names something real may nevertheless, with some justice, suggest that any term generally used only by its opponents distorts discussion from the start. If, somehow, we could ask Bateson—the man who, after all, wanted to collect his popular writings on genetics under the title *Scientific Calvinism*—whether he believed that genes invariably determine characters, he would profess amazement at the stupidity of the question and marshal umpteen passages from his published work in support of his denial. If, undeterred, we defended our question by pointing to his championing of "genetic determinism" in print, in that 1926 *Encyclopaedia Britannica* article, he would say that he had done so merely to sharpen the contrast between the debunked Lamarckian view of how adaptation comes about and what he saw as the correct view. So does Bateson count as a genetic determinist? Not if, in order to count, someone needs to believe that "genes invariably determine characters." But again, that is too stringent a standard, and too categorical, as if genetic determinism were a Mendelian unit character, either present in full or entirely absent. Better to think of it as a matter of degree: the more inclined one is to put characters down to genes for those characters, the more deterministic one is about genes. Bateson was certainly, in that sense, deterministically disposed; what is more, he went out of his way to implant that disposition in others.

Our customary intellectual tools equip us better to deal with doctrines than with dispositions. Yet a disposition is, in my view, largely what we are dealing with when we are dealing with genetic determinism: a default tendency to regard genes as overriding super-causes in the making of bodies and minds; or, looked at from the other direction, to regard bodies and minds as mainly the product of the action of "genes for" constituent characters.[4] For students who have acquired enough genetic knowledge to answer questions about it, but not so much that their guard is up against being accused of genetic determinism, "belief in genetic determinism" questionnaires can be used not just to detect this disposition, but to measure its strength.[5] For scientists whose guard is up, however, and who, if asked, would explain that their "genes for" language is just a well-understood shorthand for genotypic differences that make phenotypic differences in ordinary contexts, internal and external, something more subtle is required.

I propose what I call the "dangled prospect of simplicity" test. When news hits that an apparently complex condition may actually be the product of a single gene for it, what is the scientist's reaction? Consider the Pulitzer Prize–winning cancer physician Siddhartha Mukherjee's bestseller *The Gene: An Intimate History* (2016). It is full of passages recounting the many ways in which complex causation interposes itself between genes and characters, making for huge variability. But it is also full of passages in which characters are treated as this-or-that binaries, with the expressed character depending on whether a genetic switch at a particular locus gets thrown in one direction or the other. As a reviewer of the book, I read a pre-publication proof copy. When, in writing my review, I compared that copy with the subsequently published version, I noticed that Mukherjee had added a long footnote to his superb discussion of schizophrenia, its diverse manifestations, and the hundred-plus genes linked to it, none of which, he wrote, "stands out as the sole driver of the risk," and any one of which becomes a "gene for schizophrenia" only as context and chance allow. In the footnote, however, Mukherjee enthused about a just-published *Nature* paper purporting to show that a single gene is responsible after all: a gene whose culprit form leads to schizophrenia via over-pruning of brain synapses. In a "gene for" flash, all interactionist sensitivity was dumped, all skepticism abandoned. Thus can a disposition to genetic determinism reveal itself in those who otherwise know better.[6]

The use of "genetic determinism" as, in the main, a term of abuse nevertheless makes it less than ideal. So too does a set of crude associations—legacies from culture wars of yore—linking "genes" and "determinism" with all things bad, and "environment" and "free will" with all things good.[7] Even a little reflection will suffice to show that a cause's being nongenetic does not somehow make it intrinsically better, more desirable, less harmful, than a cause's being genetic. (Massively disrupted development is bad whether due to iodine deficiency or to the *NAA10* mutation that leads to Ogden syndrome.) More than a little reflection is required when confronting the puzzle of free will in a world of cause-leads-to-effect determinism. But one does not need to solve that puzzle in order to see that it has nothing to do with genetic determinism. The possibility of free will is just as puzzling for someone who defaults to "all in the genes" explanations as it is to someone who defaults to genes-in-their-contexts explanations.[8]

"Interaction," of course, has problems of its own. Whereas "genetic determinism" is specialist jargon, "interaction" belongs to ordinary language, and will be found wherever writers on genetics, however determinist, have addressed situations where more than one cause is in play. Reginald Punnett's *Mendelism* (1905) ended, as we saw, with a vision of future human progress as "a matter of gametes, not of training," since "the creature is not made but born." The book would go on anyone's short list of genetic-determinist texts. Yet when Punnett, in the 1907 second edition, explained how ratios outside the Mendelian norm can arise when a character is the effect not of a single allelomorphic pair, but of two pairs in conjunction, he wrote of "the interaction of factors."[9] Bateson did the same in the 1909 edition of his *Mendel's Principles of Heredity*, declaring the study of interactions to be the next great frontier: "Such interactions, as we now know, are of the greatest importance in heredity, and the progress of genetics will consist largely in disentangling the elements to which these combination-effects are due."[10]

So "interaction" talk can be found where we least expect it. Equally, it can be shunned where it might seem an eminently good fit. Richard Lewontin, arch-nemesis of genetic determinism, whose views were echt interactionist, eschewed the label. From Lewontin's dialectical-Marxian perspective, talk of "interaction" ultimately gets in the way of an appreciation of how profoundly and dynamically organisms are parts of their environments, and vice versa, by implying that there are two separate-but-interacting things, organisms and environments.[11] The evolutionary psychologist Steven Pinker is no less critical of "interaction" talk, but as a cover for dogmatic, blank-slate environmentalism, the more pernicious for appearing middle-of-the-road reasonable. As Pinker sees it, our two terms function in tandem, "genetic determinism" scaring people off from even considering gene-based explanations, "interaction" signaling that of course genes matter while ensuring that in practice—not least through what he lampoons as the "everything-affects-everything diagram" (recall that arrow-filled diagram from the Leeds Weldonian genetics course in fig. 11.4)—they rarely do.[12] Yet no one has gone further than Pinker in casting human history on the grand scale as arising from the interaction of genes and culture.[13] And when he turns to consider the future, he finds it bright in part because, in human genetics, the bad old "gene for" days are surely gone (if not long gone):

> Predictions from the 1990s that yuppie parents would soon implant genes for intelligence or musical talent in their unborn children seemed plausible in a decade filled with discoveries of the genes for X. But these findings were destined for the *Journal of Irreproducible Results*, and today we know that heritable skills are the products of hundreds of genes, each with a minuscule effect, and many with deleterious side-effects. Micromanaging an embryo's genome will always be complex and risky. Given that most parents are squeamish about genetically modified applesauce, it's unlikely they would roll the dice for genetically modified children.[14]

A different set of complications has to do with the fact that "interaction," for all its ordinariness, has acquired a narrow technical meaning *within* the semantic space covered by broadly determinism-rejecting "interaction" talk. Consider three norm-of-reaction graphs, with the y-axis representing the average value of a character of interest (plant height, say), and the x-axis representing the value of an environmental parameter of interest (nutrient level in the soil, say). Each graph has a pair of flat lines on it, representing the average character values associated with two different genotypes across the range of environmental values.

In the first graph (fig. PS.1), the lines are exactly parallel and exactly horizontal.

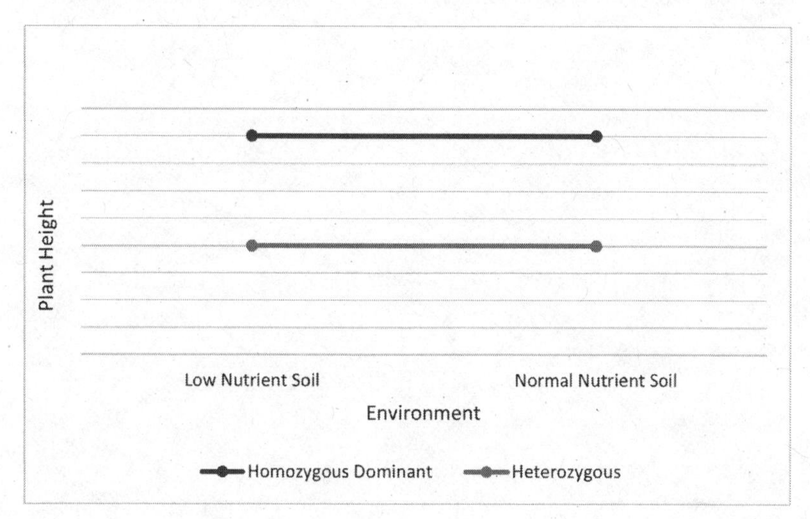

This graph shows that, in every environment, one genotype—represented by the upper line—has a greater effect on the character than the other genotype, and furthermore, that the magnitude of each genotype's effect is unaffected by the environment.

In the second graph (fig. PS.2), the two lines are also exactly parallel, but they tilt upward at an angle.

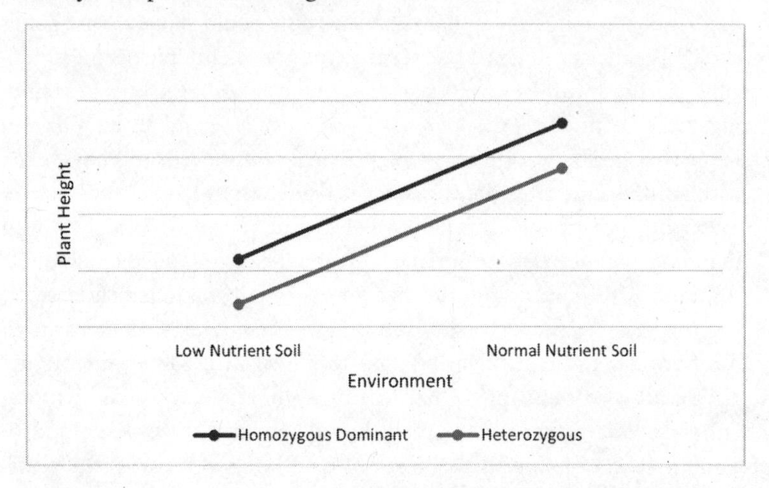

Here, the magnitude of each genotype's effect *is* affected by the environment: as the nutrient level in the soil goes up, so does plant height. But because the phenotypic "value added" is the same for each genotype, the genotype with the greater effect when, say, temperature is low continues to have the greater effect, and by the same amount, as temperature increases.

In the third graph (fig. PS.3), both lines tilt upward, but the lower line at a smaller angle.

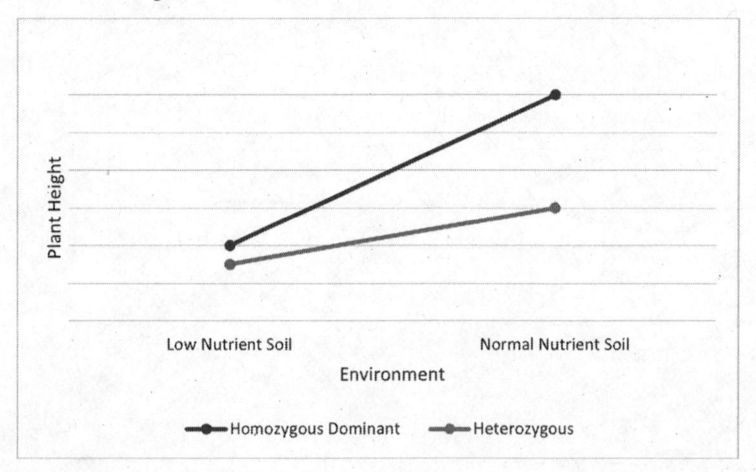

Here, it is not merely the magnitude of each genotype's effect that is affected by the environment, but the magnitude of the *difference* between their effects. In some environments, in other words, the genotypic difference makes more of a difference than in other environments.

Lancelot Hogben called that third kind of interaction "the interdependence of nature and nurture." Nowadays it is called "genotype-by-environment interaction," or "G×E" for short. If "interaction" talk is to be reserved only for such cases, then Galton's quincunx does not illustrate interaction. For one thing, the quincunx can be thought of as a blowup of just one point on just one of the lines on a G×E diagram. For another, the causes at work as the pellets bounce their way through the pegs act independently of each other. That independence is what makes their effects straightforwardly additive, as in the second graph above (a "G+E" situation). In the third, G×E graph, by contrast, we are dealing with causal interdependence, where effects are not at all straightforwardly additive. Change the environmental parameter by a little, and you can change the genotypic effect by a lot.[15]

A Weldonian lesson applies: context matters. If we want to understand norm-of-reaction diagrams, then the distinctions above should be observed strictly. But if we want to understand how norm-of-reaction diagrams came about in the first place, why they are rare in the introductory genetics courses where Punnett squares are ubiquitous, whether the situation might have been (and still might be) reversed, and why we should care, then we gain more than we lose by relaxing the restriction and allowing "interaction" talk to mark a lineage stretching back from Hogben through Weldon to Galton and Balfour.[16]

To return to my opening statement of intent in this postscript: whatever we call the cognitive legacies of Mendelian versus Weldonian organizational schemes, we all have a stake in the question of how curriculum design does or does not help students cope with the coming deluge of genetic information—about themselves, their offspring, and everyone else. "Gene for" human genetics is now largely a thing of the past; "gene for" genetics education should be a thing of the past too.[17] What is more, for all that "genetic determinism" and "interaction" are genetics-specific terms, broadly comparable concerns about cognitive disposition come up in connection with the organization of other bodies of knowledge. For other sciences too, then, history-inspired classroom experiments along the lines of the Weldonian one could help open up new options for creative updating. As a case in point, consider Hasok Chang's *Inventing Temperature* (2004) and associated work, including experimental work, on boiling point. Back in the eighteenth century, it was taken for granted that water boils over a range of temperatures, depending on the condition of the water, the container, and so on. Thermometers of the day thus represented boiling point as a spread. Nowadays, of course, boiling point is thought of as constant.[18]

That historical shift in chemistry, from sensitivity about context-dependent variability to indifference to it, has obvious affinities with the shift that exercises me in the history of genetics. Unsurprisingly, I suspect that the best way to characterize the chemical shift is not in terms of forgetting about variability—as Chang has sometimes been inclined to characterize it, prompting objections from chemists—but rather a shunting of knowledge about it to the margins. There it lies, accessible in principle without in practice being assimilated into everyday chemical thinking and doing. Likewise, I reckon that a perspicuous way to develop Chang's conviction that this shunting had consequences for chemistry and chemists (and that both are the poorer for it) is to develop and teach an experimental chemistry curriculum in which boiling point and other chemical phenomena are presented as variable. One could then see whether the students thus taught are different from students taught the traditional, constancy-emphasizing curriculum. The results may be as interesting for chemistry educators as for historian-philosophers of chemistry. More generally, this kind of endeavor presents an opportunity for a new kind of mutually beneficial collaboration between science educators and historians and philosophers of science.[19]

Even so, the human stakes are highest for genetics education. As the psychologist Steven Heine puts it:

> There's a key difference between the complexities of the atom and the complexities of our genomes: we don't make life decisions on the basis of our understanding about how subatomic particles operate. Incorrectly believing that the atom is akin to a miniature solar system has no consequences in our daily lives. In stark contrast, incorrectly believing that genes are like switches leads us to become fatalistic, and can result in increased sexism, racism, and irrational fears about our future disease risks. When it comes to our genomes, I submit that it's far better to highlight the difficulty in understanding the complex machinery of genetics than it is to give people a false sense of understanding that leads them to make potentially costly decisions in their lives.[20]

Postscript 2

A Simple Mendelian Cross Weldonized

A Mendelian cross at its simplest starts with two true-breeding varieties, differing from each other in a single character. To use Mendel's own example of seed color in the garden pea: if we cross a yellow-seeded plant from a true-breeding lineage with a green-seeded plant from a true-breeding lineage, we expect all the seeds in the first generation to be yellow, but—after the mature plants self-fertilize—we expect the next generation to contain both yellow seeds and green seeds, in a ratio of 3:1. Put more generally, when an individual from one variety and an individual from the other variety have offspring, these offspring are, with respect to that character, all like one of the parents; and when these offspring themselves have offspring, three-quarters bear that same, "dominant" character version, while one-quarter bear the other, "recessive" character version. That is the classic Mendelian pattern.

Along with that pattern comes the classic Mendelian explanation, familiar from our textbooks. It posits an underlying factor for the dominant-character version, an underlying factor for the recessive-character version, some ideas on how those factors get distributed into gametes, and a dominance rule, which states that whenever different factors end up meeting in the same organism, only the dominant-character version will be visible. As we saw in this book's introduction, these suppositions account beautifully for the pattern. They also have an intriguing corollary: in organisms showing the recessive-character version, there can be no factor for the dominant-character version present. Hence the common conviction that, in humans, where dark eye color is dominant and light eye color recessive, blue-eyed parents cannot have brown-eyed children.

From a Weldonian perspective, the simplest response to the above explanation is to say that, for a given character, something approximating the Mendelian pattern (for real patterns are never more than approximations of the theoretical one) can sometimes be deliberately engineered into being through choice of congenial breeding stocks, rigorous exclusion of other sources of internal variability from those stocks, and tight control of environmental conditions; but that nothing in general follows from the results, since other choices will yield other patterns. However, Weldon also showed that, even if those observations about the experimental design are set aside, and the Mendelian pattern accepted as revealing how heredity in nature truly and generally works, there is no compulsion, on circa 1906 biological knowledge, to accept the Mendelian explanation, because another explanation is possible.

Here is how, à la Weldon circa 1906, the Mendelian pattern can be explained on *non-Mendelian* suppositions—suppositions that, moreover, lead to Weldonian expectations about the "recessive" organisms in the 3:1 generation: namely, that some of them will harbor unexpressed factors for the dominant-character version. Again, Weldon was a chromosomal thinker, and chromosomes figure in his reasoning, in particular the notion that a character is associated with a pair of like, or "homologous," chromosomes, one inherited from the mother, one from the father.

1. Suppose first of all that, for a given character in two versions, there is not one factor per character version, but two. So, with respect to seed color in the garden pea, where the Mendelian (though not Mendel) would symbolize the hereditary constitution of the true-breeding yellow-seeded pea plants as "*AA*," and their true-breeding green-seeded pea-plant partners as "*aa*," the Weldonian would symbolize them as follows:

Yellow: *AAAA*
Green: *aaaa*

In chromosomal terms, that would mean that, on each of the two chromosomes making up a homologous pair, there is a pair of factors.

2. Suppose next, and conventionally enough, that all of the gametes of all of the yellow-seeded pea plants contain just one of the chromosomes in the homologous pair, giving each gamete an *AA* constitution. Likewise, all of the gametes of all of the green-seeded pea plants have an *aa* constitution. In that case, the plants in the first generation of offspring will all have *AAaa* constitutions.

3. Why are these hybrid plants all uniform for one of the two character versions—in the seed-color case, for yellowness? Here Weldon introduces an interesting twist on the Mendelian dominance rule. Suppose, Weldon suggests, that when *A* and *a* are present in equal numbers, yellowness will be visible; but that otherwise, whichever factor is in the majority will determine which character version is visible. So, if the plant is *AAAa*, the seed will be yellow. But if the plant is *Aaaa*, then, even though *A* is present, the seed will be green.[1]

4. Next, the plants grown from the hybrid seeds, each showing the "dominant" character version, and each having the constitution *AAaa*, produce their own gametes. At this point, Weldon introduces a hypothet-

ical process that, though he presumed it to have operated in the previous round of gamete production, could be safely ignored there because the end results were not affected. In this process, the chromosomes making up the homologous pair initially fuse, making a kind of factor magma composed of all four of the factors. The single chromosomes that get distributed into the gametes precipitate out from this composite in such a way that—with respect to the character in question, and at random—any of the three possible factor combinations may be present: *AA*, *Aa*, or *aa*. (In terms of later understandings of chromosomes, we could consider this a kind of maximal "crossing-over.")[2] Importantly, however, the three kinds of gametes will not be present with equal probabilities. Consider that when chance picks two letters at random from a four-letter word, there are six equally probable outcomes: the first letter with the second; the first with the third; the first with the fourth; the second with the third; the second with the fourth; and the third with the fourth. For *AAaa*, given these six options, only one will result in an *AA* gamete and another one in an *aa* gamete. The other four result in *Aa* gametes.[3]

5. So what happens when gamete meets gamete in the production of the next generation? To find out, we run a Punnett square, with the gamete distribution just derived represented on each axis (see fig. PS2.1).

	AA	**Aa**	**Aa**	**Aa**	**Aa**	**aa**
AA	AAAA	AAAa	AAAa	AAAa	AAAa	AAaa
Aa	AaAA	AaAa	AaAa	AaAa	AaAa	Aaaa
Aa	AaAA	AaAa	AaAa	AaAa	AaAa	Aaaa
Aa	AaAA	AaAa	AaAa	AaAa	AaAa	Aaaa
Aa	AaAA	AaAa	AaAa	AaAa	AaAa	Aaaa
aa	aaAA	aaAa	aaAa	aaAa	aaAa	aaaa

Putting these together, we get a distribution of 1 *AAAA*, 8 *AAAa*s, 18 *AAaa*s, 8 *Aaaa*s, and 1 *aaaa*. Applying Weldon's dominance rule, that translates—to return to seed color in peas again—into

9 yellow-seeds in which the *A* factor is in the majority
18 yellow-seeds in which the *A* factors and *a* factors are equal in number
9 green-seeds in which the *a* factor is in the majority

or, as a ratio (and dividing by 9):

1 yellow (majority for *A*) : 2 yellow (equal *Aa*) : 1 green (majority for *a*)

Finally, combining the yellows, we get 3 yellow: 1 green—the Mendelian pattern for the second generation.

That is the Weldonian derivation of the Mendelian pattern. In drawing attention to it, I do not, of course, mean to suggest that it is *correct*. Plainly, in light of twenty-first-century biological knowledge, it is not. Rather, for this book's purposes, three points bear emphasis. First, the Weldonian derivation is not worlds away from standard-issue chromosomal Mendelism. To go Weldonian in these matters is not suddenly to throw out everything familiar. But it is to be critically minded about the familiar—here, about whether Mendel's dominance rule is indeed, within the horizons of early twentieth-century biology, the only rule that would have worked for the Mendelian pattern, and about what happens when one tries out a different rule. (Among other things that happen, we should note in passing, is that we lose all motivation for declaring *A* the "gene for" or, more fussily, "allele for" yellowness; visible color is so much the upshot of interaction among multiple factors that, from the mere presence of a single *A*, nothing follows.) Second, the alternative rule that Weldon applied has its most interesting consequences in connection with the green-seeded pea plants in the second hybrid generation: the "extracted recessives," in the terminology of the day. On Mendelian reasoning, the extracted recessives are, in their hereditary constitution, identical to the recessives in the grandparental generation. To put the same point another way, the green-seeded plants in the second hybrid generation show no sign whatsoever of having had yellow-seeded parents. In hereditary terms, it is as if the intervening yellowness had never happened. Parental influence has gone to zero in a single generation. For reasons explored in the book, Weldon found that conclusion utterly implausible. And indeed, by the reasoning just reconstructed, most of the green-seeded pea plants—8 out of 9—harbor within them a yellow-making factor.

So how, circa 1906, would one have decided between the Mendelian suppositions and the Weldonian ones? This brings me to the third point. It is tempting to try to decide the issue empirically, by breeding from those extracted recessives. On Mendelian reasoning, since there is nothing, factor-wise, in those plants except green-making factors, the resulting lineage should be green unto the last. On Weldonian reasoning, by contrast, the expectation is that sooner or later, plants will be born with a preponderance of yellow-making factors in them, and so will be yellow-seeded. Suppose we had done the experiment, and done it well, taking due care to guard against contamination, and so on. If wholly green-seeded generation had followed wholly green-seeded generation, again and again, then, even though the Weldonian theory could not have been ruled out entirely (given the nature of chance operations), the Mendelian theory would have deserved to be considered the winner. But the reverse situation would have been much less straightforward. If yellow seeds had begun to appear, it could have been because the Weldonian theory is correct. Or, as the defender of Mendelism would have insisted, it could have been due to a spontaneous mutation among the hitherto exclusively green-making factors, or—despite all the precautions—to the experimental lineage having become contaminated with yellow-making factors. Thus has Mendelian reasoning protected itself from disconfirmation, even in the face of apparently disconfirming results.[4]

Postscript 3

From a Counterfactual Edition of the *Dictionary of Scientific Biography*

The *Dictionary of Scientific Biography* is a standard reference work for professional historians of science. Known universally as the "*DSB*," it comprises, in its now-classic first edition (1970–80), sixteen handsomely produced volumes. On pages 251–52 of volume 14, published in 1976, there is a superbly scholarly article on W. F. R. Weldon by Ruth Schwartz Cowan. It ends:

> [Weldon's] debate with the Mendelians became even more acrimonious; and a stream of critical articles flowed from his pen, to be published in *Biometrika*. In the midst of an Easter holiday devoted to biometric research Weldon collapsed and died. Many of his colleagues considered his death particularly tragic for having come when he seemed to be entering a very promising phase of his career.[1]

But if we consult a counterfactual edition of the *DSB*, from the history in which Weldon not only lived beyond spring 1906 but, not long after, published his *Theory of Inheritance*, we find a rather different ending:

> [Weldon's] debate with the Mendelians became even more acrimonious; and a stream of critical articles flowed from his pen, to be published in *Biometrika*.
>
> His criticisms reached a much larger audience after the appearance of his *Theory of Inheritance* (1906). Accessibly written and incisively argued, the book presented the debate as pivoting on a choice between two conceptions of dominance. On the Mendelian side, dominance was thought of as an absolute property of particular character versions and their germinal determinants. In seeds of garden peas, for example, yellowness was classified as dominant to greenness, and the presence of a yellowness-making determinant in a zygote was treated as a guarantor that the mature seed would be yellow. On the other side—which Weldon identified as

Galtonian, citing Galton's 1872 paper "On Blood-Relationship"—dominance was thought of as relative to contexts, internal and external. In some contexts, a yellowness-making determinant would produce yellowness in the mature seed. In other contexts, however, the character of the mature seed could end up anywhere on the color spectrum from yellow to green. In evaluating these conceptions, Weldon introduced readers to basic statistical methods, the latest findings from experimental embryology, the emerging biology of chromosomes, and his own analyses of breeding data on peas, mice, and horses. He concluded that the Galtonian conception was the more generally applicable, and furthermore, that when combined with Galton's law of ancestral heredity, the Galtonian perspective subsumed Mendelian patterns as special cases.

Although Bateson and his fellow Mendelians were predictably dismissive, *Theory of Inheritance* was soon recognized as a major contribution. Galton, to whom the book was dedicated, admired it greatly, recalling in his autobiography "the mixture of pride and astonishment which overtook me as I discovered where my own ideas ineluctably led." He praised the book fulsomely in nominating Weldon for the Royal Society's Darwin Medal, awarded to Weldon in 1908. (The award was widely interpreted as a correction on the society's part for the awarding of the medal to the anti-Darwinian Bateson in 1904.) Under the leadership of E. S. Goodrich, Weldon's zoological second-in-command, Oxford soon became the international headquarters of theoretical and empirical Darwinism, with the pioneering treatment of natural selection in Weldon's book as a basis. Outside of Britain, the book appealed especially to biologists trained in embryology. In Germany, for example, Richard Woltereck seized on Weldon's coverage of experimental work on the interaction of heredity and environment in the water flea *Daphnia* (by Weldon's former student Ernest Warren) to inaugurate research into what Woltereck enduringly called the "norm of reaction." And in the United States, an interest in testing one of the book's claims about chromosomes—that the same segment of chromosome can have different effects on a visible character depending on context—led T. H. Morgan to combine Mendelian breeding experiments with Woltereckian norm-of-reaction experiments in innovative work on the fruit fly *Drosophila*, brilliantly summarized in *The Mechanism of Weldonian Heredity* (1915), by Morgan, Calvin Bridges, Alfred Sturtevant, and Hermann Muller.

Weldon, alas, saw little of these developments. At the end of 1908, in the midst of a Christmas holiday devoted to biometric research, he collapsed and died. Many of his colleagues considered his death particularly tragic for having come when he seemed to be entering a very promising phase of his career.

Acknowledgments

This book springs from an encounter with the image that now graces the front cover. Near the end of my PhD studies at Cambridge, the sociologist Martin Richards generously invited me to sub for him on his course "Darwinism, Genetics, and the Social Sciences." As I began preparing for a lecture on the beginnings of genetics, I went off to the library in search of the first volume of *Biometrika*, so I could see for myself what Weldon found so objectionable. Leafing through, I came upon those green-to-yellow pea seeds—and I was hooked. When, the next year, I interviewed for a job in history and philosophy of science at Leeds, I showed the image by way of indicating the direction of my research.

That was over twenty years ago. Between then and now, my debts have mounted. At Leeds, I enjoyed the good fortune of taking over Bob Olby's old history of genetics course, which enabled me not only to continue my self-education but to try out my developing ideas in the company of superb undergraduate and graduate students. An invitation from the Royal Institute of Philosophy to participate in a 2003 lecture series on biology became the spur to pull together my incipient thoughts on Weldon, genetics, and the counterfactual history of science. That lecture, improved with help from Leeds colleagues and students (and, by email, Richard Lewontin), became the basis for a 2005 manifesto, "Other Histories, Other Biologies," parts of which survive in this book's chapter 11. The next year, the British Academy awarded me funds from its Small Research Grants scheme to get Weldon's "Theory of Inheritance" manuscript and associated materials microfilmed and transcribed. Luckily for me, Annie Jamieson, a developmental biologist turned historian of science who, as noted, really *got* Weldon, and then helped me to do so, did the transcribing. Just as luckily, there was enough money left over to hire Berris Charnley, a geneticist turned historian of science, to photograph the entirety of the correspondence between Weldon, Pearson, Galton, and Bateson up to Weldon's death. Most of this archival material, along with a great deal more on which my research has depended, is held in UCL Library's Special Collections, where Mandy Wise and colleagues have always been splendidly accommodating. Another London institution whose holdings proved important was the Royal Society, where Keith Moore and colleagues have provided unfailingly genial, expert support.

A new wing of the project opened in 2007–10, when Berris and I took part in the Leeds-Bristol *Owning and Disowning Invention* project, funded by the Arts and Humanities Research Council. Led by our Leeds colleague Graeme Gooday, the project offered ideal conditions for reexamining the relationship between Mendelism's theoretical and practical successes, especially in plant breeding. Another, rather different wing opened in 2012–14, when Annie and I joined with Jenny Lewis, a genetics education researcher also based at Leeds, to run the classroom quasi-experiment described in chapter 11, thanks to a funding award from the Uses and Abuses of Biology scheme of the Faraday Institute for Science and Religion. Since those two expansions, my contact with the worlds of agricultural genetics and science education has been ongoing. In agricultural genetics, my guides have included, besides Berris (whose PhD thesis on what he called "the Mendelian system" defines the state of the art, in my view), Dominic Berry, Matt Holmes, Clare Coleman, Mrinalini Kochupillai, Tina Barsby and colleagues at the National Institute of Agricultural Botany in Cambridge, and Claire Domoney and colleagues at the John Innes Centre in Norwich. In science education, Kostas Kampourakis was an early advocate for the work done in Leeds, helping Annie and me to navigate the choppy waters of publication in a new field with exemplary editorial skill and generosity. Equally helpful to us was Brian Donovan, who, with Michelle Smith, went on to turn that work into the basis for a research program whose extraordinary rigor and reach is beyond anything I could ever have imagined. It has been a joy and a privilege to collaborate with Brian and Michelle, along with the other team members in our *Honoring the Complexity of Genetics* project, funded by the National Science Foundation's Improving Undergraduate STEM Education scheme.

In retrospect, I was amazingly slow to take a serious interest in the alleged fishiness of Mendel's data. Correction began in autumn 2014, when our staff-student reading group in history and philosophy of biology at Leeds made its way through *Ending the Mendel-Fisher Controversy* (2008). Discussing Mendel's paper and its interpretation, statistical and otherwise, with Jon Hodge, John Turner, Roger White, Gabrielle White, John Grahame, and other members of the group in those months was, for me, a high point in the life of the group—maybe even in the life of the mind as I have known it so far. Invitations to speak in the months that followed, as keynote lecturer at a meeting of Norwegian historians of science (courtesy of Ageliki Lefkaditou), as the first annual John Innes Lecturer in History of Science at the John Innes Centre (courtesy of Sarah

Wilmot, who has been unstinting in help with the Innes Batesoniana), and as a presenter in the seminar series of the Department of History and Philosophy of Science in Cambridge (my graduate alma mater), allowed me to share my thoughts on the curious cultural history of the debate with exceptionally well-informed audiences. I learned much from all of them, as I did from my referees on a subsequent short paper, “Beyond the ‘Mendel-Fisher Controversy,’” published in *Science* in autumn 2015, and from the audience in Tel Aviv who heard a successor version four years later at a workshop organized by Yafeng Shan, Ehud Lamm, and Oren Harman on the historiography of genetics. For several years now, Yafeng, Charles Pence, and I have been the three musketeers of Weldon studies, and my own efforts are all the better for their companionship.

It is just about manageable, while in full academic harness, to write talks, papers, and reviews—but not a book like this one. Indispensable relief from ordinary teaching and administrative duties came in the form of a British Academy Mid-Career Fellowship (2013–14), which also funded Amanda Labbett’s crucial work transcribing the UCL letters; a Leverhulme Trust Major Research Fellowship (2017–19); and several semesters of research leave from the School of Philosophy, Religion and History of Science at Leeds, which also paid the costs associated with reproducing the book’s illustrations and funded the teacher workshops discussed in chapter 13. Overseeing the long process of drafting and revision has been Karen Darling, my editor at the University of Chicago Press. From our first conversation about the book, over coffee on a cold winter’s morning in Chicago in late 2011, I have had the glorious feeling that Karen really *got* me, and what I was hoping to achieve. That feeling has been vindicated over and over again, as I have benefited from her guidance, encouragement, and seemingly infinite patience. I am the more pleased that a story she shared with me that morning has developed into one of the more colorful threads in the narrative weave of the book. I also very much appreciate her choice of manuscript reviewers, whose sympathetically shrewd advice was just what I needed. No less improving have been the comments on chapter drafts from members of three reading groups, in Leeds and—via Zoom—Halifax (hosted by Gordon McOuat and Ford Doolittle) and Cambridge (hosted by Hasok Chang), and on parts or all of the manuscript from Alex Aylward, Stefan Bernhardt-Radu, Helen Curry, Mats Elliott, Adrienne Jessop, Kersten Hall, Jon Hodge, Kostas Kampourakis, Staffan Müller-Wille, Anya Plutynski (with Dylan Doherty), Michael Ruse, Amir Teicher, and Christian White. Alex also provided sterling support during the final submission.

To all of the above, as well as to the many more colleagues, students, archivists, and others whose diverse contributions swell my endnotes and bibliography, or who put their own work on hold in order to support me for grants and in other ways, I offer deepest thanks. I am especially grateful to my interviewees for their generosity with their time and with my preoccupations. I am also grateful to all who helped with a number of previously published writings—in the *British Journal for the History of Science* (Radick 2016a), *Philosophy* (Radick 2005a), *Philosophy of Science* (Radick 2013a), the *Studies in History and Philosophy of Science* journals (Radick 2011, 2013b, 2022b), and the *Times Literary Supplement* (Radick 2020a)—portions of which I have incorporated in this book. Portions of this book were also derived from my "Making Sense of Mendelian Genes," *Interdisciplinary Science Reviews* 45 (2020), http://www.tandfonline.com/10.1080/03080188.2020.1794387. © The Institute of Materials, Minerals and Mining. As publication drew closer, the book benefited from the graphic design work of Alex Santos, the meticulous copyediting of Norma Sims Roche, and the indexing skill of Helmut Filacchione. Many thanks to them, and to the rest of Karen's team at Chicago, including Tristan Bates, Erin DeWitt, Deirdre Kennedy, Anne Strother, and Carrie Olivia Adams. Many thanks as well to the Board of the University of Chicago Press for naming the book a recipient of support from the Susan Abrams Fund.

Outside of my professional life, friends and family have been sweetly indulgent of a project that they must surely have wondered if I would ever finish. Their many kindnesses—above all, in giving me so many opportunities to talk about it informally, in ways that have helped me keep the whole in view even as I labored away on the various parts—have been a blessing. I hope that they enjoy the book, which I dedicate, with love, to the two people who, in the most fundamental sense, got me started: my parents.

Notes

Introduction

Epigraphs: Weldon to Pearson, 23 June 1902, Pearson/11/1/22/40, PP; "The Mendelian Revelation," *Pall Mall Gazette*, 6 October 1909, clipping at G.5.d.51, BP; Bateson to Lucas, 9 November 1920 (draft), G.5.i.1, BP, quoted in B. Bateson (1928a, v–vi, epigraph quotation on vi; I have retained the emphases and punctuation from the published version).

1. On the ubiquity of talk of "'genes for' (almost) everything," see Kampourakis (2017, ch. 5). On "inheritance": for the most part, I will use it interchangeably with "heredity" because throughout most of the period that concerns me, the people that I write about used the words interchangeably.
2. Quotation from Arnold and De Saulles (2009, 31). Dallas Swallow, professor emeritus of human genetics at University College London, credits her career to such reading when she was in her mid-teens and stuck at home with flu. "Keen to find something to interest me, my mother brought a children's book about Gregor Mendel back from the library. I was hooked! I loved the segregation analyses—and had my own opportunity to try it out for myself in a project on radishes set up in school by someone from the University [of Cambridge]." Swallow (2020–21, 5–6).
3. Roarty and Bryan (2014).
4. Holladay (2004). In a survey article on which the columnist drew, Sturm and Frudakis (2004, 327) wrote, "The use of eye colour as a paradigm for 'complete' recessive and dominant gene action should be avoided in the teaching of genetics to the layperson, which is often their first encounter with the science of human heredity." I am grateful to Kostas Kampourakis for this reference.
5. Standout contributions in the earlier scholarship include Provine ([1971] 2001); MacKenzie (1981a, esp. ch. 6, with my epigraph quotation 1 on 262n20); Kim (1994); Kevles (1981, esp. 199); Olby (1989a). Other surveys—all worth consulting—can be found in Sturtevant (1965);

Carlson (1966); Froggatt and Nevin (1971a); De Marrais (1974); MacKenzie and Barnes (1974)—a widely circulated but, apart from a German translation, never published masterpiece, though MacKenzie and Barnes (1979) is a decent digest; Farrall (1975); Norton (1979)—on Weldon, Norton got there earliest of all: see postscript 2 in this book, and the citation to it in MacKenzie (1981a, 260n8); see also Norton (1973, 1975); Roll-Hansen (1980, with defending criticism in Barnes [1980]); Bowler (1989a); Depew and Weber (1995); Gayon (1998); Gillham (2001); Magnello (2004); J. Schwartz (2008); Shan (2020a, 2021); and Pence (2022a). I have also learned much from two reviews of Kim (1994): Vicedo (1995) and Barnes (1996).

6. Bateson (1908, 324). The line is used as the title of the most recent biography of Bateson by Cock and Forsdyke (2008).
7. See, e.g., Mukherjee (2016a, 69–70) and J. Schwartz (2008, ch. 7).
8. Williams (2021, quotations on 3).

Chapter One

Epigraphs: F. Burkhardt et al. (1985 13:202–4, quotations on 203, 204).

1. See Livio (2013, esp. ch. 3, 55, for the photograph). “Uncut” is Livio’s term for the pages that are still joined together, and I follow his usage. But in the technical vocabulary of the book world, as I learned from Christian White, the correct term is “unopened.” On Darwin’s reputation on heredity, see also the testimonials quoted in Vorzimmer (1970, 21) and Geison (1969, 379); nearer our own day, see, e.g., Dawkins (2010, 6). Pangenesis has found occasional, maverick champions; see, e.g., Michie (1958) and, for a recent statement and associated references, Liu and Li (2014).
2. Livio (2013, 42, emphases in original).
3. I paraphrase the account in Poczai, Bell, and Hyvönen (2014, 3–4), but see also, e.g., Cobb (2006, 957); Mawer (2006, 51); and Van Dijk, Weissing, and Ellis (2018, 351–52)—all drawing on research by Orel and Wood, most extensively presented in Wood and Orel (2001, esp. 248–77), but also in, e.g., Wood and Orel (1982, esp. 67); Orel and Wood (2000, esp. 152); Wood and Orel (2005, esp. 261, 268); Wood (2007, esp. 241); and Orel and Peaslee (2015, esp. 14). On Mendel’s life and work generally, the best of the more recent books in English are Orel (1996) and Mawer (2006). For “like engend’ring like,” see Russell (1986).
4. See, e.g., Orel (1984, 12–15, 23–26, 78–87).
5. Quoted in Wood and Orel (2001, 258). The source is Bartenstein et al. (1837, 227).
6. Darwin (1859, 12–14, quotation on 13). For Darwin on inheritance, the

best introduction is now Olby (2013). On the French medical "cradle of heredity" from which Lucas emerged, see López-Beltrán (2004), but see also Waller (2001a) on contemporary British (notably phrenological) writing along broadly similar lines. On Darwin's reading of Lucas in 1856 and its legacies, see Noguera-Solano and Ruiz-Gutiérrez (2009). On Zola's reading and its legacies for his Rougon-Macquart novels, see Lewontin (1996, quotation from Zola—from his preface to *La Fortune des Rougon*—on 31). For the history of "inheritance" talk and, latterly, "heredity" talk in English, see Radick (2013a, 715–17) and, more fully, López-Beltrán (1994) and Müller-Wille and Rheinberger (2007, 2012)—though, for reservations about the latter's overall "from generation to heredity" thesis, see Radick (2014a).

7. Darwin (1863, 23).
8. "Mendel's theory was in fact the answer to the question formulated by Abbot Napp in Brno before Mendel was accepted in the monastery": Orel 1996, 179—with Napp's question (32) quoted in a chapter titled "Heredity Before Mendel," in a book titled *Gregor Mendel: The First Geneticist*.
9. Many thanks to Helen Piel for this information.
10. The quotation from Darwin is from Notebook C (C1), in Barrett et al. (1987, 239). On Darwin's early enthusiasm for Yarrell's law, see Hodge (2009, 55, 58, 65).
11. Mendel (1866, 14, sent. 1), with commentary by Müller-Wille and Hall in Mendel (2016, and also at 7, sent. 3). Corcos and Monaghan (1993, xvi) aptly note that in its sole appearance, *vererbt* is used negatively, to characterize something *not* inherited (namely, occasional bleaching of the green color in the seeds). On *Vererbung* coming to prominence as an organizing term in Austro-German biology only from the 1880s, see Churchill (1987).
12. For a transcription of Mendel's *Origin* annotations, see Fairbanks and Rytting (2001a). In ch. 1, the only sentence to be marked (for a photograph, see Fairbanks [2020, 266]) concerns changed conditions of life inducing variability—which is also the subject of one of the most heavily marked passages in the most thoroughly annotated chapter: ch. 8, "Hybridism." For discussion, see Fairbanks and Rytting (2001b, 749–50); see also Fairbanks and Abbott (2016, itself associated with a "Darwinized" English translation by these authors of the "Versuche.")
13. See Mendel (1866, 36–37), where he scoffs at the idea that cultivation can so revolutionize a plant's physiology as to liberate it from the laws operating in open fields. Mendel does not name Darwin as the target, but the relevant passage echoes the language of the *Origin* more fully and directly than does any other in the "Versuche"; see Fairbanks and

Abbott (2016, 403–4). The issue seems to have bothered Mendel a lot: his friend Gustav von Niessl even recalled Mendel carrying out experimental transfers of plants from nature to garden, and never finding any permanent changes induced; see Iltis (1932, 102–3).

14. For the "hybrids not heredity" reading of the "Versuche," the classic sources are Olby (1979) and Brannigan (1979). Although that reading is now widely accepted among historians of science (who have, accordingly, dropped the old "neglect of Mendel" question as misguided), there is room for disagreement about virtually everything else concerning Mendel, including what, exactly, he considered his "law valid for *Pisum*" to be. On the debates and their histories, see Müller-Wille (2021).
15. The "Versuche" (Mendel 1866) exists in many editions and translations. A non-anachronistic translation into English by Staffan Müller-Wille and Kersten Hall (Mendel [2016], which I follow unless otherwise noted) offers ease of access to the German original, line-by-line scholarly commentary by the translators, and a full bibliography. An indispensable, if not infallible, companion volume is Corcos and Monaghan (1993).
16. Iltis (1932, 210); Orel (1996, 224); see also Orel (1984, 78), who notes Mendel's marking of a passage in his copy of Gärtner's 1849 book on plant hybrids encouraging the hybridizing into being of new, aesthetically pleasing ornamental plants. On Mendel the practical breeder, see Orel (1984, 78–80) and Van Dijk, Weissing, and Ellis (2018, esp. 350).
17. On Ungerian science as it mattered for Mendel, see esp. Gliboff (1999); Dröscher (2015b); and Fairbanks (2020, 265, 267), where Unger is identified as the source of Mendel's view, contra Darwin, that changing conditions do not induce new variations.
18. On Mendel's seven pairs as highlighted in the German variety lists from which he probably ordered his material, see Ellis et al. (2019, 2). The only surviving document from the backstage world of Mendel's pea experiments is an enigmatic manuscript page, called the *Notizblatt*, believed to be a record of a failed attempt by Mendel to capture seed-coat color patterns by theorizing with three, rather than two, versions of the character. See Olby (1985, 245–47, with the *Notizblatt* itself reproduced on the page facing 63). On further seed-coat-induced headaches that Mendel endured, see Corcos and Monaghan (1993, 192–95).
19. Russell (2006, 20, 24). I have not reproduced Russell's italicization. Mendel's laws, thus identified, trace back to lectures that T. H. Morgan gave in 1916, perhaps inspired by Lock (1906, 267); see Marks

(2008). But among textbook writers there was, and remains, diversity as to just how many laws Mendel discovered and what they are. For a marvelous overview, see Teicher (2020a, 28–35).

20. "*Pisum geltenden Gesetze*": Mendel (1866, 35). My gloss on the phrase is different from Müller-Wille and Hall's, though they set out the interpretive issues with admirable clarity in their commentary in Mendel (2016, 18, sent. 7). A word is needed on a German term that is notoriously untranslatable into English: *Entwicklung*. Whereas the English language, not long after Mendel wrote, recognized a division of labor between "development," meaning "change in the maturing organism," and "evolution," meaning "change in the species lineage," *Entwicklung* embraced both. Mendel announces at the start that he seeks "*ein allgemein giltiges Gesetz für die Bildung und Entwicklung der Hybriden*"—"a generally valid law for the formation and development/evolution of hybrids"; Mendel (1866, 3). He goes on later to refer to the law for *Pisum* as an "*Entwicklungs-Gesetz*"; Mendel (1866, 18). On the "Versuche" and the meanings—mathematical as well as biological—of *Entwicklung*, see Gliboff (1999, 225–27, 234–35n42).

21. Mendel (1866, 18). For an arithmetical cranking-through of the reasoning needed to get from $A + 2Aa + a$ to $2^n - 1 : 2 : 2^n - 1$, see Radick (2011, 130n6). For an elegantly algebraic reconstruction, see Teicher (2014, 191).

22. Two subtleties about the series-conforming plants need flagging. First, they were grown not from the uniformly round-and-yellow seeds that Mendel collected after hybridization, but from the mix of round-and-yellow, round-and-green, wrinkled-and-yellow, and wrinkled-and-green seeds that resulted from his planting those initial hybrid seeds. Second, the term *ABb*, for example, here represents not seeds that are round and hybrid-character yellow, but plants that, after self-fertilization, produce two kinds of seeds: round-and-yellow seeds and round-and-green seeds. For exemplary discussion of the flow of argument in Mendel's handling of his multi-character crosses, see Corcos and Monaghan (1993, 100–120); see also the helpful diagram in Orel (1996, 106).

23. It is in making the case for this crucial inference that Mendel (1866, 24, sent. 7) uses the word "*Factoren*"—ever after seized upon by Mendelian readers as equivalent to what, in the early days of Mendelism, before "gene" talk became fully established, were often called "factors"; see, e.g., Morgan et al. (1915). But in context, "*Factoren*" refers only to whatever it is in a gamete that works with another gamete to produce progeny of a particular kind. For discussion, of this term as

well as another, "*Anlage*," used in the same way as part of the same case, see Olby (1985, 250–51).

24. "*Begründung und Erklärung*": Mendel (1866, 32). Müller-Wille and Hall note that the phrase was underlined in Mendel's manuscript but apparently overlooked by the printers.
25. "*die ausserordentliche Mannigfaltigkeit in der* Färbung unserer Zierblumen": Mendel (1866, 36, emphasis in original).
26. Mendel (1866) appends an extraordinary footnote (on 41)—his only lengthy note in the whole piece—on how, far from egg cells functioning as a kind of nurse cell, pollen cells and egg cells contribute equally in the formation of the embryo, for nothing else, in Mendel's view, and as he had explained, makes sense of the return of the characters of both parents, perfectly, in the hybrid progeny.
27. Mendel's (1870) hawkweed paper—much shorter than the "Versuche"—can be read in English translation in Stern and Sherwood (1966, 49–55). In the final paragraph (55), Mendel stresses the contrast between the variable-progeny hybrid peas of his previous inquiry and the constant-progeny hybrid hawkweeds of this one. For lively discussion, see Endersby (2007, ch. 4).
28. "*sich einem Eindrucke ohne Reflexion hingeben*" (on p. 497 of Mendel's copy of *Variation*). On this annotation, see Orel (1996, 194), and for a photograph, Fairbanks (2020, 269). For discussion of this annotation, I am indebted to Daniel Fairbanks, whose translation I have adopted. On Mendel's being provoked by pangenesis into carrying out an experimental study testing whether a single pollen grain sufficed for fertilization, see Van Dijk and Ellis (2022), which also highlights the role of Mendel's worsening eyesight in bringing an end to his experimental work.
29. See, e.g., Livio (2013, esp. 38–51); Gitschier (2014, 3); Dawkins (2017, 83). Robin Marantz Henig, in a popular biography of Mendel, counterfactually imagines a meeting between Darwin and Mendel in 1862, during the latter's (factual) visit to London, where Darwin learns to his relief that inheritance is not blending (2000, 124–25). Jenkin's (1867) essay, with associated correspondence and useful commentary, can be found in Hull (1973, 302–50).
30. Fisher (1930, ch. 1). "In the future," wrote Fisher in the preface (ix), "the revolutionary effect of Mendelism will be seen to flow from the particulate character of the hereditary elements. On this fact a rational theory of Natural Selection can be based, and it is, therefore, of enormous importance." For the 1930s to 1950s as the time when Jenkin's swamping argument went from being a "celebrated diffi-

culty" for Darwin to a "most devastating criticism," see Morris (1994, 317).

31. For "completely blending . . . heritages," see Galton (1887, 401); for "particulate inheritance," see Galton (1889, 7). On the history and historiography of blending inheritance generally, see Porter (2014).
32. In an 1866 letter to Alfred Russel Wallace, Darwin wrote about the "non-blending" he had found on crossing two varieties of sweet pea, then went on to say that such cases were no more to be wondered at than "every female in the world producing *distinct* male & female offspring." See Darwin to Wallace, 6 February 1866, in F. Burkhardt et al. (1985–, 14:44–45, emphasis in original). The letter has been hailed as evidence of incipient but, alas, never fully flowering Mendelism on Darwin's part, e.g., in Livio (2013, 51–52) and in Dawkins (2003a, 35). Darwin (1868) reported the sweet-pea results in *Variation* in the chapter on crossing (ch. 15), in a subsection on the non-blending of certain characters (2:93–4; see also 1:393).
33. Darwin's Notebook M is full of examples, some close to home, e.g., M83e: "As instance of heredetary [*sic*] mind. . . . My handwriting same as Grandfather." In Barrett et al. (1987, 539). For discussion, see Radick (2018).
34. Darwin (1868, 2:70–71, 93).
35. Darwin (1868, 2:385–86).
36. "*wird in der That die Mittelbildung fast immer ersichtlich*": Mendel (1866, 10). In *Variation* (1868, 2:92), Darwin began the subsection on non-blending characters discussed in note 32 above with "When two breeds are crossed their characters usually become intimately fused together."
37. Intriguingly, Mendel does note one pea character seemingly fitting this bill. Toward the end of the section on multi-character crosses (1866, 23), he notes that the flowering time of the hybrid plants tended to be intermediate between the flowering times of the seed and pollen plants, but otherwise to behave like every other hybrid character. For discussion, see Bateson (1902a, 135–36) and Corcos and Monaghan (1993, 118–19).
38. On Mendel's materialist understanding of what he called the "material constitution and arrangement of the elements" ("*materiellen Beschaffenheit und Anordnung der Elemente*": 1866, 41), see Dröscher (2015a, 494), as well as the excellent gloss on the quoted-from sentence in Mendel (2016, sent. 6). For the case against assimilating that understanding to the later Mendelian one, see Olby (1979, esp. 65–66). For the fluid-mixing reading, see Meijer (1983, esp. 139–42). The single

point in the "Versuche" where Mendel writes *aa* is in his discussion of *Phaseolus* (1866, 35), where the notation arises as an artifact from his complexly reinterpreted combination-series mathematics, as Corcos and Monaghan (1993, 151) explain.

39. On Darwin's response to Jenkin, the best introduction remains Vorzimmer (1963), later expanded in Vorzimmer (1970) and popularized in Gould (1991). Valuable treatments since include Morris (1994); Gayon (1998, ch. 3); Bulmer (2004a); and Lewens (2010, 831–32).
40. Darwin owned two German publications including brief treatments of Mendel's work: an 1869 brochure by the Giessen botany professor Hermann Hoffmann on what distinguishes species from varieties; and an 1880–81 book on plant hybrids (photographed in Livio [2013]) by the Bremen physician Wilhelm Focke. For comprehensive quotation and discussion, see Olby (1985, 222–24, 227–30). On Hoffmann and Focke in their own rights, see Roberts (1929, 204–18).
41. Mendel (1866, 43–44); Darwin (1868, 2:88–89). Like Mendel, Darwin revered Kölreuter and Gärtner, referring to them often and annotating their books heavily (see Di Gregorio [1990, 248–97, 458–71]). On Kölreuter and Gärtner, see Roberts (1929, 34–61, 164–78).
42. But see Olby (2009, 43–44) on how productive it might nevertheless have been for Darwin to respond to Mendel's critique, just as it proved productive for him to respond to Jenkin's critique.
43. Darwin (1868, 2:70–71, 93, quotation on 71). That Darwin would have seen in the "Versuche" at most another set of observations on hybrids to add to the pile has long been, and remains, the conventional view: see, e.g., Huxley (1960, 30–31); De Beer (1964, 215); Lorenzano (2011, 40–41). Even Livio (2013, 57) accepts it. For Darwin's snapdragon cross as a Mendelian near miss, see, e.g., Iltis (1932, 127–28).
44. Laxton (1866a). Darwin owned and (lightly) annotated these proceedings; see Di Gregorio (1990, 428). On Laxton, see Roberts (1929, 104–10).
45. For a transcription of Darwin's 1865 manuscript with commentary, see Olby (1963). On Darwin's ideals of scientific argumentation as reflecting the philosophy of science of his place and period, see Hull (2009, esp. 175–85). On the *Origin* as structured in conformity with those ideals, see Hodge (1977).
46. Darwin (1868, 1:8–14; cf. the 1860 letter from Darwin discussed in Hull [2009, 184–85]); Darwin (1868, 2:357). Contemporary reviewers readily connected the dots between natural selection and pangenesis; see Holterhoff (2014, esp. 665–66). On how and why Darwin found pangenesis *so* compelling, see Stanford (2006, esp. 65–66).
47. On the deep, Grantian and then Müllerian roots of pangenesis in Darwin's thinking, see Hodge (1985), as well as the further reflections by

Hodge (1989, 2010). On the development of Darwin's ideas in the first half of the 1860s in response to publications by Herbert Spencer and the French hybridist Charles Naudin, among others, see Geison (1969). For the manuscript citation to Müller, see Olby (1963, 253). On Grant's "gemmules," see Hodge (1985, 210, drawing on Sloan [1985, 80]).

48. Huxley was put on notice about the thrust of the theory no later than 1857, in a letter from Darwin noting his view that the power responsible for parthenogenesis also produced, "for instance, nails on the amputated stump of a man's fingers, or the new tail of a Lizard," and that "propagation by true fertilisation, will turn out to be a sort of *mixture* & not true *fusion*, of two distinct individuals, or rather of innumerable individuals, as each parent has its parents & ancestors:—I can understand on no other view the way in which crossed forms go back to so large an extent to ancestral forms." Darwin to Huxley, before 12 November 1857, in F. Burkhardt et al. (1985–, 6:484–85, quotation on 484, emphases in original). Cf. Fisher (1930, 1–2), where the letter is quoted as a sign that Darwin early on sensed the problem with blending inheritance.

49. Laxton (1866b). Laxton's little report begins with praise for Darwin, which might have caught his eye. In a previous issue of the *Gardeners' Chronicle*, Darwin had described what Laxton (900) called a "very ingenious" method for tricky crosses, but which Laxton had not made use of with his peas.

50. In peas, the phenomenon became sufficiently well established that in 1902, William Bateson misinterpreted the "Versuche" as reporting evidence of direct paternal influence, in a passage on the increase in the purply-spottedness of seed coats after crossing—a misinterpretation that Bateson, on the quiet, later corrected via the insertion of a new, clarifying footnote into the English translation of Mendel's paper. See Bateson (1902a, 51 [for the passage sans footnote], 139–40, 161); Bateson (1909, 343 [where a new footnote now insists that the increase be interpreted as taking place not in the maternal seed coats but in the seed coats of the hybrid offspring]).

51. Darwin to Hooker, 2 October 1866, in F. Burkhardt et al. (1985–, 14:337–38, quotation on 337); Darwin (1868, 1:397–98; 2:357–58 [on "pangenesis"], 365 [on Laxtonian phenomena in peas]). At Darwin's request, Laxton later sent him his collection of pea seeds; Darwin to Laxton, 31 October 1866, in F. Burkhardt et al. (1985–, 14:365–66). A more famous though less straightforward example from this class of fact was the persistent influence of a male quagga on later offspring from a mare of Lord Morton's; see Darwin (1868, 1:403–4, 2:365–66). For discussion, see R. Burkhardt (1979).

52. Darwin (1868, vol. 2, ch. 27; on functional independence, see 368–71; on development, see 366–68). I take my list of topics from the one that heads the lightly but comprehensively modified version of the chapter in the second edition (1875).
53. See Darwin (1868, 2:371–73, quotation on 372; and 1:284–87), for his studies of the effects of use and disuse on duck limbs. From the first edition of the *Origin* until the end of his publishing life, Darwin accepted and indeed called upon Lamarckian inheritance. The best treatment of Darwin's Lamarckism to date is in Hoquet (2018). Curiously, Yarrell's law, which meant so much to Darwin in the notebook period, and which he affirmed in the 1842 *Sketch* and the 1844 *Essay*, is nowhere discussed under that name in the *Origin* or *Variation*, though he addressed the topic and its complexities in one of the inheritance chapters (1868, 2:62–64). Many thanks to Jon Hodge for discussion on this point.
54. Darwin (1868, 2:374–77, 395).
55. Darwin (1868, 2:377–83).
56. Darwin (1868, 2:398–402, quotation on 400, "pure" and "hybridised" gemmules on 400–401). On reversion as a long-running preoccupation for Darwin in his studies of domesticated animals, see Bartley (1992, esp. 323–27).
57. Darwin (1868, 2:404).
58. Darwin (1859, 25–27).
59. Mendel discussed the beetle—the subject of his first scientific publication—and its effects in the "Versuche" (1866, 9–10).
60. Darwin (1881).
61. Darwin (1868, 2:398 for mutilations, 401 for pigeons).
62. Darwin (1868, 2:357 [Whewell] and 404).

Chapter Two

Epigraphs: Pearson (1914–30, 3B:542); Pearson (1914–30, 2:201).

1. Introductions to the quincunx, including virtual versions of it in action, are easily found online.
2. The classic biographical sources on Galton are Galton (1908) and the monumental Pearson (1914–30). Of the more recent biographies, the best is Gillham (2001). Some very helpful recent expositions of Galton's work on heredity are available, notably Gayon (1998, chs. 4 and 5); Bulmer (2003); and J. Schwartz (2008, chs. 1 and 2).
3. Mehler (1996, 261).
4. See the *OED Online*, entry for "hereditarian," citing Papillon (1873, esp. 61).

5. For Galton's certificate as a candidate for election to the Royal Society, see EC/1860/10, RS. For his remarking on the shared birth year with Mendel, see Galton to Pearson, 16 March 1908, in Pearson (1914–30, 3A:335).
6. Galton (1869, v); Waller (2002, 38–39). On Galton's time in Africa and its afterlife in his later scientific work, see Fancher (1983a). Waller (39) suggests that Galton's shift toward the family line may have been influenced by his reading of G. H. Lewes's *The Physiology of Common Life* (1859–60)—at that moment the most comprehensive and up-to-date book on inheritance in English. On Darwinians at the Ethnological Society in the early 1860s, see Desmond and Moore (2009, 341).
7. Galton (1865, quotation on 321; 1869, 336–50). A letter from Galton, published in the *Times* on 26 December 1857 under the title "Negroes and the Slave Trade" opens, "Sir,—I do not join in the belief that the African is our equal in brain or in heart; I do not think that the average negro cares for his liberty as much as an Englishman, or even as a serf-born Russian; and I believe that if we can, in any fair way, possess ourselves of his services, we have an equal right to utilize them to our advantage as the State has to drill and coerce a recruit who in a moment of intoxication has accepted the Queen's shilling, or as a shopkeeper to order about a boy whose parents had bound him over to an apprenticeship." He goes on to say that just as the State's and the shopkeeper's actions are non-tyrannical only so far as they elevate, rather than degrade, those made to serve, so Britain's use of African laborers in the colonies can be legitimate only if Africans end up better off as a result. To reintroduce the slave trade would, in Galton's view, be sure to make them worse off. No, Africans should freely choose to become immigrants, through skillful recruitment, and then be treated "justly."
8. Galton (1908, 287–88).
9. Galton (1865, 157–59, 165, quotation on 157). For the domestication of animals by savages—a subsidiary theme of the *Origin*'s first chapter—as Galton's intellectual bridge from geography to heredity, see Pearson (1914–30, 2:70–75).
10. Galton (1869, 31–32). A little earlier in the book, we read: "There is a definite limit to the muscular powers of every man, which he cannot by any education or exertion overpass. This is precisely analogous to the experience that every student has had of the working of his mental powers" (15).
11. Galton to Darwin, 24 December 1869, in F. Burkhardt et al. (1985–, 17:532), and in facsimile in Pearson (1914–30, vol. 1, insert between 6 and 7). I have tidied up the spelling a little. On Galton as someone

"whose intellectual life utterly hinged on Darwin's work," more so than any other Victorian's did, see Bynum (1993, 42).

12. Geddes (1889).
13. See, e.g., MacKenzie (1981a, 12, 29–36, 51–56, 71–72). For further references and discussion, see Waller (2004, esp. 142, 162–63).
14. See, e.g., Forrest (1974, 85); Cowan (1977, esp. 152–54); Fancher (1998); Waller (2004); and Mukherjee (2016a, 67). On Galton from the 1860s as responding to a perceived crisis in the established social order and its guiding science, classical political economy, see Renwick (2012, chs. 1 and 2, esp. 49–51).
15. See the extract from Galton's 1886 Royal Society gold medal acceptance speech, quoted in Pearson (1914–30, 2:201); see also Galton (1908, 288).
16. Galton (1908, 288); Herbert Spencer (1864, 1:238 [ch. 8] and 273 [ch. 10]); Galton (1869, 2; 1889, 1). In a letter Galton received in 1866, at the start of a serious mental breakdown (hence the gap between the articles and the book), his sister Bessy advised a long holiday, adding, "Your heredity will also be better by returning to it with a fresh eye and refreshed head." E. A. Wheler to Galton, 15 February 1866, quoted in Forrest (1974, 86).
17. Galton (1865, quotations on 165 and 166).
18. Galton (1865, 326–27); a similar view is expressed in Galton (1869, 64).
19. Galton (1869).
20. Galton (1865, 157, 160).
21. There are several standard histories of the rise of statistics (in both senses). The liveliest is Hacking (1990; on Galton, see ch. 21). On eugenic ambitions as stimulating and shaping the development of statistics in Britain from Galton to R. A. Fisher, see MacKenzie (1981a; on Galton, see ch. 3).
22. Galton (1869, 6–26).
23. Galton (1869, 26–49, esp. 26–33, quotation on 26). On Galton's using "normal curve" regularly by the late 1880s, see Hacking (1990, 184).
24. Darwin (1868, 2:7). Galton (1869, 2) referred to Darwin's puff at the start of *Hereditary Genius*, suggesting that, if the comparatively small 1865 data set had been sufficient to persuade Darwin that genius is hereditary, "the increased amount of evidence submitted in the present volume is not likely to be gainsaid."
25. Galton (1869, 363–73, quotation on 373). On Galton's law-of-error mathematizing of pangenesis, see J. Schwartz (2008, 11–12).
26. Galton (1869, 371–73, quotations on 373).
27. On the book's reception, see Gökyiğit (1994).
28. Darwin to Galton, 23 December 1869, emphasis in original, in F. Burk-

hardt et al. (1985–, 17:530–31), and in facsimile in Pearson (1914–30, vol. 1, insert between 6 and 7); Darwin (1871a, 1:104, 111, 168, 171, 173, 177–79). Note that in discriminating zeal and hard work from intellect, Darwin was echoing Galton (1869, 84), who saw ability as made up of those three components; the two men were not as far apart on "hereditary genius" as might appear.

29. Galton (1871a); Darwin (1871b). Galton (1871b) ingratiatingly suggested that Darwin was like the leader-monkey and Galton a brave but misguided follower-monkey out of the origin-of-language scenario in the *Descent*. Pangenesis letters from a number of people featured regularly in *Nature* throughout May 1871. For the whole of the Darwin-Galton correspondence on the transfusion experiments, see Pearson (1914–30, 2:156ff). On Galton's subsequent experiments, see Galton (1875a, 90).

30. Galton (1865, 321–22, quotation on 322, emphasis in original, and see 319 for "mongrel"). Cf. Galton (1869, 349, 370ff). Pearson (1914–30, 2:169–70) read Galton's shifts roughly as I do. Galton's annotations on his copy of *Variation* show his querying attitude toward the inheritance of acquired characters as well as his admiration for pangenesis; see Cowan (1969, 63, 96–99; 1977, 167–68). In the Victorian period, that attitude was a rarity (see Lidwell-Durnin 2020); besides Galton, the only other notable men of science who shared it were Alfred Russel Wallace and the ethnologist J. C. Prichard. Many thanks to Jon Hodge for helpful discussion of this point.

31. Galton (1865, 158–59, 318–19).

32. In the February 1877 Royal Institution lecture discussed in the next section, Galton (1877, 493) mentioned having used the device in that same venue, to illustrate different points, three years before. The lecture he gave there in February 1874 was titled "Men of Science, Their Nature and Nurture"; the write-up in *Nature* told of his illustrating the law of error "by many experiments" [Galton] (1874b, quotation on 345). For a transcription of the writing around the original 1873 quincunx in fig. 2.2, now held at University College London, see Stigler (1986, 277).

33. In his 1877 lecture (494, 512), Galton compared the device to a gardening tool called a harrow—a board with rows of little spikes sticking out, used to smooth soil. The name "quincunx" attached itself sometime after Galton, in *Natural Inheritance* (1889, 64), referred to the four-corners-plus-one-in-the-middle "quincunx fashion" in which the pins were arranged. I learned from John Turner that, like harrows, quincunxes were commonly encountered in the English countryside, where they were a favored pattern for arranging trees in an orchard.

34. Even in the specialist scholarly literature, Galton's hereditarianism tends to be assumed and then treated as in need of explanation; see, e.g., Waller (2002) and Bulmer (2003, 67–71).
35. Galton (1872a, quotation on 395–96, emphasis in original; 1873a; 1875a [see 81 for "stirp" defined]; 1876 [almost the same as the 1875 paper, but with some modifications and additions]). Cowan (1969, 120–43) is still the best guide to these papers, though Cowan (1977) includes an interesting supplemental diagram (175). Galton had used the representative government analogy in *Hereditary Genius* (1869, 367–68), also in connection with patent and latent elements, but to a somewhat different purpose. It must be said that Darwin never found Galton's physiological views intelligible, let alone convincing. In correspondence in late 1875 over Galton's latest paper, they largely talked past each other, though one of Galton's letters included a passage—on the various ways in which gemmules for blackness and whiteness might combine to produce cells that are gray, or black, or white, or any other tint in between—that went on to have an "anticipation of Mendel" afterlife. See Pearson (1914–30, 2:187–91; Lints and Delcour (1968); Olby (1985, 55–57).
36. Galton to Candolle, 27 December 1872, in Pearson (1914–30, 2:135–36, quotation on 135); also in Fancher (1983b, 348). Galton was sent a pre-publication copy of Candolle's book (1873). The fullest discussion of that book and Galton's response to it—which included a review (1873b)—remains Fancher (1983b). The Candolles had featured on Galton's list of eminent lineages in *Hereditary Genius* (1869, 210). In fairness to Galton, he had there boasted that his research, in undermining the idea that "each man is an independent creation," revealed each instead to be "a mere function, physically, morally, and intellectually, of *ancestral qualities and external influences*" (1869, 305, emphasis added).
37. Maxwell (1873, quotation on 822). On Maxwell's lecture as one of a series of responses by him to Galton's work in the 1870s, see Radick (2011, 130–32).
38. Galton (1874a, quotation on 260). For a detailed study of this book and the research it reports, see Hilts (1975). On Galton and the professionalization of British science, see Waller (2001b).
39. Galton (1874a, 30, emphasis in original). Indeed, quincuncial thinking can help in making sense of that curious phrase "what would have been the case." Given a quincunx with *n* pellets at rest, piled up at the bottom in the shape of a bell, the pellets represent—on the interpretation ventured here—the aggregate result of *n* separate developmental trajectories from initial embryonic averageness. For

Galton, then, an ordinarily bell-curvy population was as if produced by an enormously prolific but otherwise average couple, whose germinal contributions for the character in question destine their offspring for averageness; for only in that situation might the differences in outcome be put down to causes beyond the initiating one. As Galton wrote (1877, 512, emphasis in original), describing the 1874 vintage quincunx, "The essence of the law is that differences should be wholly due to the collective actions of a host of independent *petty* influences in various combinations, as was represented by the teeth of the harrow, among which the pellets tumbled in various ways."

40. Galton (1874a, 12, 39).
41. Galton (1873a, 116). On how Galton came up with "nature and nurture"—and the lack of evidence for any inspiration from *The Tempest*, notwithstanding Shakespeare's use of "nature" and "nurture" in close proximity in describing Caliban (4.1.179–80), as well as Galton's high opinion of Shakespeare on twins—see Fancher (1979). For a pre-*Tempest* "nature"-"nurture" conjoining by the English schoolmaster Richard Mulcaster in 1581, see Pinker 2004, 5.
42. Galton (1874a, 12–16, quotations on 12, 14). Galton had briefly discussed twins in "Hereditary Talent and Character" (1865, 323).
43. Galton (1875b, quotation on 576). On Galton's twin research, including the surprise bordering on mistrust with which he regarded his results, see Burbridge (2001, esp. 325–30). The more familiar idea of establishing nature-nurture relations by looking at how a given inborn constitution develops under different environmental circumstances can be found in *Hereditary Genius* (1869, 38), where Galton asserted that, even if they had been "changelings," most of the men who rose to eminence in a given period would, had they lived long enough and stayed healthy, still have risen to eminence.
44. Darwin to Galton, 10 November 1875, in Pearson (1914–30, 2:188–89, quotation on 188).
45. For outstanding technical treatments of this period in Galton's work—routinely described as one of "breakthrough" in his statistical innovations—see MacKenzie (1981a, 56–68) and Stigler (1986, 273–99). Once the quincunx was around, it became available for new uses, and never more dramatically so than in Galton's 1877 lecture, now celebrated as inaugurating not merely a statistical approach to selection (see Gayon [1998, 103, 131, 154–61]) but statistical explanations generally (Hacking [1990, ch. 21]; Stigler [2010]; Ariew, Rohwer, and Rice [2017]).
46. Galton (1877, quotations on 492).
47. Galton (1877, esp. 512–13). On Galton's sweet-pea experiments, see

Pearson (1914–30, 3A:3–7); on Darwin's participation in 1875, see Pearson (1914–30, 2:180–81, 187, 189).

48. Galton (1869, 364–68, quotation on 367).
49. Galton (1869, 368–70, quotations on 369).
50. Galton (1883, quotation on 14, referring to the frontispiece). For his recounting of the history of composite portraiture up to that point, see Galton (1883, 339–48). Galton's claim to independent invention was contested; see Tucker (1997, 397).
51. Galton (1883, 24–25). For "viriculture," see Galton (1873a, 119). Curiously, "stirpiculture," though seemingly an obvious coinage for post-1875 Galton, was in use by 1865 by the American religious radical John H. Noyes. Although he developed and named the idea independently of Galton, Noyes was inspired by *Hereditary Genius* to put his human stockbreeding ambitions into practice in his community at Oneida, New York; see Gayon (1998, 429–30n45). Like "heredity" in the 1860s, "eugenics" in the 1880s was not absolutely new to English, though it may as well have been.
52. Galton (1883, 305–6).
53. Galton (1883, 307, 181–82).
54. For the eugenic remarks in *Natural Inheritance*, see Galton (1889, 197–98, quotation on 198); see also his remarks on heredity and environment (esp. 8–11, 195) and on regression (esp. 95–110, 194–95). Valuable *explications des textes* include Roberts (1929, 241–50) and Gillham (2001, 258–67).
55. Galton (1889, 18–27, quotations on 20–21). For a superb discussion of this chapter in the context of Galton's writings on organic stability generally, see Gould (2002, 342–51).
56. Galton (1889, 27–28).
57. Galton (1889, 28–34). On Galton's Darwinism and the ambiguous status of natural selection within it, see Bowler (2014, esp. 276–79).
58. Galton (1908, 295–96, reflecting on his 1884 paper). For this paper as a parting shot in the exchanges with the by-then-deceased Maxwell, see Radick (2011, 132).
59. Galton (1889, 9). See also Galton (1885, 1213–14).

Chapter Three

Epigraphs: B. Bateson (1928a, 20), also in G.1.a.79 and G.1.b.26, BP; Weldon, MS "Clip 8," in Pearson/5/2/9/8, PP. Versions of the latter can be found in Weldon (1901a, 1) and Weldon (1902c, 633–34).

1. Darwin (1859, 434–39, quotation on 434–35). For morphology from

Goethe to Darwin, an excellent brief guide is Richards (2008, appendix 1). For morphology after Darwin, see Bowler (1996).

2. Letter from Weldon to W. Weldon, 18 March 1880, in Pearson/5/1/3/4, PP.
3. Bateson (1922, 389).
4. Much more has been written about Bateson than about Weldon. The standard source for Weldon's life and work is Pearson (1906). The best modern treatments are Farrall (1969, ch. 3) and Olby (1989a, also outstanding on Bateson). The indispensable Bateson biographies are B. Bateson (1928a, 1–160) and Cock and Forsdyke (2008). On Bateson's scientific research up to the mid-1890s, see also Peterson (2008); Richmond (2008, 214–23); and Bowler (1992).
5. Letter from Weldon to "the most noble . . . ," 6 March 1885, SZN. On the Stazione Zoologica in Naples and the scientific culture it begat, see De Bont (2014) and the contributions to a January 2015 special issue of *History and Philosophy of the Life Sciences* (Dröscher 2015b).
6. On Cambridge before the Maxwell-Foster era, see Garland (1980). On Foster's reforms in biology, see Geison (1978). On the heyday of Balfourian morphology, see H. Blackman (2004, 2007a,b).
7. A superb discussion of Balfour's Darwinism can be found in Bowler (1989b, 290–92). The key Balfourian paper is Balfour (1875, esp. 112–13). On Balfour's life and work generally, see B. K. Hall (2004).
8. Balfour (1880, 637). A nearly identical passage appears at the start of the first volume of his *Treatise on Comparative Embryology* published that same year. See Balfour (1885, 2:2–3).
9. Darwin to Balfour, 4 September 1880, available on the Darwin Correspondence Project website (letter 12706). As a student friend of one of Darwin's sons, Balfour had visited Down House in late 1872 and told Darwin about some pangenesis-testing skin transplantation experiments he had been conducting with brown and white rats. Darwin was hugely impressed, judging Balfour "very clever and full of zeal for [Biology]": Darwin to Galton, 30 December 1872, in Pearson (1914–30, 2:176).
10. W. F. R. Weldon to Anne Weldon, 27 April 1878, 2 March 1879, 28 May 1879, in Pearson/5/1/3/4, PP; W. F. R. Weldon to Dante Weldon, 9 March 1879, in Pearson/5/1/3/5, PP. On Garrod's teaching and possible influence, see Pearson (1906, 7–8). Garrod was the older brother of the much more famous Archibald Garrod, a medic and physiologist who, as we shall see, became important for Mendelism and medicine.
11. W. F. R. Weldon to Anne Weldon, 9 February 1880, 29 February 1880, and 4 April 1880, in Pearson/5/1/3/4, PP; Pearson (1908, 7–8). There were, confusingly, two Adam Sedgwicks active in Cambridge science

in the nineteenth century: a geologist, who mentored the student Darwin in the 1820s and died in 1873; and a morphologist, who came to Cambridge as a student in 1874 and was great-nephew to the geologist.

12. Weldon to Anne Weldon, 4 April 1880, in Pearson/5/1/3/4. Cf. B. Bateson (1928a, 13) on the Cambridge by-election of 1882 as, supposedly, the only one in which Bateson involved himself—a view largely adopted by Cock (1983, 28—though he did note the evidence to the contrary).
13. Bateson to G. H. Fowler, 24 June 1906, G.5.b.16, BP; quoted in B. Bateson (1928a, 103).
14. Documents from these years can be sampled in B. Bateson (1928a, 3–10), and more extensively in G.3.g, BP.
15. On Rugby School's culture and its influence on Victorian Britain generally, see Heffer (2013, esp. 1–30).
16. *Report of the Fiftieth Meeting of the BAAS* [. . .] (1880, x, and in the List of Members, 78 and 83). Many reports of BAAS meetings are archived at the Biodiversity Heritage Library, https://www.biodiversitylibrary.org/item/93735#page/3/mode/1up. See also the chronology of W. F. R. Weldon's life penned by Florence: Pearson/7/29/3, PP, 2. She had also accompanied the Weldons to the 1877 BAAS meeting in Plymouth (1). On BAAS meetings in the nineteenth century from the perspective of the women who attended, see Higgitt and Withers (2008). On Weldon's parents, Walter and Anne Weldon, see "Obituary" (1885) and F. W. R. (1889).
17. Weldon to Dohrn, 2 February 1882, SZN; Weldon to Balfour, 3 June 1882, in the Balfour Family Papers, National Archives of Scotland, microfilm GD433/2/103B/97, cited in H. Blackman (2007a, 418–19n20. For Balfour's wall-lizard research, see Balfour (1879); for Weldon's, see Weldon (1883). This work led Weldon for a time to specialize in the comparative morphology of the vertebrate kidney; see Weldon (1884a,b, 1885). For SZN letters, I thank Maurizio Esposito.
18. Lankester to an unnamed correspondent, 1 August 1882, quoted in Lester (1995, 103). On the reaction to Balfour's death generally, see H. Blackman (2004).
19. On the studentship, see R. MacLeod (1994). On the new posts and their occupants, see Geison (1978, 378–79). On the laboratory, see Richmond (1997, with an 1884 testimonial from Weldon quoted on 444). On the expansion of morphological teaching at Cambridge through the 1880s, see H. Blackman (2007b, esp. 89–99). For a list of Weldon's Cambridge students, see Pearson (1906, 15n).
20. Weldon to Thompson, 25 April 1885, TP; Pearson (1906, 11–12). According to Pearson, Weldon was sometimes an effective speaker for the

general public, and sometimes not. The novelist and writer on evolution Samuel Butler attended this lecture and found it "very dull," according to H. F. Jones (1919, 2:5, quoted in Cock and Forsdyke [2008, 101]). But Butler was only one of about 250 people attending, and the fact that Weldon was invited back (on 9 February 1894) suggests he did not disgrace himself. Many thanks to Frank James for information and discussion, and Maurizio Esposito for the letter.

21. Bateson (1913, 39–40). For further discussion, see ch. 5 of this book.
22. On the morphologists on *Balanoglossus*, see B. K. Hall (2005, esp. 1–4). For Balfour's treatment, see Balfour (1885, vol. 2, ch. 21). On the vertebrate origins debate generally, including superb coverage of the contributions of Balfour, Bateson, and Brooks, see Bowler (1996, ch. 4). On Weldon's role in Bateson's investigations, see B. Bateson (1928a, 102); Bateson (1885b, 116).
23. On Bateson's time with Brooks, the main source is Bateson (1910), but see also B. Bateson (1928a, 17–18), extracting his letter to his mother from Hampton on 9 and 20 July 1882, the complete version of which can be found at A.1.a.1, BP. I follow Bateson himself in crediting Sedgwick, though his wife credited Weldon with "help and encouragement" in applying to Brooks. On Brooks and his significance, see, for contemporary testimonials, "William Keith Brooks" (1910) and Conklin (1913), and for historical assessments, Benson (1985, 2010) and McCullough (1969).
24. Agassiz (1881, esp. 413 ["The time for genealogical trees is passed"]); Brooks (1882–83). On Agassiz's address—and Darwin's reply—see Bowler (1996, 81–82). So far as I know, Bateson's skepticism about evolutionary genealogies has not before been traced to Agassiz's.
25. Brooks (1883; for "saltatory evolution," see 83, 296 [discussing Galton's stone]). Brooks attributed the expression to the Harvard-trained malacologist William Healey Dall, who coined it in an 1877 paper; see Lindberg (1998). An excellent discussion of Brooks's book can be found in Robinson (1979, ch. 5).
26. Bateson (1886b, 1). Bateson's closing words are a rendition of Brooks's closing in his "Speculative Zoology" (1882–83, 380).
27. Bateson (1886b, 2–14, quotation on 3). Cf. Brooks (1883, 307–11). Bateson tried out a version of the argument in a talk in Cambridge in November 1885 (1885c). For his *Balanoglossus* publications, see Bateson (1883, 1884a–c, 1885a–c, 1886a,b).
28. W. Bateson to B. Bateson, 16 April 1906, quoted in B. Bateson (1928a, 102); Punnett (1950, 2); Weldon chronology, 3–4, Pearson/7/29/3, PP. Bateson referred to biological discussions with Weldon in a note to D'Arcy Thompson, 13 January 1886, TP.

29. Weldon was elected a fellow in November 1884 and Bateson in November 1885; see *Nature* (13 November 1884), 46; (5 November 1885), 22.
30. On the aims of the steppe expedition, see Bateson's funding application to the Royal Society for £200, 25 March 1886, *Government Grant Applications and Reports 1884–1896*, 2:784–85, RS.
31. Brooks (1883, 89–93, esp. 91–92). Useful summaries of Schmankewitsch's papers, with citations, can be found in the *Annals and Magazine of Natural History* 17 (1876): 256–58; Bateson (1894, 96–97); and Vernon (1902, 271–75). I am grateful to Elena Aronova and Alexei Kouprianov for information on Schmankewitsch.
32. The Cambridge board that awarded Bateson £50 for 1886 and, if needed, £50 for 1887 noted the "great value" of his well-known *Balanoglossus* work; *Nature* (6 May 1886), 19.
33. W. Bateson to Anna Bateson, 22 November/4 December 1886, in B. Bateson (1928b, 128–33, quotation on 132 [though I have modified the spelling]); also G.1.a.37 and G.1.b.27, BP.
34. W. Bateson to his mother, mid-September 1886, in B. Bateson (1928b, 102–6); Bateson (1889).
35. W. Bateson to his mother, 27 September/7 October 1886, in B. Bateson (1928b, 107–10, quotation on 109). In the letter cited in note 33 above, he admonished Anna (132): "Don't you jeer at my 'geometrical biology'!"
36. W. Bateson, 10/22 November 1886, in B. Bateson (1928a, 19–21).
37. Weldon (1887a, quotation on 149–50, emphases in original).
38. As Weldon put the point in a grant application submitted to the Royal Society that March for funds to continue his Bahamian research: "The importance of larval variation of this kind is great, as bearing on the influence of external stimuli on the transmission of hereditary character, and also as possibly affording a clue to the laws and limits of larval variation in general." *Government Grant Applications and Reports 1884–1896*, 2:981–82, RS.
39. Weldon to Mayer, 25 May 1887, SZN.
40. Weldon (1887b).
41. Weismann (1889–92) presents his 1880s papers in English translation, including the germ-plasm continuity paper (ch. 4). For the previously appearing abstracts in *Nature* and elsewhere, see xii. On Weismann's theorizing, see Robinson (1979, esp. chs. 7 and 8).
42. On the Manchester discussion, see the editors' preface to Weismann (1889–92, ix) and Poulton (1937, 400–402). For Weismann as no Weismannian when it came to the conditioning effects of the environment, see Winther (2001).

43. The best survey of the long run of debate over Lamarckian inheritance remains Bowler (1983, ch. 4).
44. See *Report of the Fifty-Seventh Meeting of the BAAS* [. . .] (1887, 736, 740, 755).
45. W. Bateson, 10/22 November 1886, in B. Bateson (1928a, 19–21, quotation on 20, emphasis in original). The hydrangea case remains a textbook example of how differences in environment can elicit different phenotypes from a single genotype (see, e.g., Campbell [1993, 271]), illustrating the norm-of-reaction concept discussed later in this book.
46. In the sole publication from the steppe expedition, Bateson (1889, 34) wrote that he expected cockle structure to track salinity levels back and forth, but did not know "the length of time and the number of generations necessary to effect these changes"; cf. Provine (2001, 39).
47. W. Bateson to Anna Bateson, 22 December 1886, G.1.a.38, BP. Earlier letters instructed Anna to thank Weldon for his letters (5 June 1886, G.1.a.35, BP) and to spell his name properly in her own (22 November/4 December 1886, G.1.a.37, BP).
48. W. Bateson to M. Bateson, 25 January 1887, G.1.a.50, BP.
49. W. Bateson to his mother, 21 January 1887, G.1.a.82, BP, and 20 March 1887, G.1.a.85, BP.
50. W. Bateson to his family, 30 May/11 June 1887, G.1.a.88, BP. Around this time, Bateson also wrote about his sadness on hearing some bad, presumably medical, news about Mrs. Weldon: Bateson to M. Bateson, 23 June 1887, G.1.a.51, BP.
51. Bateson (1888); Weldon to Bateson, 7 March 1888, G.4.g.4, BP. For "Is everything adapted?," see Eimer (1890, 63).
52. Weldon to Bateson, 26 June 1888, G.4.g.5, BP.
53. W. Bateson to Anna Bateson, 2 September 1888, in B. Bateson (1928a, 38–39).
54. *Report of the Fifty-Eighth Meeting of the BAAS* [. . .] (1888, 692–94, quotation on 692). There is no evidence that Bateson or Weldon attended, though nothing would have been more natural. Again, Weldon was, like his father, and with Florrie at his side, a BAAS regular; while for Bateson, a member since 1884, the meeting would have represented a matchless chance for reentry into the British scientific scene.
55. Weismann (1888, esp. 431–34).
56. Weldon to Bateson, 28 September 1888, B.13, BP. For a somewhat different gloss on this letter, see Olby (1989a, 309–10). Although Pearson of course had no knowledge of this letter, he reported Weldon's working on the laws governing larval-adult evolution in 1888 because Florrie recalled Weldon telling her then about his ideas; see Pearson

(1906, 13) and the Weldon chronology, 6, Pearson/7/29/3, PP, with discussion in Gayon (1998, 198–99).

57. On the establishment of the Plymouth laboratory—which was largely the work of Lankester—see Lester (1995, ch. 9), and the first issue (August 1887) of the *Journal of the Marine Biological Association.*
58. Weldon was one of the signatories to an 1886 MBA-related memorandum (*Nature* [24 June 1886], 179–81) calling on the government to create a new, scientifically oriented and managed Fishery Department. He came to the under-construction laboratory in November 1887, putting in several months of work on the Royal Society–funded crustacean studies before the summer was over; see the *Journal of the Marine Biological Association*, o.s. 1, no. 2 (August 1888), 117, 144.
59. *Athenaeum* (21 April 1906): 485. For the Weldons' trips to France in the later 1880s "on chlorine business," see the Weldon chronology, 4–5, Pearson/7/29/3, PP.
60. Not long after his father's death, Weldon wrote to Thompson about wanting to "begin seriously with some Crustaceans"; Weldon to Thompson, 7 November 1885, TP.
61. Notebook, "Decapod Crustacea of Plymouth Sound, 1888. Records of Prof. W. F. R. Weldon," PWE1, NMBL. Despite the year on the cover page, entries go back to 1887 (even before November) and forward to 1889.
62. Notebook, "Measurements of Crustacea (Palaemon & Pandalus)," Pearson/5/2/1/1, PP. The first entry is dated 5 November 1888.
63. See, e.g., Wallace (1889, ch. 3), which surveyed Darwinism-inspired quantitative studies of variation in nature, going back decades. Galtonian inspiration cannot be ruled out entirely, however. Galton was very active at the 1880 Swansea BAAS meeting, which Weldon in later life recalled as where he had first seen Galton in action; Pearson (1906, 8). And in 1886, the Royal Society awarded Galton its gold medal for his services in bringing together statistics and biology; Pearson (1914–30, 2:201).
64. Bourne (1906, 112); Weldon chronology, 5–7, Pearson/7/29/3, PP.
65. Galton (1889, 136–37).
66. Weldon to Galton, 11 June 1889 and 16 November 1889, Galton/2/5/1/6, GP.
67. Weldon's paper is not preserved; the reconstruction here is based on remarks in the correspondence between Weldon and Galton in January–March 1890 and the subsequently published revision of the paper.
68. In a February 1890 letter to Galton, Weldon wrote that his expanded data set undoubtedly bore out "the truth of your remarks on Natural

Selection": Weldon to Galton, 16 February 1890, Galton/3/3/22/17, GP; see too note 6 in the next chapter.

69. According to the Plymouth lab's journal, Weldon was there 20 March–24 April and then more or less continuously from 19 June, with Bateson there from 1 April. Although Bateson's appointment finished at the end of September, he returned in October and stayed for another month. *Journal of the Marine Biological Association* 1, no. 2 (October 1889): 118, and 1, no. 3 (April 1890): 222. Bateson's sense perception publications (1890a,b) became well regarded in the emerging field of comparative psychology; see, e.g., Hobhouse (1901, 65, 96–97).
70. Bateson (1889); Galton to Bateson, 14 October 1889, A.9.c.4, BP; Bateson to Galton, 14 October 1889, Galton/3/3/2/6, GP; Galton to Bateson, 16 October 1889, A.9.c.5, BP. Although the month of publication of the 1889 paper is not printed there, Bateson also sent a copy to a Plymouth colleague, Walter Garstang, in early October: W. Garstang to Bateson, 5 October 1889, G.4.g.1, BP. In correspondence with his family from the steppe, Bateson referred often to *The Art of Travel*, at one point even taking Galton's advice about not washing: W. Bateson to his mother, 27 September/7 October 1886, in B. Bateson (1928b, 107–10, 109); Galton (1872b, 122).
71. Bateson to A. Newton, 8/9 March 1889, A.1.b.6, BP. Bateson belatedly published the negative data from the steppe (1894, 96–101).
72. Bateson (1890c).
73. I quote from Bateson's unsuccessful 1890 application for the Linacre Professorship, printed in B. Bateson (1928a, 30–37, quotations on 36).

Chapter Four

Epigraphs: Lankester to Newton, 4 February 1899, Add.9839/1L/7–24, Newton Papers, Cambridge, emphasis in original; Weldon to Galton, 28 November 1899, Galton/3/3/22/17, GP; Pearson to F. Weldon, 19 October 1906, Pearson/5/3/8/1, PP.

1. On the Royal Society in the nineteenth century, see M. B. Hall (1984). On the criticisms of Babbage and others in 1830, see Babbage (1830) and Hall (1984, 45–51).
2. These and other reforms by no means dispelled all dissatisfactions with the Royal Society. For a complaint about the society's isolation, narrowness, and inactivity, see *Nature* 74 (6 September 1906): 466–68. On the role of medals in the emerging "reward system" for science, see R. MacLeod (1971).
3. For the quoted phrase, see the anonymous *Nature* review cited in the previous note (466).

4. Certificate of Election and Candidature [Weldon], EC/1890/16, RS; "The Ladies' Conversazione of the Royal Society," *Nature* 42 (26 June 1890): 210–11. On the society's election procedure, see M. B. Hall (1984, 81–82).
5. Galton (1889, 119–24). Galton's pages on natural selection come in the middle of a chapter about the data on stature.
6. See the letters from Weldon to Galton, 7 January 1890–16 February 1890 ("The curves must be taken as a complete demonstration of the truth of your remarks on Natural Selection [pp. 119–121 of Natural Inheritance]"), Galton/3/3/22/17, GP. On Galton's Royal Medal, see Pearson (1914–30, 2:201, 3B:476).
7. Galton (1889, quotation on 120); Weldon to Galton, 26 February 1890–16 March 1890, Galton/3/3/22/17, GP; Weldon (1890a, quotation on 451).
8. Weldon to Galton, 21 April 1890 and 14 May 1890, Galton/3/3/22/17, GP. Weldon's selection as a candidate for the fellowship was reported in the Royal Society council's minutes for 24 April 1890: *Royal Society Minutes of Council* 6 (1884–92): 302–4, 304, RS.
9. Galton (1888). For Galton on correlation, see Stigler (1986, 297–99). For Darwin on correlation, see, e.g., Darwin (1859, 11–12, 143–50; 1868, vol. 2, ch. 25).
10. Weldon to Galton, 19 February 1890, Galton/3/3/22/17, GP. On the evidence of this letter—the first in the surviving correspondence to mention correlation—Weldon was well on his way to completing the research published in his 1890 *Crangon vulgaris* paper before he had even heard of Galton's new test (which, again, did not appear in *Natural Inheritance*). Cf. Provine (1971, 30); Olby (1989a, 309); and Gayon (1998, 198–200), which, following Pearson (1906, 13–14), treat correlation as the bridge between Weldon's morphological and statistical work.
11. Weldon to Galton, 2 March 1890, Galton/3/3/22/17, GP. Later that year, Weldon and a Plymouth colleague published their lobster-rearing findings in the *Journal of the Marine Biological Association:* Weldon and Fowler (1890).
12. Weldon (1890a, 453).
13. Weldon (1892; 1893, quotation on 329).
14. On Pearson's "pre-statistical" life and thought, and the connections between them (and the times generally), Porter (2004) is indispensable. His extracurricular writings of the 1880s can be sampled in Pearson (1888; see 430–31 for a concise statement of his Darwinian beliefs). On Darwinian-socialist London, see Hale (2014, chs. 4–7). For Pearson's lecture on *Natural Inheritance*, see Karl Pearson, "On the

Laws of Inheritance According to Galton," Pearson/1/5/19/1, PP. On the University of London reform campaign that brought Pearson and Weldon together, see Porter (2004, 218); for the complex institutional backstory, see Barton (2018, 342–43). For Weldon's acknowledgement in the 1893 paper, see Weldon (1893, 324).

15. They knew each other well enough by November 1892 that Weldon's letter reporting his breakthrough in decomposing the Naples curve ended, "If you scoff at this I shall never forgive you"; Weldon to Pearson, 27 November 1892, Pearson/11/1/22/40, PP. In spring 1893 Pearson involved Weldon (and tried to involve Galton) in Pearson's Gresham Lectures on the laws of chance; see Magnello (1996, 51–52).
16. Pearson (1894a, republished along with many other papers from the series in Pearson [1948]). Up to 1904, Pearson gave numbers only to the parts published in the *Philosophical Transactions*. By then, an additional eight had been published in the *Proceedings* alone. Number 19 appeared in 1916. (17 never appeared.)
17. Pearson's article (1894b) was published in the February issue of the *Fortnightly Review*): Weldon to Galton, 4 February 1894, Galton/3/3/22/17, GP; see also the undated letter with data from Weldon to Pearson in Pearson/5/2/1/3, PP, and Weldon to Bateson, 15 February 1894, B.13, BP. On Pearson's Monte Carlo empiricism, see Porter (2004, 253–54). On "Weldon's dice," see Labby (2009). Weldon told Galton that the results were for use in an upcoming Royal Institution lecture, which had taken place by the time he wrote to Bateson.
18. Weldon to Galton, 16 February 1890, Galton/3/3/22/17, GP.
19. Weldon to Galton, 19 February 1890, Galton/3/3/22/17, GP.
20. Weldon to Galton, 18 September 1890, Galton/3/3/22/17, GP; Weldon (1890b).
21. Certificate of Election and Candidature [Bateson], EC/1894/05, RS; B. Bateson (1928a, 29–37, 42). Bateson consulted extensively with Weldon and many others before applying for the Oxford position; see the correspondence preserved in G.3.m, BP.
22. Bateson to Sedgwick, 8 October 1890, and Sedgwick to Bateson, 9 October 1890, G.3.i.1–2, BP.
23. B. Bateson (1928a, 51); Certificate of Election and Candidature [Bateson], EC/1894/05, RS; Royal Society council's minutes for 10 May 1894: *Royal Society Minutes of Council* 7 (1892–98): 91, RS. The quotation is from the report of Bateson's election in the *Star*, 7 June 1894, G.3.p.13, BP.
24. Bateson and Bateson (1891, quotation on 157). No abstract or other text was printed from the reading of the paper at the 10 November 1890 meeting in Cambridge; see *Proceedings of the Cambridge Philosophi-*

cal Society 7 (1889–92): 96. The Linnean Society reading took place on 2 April.

25. "The Royal Society Conversazione," *Nature* 44 (25 June 1891): 187–88, 188. On what came to be called "Bateson's Rule," see G. Bateson (1972, quotation on 380).
26. Bateson (1891; 1892a, quotation on 163; 1892b–d). The book that incurred Bateson's wrath was Poulton's *The Colours of Animals* (1890). The ill-tempered correspondence between Poulton and Bateson from October–December 1892 can be found in G.3.n, BP.
27. Bateson (1892e, esp. 181–82, 192).
28. Bateson and Brindley (1892, quotations on 197, with dimorphism [a long-established term in natural history] characterized in terms of a two-peak error curve on 194).
29. A great deal of Bateson's scientific correspondence while he was preparing *Materials* can be found in G.3.q, BP, esp. items 37 ff. For his correspondence with Macmillan, see G.7.l, BP.
30. Bateson (1894, esp. 1–80, with "merism" defined on 20). The preface is dated 29 December 1893. The date of publication was 6 February 1894; see the letter from Macmillan to Bateson, 6 February 1894, G.7.1.21, BP. The most thorough and thoughtful commentary on the book is in Gould (2002, 398–409).
31. Bateson (1894, 70, emphasis in original).
32. Another, now more famous book with the same mission of forging a Darwinism-free, geometrized, forces-and-motion biology is *On Growth and Form* (1917), by Bateson's Cambridge contemporary D'Arcy Thompson. I learned from Matt Holmes that Thompson's first public expression of skepticism about Darwinism was at the 1894 BAAS meeting—the first such meeting after *Materials* came out.
33. Bateson wrote his recipient wish list on the 6 February 1894 letter from his publisher, who in reply confirmed the number to be sent out: G.7.l.21–2, BP. Also on Bateson's list were Huxley, Lankester, and the physiologist Charles Sherrington, who had responded enthusiastically to a draft of the introduction, praising the exposure of "the fallacies of the Adaptation priesthood"; Sherrington to Bateson, 14 August 1892, A.9.(a).6, BP.
34. See, e.g., Weldon to Bateson, 23 December 1890, G.4.k.4, BP, sending a drawing of the insides of a hagfish he had found with an extra gill pouch (reproduced, with credit, in Bateson [1894, 173]); and undated notes from Weldon, G.4.g.7 and G.4.g.21, BP, with translated extracts and an image comparing normal and parasitized crab abdomens, drawn from the work of the French naturalist Alfred Giard (the image is reproduced, with no mention of Weldon, in Bateson [1894, 95]).

35. "In the last few evenings I have wrestled with a double humped curve, and have overthrown it." Weldon to Pearson, 27 November 1892, Pearson/11/1/22/40, PP. The 1892 *Proceedings of the Zoological Society of London* indicate that the dimorphism paper was read (and received) on 15 November. The very possibility that an asymmetrical curve masked two peaks would not have been news to Pearson, who raised it in his 1889 lecture on *Natural Inheritance:* MS p. 19, Pearson/1/5/19/1, PP.
36. Weldon (1893, 329); Bateson and Bateson (1891, 126–27).
37. Weldon to Bateson, 15 February 1894, B.13.6, BP; Weldon (1894a).
38. Weldon to Galton, 27 November 1892, Galton/3/3/22/17, GP.
39. I have synthesized points made across Weldon's review and letters, which, in addition to the 15 February 1894 letter cited above, include letters to Bateson on 13 February 1894, G.3.q.8, BP, and on 1 and 6 March 1894, B.13.3 and B.13.5, BP—the latter responding to a reply from Bateson on 4 March 1894, B.13.4, BP—as well as a letter to Sedgwick on 23 February 1894, MS Add.9967/93, Correspondence of Adam Sedgwick, Cambridge University Library. Many thanks to Matt Holmes for discovering (and copying) this last letter.
40. Over thirty print notices of the book are preserved in G.3.p, BP. The *Westminster Review* review (July 1894) is item 17; the *Science* review (4 January 1895) is item 31. Within British periodicals, the most critical "first wave" review, apart from Weldon's, was the zoologist Peter Chalmers Mitchell's in the May 1894 *Natural Science* (item 11). Sales records can be found in G.4.b, BP.
41. Galton (1894, quotations on 362, 368, 372). Galton's reading notes on the book are in Galton/2/5/4/1/5, folio pp. 5–8, GP. Galton was better prepared for the endorsement of his views than he let on, as Bateson had sent him a copy of the earwigs paper the previous year, and Galton in reply had proposed a selective breeding experiment to test whether the high and low forms really were non-blending varieties: Galton to Bateson, 22 April 1893, A.9.c.7, BP.
42. Weldon to Galton, 13 July 1894, Galton/3/3/22/17, GP.
43. The main sources for reconstructing the committee's prehistory and early history are Weldon's letters to Galton, from 18 November 1893 onward, Galton/3/3/22/17, GP; the minute book for the committee, CMB/65, RS; and "Miscellanea" from the committee, MM/15/72–103, RS. Poulton and the Cambridge medic-mathematician Donald Macalister—who helped Weldon with statistics pre-Pearson (MacKenzie 1981a, 235)—were also founding members, but were not present at the first meeting, whose minutes record nothing about Weldon's research on selective destruction in crabs (CMB/65/1, RS). On the beginnings of that project, see Weldon to Galton, 14 July 1893,

Galton/3/3/22/17, GP ("As I told you, I hoped to measure the selective death-rate in Crabs during this summer. I have begun to do so: and the results are extraordinary")—a letter that puts paid to the claim in Hale (2014, 319) that the crab selection study was Weldon's response to an exchange of fire with George Romanes in *Nature* in late spring 1894. On this exchange, see note 72 below.

44. Weldon (1895a). On the paper's mathematical reasoning, see Norton (1973, 301–2). Weldon was fascinated by a parallel he saw between his new selection mathematics and stimulus-response mathematics in physiology and psychophysics; see Weldon (1895a, 371–74), as well as a number of letters to and from Pearson, e.g., Weldon to Pearson, 27 September 1893, Pearson/11/1/22/40, PP; Weldon to Pearson, 29 September 1893, Pearson/5/2/1/2/1, PP; Pearson to Weldon, circa early February 1895, and Weldon to Pearson, 12 February 1895, Pearson/5/2/4/1–2, PP.
45. For a perceptive appreciation of Weldon's achievement in the paper, see Gayon (1998, 211–15).
46. For the minutes of the 1893–94 Procedure Committee, along with the initiating letter from Foster (the Royal Society's senior secretary) and its final report, see *Committees* [. . .] *Procedure*, CMB/43/3, RS. For the council's subsequent actions, see *Royal Society Minutes of Council* 7 (1892–98): 114–15, 137, 140, 142–43, RS.
47. Weldon to Galton, 7 February 1895, Galton/3/3/22/17, GP; *Royal Society Minutes of Council* 7 (1892–98): 143, 146, RS.
48. "Science Gossip," *Athenaeum*, 9 March 1895, cutting in G.3.n.9, BP.
49. Weldon (1895b).
50. Vernon (1895); for Weldon's report on it, see 30 January 1895, RR.12.399, RS. On Vernon, see Bedford (1951). For Weldon's ill-fated water flea (*Daphnia magna*) work, see the entries for October 1894 in the notebook titled "*Daphnia* 1894–5," Pearson/5/2/1/6/1, PP; Weldon to Galton, 29 October 1894, 12 and 13 November 1894, 23 January and 28 March 1895, Galton/3/3/22/17, GP; Weldon to Pearson, 22 and 30 December 1894, Pearson/11/1/22/40, PP.
51. See, e.g., Weldon on how the appearance of right-dentary-margin sports goes away once one looks more closely and/or knows more about how crab parts break off and regenerate (1895a, 376–78); on how no speculations about adaptation were used in the making of the new knowledge about natural selection (1895a, 379; 1895b, 381); and on the link between museum-going and a stress on sports (1895b, 380).
52. Thiselton-Dyer (1895, quotations on 461). On Thiselton-Dyer, see Geison (1976).
53. Bateson (1895a).

54. See the letters published in *Nature* 52 (1895) on 2 May (Thiselton-Dyer), 9 May (Bateson), 16 May (Weldon and the botanist William Hemsley), 23 May (Thiselton-Dyer), 30 May (Bateson and Weldon), and 6 June (Thiselton-Dyer and Weldon). The *Gardeners' Chronicle* reported on the debate in its 11 May 1895 issue (588). An indirect contribution on the Darwinian side came from Alfred Russel Wallace (1895), who, at just this moment, published a two-part attack on Bateson's and Galton's views.
55. The quotation is in a letter from Weldon to Galton, 31 May 1895, Galton/3/3/22/17, GP. The Weldon-Bateson correspondence throughout and after the *Cineraria* controversy is in B.10, BP. Despite some inaccuracies, the coverage of the controversy in Provine (1971, 45–48) remains valuable.
56. Pearson (1892, quotations on 33, with discussion of ether-squirts on 318–23 and Weismannian metaphysics on 397–99). In his 1889 lecture on *Natural Inheritance*, Pearson had noted that Galton's methods "do not take us into the metaphysical regions of germ plasm," and that his conclusions sat as comfortably with Darwin's pangenesis as with Weismann's theory; MS p. 2, Pearson/1/5/19/1, PP. In *The Grammar of Science*, there is no mention of Galton, except an indirect and disobliging one, when Pearson snickers at the idea that anyone ever thought that they could dispatch pangenesis—so openly hypothetical and imaginatively economical—by failing to find gemmules in blood (397). On Pearson's philosophy of science in principle and practice, see Gayon (2007) and Porter (2004, esp. ch. 7). Invaluable for the study of Pearson are Porter's book, the biographical E. Pearson (1938), and the bibliographic Morant (1939).
57. Pearson (1894a, quotation—in the 1948 reprint—on 29, emphasis in original, citation of Bateson on earwigs on 30).
58. George Darwin, report on Pearson's "Contributions," 9 November 1893, RR.12.11, RS.
59. Francis Galton, report on Pearson's "Contributions," 22 November 1893, RR.12.12, RS. Stephen Stigler (1986, 332) describes the memoir's algebra as "frightening." It nevertheless introduced two enduring innovations: the use of the term "standard deviation" for the *x*-axis value marking the halfway point in normal-curve variation—a measure of curve width (see Stigler [1986, 328] for Pearson's earlier use of the term in his lectures); and what became Pearson's signature method of curve fitting, the "method of moments," creatively importing techniques from mechanics. For discussion, see, in addition to Stigler, Yule and Filon (1936, 80); Magnello (1993, 72–80); and esp. Porter (2004, 239–41), who brings out the links between the method

of moments and Pearson's earlier interest, as a teacher of engineers, in graphical statics as a means for making mathematics more visual.

60. Pearson's first statistical publication was an October 1893 letter in *Nature* sketching part of the more general treatment of asymmetrical curves that he went on to publish (1895). On his 1893–94 Gresham Lectures as launching the new evolutionarily directed statistics, see Magnello (1996, 59–61).
61. Pearson (1895a, quotation—in the 1948 reprint—on 43, with analysis of Weldon's best-case data set on 82–83). On this memoir, see Stigler (1986, 333–36); Magnello (1993, 100–108); and Porter (2004, 244–45).
62. Galton (1895). A note at the bottom indicates that Pearson's memoir was read on 24 January 1895.
63. Certificate of Election and Candidature [Pearson], EC/1896/14, RS. It is dated 24 January 1895.
64. Pearson to G. Udny Yule, 25 January 1895, Pearson/11/2/20/1, PP; Weldon to Galton, 27 January 1895, Galton/3/3/22/17, GP. For the *P. varians* work, see Pearson (1895a [1948 reprint], 101–2).
65. Quotations from Weldon to Galton, 11 February 1895, Galton/3/3/22/17, GP. Extracts in a similar spirit from Weldon's letters to Galton between January and March 1895 are printed in Stigler (1986, 336–38). For Pearson's commentary, Weldon's response (dated 12 February 1895), and associated materials, see Pearson/5/2/4/1–5, PP.
66. After the Royal Society skew-curves discussion, Pearson wrote to his assistant Yule about Weldon: "He is such a splendid fellow really, that I hate having rows with him & yet am always dropping into them. I was quite unconscious of it, but I fear he considers that we are poaching on his preserves. I would retire at once, but I am sure only the mathematician can bring down the game upon them." Pearson to Yule, 25 January 1895, Pearson/11/2/20/1, PP. On their spending "a delightful afternoon, abusing each other in a friendly way about [skew curves] for some hours," see Weldon to Galton, 6 March 1895, Galton/3/3/22/17, GP, quoted in full in E. Pearson (1965, 17). For the measurement project, see, e.g., Pearson to Galton, 6 June 1895, Galton/3/3/16/9, GP. The texture of Pearson's friendship with Weldon is captured beautifully in Kevles (1995, 29–30).
67. Weldon to Galton, 23 June 1895, Galton/3/3/22/17, GP, referring to Pearson (1895b). From the start of the Royal Society committee, Weldon had envisioned the experimental study of heredity, understood as a "special case of correlation," and taking the form of breeding studies with plants and animals (such as his own with *Daphnia*), as falling within its remit. See his two-page note "Suggested Functions of Committee," MM/15/72, RS.

68. "University College, London," *Times* (London), 3 October 1895, 8; Weldon to Galton, 6 October 1895, Galton/3/3/22/17, GP ("I was glad of an opportunity of saying how much I really value [Pearson's] work: because in talking of it, both to you and to him, I spend so much time in trying to criticise points of detail, that I fear I sometimes seem to treat it unfairly").
69. Weldon to Galton, 5 January 1896, Galton/3/3/22/17, GP, discussing Francis Edgeworth; for Edgeworth's interactions with Weldon and Pearson, see E. Pearson (1965, 10–15); Stigler (1986, 327–44); and Gillham (2001, 276–79).
70. Pearson (1896a). Mathematically, the memoir is esteemed for advancing the theory and practice of correlation: see Stigler (1986, 342–44) and Haldane (1957, 308). Biologically, its contents are forgotten save for the framing of a "fundamental theorem in selection"—and that only because Ronald Fisher (1930) later framed a much more famous (and completely different) version; see Gayon (2007) and, on the memoir's general treatment of selection and heredity, Gayon (1998, 230–39). Weldon admired the memoir despite the damage done to his contention that correlation coefficients are constant across local races (Pearson [1896a, 126–28, in the 1948 reprint]). On panmixia, see Weismann (1886, definition on 21).
71. On the vogue for degenerationism, see Pick (1989); on its panmictic front, see Hale (2014, ch. 7). The "boom" was specifically a "'Nordau' boom," after Max Nordau, German author of *Degeneration*, published in English in 1895; see *Review of Reviews* 12 (1895): 183; 13 (1896): 565. I owe these periodical references to the SciPer project website (www.sciper.org.uk), and the *Punch* cartoon (fig. 4.6) to Klette (2015).
72. Weldon (1894b) made this argument in his reply to a *Nature* letter on panmixia by the Darwinian physiologist and psychologist George Romanes; Pearson (1894c) did so in reviewing Benjamin Kidd's *Social Evolution* (1894), which argued that, because of the deleterious effects of panmixia, socialism would bring about evolutionary decline.
73. Pearson (1896a [1948 reprint], quotation on 114, with discussion of panmixia on 167–77); Galton to Pearson, 9 May 1896, Galton/3/3/7/20, GP. When Galton wrote to Pearson, the Royal Society's "selected candidates" had already been announced: *Nature* 54 (7 May 1896): 9–12. The election took place on 4 June 1896: Certificate of Election and Candidature (Pearson), EC/1896/14, RS.
74. "The Ladies' Conversazione of the Royal Society," *Nature* 54 (18 June 1896): 159–61; Pearson to Yule, 30 December 1895, Pearson/11/2/20/1, PP. On reproductive selection, see Pearson (1896a [1948 reprint], 118–19; 1896b; 1897a, 1:63–102).

75. For the committee's history into 1900, see, in addition to the sources listed in note 43 above, Weldon to Pearson, 26 November 1896, Pearson/11/1/22/40, PP; circulars and related documents from the 1896–97 surge, in Galton/2/5/4/1, GP; letters to Galton from the same period on the Down House proposal, in Galton/2/5/4/2, GP; Bateson's letters to Galton, in Galton/3/3/2/6, GP; and the committee materials that Bateson preserved, mainly in C.15.b–f and G.3.0, BP. I take the phrase "Darwinian Institute" from Pearson to Galton, 27 November 1896, Galton/3/3/16/9, GP, though the name and the idea were older, as Galton noted (on a page in Galton/2/5/4/1/1, GP). For Bateson on substantive variation, see Bateson (1894, 23–25, 36–60).
76. On this correspondence—which started with an *esprit-de-l'escalier* letter from Lankester after Weldon had got the better of him in a discussion at the Linnean Society in June (at a lecture by Wallace)—see Pence (2011, 480–81). For the letters, see *Nature* 54 (1896), 16 July (Lankester), 30 July (Thiselton-Dyer, Weldon, and Cunningham), 20 August (Lankester), 3 September (Weldon), 17 September (Pearson), 1 October (Thiselton-Dyer and Cunningham), 22 October (Meldola), and 12 November (Bather and Meldola). The Lamarckian zoologist J. T. Cunningham was the most implacable critic of Weldon's crab research, from first (*Nature* 51, 28 March 1895) to last (*Nature* 130, 24 September 1932).
77. Weldon to Pearson, 8 and 13 August 1896, Pearson/11/1/22/40, PP. (Weldon read the *Critique* in the original German.) To Galton earlier in the summer, Weldon had expressed his annoyance at his colleagues having apparently slept through schoolboy lessons on Hume on cause and effect being nothing but universal correlation: Weldon to Galton, 30 June 1896, Galton/3/3/22/17, GP. Weldon's backstage comments reveal his eye-rolling attitude to the front-of-stage philosophizing, his own included (cf. Gayon [1998, 216]). Pearson, however, found the *Nature* discussion very stimulating, as did Galton. In responding to it that autumn, in dialogue with each other, they made the first serious attempts to solve what remains an outstanding problem for statistical science: how to distinguish causal from noncausal correlations.
78. Bateson to Galton, 15 October 1896, in Galton/2/5/4/3, GP. Many of the documents relating to this episode can be found in this folder and also in C.15.b–c, BP.
79. Quotation from Weldon to Galton, 22 October 1896, Galton/3/3/22/17, GP—a cover note for the longer, formal letter with the same date now found (in draft and final versions) in Galton/2/5/4/3, GP. On the envisaged testing of whether Plymouth shore crabs show greater deviations as they grow, see Weldon (1895a, 367–68), though the more

general ambition of measuring molted shells in captive crabs had already been set out in a 27 November 1893 letter from Weldon to Galton, in MM/15/85, RS. For the phrase "law of growth," see, e.g., Weldon (1895b, 381). Cf. J. Schwartz (2008, 56–57), which describes the "great shock" with which Galton received Weldon's "frank confession" of the screw-up now exposed by Bateson.

80. Galton to Bateson, 20 November 1896, C.15.b.32, BP.
81. Galton to Weldon, 17 November 1896, Galton/2/5/4/3, GP; Weldon to Galton, 17 and 26 November 1896, Galton/3/3/22/17, GP.
82. Weldon to Galton, 16 January 1897, Galton/3/3/22/17, GP. See also the minutes for the 14 January 1897 meeting, CMB/65, RS.
83. Weldon to Galton, 5 January 1896, Galton/3/3/22/17, GP; Weldon to Pearson, 17 May and 25 August 1896, Pearson/11/1/22/40, PP; H. Thompson (1896). Weldon's most energetic critics went after Thompson too, Bateson privately (see his October–November 1896 correspondence with Thompson in C.15.b, BP) and Cunningham publicly (see his October 1896–January 1897 correspondence with Thompson and Weldon in *Nature*).
84. Weldon (1898). Weldon's notebooks from his summer crab studies between 1895 and 1898 are held in Pearson/5/2/1/1–2, PP, though the regular updates he sent to Galton and Pearson in these years (along with a one-off letter to Bateson, 29 October 1897, C.15.c.10, BP) are more instructive for following his progress. The most comprehensive historical discussion of this work, including responses to it, is in Magnello (1993, ch. 4).
85. Weldon 1898, esp. 892–97. The previous summer, near the start of a somewhat critical review of Pearson's *The Chances of Death*, Weldon (1897a, 50) wrote, "With the exception of Mr Francis Galton . . . no one has done so much as Professor Pearson to make the systematic investigation of animal statistics possible."
86. Weldon to Galton, 27 May 1898, Galton/3/3/22/17, GP; Weldon, Nomination "Prof. Karl Pearson" (Darwin Medal), *Medal Claims 1873–1909*, RS, cited in Magnello (1996, 45n16). The citation stressed the role of Pearson's work in making possible "the direct demonstration of a selective death-rate" beyond the simplest cases—such a demonstration being "the natural and logical sequel to the work of Darwin himself": *Nature* 59 (8 December 1898): 139.
87. On Bateson's funded research project, see his application letter to Galton, 11 November 1897, MM/15/93, RS, along with a 1 February 1897 letter from Bateson referred to there (now in Galton/2/5/4/3, GP) and the minutes for the committee's 16 November 1897 meeting, CMB/65, RS.

88. See Weldon to Galton, 23 February, 9 and 10 March 1897, Galton/3/3/22/17, GP.
89. Weldon to Galton, 3 June 1899, Galton/3/3/22/17, GP.
90. Weldon to Galton, 28 November 1899, Galton/3/3/22/17, GP. Weldon wrote "absurd" after scratching out "nonsense."
91. See the letters to Foster, the society's president, from Galton (14 December 1899), Weldon (12 January 1900), Meldola (22 January), Pearson (23. January), and Thiselton-Dyer (23 January), MM/15/96–101, RS.
92. See the meeting minutes in CMB/65, RS. Foster came down on the side of reappointment after discussing the matter with Bateson; see Foster's notes on the subject in MM/15/81 and MM/15/94, RS.

Chapter Five

Epigraphs: Pearson (1900a, 486); Thomson (1900a, 334).

1. On the nineteenth-century spread of Mendel's publications and references to them, see Olby (1985, 216–34). For the Cambridge University holding, see Olby (1987, 404–5, 420). The "Versuche" offprint order is noted in the upper left-hand corner of the first page of Mendel's manuscript. Stomps (1954) transcribes Beijerinck's cover letter and gives the date as 1900, though the letter itself has since been lost; see Meijer (1985, 194).
2. On Beijerinck's hybridization work (mainly done at Wageningen) and its agricultural improvement context, see Zeven (1970). As Zeven notes (265), a linguistic convention widely followed at the time distinguished "crossing," involving different varieties of a species, from "hybridization" or "bastardization," involving different species. Mendel's "Versuche" rode roughshod over this distinction, however, and, like most historians of Mendelism before me, I shall do likewise.
3. De Vries (1889, esp. the preface and ch. 1; I have relied on the 1910 English translation). On how and why, through the 1880s, heredity took over from physiology as de Vries's main research preoccupation, see Stamhuis (2015, 33–34) and Theunissen (1994). On his hybridization experiments of the 1890s as defending and extending the program of the 1889 book, see Stamhuis, Meijer, and Zevenhuizen (1999) and Lenay (2000).
4. For the *Lychnis* crosses, see de Vries (1900a), where he described the ratio among the offspring of the hybrids as ¾ to ¼, though—following others—Meijer (1985, 215) points out that ⅔ to ⅓ would have been a more accurate approximation of de Vries's own numbers, and indeed were what he had reported before. For the poppy (*Papaver som-*

niferum) crosses, and the plate illustrating them, see Heimans (1978, 473) and Darden (1985). Whenever the plate dates from—and there is room for doubt that it is from before 1900 (see Kottler [1979, 530–33] and Meijer [1985, 195–97])—the general pattern it shows was one that de Vries thought he had found sufficiently often to write in 1896 of a "1.2.1 law," as discussed more fully later in this chapter.

5. For that title—which differed from the title of the published version ("Hybridising of Monstrosities"), see, e.g., the conference report in *Nature* 60 (27 July 1899): 305–7. Many thanks to Clare Coleman for this reference. On de Vries and Beijerinck, see Meijer (1985, 194, 211, 214–15) and Bos (1999, 678).
6. A photograph of the offprint appears in Zeven (1970, 273). For the case (which I find persuasive) for de Vries as having read the "Versuche" some years earlier, after following up a citation to it in a survey by the American breeder L. H. Bailey, but as not appreciating its interest at the time, see Meijer (1985, 193–95).
7. According to the botanist T. J. Stomps, who had been a junior colleague of de Vries, he wrote the German paper first, sending it off on 14 March, and then decided a few days later to translate it in abridged form for a French readership. See Stomps (1954).
8. De Vries (1900b). The paper was read on 26 March.
9. Weldon to Galton, 24 May 1895, Galton/3/3/22/17, GP. The first entry in Weldon's notebook for his Shirley poppies work is dated 25 March 1900: Pearson/5/2/1/8, PP.
10. Pearson to Weldon, 3 August 1898, Pearson/5/3/9, PP, emphasis in original. On Lankester's appointment, see Lester (1995, 127–31).
11. Pearson to Galton, 16 November 1898, Galton/3/3/16/9, GP.
12. Pearson to Weldon, 28 February 1899, Pearson/5/3/9, PP.
13. Weldon to Pearson, 13 October 1899, Pearson/11/1/22/40, PP. Wells's *War of the Worlds* was published in serialized form in 1897 and as a book in 1898.
14. Weldon to Galton, 5 November 1899 and 19 February 1900, Galton/3/3/22/17, GP; Weldon to Pearson, 11 November 1899 and 20 March 1900, Pearson/11/1/22/40, PP. The envisaged textbook was never published, but surviving MS chapters and lecture scripts can be found in Pearson/5/2/9, PP.
15. Weldon to Galton, 19 February 1900, Galton/3/3/22/17, GP. On the disdain for evolution—associated above all with Herbert Spencer—among the Kantian-Hegelian idealists who held sway in Oxford philosophy from about 1880 to 1910, see Hobhouse (1913, xv–xviii).
16. On the homotyposis project, see Magnello (1998, 58–62) and Provine (1971, 58–64).

17. Weldon to Pearson, 3, 20, and 26 March 1900, Pearson/11/1/22/40, PP.
18. Thomson (1900a,b). On Thomson as popularizer, see Bowler (2009, 233–40).
19. Thomson (1900a, 331).
20. Thomson (1900a, 331–32, quotations on 332). For the state of the art circa 1900, the universally esteemed statement—which Thomson drew on—was the Columbia zoologist E. B. Wilson's *The Cell in Development and Inheritance*, published in a revised second edition in January that year (Wilson 1900). On the historical emergence of the new knowledge of cell structure and function in the latter nineteenth century, mainly in the German lands, see W. Coleman (1965). Chromosomes got their name in 1888, but observations of them went back at least to the 1840s.
21. Thomson (1900a, 332).
22. Thomson (1900a, 332–33, quotations on 333); cf. Pearson (1900a, 452).
23. Thomson (1900a, 333–34, quotation on 334); cf. Pearson (1900a, 486–89, quotation on 456). To regard regression as having a biological (and more generally a causal) explanation, rather than a mathematical one, has come to be known as "Galton's fallacy": see Holt (2005, 89). For Weldon, regression was purely a chance phenomenon, which he hoped his illustrative use of dice-roll tables would bring home to his audiences; see Weldon to Pearson, 16 August 1898, Pearson/11/1/22/40, PP.
24. Thomson (1900a, 334; 1900b, 264; 1899, 161–62, quoting Pearson 1898, 412, emphasis in original). In the latter 1890s, the key publications were Galton (1897a), glowingly and lengthily précised under the title "A New Law of Heredity" in *Nature* 56 (8 July): 235–37; Galton (1897b), in which he insisted—to no avail—that the law applied only to intravarietal crosses and so had nothing to do with hybridism; Galton (1898), which introduced the diagram; and Pearson (1898), which gave the law its name (in capital letters), along with a brief account of how Pearson came to accept it (having, he realized, earlier misunderstood Galton's position). On the law's history up to 1900, see, e.g., Pearson (1914–30, 2:84; 3A:34–45); Swinburne (1965); Froggatt and Nevin (1971a, 5–8; 1971b, 5–10); Provine (1971, 179–85); Gayon (1998, 109–10, 132–34); and Bulmer (2003, 238–57). On the debt that the basset hound breeder, Sir Everett Millais, owed to Galton's earlier statements of the law in developing the breed, see Worboys, Strange, and Pemberton (2018, 167–69).
25. Pearson to Galton, 17 November 1899, Galton/3/3/16/9, GP.
26. Pearson (1900a, 475–81, 486–96; 1900b). The former was published in January 1900 and the latter read on 25 January.
27. See the "Diary of Societies" listings in *Nature* 61 (29 March 1900): 532.
28. Pearson to Galton, 13 February 1895, Galton/3/3/16/9, GP.

29. Pearson to Galton, 12 February 1897, Galton/3/3/16/9, GP.
30. "As a rule nobody takes any interest in my papers when read at the meetings unless you are there": Pearson to Galton, 17 November 1899, Galton/3/3/16/9, GP.
31. Ewart's paper, titled simply "Variation," was not published in the *Proceedings*. But his paper on his telegony-exploding experiments (1899) had been published there less than a year before. For Pearson's plug in the new *Grammar* edition, see Pearson (1900a, 454). Ewart (1901) later surveyed his then-recent studies of variation, including the experiments discussed by Thomson. On Ewart historically considered, see Button (2018). On telegony—as Weismann named it in 1892—historically considered, see R. Burkhardt (1979).
32. Vernon (1900).
33. Pearson (1900c).
34. Pearson and Lee (1900a,b). After the 16 November 1899 reading, Pearson decided not only to revise the memoir, but to break it up into two separate memoirs: one on aspects of the theory and practice of correlation mathematics (read on 1 March 1900: Pearson [1900d]); and the one co-authored with Lee (Pearson and Lee [1900b]). On Lee's life and work, see Renwick (2018).
35. Pearson to Yule, 3 April 1900, Pearson/11/2/20/1, PP. Pearson wrote to Galton the previous spring: "The talk with a man, who understands what you are aiming at and sympathises & suggests, is invaluable & I have lost both Weldon and Yule at once." Pearson to Galton, 17 May 1899, Galton/3/3/16/9, GP.
36. Bateson (1894, 573–75).
37. Bateson (1895b).
38. Although nothing much came of it, Bateson's butterfly research more nearly met his own ideal methodologically, as the two pairs of varieties he worked with were collected from the wild. On this research, see, e.g., Bateson to Galton, 1 February 1897, Galton/2/5/4/3, GP; and 11 November 1897, MM/15/93, RS; Bateson (1898a, 365–67 and 1898b, with a description of the experimental setup in his garden); and Bateson and Saunders (1902, 12–13). On the poultry research, see, in addition to the same letters, Bateson and Saunders (1902, 89–124).
39. See B. Bateson (1928a, 58, 61). After John came Martin (1899) and—with a Mendel-honoring name—Gregory (1904).
40. Bateson (1897, esp. 349–53; 1898a, 367).
41. See esp. Bateson to Galton, 1 February 1897, Galton/2/5/4/3, GP. The study of transmissibility, prepotency, and inbreeding was item 4 on Bateson's wish list for a Down House experiment station when he was asked about it in July 1899; see B. Bateson (1928a, 71–72).

42. See Saunders (1897), reporting that work begun in August 1895. On Saunders and her collaboration with Bateson, see Richmond (2001, esp. 60–63; 2006, esp. 569–71).
43. Bateson mentioned the allotment in his successful funding-application letter: Bateson to Galton, 11 November 1897, MM/15/93, RS. For the division of labor between Bateson and Saunders, see their draft reports on their 1898 and 1899 breeding experiments, G.3.o.19 and G.3.o.25a, respectively, BP.
44. On this research and de Vries's role in it, see the 1897 funding-application letter cited in the previous note; Bateson (1900, 65); de Vries (1900a, 71); and Bateson and Saunders (1902, 14–15).
45. W. Bateson to B. Bateson, 8 July 1899, in G.3.g.56, GP. For further comments on de Vries and Saunders from Bateson's letters at this time, see Cock and Forsdyke (2008, 186–87).
46. Bateson (1900, quotation on 62).
47. Hybrid Conference Report 1900, *Journal of the Royal Horticultural Society* 24:44–46, quotation on 45.
48. At the 1899 conference, an orchid breeder, C. C. Hurst (1900, esp. 103–4, 112)—later a staunchly loyal Mendelian—sought to apply Galton's law to hybrid data, but without much success.
49. Masters to Bateson, 15 March 1900, G.3.o.32, BP; see also W. Wilks (the RHS secretary) to Bateson, 10 April 1900, G.3.o.36, BP. Bateson's 8 May lecture was actually his second appearance at the society after the diplomatic needling. On 24 April, he attended a meeting of the RHS scientific committee, where he affirmed the message of his 1899 lecture on how more thorough record keeping by breeders would aid scientific understanding; see Olby (1987, 409). On the collaboration between Bateson and the society, see Olby (2000a,b); see also Fletcher (1969, 248–50).
50. Bateson (1901a); B. Bateson (1928a, 73).
51. Mendel's work with hawkweed was mentioned by the Kew botanist R. Allen Rolfe, in a lecture on hybridization and systematic botany (1900, 187).
52. The report, from the 12 May 1900 issue, is quoted in full (400–401) by Olby (1987) as part of a masterful case for Bateson's having first read de Vries, not Mendel.
53. Bateson and Saunders (1902, 157–59).
54. For this interpretation of de Vries on Mendel, I am indebted to Stamhuis, Meijer, and Zevenhuizen (1999, 249–55), summarizing an analysis set out more extensively in Zevenhuizen (1998).
55. For Bateson's animadversions on pangenesis, see Bateson (1894, 75). For Pearson's derivation of Mendel's law from Galton's, see Pear-

son (1914–30, 2:84n2). Alongside the detective work that uncovered de Vries, not Mendel, as read on the train, Olby (1987) furnished a pioneering treatment of "the transition from Galton to Mendel" (408) in Bateson's research circa 1900—but without considering the significance of de Vries's distinctive way of stating Mendel's law.

56. See de Vries (1900c, the longer, German, Mendel-mentioning version of 1900b, and 1900d, a characters-as-units-oriented French version, written and submitted around the same time as 1900b, and more fulsome in its praise of Mendel); Correns (1900a–c); Tschermak (1900a–b). On Correns's crossing experiments with *Pisum*, from 1896 (when he first read Mendel) to 1899, see Rheinberger (2003). On Tschermak's similar experiments, from 1898 to 1899 (when he first read Mendel), see Gliboff (2015). For Correns, the end in view was understanding the effects of foreign pollen ("xenia": yet another name for Darwin's "direct action of the male element on the female"), while for Tschermak, it was understanding the effects of inbreeding. Although a little out of date, the best overview remains Olby (1989b).
57. Bateson to Galton, 9 August 1900, Galton/3/3/2/6, GP, emphasis in original.
58. Bateson (1901a). For "rediscovery" in quotation marks (something started by Correns), see 60.
59. Correns (1900a, 164 [127 in the English translation]). In his basset-hounds paper, Galton (1897a, 403) adduced this gamete-level halving as evidence in favor of the halving of ancestral influence.
60. On Bateson on chromosomes, see W. Coleman's (1970) classic paper, the critical response in Cock (1983), and the more recent overviews in Cock and Forsdyke (2008, ch. 13) and Rushton (2014). For a statement of Bateson's views as the chromosomal gene theory began hitting its stride, see Bateson (1909, 270–72).
61. Bateson (1901a, 55, emphasis added).
62. For an overview of the vortex theory by a historically minded contemporary, see Merz (1965, 2:56–66). It is to Merz that I owe the phrase "mechanical view of nature." For an overview by a culturally minded historian of physics, emphasizing the vortex theory's status as a "Victorian theory of everything," see Kragh (2002). The quotation is from Maxwell (1875, 472).
63. William Coleman (1970) pieced together the documentary evidence in his pioneering article from half a century ago. Although it remains an indispensable guide, its construal of Bateson's attitude to biological form as part of a more general conservatism is problematic, for reasons I explore at length in ch. 10 of this book.

64. Bateson to F. B. Borradaille, 28 January 1924, cited in W. Coleman (1970, 270) and quoted in Schaffer (2000).
65. Bateson (1924, 408).
66. See esp. Bateson's famous letter to his sister Anna of 14 September 1891, about what he regarded as "the best idea I ever had or am likely to have," in B. Bateson (1928, 42–43). The other letters collected on 44–46, together with the editorial comments, are also instructive, as is Bateson's undated and unpublished manuscript "A 'Vibratory Theory' of Linear and Radial Segmentation as Found in Living Bodies," which exists in two slightly different copies in A.9.c/d, BP. Bateson was sensitive to the limitations of thinking with Chladni patterns; see Bateson (1913, 60) on how heaped sand does not go on to differentiate chemically as would divided living tissue. For a marvelously clear discussion of Chladni patterns as models of Batesonian morphogenesis, see Newman (2007, esp. 88–90).
67. Bateson 1917, 209.
68. On the concurrent empirical work, see, e.g., Weldon to Pearson, 11 July 1900, Pearson/11/1/22/40, PP. In a letter to Pearson of 22 July, Weldon dismissed the "unhappy Chap. 1" of his textbook as "rubbish." For Warren's *Daphnia* research, see Warren (1899, reporting results affirming Galton's new law, and 1900).
69. Pearson (1901a). Quotations from Weldon to Pearson, 11 July and 12 August 1900, Pearson/11/1/22/40, PP. For the planned use of artificial fertilization, see Weldon to Pearson, 14 November 1899, Pearson/11/1/22/40, PP.
70. Weldon to Pearson, 16 October 1900, Pearson/11/1/22/40, PP.
71. Weldon to Pearson, 25 October and 16 November 1900, Pearson/11/1/22/40, PP. Bateson taught regularly in Cambridge from 1899, when he was appointed as deputy to Newton. See B. Bateson 1928a, 62. The printed syllabus for his 1899 lectures on the "practical study of evolution," over two terms, can be found in G.5.a.3, BP. For Sollas *père* on Weldon, see Sollas (1900, 486).
72. Weldon to Pearson, 16 November 1900, Pearson/11/1/22/40, PP. In a letter later that year (5 December), Weldon recalled Bateson confirming in conversation what Weldon had understood from his first reading of Mendel's paper. So they seem to have discussed Mendel sometime between mid-October and mid-November. Another letter (6 December) mentions finding "an old letter from Sedgwick, about Mendel," with mention, too, of Sedgwick having discussed Mendel with Bateson. So perhaps it was Sedgwick from whom Weldon first heard about Mendel's paper.
73. Pearson to Galton, 13 December 1900, Galton/3/3/16/9, GP.

74. For Bateson's linking of an appetite for "statistical refinement" to the reluctance to give due attention to discontinuous variation, see Bateson (1897, 347).
75. Weldon to Pearson, 7 November 1900, Pearson/11/1/22/40, PP.
76. See esp. Weldon to Pearson, 3, 10, and 12 December 1900, Pearson/11/1/22/40, PP.

Chapter Six

Epigraphs: Pearson to Weldon, 30 July 1900, Pearson/5/3/9, PP; Weldon to Pearson, 15 December 1901, Pearson/11/1/22/40, PP; Weldon (1902a, 242, emphasis added).

1. Weldon (1902a).
2. See Pearson (1903a, 213–14, esp. the note on 213) for his role as source for the eye-color passage in Weldon (1902a). In earlier correspondence they had puzzled at length over human eye color and Galton's treating it as an instance of "alternative" (i.e., either/or—in Pearsonese, "exclusive") inheritance. See Weldon to Pearson, 12 April 1899, Pearson/11/1/22/40, PP, with a helpful commentary in Bulmer (2003, 243–44).
3. Pearson (1901b, quotations on 15, 17); Yule to Pearson, 9 September 1900, Pearson/11/2/20/1, PP.
4. Weldon to Pearson, 14 November 1900, Pearson/11/1/22/40, PP.
5. Weldon to Pearson, 21 January 1901 [misdated by Weldon 1900], Pearson/11/1/22/40, PP; see also Weldon to Pearson, 29 December 1900, Pearson/11/1/22/40, PP.
6. Galton to Pearson, 7 January 1901, Galton/3/3/7/20, GP.
7. Pearson to Galton, 10 January 1901, Galton/3/3/16/9, GP.
8. Weldon to Pearson, 16 November 1900, Pearson/11/1/22/40, PP. In fact Pearson had been thinking and consulting with colleagues about starting a new statistical journal for some time; see Pearson to Yule, 8 December 1900, Pearson/11/2/20/1, PP; and Yule to Pearson, 15 May 1900, Pearson/11/1/23/7, PP.
9. Davenport (1899, 1900a). Weldon regarded his own textbook as improving on Davenport's "very useful" one by going further in explicating the mathematics behind the methods; see Weldon to Pearson, 12 November 1899, Pearson/11/1/22/40, PP. On Davenport's life and work, see Riddle (1947) and Kevles (1995, ch. 3).
10. Weldon to Pearson, 2 December 1900 (source of the "best American" quotation) and 22 January 1901, Pearson/11/1/22/40, PP; Pearson to Galton, 13 December 1900, Galton/3/3/16/9, GP.

11. The largest cache of documents relating to the homotyposis controversy through February 1901 is in C.18, BP (typed transcriptions of most of which can be found in G.5.m, BP), but see also esp. the February 1901 Bateson-to-Pearson letters in Pearson/11/1/2/23, PP; Pearson to Yule, 8 December 1900, Pearson/11/2/20/1, PP; Pearson to Galton, 13 December 1900 and 1 February 1901, Galton/3/3/16/9, GP; and Weldon to Pearson, 25 February 1901, Pearson/11/1/22/40, PP.
12. Bateson (1901b).
13. Pearson to Bateson, 17 February 1901, C.18.6, BP; see also Pearson to Weldon, undated but clearly a response to Weldon's letter of 25 February 1901 (in which Weldon told Pearson about writing a letter to the council complaining about the society's handling of the matter), Pearson/5/3/9, PP.
14. Bateson to Pearson, 18 February 1901, C.18.7, BP, and also as an enclosure with Weldon to Pearson, 25 February 1901, Pearson/11/1/22/40, PP; Pearson to Bateson, 19 February 1901, C.18.9, BP.
15. Pearson to Galton, 18 April 1901, Galton/3/3/16/9, GP.
16. Galton to Pearson, 23 April 1901, Galton/3/3/7/20, GP.
17. Pearson to Galton, 27 April 1901, Galton/3/3/16/9, GP; Weldon to Pearson, 29 April 1901, Pearson/11/1/22/40, PP.
18. Pearson to Weldon, 3 May 1901 (I have tidied up the tired Pearson's spelling a little), Pearson/5/3/9, PP, transcribed in full in E. Pearson (1938, 51–52); Davenport (1900b); von Guaita (1898, 1900). On von Guaita's experiments, see Gliboff (2015, 106–7).
19. Weldon to Pearson, 8 May 1901, Pearson/11/1/22/40, PP.
20. Davenport (1901), published in the June issue of the *Biological Bulletin*, house journal of the premier American marine biological laboratory at Woods Hole, Massachusetts.
21. Farmer (1901, quotation on 438). "Meiosis," the now-familiar term for the gamete-producing process, is Farmer's coinage (circa 1904) and main claim to fame. For Thomson's reliance on Wilson, see Thomson 1900a, 332.
22. Weldon to Pearson, 9 May 1901, Pearson/11/1/22/40, PP.
23. Weldon to Pearson, 10 May 1901, Pearson/11/1/22/40, PP; see also his letter to Pearson of 19 May for Weldon's ambitions for the bibliography and his working methods. Weldon's abstracts from his reading can be found in a notebook titled "Variation—Vol 1—Inheritance," Pearson/5/2/10/1, PP. The entries are undated, but the one on Correns (1900a–c) is on 17–28 and the one on Tschermak (1900b) on 64–70.
24. Weldon to Pearson, 10 May 1901, Pearson/11/1/22/40, PP; Weldon, "Variation—Vol 1—Inheritance," 28, in Pearson/5/2/10/1, PP, empha-

sis in original ("not" is underscored twice). Cf. Correns (1900a, 121–22; 131–32 in the English translation).

25. Weldon to Pearson, 29 May 1901, Pearson/11/1/22/40, PP, emphases in original; Standfuss (1896); Bateson (1897, 352–53); Weldon to Bateson, 13 March 1898, C.15.f.16, BP. The 22-page abstract Weldon made for *Biometrika*, referred to in the letter to Pearson, can be found in Pearson/5/2/2/7, PP.
26. Weldon to Pearson, 22 May 1901, Pearson/11/1/22/40, PP. For Weldon's abstract of von Guaita, see Weldon, "Variation—Vol 1—Inheritance," 49–55, in Pearson/5/2/10/1, PP.
27. Weldon to Pearson, 29 May 1901, Pearson/11/1/22/40, PP.
28. Weldon to Pearson, 20 June 1901, Pearson/11/1/22/40, PP. Assembled by the first Linacre Professor, George Rolleston, the skull collection has since been absorbed into the Oxford Museum of Natural History. On the skullful group photograph in fig. 6.2, see Weldon to Pearson, 25 July 1901, Pearson/11/1/22/40, PP. Many thanks to Roger Wood for the tip-off about this photograph's existence and whereabouts.
29. Weldon to Pearson, 18 August and 22 September 1901, Pearson/11/1/22/40, PP.
30. Weldon (1901b); Weldon to Pearson, 18 August 1901, Pearson/11/1/22/40, PP. At the start of the summer, Weldon guessed that the goal of making *Biometrika* a success would help him work his way out of his slump, to which end he also gave up booze (which had become a problem) and cigarettes (which he missed much more). Even so, he was saved from making mathematical errors in the snails paper only thanks to Alice Lee's backstage corrections, for which he was immensely grateful. See his letters to Pearson of 25 June, 16 August, and 11 September 1901, Pearson/11/1/22/40, PP.
31. Weldon (1901b, esp. 122).
32. Weldon (1901b, 119); Pearson (1900a, 481–86). For a fine historical appreciation of the paper, see Gayon (1998, 223–24). The snail data can be found in Pearson/5/2/1/9, PP.
33. Weldon (1901a,c); Weldon to Pearson, 22 September 1901, Pearson/11/1/22/40, PP.
34. Brautigam (1993, 69–70).
35. Galton to Pearson, 21, 23, and 25 October 1901, Galton/3/3/7/20, GP.
36. Galton (1901, quotations on 659, 663–64); Pearson (1914–30, 3A:235).
37. *Times* (London), 30 October 1901, 3, 7; Pearson to Galton, 30 October 1901, Galton/3/3/16/9, GP. According to Pearson, the lecture "attracted more attention and bore ampler fruit" in the United States, where it was printed in the *Report of the Smithsonian Institution:* Pearson (1914–30, 3A:235).

38. Weldon to Pearson, 31 October 1901, Pearson/11/1/22/40, PP. In that era, the Prussian general Helmuth von Moltke (1800–91) was a byword for military prowess. In that week, Buller's sacking—and general scapegoating for British Army failings in South Africa—made him a byword for the opposite.
39. Mendel (1901). In a celebratory note about the new translation in late September 1901 in *Nature* (64:530–31), Bateson was mistakenly identified as the translator. It was in fact the fern specialist Charles Druery. On the Englishing of Mendel, see Hall and Müller-Wille (2013).
40. Bateson (1901c, quotation on 1–2, emphasis in original).
41. Weldon to Pearson, 13 November 1901, Pearson/11/1/22/40, PP.
42. Weldon (1902a, 232–35). For Weldon's probable-error formula, see 233 (cf. Davenport 1899, 14; Harris 1912, 741–42). For Pearson's now-famous test, see Pearson (1900e).
43. Weldon to Pearson [undated but, on internal evidence, between 21 and 25 November 1901], Pearson/11/1/22/40, PP, crossing out and underscoring in original. Parts are quoted in Magnello 2004, 22–23.
44. Weldon to Pearson, 21 November 1901, Pearson/11/1/22/40, PP; Laxton (1866a,b).
45. Weldon (1902a, 238); Weldon to Pearson, 28 November 1901, Pearson/11/1/22/40, PP. Three letters from Weldon to Tschermak (1901–2) are preserved in the Tschermak Papers, box 4, folder 84, Archive of the Austrian Academy of Sciences, Vienna. Many thanks to Sander Gliboff for alerting me to them and sharing his transcriptions. In a letter of 21 November 1901, Weldon replied to Tschermak, "I am especially interested to find that you now think Mendel's laws only special cases of something much wider and more general . . . I had felt bound to form this conclusion myself, as the only way of reconciling the published statements; but it is very difficult for one who is not familiar with the phenomena described to trust his judgment of the evidence; so that I was very glad to read your statement of your latest work."
46. Weldon to Pearson, 25 November 1901, Pearson/11/1/22/40, PP, emphasis in original.
47. Weldon to Pearson, 28 (?) November 1901, Pearson/11/1/22/40, PP, emphasis in original. (The question mark in the date is Weldon's.)
48. Weldon to Pearson [undated but, on internal evidence, around 25 November 1901], Pearson/11/1/22/40, PP, emphasis in original. At the top of the letter Weldon added, "I don't know if I have told you that in Telephone crosses green and wrinkled are often fairly 'dominant.'" See also Weldon to Pearson, 28 November and 1–4 December 1901, Pearson/11/1/22/40, PP. Letters to Weldon from Vilmorin (27 November 1901) and Carters (2 December 1901) are in Pearson/5/2/10/8–9.

49. Weldon to Pearson, 25 November 1901 (emphasis in original) and an undated letter from around that date (see previous note), Pearson/11/1/22/40, PP.
50. Cf. Bateson's introduction to the new English "Versuche," in which he retained the singular "law" but characterized it in terms of the purity of the hybrid's gametes: "In respect of certain pairs of differentiating characters the germ cells of a hybrid, or cross-bred, are *pure*, being carriers and transmitters of either the one character or the other, not both." Bateson (1901c, 1, emphasis in original).
51. Weldon (1902a, 228–29).
52. Weldon (1902a, 229–35, quotations on 235).
53. Weldon (1902a, 236–44, quotation on 244), citing Bateson (1897, 1898a).
54. Weldon (1902a, 244–49, quotation on 248).For *Telephone* as "deep green," see the 1879 advertisement in Charnley and Radick (2013, 226). At the time of this writing, *Telephone* seeds are for sale from a British heritage-seeds firm, The Real Seed Catalogue.
55. Weldon (1902a, 249–50).
56. Weldon (1902a, 246).
57. Weldon (1902a, 251–52).
58. On the color-plate-induced delay, see Weldon to Pearson, 15 December 1901, Pearson/11/1/22/40, PP, and Pearson to Galton, 10 and 28 January 1902, Galton/3/3/16/9, GP. Even worse disappointment was avoided, albeit late in the production process. In his 28 January letter to Galton, Pearson wrote that Weldon's peas "gave a beautifully continuous scale of colour from pure green to pure yellow, but the colour lithographers found it easier to give one half the peas a yellow shade & the other half a green! Thus the plate effectually demonstrated the truth of Mendel's Law instead of disproving it by continuity of shading!!" Discussed in MacKenzie 1981a, 261n9.
59. Weldon to Pearson, 15 December 1901, Pearson/11/1/22/40, PP.
60. Weldon to Pearson, 15 and 18 December 1901, Pearson/11/1/22/40, PP. I have learned much about re-hybridizing in late nineteenth-century British biology from Clare Coleman's Leeds PhD thesis (2021).

Chapter Seven

Epigraphs: Galton to Pearson, 28 February 1902, Galton/3/3/7/20, GP; Weldon to Pearson, 28 May 1902, Pearson/11/1/22/40, PP; Pearson to Yule, 27 August 1902, Pearson/11/2/20/1, PP; Bateson to B. Bateson, 3 October 1902, G.3.d.5, BP, with typescript at G.8.g.1d, BP.

1. Bateson to Pearson, 13 February 1902, Pearson/11/1/2/23, PP.
2. Bateson to Pearson, 10 February 1902, in the Weldon-to-Pearson corre-

spondence, Pearson/11/1/22/40, PP, with other versions at C.18.14 and G.5.m.17, BP; Pearson to Galton, 11 February 1902, Galton/3/3/16/9, GP.

3. See "Bateson and Pearson: Controversy over Paper 1902," folder CD.260–77, RS; see also relevant correspondence in C.18 and, in typescript, G.5.m, BP.
4. Pearson to Yule, 27 August 1902, Pearson/11/2/20/1, PP.
5. Bateson (1894, vii, 76, and 70 [also 89], respectively); cf. Pearson (1892, vii [and throughout], 32, and chs. 7–9, respectively).
6. Pearson to Bateson, 30 December 1895 and 3 January 1896, G.5.m.3–4, BP; Bateson to Pearson, 1 January 1896, Pearson/11/1/2/23, PP. On Bateson and women's education, see B. Bateson (1928a, 59–60). For Pearson on that topic, see Porter (2004, 159–61).
7. Bateson (1902a, 16–17) cited Pearson's writings on Galton's law, including the second edition of Pearson's *Grammar of Science* (1900a) and the full memoir version of his 29 March 1900 Royal Society talk (Pearson and Lee 1900b).
8. Pearson to Bateson, 12 October 1901, C.18.11, typescript in G.5.m.15, BP.
9. Bateson and Saunders (1902, 158).
10. Bateson to Pearson, 13 February 1902, Pearson/11/1/2/23, PP.
11. Pearson to Bateson, 15 February 1902, C.18.19, typescript in G.5.m.21, BP. For fuller quotations from this exchange, see Provine (1971, 63–64).
12. Weldon to Pearson, 12 and 16 February 1902, Pearson/11/1/22/40, PP. Between these dates Weldon corresponded briefly with Bateson about the *Biometrika* article, sending him a copy and a set of peas: Bateson to Weldon, 14 February 1902, C.18.18, typescript in G.5.m.20, BP.
13. Galton to Pearson, 25 January 1902, Galton/3/3/7/20, GP; Weldon to Pearson, 16 and 18 February 1902, Pearson/11/1/22/40, PP.
14. Bateson (1901b, 416–17) introduced the new terminology in a note added to the manuscript of his homotyposis commentary in November 1901, after Pearson's original paper had been published.
15. Pearson (1902).
16. Weldon (1902b); de Vries (1901; for the topics mentioned, see, in the 1910 English translation, 47–53, 71–129, quotation on 71). Although de Vries's mutationism is closely associated with the evening primrose, *Oenothera lamarckiana*, he also commended another plant, the grass *Draba verna*, as exemplifying his theory, and Weldon studied it excitedly through February and March 1902, reporting on his progress—or more often lack of it—periodically in his letters to Pearson into the autumn.
17. Weldon (1902b, 367–68). Up to this time, Weldon's only public expression of interest in the new experimental embryology out of Germany and France was an 1897 review in *Nature* of books in the field, includ-

ing one by an American exponent: *The Development of the Frog's Egg*, by T. H. Morgan. See Weldon (1897b) and, on the era's *Entwicklungsmechanik* generally, see Hopwood (2009, 298–301). For de Vries's unimpressed response to Weldon's article, see Provine (1971, 68). (Provine's account of the article misleadingly ignores the points about development and environment stressed here.)

18. Weldon to Pearson, 23 April 1902, Pearson/11/1/22/40, PP; Cuénot (1902), reported in *Nature* (65 [17 April 1902]: 580) as a first attempt to check whether Mendel's law holds for animals as well as for plants. On Cuénot's status as not just the first but, until the Great War, the only card-carrying Mendelian in France, see Burian, Gayon, and Zallen (1988). On Mendelism in France generally, see Gayon and Burian (2000).
19. Weldon to Pearson, 27 April and 16 May 1902, Pearson/11/1/22/40, PP.
20. Bateson and Saunders (1902); Bateson (1902a). Apropos of Weldon's limerick, Bateson belonged to the Cambridge Botanic Garden Syndicate, which he thanked on the first page of *Report 1* for rent-free gardening space.
21. Bateson and Saunders (1902, quotation on 4). At this point, the Evolution Committee functioned more or less exclusively to fund and publish the work of Bateson and his associates. On Saunders's experimental work in *Report 1*, see Richmond (2001, 66–67); on Bateson's experimental work, and how it changed with Mendelization, see Cock (1983, 3–7).
22. Bateson and Saunders (1902, 126, 159–60); Galton (1869, 50–53).
23. Bateson and Saunders (1902, e.g., 18–19 [for 1:1] and 26 [for 9:3:3:1]). For the phrase "Mendel numbers," see Weldon to Pearson, 22 December 1904, Pearson/11/1/22/40, PP; see also Porter (2018, 282, quoting Wilhelm Weinberg). The 9:3:3:1 ratio can be calculated easily by combining two 3:1 ratios (3 × 3, 3 × 1, 1 × 3, 1 × 1); Mendel's not going there underscores the extent of his investment in 1:2:1 as "the law valid for *Pisum*."
24. Bateson and Saunders (1902, 11–12, 129–32, 152).
25. In the final part, the second-last section is headed "Non-Mendelian Cases." Bateson divided up his poultry results into two groups, Mendelian and non-Mendelian.
26. Bateson and Saunders (1902, 119, 86–87, and 23, respectively).
27. Bateson and Saunders (1902, 12, 139 and 147, and 152, respectively).
28. Bateson and Saunders (1902, 157–59, quotation on 157). The contrast between the two laws is presented as maximally sharp here, but elsewhere in the report even gametic or "germ cell" purity is considered negotiable (see 127).

29. Bateson and Saunders (1902, 6, 125). On the theme of Mendelian "retrodictions," see Buttolph (2015).
30. Bateson and Saunders (1902, 132–36, 139–41, quotation on 134).
31. Bateson and Saunders (1902, 133–34). The case of alkaptonuria comes up in a discussion of other phenomena that might be explained by the view that recessive homozygotes are generally weak: why in every generation there are always weaker individuals, despite regular culling; and why the idea that variation is greater on the farm than in the wild became entrenched (in the wild, recessive homozygotes appear just as frequently, but never live long enough to be noticed). For Bateson's correspondence with Garrod through the first half of 1902, see G.7.h, BP. For Garrod's use of Bateson's analysis, see Garrod (1902, 1618).
32. [Notes from Saunders, March 1902], G.4.d.3, BP.
33. See the correspondence from mid- to late February 1902 in G.5.n.1–4, BP.
34. Richard Wright to Bateson, 28 April 1902, G.5.n.20, BP; B. Bateson to W. Bateson, 1 May 1902, G.5.n.24, BP.
35. Bateson (1902a, 198 ["the Mendelian"], 115 ["dazzling"], 136–37 [on dominance and Mendel's discovery]).
36. Bateson (1902a, esp. 55n, 106, 117–20, 130). Bateson at one point compares the relationship between the hybrid character and its constituent gametes to the relationship between salt's properties and the properties of sodium and chloride (23).
37. Bateson (1902a, esp. 16–35, 108–16, 135–36, 204).
38. Bateson (1902a, 26–30).
39. Bateson (1902a, quotations on 18 and 32, respectively). At the very start of the book, Bateson presented Mendel as the man who, confronting "the jungle of phenomena . . . cut a way through" (v).
40. Bateson (1902a, 193–200).
41. Bateson (1902a, 119–68, 178–83). According to Bateson (139–41), the phenomenon that attracted Darwin to Laxton—the influence of pollen on maternal characters such as seed coat ("xenia")—had since been widely recognized, and meant that reports of, e.g., the hybrids in a yellow-green cross all going green could not be trusted unless one knew for sure that the observer was alert to the possibility that the greenness was in the seed coat, perhaps with yellow cotyledons within. See too ch. 1, note 50, in this book.
42. Bateson (1902a, 186 ["special pleading"], 119 [on diversity in how varieties conform to Mendelian expectations], 203–4 [imagining a variety like *Telephone* becoming universal]). For follow-up experiments with peas, see Lock (1904, 1905, 1908).

43. Bateson (1902a, 12).
44. Bateson (1902a, v, 115). Bateson inserted a passage on the 9:3:3:1 ratio into the reprint of his first Mendelian paper included in the *Defence* (11–12) and added many new notes to the English "Versuche," including one to the "reproductive cells" section flagging the paragraphs that "contain the essence of the Mendelian principles of heredity" (67).
45. Bateson (1902a, 208).
46. Saunders found that, in some *Matthiola* crosses, green-seededness correlated with hairiness while brown-seededness correlated with smoothness; see Bateson and Saunders (1902, 81). For discussion, see J. Schwartz (2008, 114).
47. Bateson (1902a, vi and x–xi). Bateson managed to imply that these vices flowed from a myopic concentration on "numerical precision" and "statistical nicety" as the only forms of "exactness" that mattered.
48. Weldon to Pearson, 3 June 1902 (where Weldon offered to be let go), Pearson/11/1/22/40, PP.
49. Weldon to Pearson, 4 June 1902, Pearson/11/1/22/40, PP.
50. Bateson and Saunders (1902, 15–19).
51. Weldon to Pearson, 6 June 1902 ff, Pearson/11/1/22/40, PP. About the identity of extracted dominants and recessives with the grandparental forms, Weldon wrote to Pearson on 6 August 1902, "This is simply a lie, and it really should be pilloried as such."
52. Weldon to Pearson, 23 June 1902, Pearson/11/1/22/40, PP; see also Weldon to Galton, 1 July 1902, Galton/3/3/22/17, GP.
53. Pearson to Bateson, 9 June 1902, Pearson/11/1/2/23, PP, and also in G.5.m.32, BP.
54. Galton to Bateson, 13 June 1902, G.5.n.12, BP; Galton to Pearson, 12 June 1902, Galton/3/3/7/20, GP.
55. Tschermak's note (in German) is included in Weldon to Pearson, 17 June 1902, Pearson/11/1/22/40, PP. Tschermak's review—sympathetic to Weldon—appeared not in *Biometrika* but in a German journal; see Šimůnek et al. (2012, 248–49).
56. D[ixey] (1902).
57. Yule (1902, 194).
58. Although Bateson's group tends to be thought of as "Newnham College Mendelians" in the first instance, my impression is that men and women got involved in roughly equal numbers from roughly the same time. On the membership of that group, see Richmond (2006, esp. 576–81; 2001, 64–72).
59. Bateson (1902a, 119). On Killby, see Richmond (2001, 69).
60. W. Bateson to B. Bateson, 11 April 1902, G.3.b.9, BP. For Lock's letters to

Bateson from Ceylon, starting in October 1902, see D.23, BP. On Lock, see Edwards (2013). On Bateson's lectures and their influence, see Richmond (2001, 65–66). Syllabi and notes can be found in G.5.a, BP.

61. Biffen to Bateson, 4 June 1902, G.5.n.6, BP. On Biffen and his role in the "Mendelian system" that emerged in Britain and beyond, see Charnley (2011; on Biffen and Wood from 1902, see 41).
62. L. Doncaster to W. Bateson, 7 and 21 June 1902 (quotation from latter), Letters, vol. 11, nos. 1414–74, BP. On Doncaster, see Bateson's obituary, G.7.n.15, BP.
63. Hurst (1900); Bateson and Saunders (1902, 157). On Hurst, see Cock and Forsdyke (2008, ch. 10, "Bateson's Bulldog"), and more extensively, R. Hurst (1971). In autumn 1900, Hurst collected data on the head sizes of local schoolchildren for Pearson, who, in a letter of 25 October telling Hurst that a head spanner and data forms would be arriving soon, mentioned Mendel. R. Hurst (1971, 522 [quotation from Pearson letter], 525–26).
64. R. Hurst (1971, 501–2); Weldon to Pearson, 16 February 1902, Pearson/11/1/22/40, PP. For discussion of Bateson's un-Pearsonian opposition to war and imperialism, see MacKenzie (1981a, 148).
65. Hurst (1902, quotations on 689 [emphasis in original], 690, 694). It was undoubtedly this paper—not reprinted in Hurst's collected genetics papers (1925) and passed over in silence in his wife's Hurst-centric history of genetics (R. Hurst [1971])—that Bateson saw advance notice of in the *Gardeners' Chronicle* and referred to in his first letter to Hurst (see next note). Hurst was tipped off about Weldon's paper by a fellow orchid specialist, the Kew-based Robert Allen Rolfe, in a letter of 6 April ("It voices several of my difficulties"); R. Hurst (1971, 590).
66. On Bateson's visit, which Hurst soon reciprocated, see R. Hurst (1971, 560–64); see also Bateson to Hurst, 24 and 27 March 1902, typescripts in D.21.a, BP. Bateson's preserved correspondence with Hurst is more voluminous than with anyone else: Cock and Forsdyke (2008, 265).
67. Hurst (1904, 11–13, quotations on 11–12).
68. Bateson to Hurst, 13 September 1902, typescript in D.21.a, BP.
69. W. Bateson to B. Bateson, 27 August 1901, G.3.b.8, BP, reprinted in Bateson (1928a, 65–66).
70. On the American breeders and their world, see Paul and Kimmelman (1988, see 286 for the Mendel-disseminating activities of the USDA in 1901–2) and, more extensively, Kimmelman (1987, esp. chs. 1–3).
71. Bateson (1904a, 1).
72. Bateson (1904a, 2–3).
73. Bateson (1904a, 3–4).
74. Bateson (1904a, 3–4).

75. In Bateson (1904a, 9).
76. In Bateson (1904a, 8–9). On Bailey's citing of Mendel, see Olby (1985, 231). For Bailey's life and career up to this point, see Kimmelman (1987, 112, 189–96).
77. In Bateson (1904a, 9); Spillman (1902a–c). On Spillman, see Kimmelman (1987, 141, 169n178).
78. In Bateson (1904a, 9). On Wood, see the biographical note in the online finding aid for his papers at https://archives.tricolib.brynmawr.edu/repositories/5/resources/9876 (accessed 18 April 2019).
79. Spillman to Hurst, 21 January 1903, quoted in Cock and Forsdyke (2008, 271).
80. De Vries (1904). In late November, Bateson further developed his own version of compound character analysis/synthesis in a paper read to the Cambridge Philosophical Society: Bateson (1902b).
81. In Lynch (1904, 30).
82. Hays (1904, esp. 61). Burbank's own theorizing of his celebrated successes—which made him the Thomas Edison of plant varieties—never made room for Mendel, de Vries, or Weismann; see, e.g., Palladino (1994, 411–19); Gould (1996).
83. Beach (1904, 67).
84. Cannon (1904, quotations on 89–90, emphases in original).
85. In Hill (1904, 114).
86. Bailey (1904, 120–22).
87. Bailey (1904, 122–24).
88. In Ward (1904, 155). The speaker, C. W. Ward, ended his presentation by saying that, having never heard of Mendel before the meeting, he now intended to get hold of his writings, "for if there is a fixed rule by which I can produce six inch carnations on four foot stems I certainly wish to learn that rule" (154).

Chapter Eight

Epigraphs: Weldon to Pearson [late July 1902], Pearson/11/1/22/40, PP; Hurst to Bateson, 27 May 1903, D.21.c.1, BP (slightly differently transcribed in R. Hurst [1971, appendix A, 4]); Weldon to Pearson, 31 December 1903, Pearson/11/1/22/40, PP; Punnett (1950, 7–8).

1. Weldon to Pearson [late July 1902], Pearson/11/1/22/40, PP. Weldon's evening discourse on inheritance took place at the Grosvenor Hall in Belfast on 15 September 1902: *Report of the Seventy-Second Meeting of the BAAS* [. . .] (1902, lxxvi). For discussion of what seems to be Weldon's draft of this lecture, see ch. 11, esp. note 32.
2. Weldon to Pearson, 2 August 1902, Pearson/11/1/22/40, PP. Weldon went

on, "Where reason failed, this piece of foolery seems to have succeeded."

3. For the debut of "Mendelism," see Bailey (1903, 451), discussed in Paul and Kimmelman (1988, 295, 304n18) and in Olby (2000a, 1048). For the new identification of a chromosomal basis for Mendelian patterns, the classic papers are Boveri (1902) and W. Sutton (1902, and esp. the masterfully synthetic 1903), along with Cannon (1902) and McClung (1902). For snapshots of the state of discussion before and after these papers, see V. H. Blackman (1902, 104–5) and Wilson (1902), respectively. Excellent historical overviews include R. Hurst (1971, 617–19); Robinson (1979, ch. 9); and J. Schwartz (2008, ch. 8). On Wilson's "Sutton-Boveri hypothesis" label for the chromosome theory of Mendelian heredity as misleadingly exaggerating Boveri's Mendelism, see Martins (1999). On the role of Mendelian theory in bringing about consensus on what the cytological facts were, see Baxter and Farley (1979).
4. Spillman (1903).
5. On the ABA, launched in 1903, see Kimmelman (1983; on Hays, see 168–71).
6. Davenport (1904a [reporting his data on coat-color inheritance in mice]; 1904b [assessing the fit with Mendelism for data from others on "wonder horses," polydactylism, and deaf-mutism]); Riddle (1947, 82).
7. Weldon to Pearson, 18 March 1903, Pearson/11/1/22/40, PP.
8. On the Inter-departmental Committee on Physical Deterioration (to give the committee its official name), see Hynes (1968, 22–24; on the Scouts movement as a response to it, see 27–29). See also Soloway (1995, 43–47).
9. Galton (1903).
10. Weldon to Pearson, 29 April 1904, Pearson/11/1/22/40, PP.
11. Weldon to Pearson [late July 1902], Pearson/11/1/22/40, PP.
12. Weldon (1902d, 44–45).
13. Weldon (1902d, 46).
14. Weldon (1902d, 47–49); see also ch. 5, note 4, and J. Schwartz (2008, 119–21, esp. on Weldon's calculation of a 1 in 17 chance of de Vries getting numbers conforming to a 2:1 ratio given a 3:1 reality).
15. Weldon (1902d, 49–55, quotation on 53).
16. Weldon to Pearson, 2 November 1902, Pearson/11/1/22/40, PP.
17. Bateson to Hurst, 17 January 1903, typescript in D.21.a.1, BP. In a May 1904 letter to Weldon, de Vries agreed that it was difficult to draw any generalizable conclusion from his *Lychnis* hybrid experiment, since

he had never managed to repeat it, despite trying. De Vries to Weldon, 11 May 1904, Pearson/5/3/5, PP.

18. Weldon to Pearson, 6 August 1902, Pearson/11/1/22/40, PP, emphasis in original; Bateson (1902a, 173–74); Darbishire (1902, 102). On Darbishire's life up to this point, see Ankeny (2000, 317) and Wood (2015, 17). Both tell of how Darbishire gave a lecture on Mendel's discovery in autumn 1901 at the Oxford University Junior Scientific Club, where his audience members included the eight-year-old J. B. S. Haldane.
19. Weldon to Pearson, 25 September 1902, Pearson/11/1/22/40, PP, emphasis in original; Darbishire (1902, 101).
20. Weldon to Pearson, 30 October 1902, Pearson/11/1/22/40, PP; Schuster (1905, 1).
21. Weldon to Pearson, 11 December 1902, Pearson/11/1/22/40, PP; Darbishire (1902, 102, 104).
22. Darbishire (1902).
23. Darbishire (1903a).
24. Darbishire (1903a).
25. Darbishire (1903a, 173).
26. Weldon to Pearson, 15 October 1902, Pearson/11/1/22/40, PP. (There are two letters with this date.)
27. Weldon to Pearson, 29 January 1903, Pearson/11/1/22/40, PP, emphases in original.
28. Weldon to Pearson, 22 February 1903, Pearson/11/1/22/40, PP.
29. Bateson (1903a). Castle and Allen (1903, 612–13) proposed more or less the same explanation independently.
30. Weldon to Pearson, 24 March 1903, Pearson/11/1/22/40, PP, with comments on Bateson's rhetoric in Weldon to Galton, 22 May 1903 (one of two letters with this date), Galton/3/3/22/17, GP.
31. See the letters published in *Nature* 67 and 68 (1903) on 2 April (Weldon), 23 April (Bateson), 30 April (Weldon), and 14 May (Bateson and Weldon; quotation on 34), along with a siding-with-Darbishire anonymous notice of the latest *Biometrika* on 9 April. Bateson's letters are republished in Bateson (1971, 2:112–14). For a discussion picking up on other dimensions of the correspondence, see J. Schwartz (2008, 122–24).
32. Weldon (1903, quotations on 289, 297); Darbishire (1903b).
33. Weldon to Pearson, 16 April 1903, Pearson/11/1/22/40, PP, emphasis in original.
34. Weldon to Pearson, 26 April 1903, Pearson/11/1/22/40, PP.
35. Weldon to Galton, 22 May 1903 (the other of two letters with this date—see note 30 above), Galton/3/3/22/17, GP. See also note 74.

36. Weldon to Pearson, 16 and 17 May 1903, Pearson/11/1/22/40, PP.
37. Bateson (1903b). For accounts of the meeting, see *Nature* 68 (25 June 1903): 19; Mitchell (1937, 167); Weldon to Galton, 29 May 1903, Galton/3/3/22/17, GP.
38. Weldon to Pearson, 27 May 1903, Pearson/11/1/22/40, PP.
39. Weldon to Pearson, 21 June 1903, Pearson/11/1/22/40, PP, reporting that two people he had met subsequently "professed to consider that Bateson had been utterly destroyed."
40. Hurst to Bateson, 27 May 1903, D.21.c.1, BP, transcribed nearly in full in Cock and Forsdyke (2008, 229).
41. Bateson (1903b, 104). To the best of my knowledge, this paper marks the beginning of Bateson's referring to the hybrid generation as "F_1" rather than "F."
42. Bateson (1903b, 106). The papers by Darbishire (1903b) and Weldon (1903) appeared in the June 1903 issue of *Biometrika*.
43. Coat color in mice rapidly became a site of further, virtuosic explanatory extension (and exculpation) among the Mendelians. For a glimpse of the new possibilities, see Cock and Forsdyke (2008, 232). On Bateson as once again pointing the way, see J. Schwartz (2008, 123).
44. Johannsen (1903a); Weldon and Pearson (1903); Pearson to Galton, 30 August 1903, Galton/3/3/16/9, GP. In an article that same year on Darwinism and heredity, Johannsen summed up his experiments, with barley and peas as well as beans, as showing that "individual variations are *not* heritable. Or at least so says *all my material*!" Johannsen (1903b, 11, emphases in original). On Johannsen's awareness of Mendel's work at this time, see Müller-Wille and Richmond (2016, 368, 378). On Johannsen and his experiments, see, e.g., Provine (1971, 92–97); Roll-Hansen (1989); Gayon (1998, 260–71); Meunier (2016).
45. Castle (1903a); Castle and Allen (1903, 606).
46. Bateson (1903b, 82).
47. Garrod (1903, 413–14, 417)—a restatement (in German) of material already presented in Garrod (1902, 1618–19).
48. Weldon (1904a). Weldon was freshly returned from a snail-hunting holiday with his wife in Sicily when the brief remarks on Sicilian albinism in Garrod's 1903 paper caught his attention; Weldon to Pearson, 3 February 1904, Pearson/11/1/22/40, PP. As Galton was just then in Sicily, Weldon asked him to keep his eyes peeled for albinos: Weldon to Galton, 14 February 1904, Galton/3/3/22/17, GP.
49. Weldon to Pearson, 12 June 1903, Pearson/11/1/22/40, PP. The approach was made by the Cambridge physicist Joseph Larmor, a St. John's colleague of Bateson's who, not long before, had suggested to Bateson that, as Bateson recalled, "a jury of good men & true should be em-

paneled to try the Mendelian issue." Bateson to Larmor, 28 May 1903, Larmor Collection, Special Collections, Library, St. John's College, University of Cambridge (slightly differently transcribed in Cock and Forsdyke [2008, 226]).

50. Weldon to Pearson, 9 and 20 February 1904, Pearson/11/1/22/40, PP; *Nature* 69 (10 March 1904): 454; Cock and Forsdyke (2008, 231–32).
51. Weldon to Pearson [undated, but circa 20 March 1904], Pearson/11/1/22/40, PP. Indeed, Arthur Sutton, who was at the Linnean Society meeting, wrote to Bateson afterward, "I had no idea myself that there was any point in dispute between Professor Weldon and yourself"—though, he went on, since Weldon plainly accepted some Mendelian results, and Bateson did not entirely deny a role for ancestry, the exact nature of the dispute was a little elusive. Sutton to Bateson, 20 February 1904, G.5.c.22, BP. In an incomplete short MS on "Messrs. Sutton's Cross-Bred Primulas and Mendel's Laws," Weldon reported that the nursery workers who looked after Bateson's experiments recognized that when it came to the often pink- or lavender-flushed or -flaked white plants, "Mr. Bateson's description of the plants as 'albino' was simply inaccurate." Pearson/5/2/8/6, PP.
52. Galton (1905, 47, emphasis in original). On this meeting and its significance for the Sociological Society, see Renwick (2012, 131–39).
53. Bateson (1905). For related correspondence, see G.6.a.1–11, BP; Pearson (1914–30, 3A:220–21); Bateson to Galton, 22 June 1904, Galton/3/3/2/6, GP.
54. Weldon (1905, 57–58).
55. Weldon to Pearson, 11 August 1903, Pearson/11/1/22/40, PP. The letter makes plain that Weldon was responding to a passage in the *Defence* (1902a, 115; see also 3) where Bateson commended "to those who are more familiar with statistical method, the consideration of this question: whether dominance being absent, indefinite, or suppressed, the phenomena of heritages completely blended in the zygote, may not be produced by gametes presenting Mendelian purity of characters." In both passages, Bateson used stature as an example and allowed that sensitivity to environmental differences could make the distribution even more smoothly continuous. See also Bateson and Saunders (1902, 152–53) and, for discussion, Stoltzfus and Cable (2014, 520–21).
56. Johannsen (1903a, 213–14).
57. Yule (1902, 236).
58. Shortly after the publication of *Report 1* and the *Defence*, Yule wrote to Pearson about Bateson: "The man seems to have clean lost his head from the way he talks about Mendel: 'wet dish clouts' are rather good treatment for such style. He does not seem to realise that high falu-

tin' of that kind only tends to make one the more critical." Yule to Pearson, 14 June 1902, Pearson/11/1/23/7, PP.

59. Yule (1902, 196–207). For an outstanding discussion of Yule's paper in its context, see Tabery (2004).
60. Yule (1902, 201–7).
61. Yule (1902, 222–27, quotations on 225–27, emphases in original). Tabery (2004) characterizes what Yule did as a "reduction" of Mendel's laws by the law of ancestral heredity, but that seems to me a poor fit, since, ordinarily, the theory at the end of a derivation is regarded as the reduced one (cf. the relationship between Newton's laws and Kepler's laws).
62. Yule (1902, 227–32, quotation on 228, emphasis in original).
63. Yule (1902, 228). On the role of heredity in the incidence of tuberculosis as a topic of wide interest at this time, see Olby (1993, 427–30).
64. Yule (1902, 232–37).
65. Lock to Bateson, 25 January 1903, D.23.8, BP. Bateson's own impatience with talk of the Galtonian and Mendelian theories being parts of a larger whole can be gauged from an unpublished letter he had written to *Nature* a few months earlier, responding critically to a letter from the York science teacher Hugh Richardson. Against the supposed antagonism between the two theories as described in *Nature*'s review of *Report 1* and the *Defence*, Richardson reported that his own work with sweet-pea hybrids suggested that dominance was a matter of degree, increasing as a character version became established in a lineage. See *Nature* 66 (23 October 1902): 630–31; Bateson, "Mendelian and Galtonian Heredity: The Case of Sweet Peas," G.5.n.27, BP.
66. Castle (1903b, 232–34). Independently of Castle (and a century later), Tabery (2004, 88–89) also judged Yule to have made a mistake. On any mistake being not Yule's but Castle's, see Edwards (2008, 1149–50).
67. Weldon to Pearson [undated, but circa 20 January 1903], Pearson/11/1/22/40, PP.
68. Pearson (1903a, 228–29).
69. Pearson (1904a [read at the Royal Society on 26 November 1903], terms on 54, quotation on 73). Although Weldon later adopted Pearson's "proto-" and "allogenic" terminology, there is no sign of it in any writing that can be confidently dated before autumn 1903 (cf. Sloan [2000, 1076]). On Pearson's memoir as more conspicuously (albeit tacitly) indebted to Yule's article than to anything by Weldon, see Edwards (2008, 1144–45). Both Yule's article and Pearson's memoir figure in the prehistory of what would come to be known as the "Hardy-Weinberg theorem," stating that, when a freely interbreeding population is insulated from natural selection and other disturbing influences, the

frequencies of dominant homozygotes, recessive homozygotes, and heterozygotes will remain stable. As ever, the ur-statement was in *Report 1*. See Bateson and Saunders (1902, 130); Yule (1902, 225); and Pearson (1904a, 60). "Hardy" was the Cambridge mathematician G. H. Hardy; "Weinberg" was the German physician Wilhelm Weinberg. Each published his version of the theorem in 1908. In the weeks after the reading of Pearson's memoir, Pearson and Yule skirmished publicly over Johannsen on pure lines. See Pearson (1903b); Yule (1903); and the letters in *Nature* (7 January 1904): 223–24.

70. Pearson (1903c, 1904b, quotations from the latter on 154–55).
71. Weldon to Pearson, 18 October 1903, Pearson/11/1/22/40, PP. In October 1896, Galton, as referee for a Royal Society paper of Pearson's examining the evidence for telegony in humans, had tried without success to persuade Pearson to take into account the possible variability of the effects of nurture within the same family due to birth order: "The fortune & physical well-being of the parents may be (1) reduced, (2) remain the same or (3) be improved by the time the younger children are born. If reduced, there is less air space, less food and a worn out mother—if the same, the same occurs but in a less degree. If improved, there is the change from a street to a suburban villa & country air, with better food, doctoring etc." Galton to Pearson, 8 October 1896, Galton/3/3/7/20, GP; see also Pearson to Galton, 10 October 1896, Galton/3/3/16/9, GP; Pearson (1897b, 277).
72. When it came to natural selection's existence and power, "mere drudging proof of it time after time is not wasted, I am sure," Weldon explained to Pearson earlier that year: Weldon to Pearson, 17 June 1903, Pearson/11/1/22/40, PP. The paper Weldon finally published in 1904—his last on *Clausilia*—in fact reported a failure to find the expected narrowing, though a similar study undertaken that summer by a student, A. P. Di Cesnola, found it for a different species of snail. See Weldon (1904b); Di Cesnola (1907); and, for discussion, Hutchinson (1990). Many thanks to Juan Escalona for alerting me to Di Cesnola's Weldonian selection work, on snail shells and also on praying-mantis coloration (1904).
73. Weldon to Galton, 25 October 1903, Galton/3/3/22/17, GP; Galton to Weldon, 7 November 1903, Galton/3/3/7/28, GP, emphasis in original—though the next, very long letter in the sequence, cited in the next note, is indispensable for the interpretation ventured here as to what was on Weldon's mind.
74. Weldon to Galton, 8 November 1903, Galton/3/3/22/17, GP; Darwin (1868, 1:112–15); Weldon to Pearson, 12 December 1900, Pearson/11/1/22/40, PP ("Have you read about the Porto Santo rabbits lately? I do not at

present see any way of explaining Darwin's statements other than by saying that the embryonic 'mechanism' for producing black hair, under a suitable external set of conditions, has been inherited generation after generation for 500 years, and suppressed for lack of the external factor. How else can a wild Porto Santo rabbit redevelop the black points of a European Rabbit in the zoo? I must try to get some home alive!").

75. Herbst (1901); Weldon (1902b, 367); Weldon to Galton, 13 November 1903, Galton/3/3/22/17, GP. Weldon would have known Herbst and Driesch from the Naples station. On Herbst and the background to his 1901 book's picture of development as a process whereby external stimuli release internal potencies, see Oppenheimer (1970). On Driesch's work in that same period of close collaboration with Herbst, see Churchill (1969).
76. Weldon to Galton, 8 November 1903, Galton/3/3/22/17, GP.
77. Weldon to Pearson, 4 December 1903, Pearson/11/1/22/40, PP, referring to Laurence Sterne's *Tristram Shandy* (1760–67, book 2, ch. 19).
78. Weldon to Pearson, 8 December 1903, Pearson/11/1/22/40, PP.
79. Pearson (1904c); Weldon (1904a); Pearson (1904d), responding to Castle (1903b).
80. Weldon to Pearson, 18 March 1904, Pearson/11/1/22/40, PP. For the diagram, see Pearson (1904a, 83). For discussion, see Sloan (2000, 1076–77) and Bulmer (2003, 263–64).
81. See esp. Weldon to Galton, 6 and 20 March and 2 April 1904, Galton/3/3/22/17, GP; Weldon to Pearson, 3 and 11 March and 24 April 1904, Pearson/11/1/22/40, PP. Darbishire (1904) published a long fourth report on the mouse-crossing work in the January 1904 *Biometrika*.
82. Weldon to Pearson, 29 April 1904, Pearson/11/1/22/40, PP.
83. Doncaster (1904)—a paper inspired by the earlier work of Vernon along similar lines.
84. Weldon to Pearson, 8 May 1904, Pearson/11/1/22/40, PP, emphasis added.
85. Weldon to Pearson, 10 May 1904, Pearson/11/1/22/40, PP, emphasis in original.

Chapter Nine

Epigraphs: Inter-departmental Committee on Physical Deterioration (1904, 2:31); Bateson (1904b, 589); Punnett (1905, 59–60); Weldon, "Theory of Inheritance," MS ch. 5, 24, Pearson/5/2/10/4, PP.

1. On the committee and its findings, see Heggie (2008); see also Solo-

way (1995, 43–47) and Hynes (1968, 22–24). For "hereditary taint," see Inter-Departmental Committee (1904, 1:46–47).

2. Bateson (1904b, quotations on 589). A generous summary, including the epigraph quotation, was published two days later in the *Times* (London), 20 August 1904, 12.
3. "The British Association," *Times* (London), 20 August 1904, 8. See also two subsequent accounts, "Section of Zoology: Heredity and Variation," *British Medical Journal* 2 (27 August 1904): 448–49; and "Zoology at the British Association," *Nature* 70 (29 September 1904): 538–41. For the full list of papers (with some abstracts) read, though not always by their authors—Lock, for instance, was still in Ceylon—see the *Report of the Seventy-Fourth Meeting of the BAAS* [. . .] (1904, 590–95). On Darbishire's public move toward Mendelism, see Ankeny 2000, esp. 331–38. After the meeting, Pearson used the *Nature* Letters page both to query the closeness of fit between Darbishire's results and Mendelian predictions (*Nature* 70 [29 September 1904]: 529–30) and to reiterate how overblown talk of incompatibility between rival schools was, given Pearson's own Mendelian theorizing (*Nature* 70 [27 October 1904]: 626–27).
4. "P[unnett] agrees with me that the silence of *Nature* is significant, and I no longer feel much doubt that 'biometry' is damaged heavily." W. Bateson to B. Bateson, 27 August 1904, G.3.b.14, typescript in G.3.e.02.d, BP.
5. Weldon to Galton, 24 August 1904, Galton/3/3/22/17, GP. Weldon reported that when, in connection with the Mendelian claim that combs on the heads of fowl follow Mendel's principles, with extracted recessives indistinguishable from grandparental recessives, he asked Punnett for more precise numbers for the comb tubercles, Punnett "sneered, saying they left such unimportant statistical details to us."
6. "The British Association," *Times* (London), 20 August 1904, 8; "Zoology at the British Association," *Nature* 70 (29 September 1904): 539.
7. Weldon to Pearson, 7 and 10 November 1904, Pearson/11/1/22/40, PP.
8. *Nature* 71 (1 December 1904): 110. Bateson was proposed for the medal by Adam Sedgwick; see *Medal Claims 1873–1909*, 263–64, RS. (Two years previously, Bateson had successfully proposed Galton; see, in the same volume, 244–45.)
9. Bateson et al. (1905). For discussion, see Cock and Forsdyke (2008, 274–75). Three more reports to the committee followed, completed in 1906, 1908, and 1909: see Bateson et al. (1910).
10. Punnett 1905.
11. On "genetics" in 1905–6 and earlier, see Cock and Forsdyke (2008, 248).

12. Bateson (1907a, 91). Also included was a biographical essay on Mendel, including a facsimile letter and a photograph of Mendel's monastery (85–89).
13. Weldon acknowledged Bateson's stimulus in an April 1905 letter, when he was drafting what became the Galton section of a chapter comparing and contrasting Galton's and Mendel's theories. "I have been writing up a long account of the 'stirp,' because I think it wants doing, in answer to Bateson's statement that the Law of Ancestral Heredity is incomplete, because it does not give a theory of the constitution of gametes." Weldon to Pearson, 17 April 1905, Pearson/11/1/22/40, PP.
14. See ch. 2 in this book.
15. See ch. 8 in this book, note 71.
16. See ch. 8 in this book.
17. Galton (1905, 45).
18. Weldon to Galton, 24 August 1904, Galton/3/3/22/17, GP.
19. Indeed, when Weldon and Pearson in 1899 corresponded about how to make sense of Galton on eye color in *Natural Inheritance*, Weldon wrote of Galton's "habit of thinking in terms of Weismannian determinants: and I imagine him to think that eye-colour is as a rule determined by one complex determinant, which must [therefore] come from only one parent: but that in rare cases the determinant may become broken up into its constituent units during fertilisation, under which circumstances the new determinant may be formed from elements provided partly by one parent, partly by the other." Weldon to Pearson, 12 April 1899, Pearson/11/1/22/40, PP.
20. Just before the passage quoted at note 18, Weldon wrote to Galton (who was also reviewing the second report): "The whole simplicity of Mendelism depends on finding a category so wide that all these minor differences [e.g., concerning the number of tubercles in fowl combs], which can alone decide between your reversionary series and Mendel, are concealed. The moment you get closer descriptions, you find strange stories of the odd behaviour of extracted forms, such as those in the report of the Evolution Committee you have just read." Weldon to Galton, 24 August 1904, Galton/3/3/22/17, GP.
21. Weldon to Galton, 8 October 1904, Galton/3/3/22/17, GP.
22. Weldon to Pearson [9 October 1904], Pearson/11/1/22/40, PP.
23. Weldon to Pearson [9 October 1904], Pearson/11/1/22/40, PP.
24. Weldon to Pearson, 11 October 1904, Pearson/11/1/22/40, PP; Pearson (1906, 44).
25. Weldon to Pearson, 21 October 1904, Pearson/11/1/22/40, PP; see also the letter of 7 November, quoted in Pearson (1906, 45).
26. "Hooray! Let 'em all come! . . . Bateson must come, if he wants to, and

it is a sort of compliment, because it shows he is afraid." Weldon to Pearson, 11 November 1904, Pearson/11/1/22/40, PP.

27. Weldon to Pearson, 16 October 1904, Pearson/11/1/22/40, PP.
28. Weldon to Pearson, 16 October 1904, Pearson/11/1/22/40, PP.
29. Weldon to Pearson, 16 October 1904, Pearson/11/1/22/40, PP. Weldon responded directly and at length to Galton's annoyance in Weldon to Galton, 10 October 1904, Galton/3/3/22/17, GP.
30. Weldon to Pearson [October 1904], Pearson/11/1/22/40, PP. "Machine" here harks back to "Theories of the 'mechanism' of inheritance," as Weldon put it in his draft lecture outline. He eventually recommended that *Report 2* be published as it was, whereas Galton complained that it was badly written. RR.16.171–80, RS. (Hurst wrongly suspected that Pearson was the complainant—a suspicion that cannot have made relations any easier; Cock and Forsdyke [2008, 274–75].)
31. Pearson to Galton, [21] November 1904, Galton/3/3/16/9, GP. Pearson feared the room would be inhospitably cold, and Weldon confirmed it was "hardly fit to support human life at the best of times." Weldon to Galton, 27 November 1904, Galton/3/3/22/17, GP.
32. Bateson's admission card and extensive (but scrawled) notes can be found in G.5.b.1–7, BP. Pearson also took notes, preserved—along with the published syllabus for the lectures—in Pearson/5/2/6, PP. Pearson's notes peter out after lecture 5, and Bateson's are detailed mainly for lectures 6–8.
33. Hurst to Bateson, 9 December 1904, D.21.c.4, BP. I am not sure about "sighs."
34. Bateson to Hurst, 5 February 1905, D.21.a.54, BP.
35. Lecture-by-lecture abstracts appeared fortnightly and then weekly in the *Lancet* from early January to late March 1905. Along the way, the *British Medical Journal* published a two-part digest of the series overall, noting at the start that the "medico-biological importance of these lectures and the large audiences they have drawn showed that the subject of heredity as expounded by Professor Weldon commanded interest and attention." See "Current Theories of the Hereditary Process," *Lancet* 1 (7 January 1905): 42; (21 January 1905): 180; (4 February 1905): 307–8; (25 February 1905): 512; (4 March 1905): 584–85; (11 March 1905): 657; (18 March 1905): 732; (25 March 1905): 810; "Theories of the Hereditary Process," *British Medical Journal* 1 (25 February 1905): 441–42, quotation on 441; (4 March 1905): 505–6.
36. "Current Theories of the Hereditary Process," *Lancet* 1 (7 January 1905): 42.
37. "Current Theories of the Hereditary Process," *Lancet* 1 (21 January 1905): 180.

38. "Current Theories of the Hereditary Process," *Lancet* 1 (4 February 1905): 307–8.
39. "Current Theories of the Hereditary Process," *Lancet* 1 (25 February 1905): 512. The version in Pearson's notes is more pungent: "Every character at the same time is inherited & is also what is called acquired. Weismann [division] into germinal & somatic characters is not real & did harm. No purely somatic characters, no character, grows to same condition under any environment. No environmental condition which affects all animals in same way; every character is germinal & environmental." Pearson/5/2/6, PP, emphases in original. Weldon blamed Weismann for the distinction: see, e.g., Weldon to Galton, 10 October 1904, Galton/3/3/22/17, GP.
40. Weldon to Pearson [20 January 1905], Pearson/11/1/22/40, PP. The documentary material relating to Weldon's never-settled views on chromosomes is not easy to interpret. Beyond the lecture 5 summaries, published and unpublished, sources include a long letter to Galton on 6 June 1896, in Galton/3/3/22/17, GP; a long notebook entry, "About Mitosis," dated 2 January 1905, Pearson/5/2/10/3, PP; an untitled and undated manuscript using Pearson's "protogenic"/"allogenic" terminology (absent from Weldon's lectures and his extant book manuscript), in Pearson/5/2/8/2, PP; a set of reflections on chromosome individuality and its consequences, in Pearson/5/2/10/6, PP; and several letters from Weldon to Pearson between December 1904 and February 1905, notably one on chromosome individuality, 8 February 1905, Pearson/5/2/10/10, PP. After Weldon's death, Pearson drew on this material for what he called "a mathematical theory of determinantal inheritance from suggestions and notes of the late W. F. R. Weldon" (Pearson 1908). Although untrustworthy as a guide to what Weldon believed, it is invaluable for understanding the surviving Weldoniana as attempts to see where various—sometimes contradictory—suppositions led. Pearson's paper is itself hard going, though on Bernard Norton's gloss it includes an interesting demonstration of how, with the right assumptions about chromosomal-unit exchanges and dominance relations, most of the "pure" dominants and extracted recessives in a Mendelian F_2 generation will turn out to harbor opposite-character determinants: Mendelism without gametic purity. See Norton (1979, 191–93) and postscript 2 in this book. Many thanks to Charles Pence for illuminating correspondence and discussion on this topic (see also Pence [2022b]).
41. Saunders to Bateson, 25 January 1905, G.5.b.4, BP; Hurst to Bateson, 9 December 1904, D.21.c.4, BP. The question of how nuclei divide was separate from the question of whether inheritance is exclusively nu-

clear, and Weldon, in common with a number of biologists, believed it was not. See "Current Theories of the Hereditary Process," *Lancet* 1 (4 February 1905): 307–8, and, on the wider interest in cytoplasmic inheritance in that period, Harwood (1993, 61ff.) and Sapp (1987, 14–16).

42. "Current Theories of the Hereditary Process," *Lancet* 1 (4 March 1905): 584–85, quotation on 585. On Weismann's position and the criticisms it had already attracted from T. H. Morgan, see Esposito (2013a).
43. "Current Theories of the Hereditary Process," *Lancet* 1 (11 March 1905): 657; (18 March 1905): 732; (25 March 1905): 810. Some advertised topics got left out, notably Johannsen's pure lines. What appears to be a nearly complete draft text for the lecture on Mendel can be found in Pearson/5/2/9/12, PP. It is quoted from in Martins (2020, 19–21), though misidentified as coming from the earlier, abandoned book on evolution.
44. Bateson to Hurst, 5 February 1905, D.21.a.54, BP, emphases in original.
45. Galton to Weldon, 16 January 1905, Galton/3/3/7/28, GP.
46. Galton (1906, 16).
47. On Weldon's "astonishing energy," physical and intellectual, see Galton to Millicent Lethbridge, 6 September 1904, in Pearson (1914–30, 3B:528); Pearson (1906, 44).
48. Weldon's notes on his (soon abandoned) silkworm experiments can be found in Pearson/5/2/1/11, PP. For Pearson's recollection of this period, see Pearson (1906, 45–47).
49. Weldon to Pearson, 3 and 17 April 1905, Pearson/11/1/22/40, PP. The quotation is from the latter letter. The former letter is partially quoted in Pearson (1906, 45).
50. Bateson to Hurst, 3 July 1905, D.21.a.68, BP, quoted in R. Hurst (1971, 843). The likely conduit was George Darwin; in a letter to Pearson dated 2 July, Weldon wrote that Darwin had been at Oxford the night before, "raging against the way in which Bateson was persuading the Cambridge biologers to neglect your statistics, because W. B. himself could not follow what you write." Weldon to Pearson, 2 July 1905, Pearson/11/1/22/40, PP.
51. Weldon to Galton, 12 September 1905, Galton/3/3/22/17, GP.
52. Weldon, "Theory of Inheritance," MS ch. 1, Pearson/5/2/10/4, PP. The manuscript also includes what is plainly an earlier draft of the chapter. On the structure and contents of the manuscript, see Edgar Schuster's "Notes on Weldon's 'Inheritance Manuscript,'" Pearson/5/2/10/5, PP.
53. Weldon, "Theory of Inheritance," MS ch. 1, 4–5, Pearson/5/2/10/4, PP. On the history and significance of the discovery of the variation of latitude, see Carter and Carter (2002).

54. Weldon (1906a, 85–93); Turner (1904, ch. 6). In the mathematical appendices following the first chapter of Weldon's "Theory of Inheritance," Weldon thanked Turner—the Savilian Professor of Astronomy at Oxford and director of the Radcliffe Observatory—for help with one of the proofs.
55. Argon's discovery story got a fresh retelling in 1905 with the appearance of the third edition of William Ramsay's popular book *The Gases of the Atmosphere*; see Ramsay (1905, ch. 5).
56. Weldon, "Theory of Inheritance," MS ch. 1, 5, Pearson/5/2/10/4, PP; Weldon (1906a, 94).
57. Weldon, "Theory of Inheritance," MS ch. 1, 5, Pearson/5/2/10/4, PP.
58. Weldon (1906a, 94–109, quotation on 97); Weldon, "Theory of Inheritance," MS ch. 1, 6–21, Pearson/5/2/10/4, PP.
59. Weldon, "Theory of Inheritance," MS ch. 1, 1–2, Pearson/5/2/10/4, PP.
60. Weldon, "Theory of Inheritance," MS ch. 1, 20–21, Pearson/5/2/10/4, PP; see also Weldon (1906a, 107).
61. Weldon, "Theory of Inheritance," MS ch. 1, 2, Pearson/5/2/10/4, PP. Weldon (1902c, 637) had earlier called for cooperative work to overcome this standard division of labor.
62. A latish change along these organizational lines would explain why, within the manuscript, there is a draft version of ch. 2 (missing Galton's corrections) that is labeled as ch. 7, and which describes chs. 2 through 4 as dealing with the effects on tissues when the normal relations between parts are disturbed (28).
63. Weldon, "Theory of Inheritance," MS ch. 2, 1, Pearson/5/2/10/4, PP, emphases in original. Galton added in the final sentence; see the first of the pasted-in scraps on the manuscript in Pearson/5/2/9/13, PP. Weldon sent the draft in mid-April and received Galton's corrections in early May; see Weldon to Galton, 19 April and 5 May 1905, Galton/3/3/22/17, GP. Weldon was especially pleased to discover that, more than he had guessed, Galton was prepared to accept that determinants might vary in passing from one generation to the next.
64. Weldon, "Theory of Inheritance," MS ch. 2, 2–8, quotation on 2, Pearson/5/2/10/4, PP.
65. Weldon, "Theory of Inheritance," MS ch. 2, 9–15, quotation on 9, Pearson/5/2/10/4, PP.
66. Weldon, "Theory of Inheritance," MS ch. 2, 11, 16, Pearson/5/2/10/4, PP.
67. Weldon, "Theory of Inheritance," MS ch. 2, 16–23, quotation on 21, Pearson/5/2/10/4, PP. Notice that, on a Galtonian reading of Mendelian work, its centerpiece "allelomorphs" are not genuinely non-blending, like basset-hound coat colors, which stay non-blending under conditions of random breeding, but pseudo-non-blending, since they de-

pend for their existence on the artificial purging of all the variability that would otherwise make the characters look blending. End the regime of extreme selectivity, and yellowness will shade into greenness, roundness into wrinkledness, hairiness into hairlessness, beardlessness into beardedness, and so on.

68. Weldon, "Theory of Inheritance," MS ch. 2, 22–23, quotation on 22, Pearson/5/2/10/4, PP.
69. Ch. 3, mainly on regeneration in *Stentor* and planaria, exists in two versions, one labeled "ch.2." Ch. 4 is mainly about regeneration in sea urchins. The next chapter, on regeneration in earthworms on up, should be ch. 5, but is also labeled "ch. 4." The last chapter in the set, on the developmental effects of environmental changes, should be ch. 6, but is labeled "ch. 5."
70. Weldon, "Theory of Inheritance," MS ch. 5, 27–28, Pearson/5/2/10/4, PP, emphasis in original.
71. Weldon, "Theory of Inheritance," MS ch. 2, 3 (quotation) and ch. 5, 4, Pearson/5/2/10/4, PP. For further details on the Galtonian-chromosomal explanation of the Mendelian pattern, see postscript 2 in this book.
72. Weldon, "Theory of Inheritance," MS ch. 2, 7, and ch. 8 (incomplete, with several pages on peas and several on "Cuénot on Darbishire"), Pearson/5/2/10/4, PP; Weldon to Pearson, 18 October 1905, Pearson/11/1/22/40, PP. Weldon told Pearson he was also photographing the mice in order to include a "scale of whiteness" in the book. Weldon's mouse-breeding records were published posthumously in *Biometrika:* Weldon (1907); [Weldon] (1915, which included a finely graded photographic scale of piebaldness in the mouse coats). The original records can be found in Pearson/5/2/1/12, PP.
73. Weldon, "Theory of Inheritance," MS ch.1, 2, and ch. 2, 3–4 and 20, Pearson/5/2/10/4, PP. Weldon connected the dots between his emphasis on ancestry-environment interactions, Pearson's version of Galton's law, and the efficacy of selection most fully in the manuscript of what is possibly the 1902 Belfast address, in Pearson/5/2/10/7, PP, building on the statement in Weldon (1902c).
74. Quotation from Weldon to Galton, 13 October 1905, Galton/3/3/22/17, GP. Weldon's correspondence suggests that he came to think Darbishire neither bright enough nor assiduous enough to do first-rate scientific work, which made him vulnerable—as Herbert Thompson in the 1890s had not been—to Bateson's needling attentions. But Darbishire never fully accepted Bateson's position nor rejected Weldon's, even carrying out dice-rolling experiments after the latter's death. For a comprehensive study of Darbishire's work, see Wood (2015). For

the early publishing history of Punnett's book, which first appeared in May 1905, see Punnett (1910, iv). On the prehistory of the Punnett square—introduced to a wider audience in the 1907 second edition of *Mendelism*—see Edwards (2016, 2012a). On Punnett's life and work generally, see Edwards (2012b).

75. Bateson to Hurst, 26 May and 3 July 1905, typescripts, D.21.a.66 and D.21.a.68, BP.
76. Quotations from Bateson to Hurst, 18 April and 9 May 1905, typescripts, D.21.a.60 and D.21.a.65, BP. It seems that Bateson and Florence Durham were already at work on a histological study of color in horse hair when Hurst started in on a genealogical study of color in horse coats. A great deal of the Bateson-Hurst correspondence on horses and associated controversies in 1905–6 is quoted at length in R. Hurst (1971, 810–917). For a more concise version, see J. Schwartz (2008, 135–43). For Bateson's sense that every "exception" (his quotation marks) would be a hostage to fortune, see Bateson to Hurst, 2 December 1905, typescript, D.21.a.81, BP.
77. Pearson to Weldon, 19 November 1905, Pearson/5/3/9, PP.
78. "University of London: Advanced Lectures in Botany: A Course of Six Lectures on 'The Facts of Heredity,'" G.5.a.57, with notes at G.5.a.34, BP; Bateson to Hurst, 24 November 1905, typescript, D.21.a.79, BP, quoted in R. Hurst (1971, 865–66).
79. The advance schedule, published in *Nature* 73 (30 November 1905): 120, listed, in addition to Hurst's paper, two statistical papers by Yule and a biometric study of paramecia by a new American recruit to the biometric methods, Raymond Pearl. Bateson's understanding was that a notice would appear in the *Times* advertising "a general discussion on statistical methods"; Bateson to Hurst, 2 December 1905, typescript, D.21.a.81, BP, quoted in R. Hurst (1971, 868).
80. Bateson, Saunders, and Punnett (1906). Although hastily put together, this paper, analyzing crosses with sweet-pea varieties, reflects a number of then-recent innovations: in its use of a new language of "factors" as what get inherited; in its attention to character versions getting inherited together (what came to be called "linkage"); in occasioning the first public presentation of a diagram ancestral to the Punnett square (see Punnett [1950, 8–9]; Edwards [2012a, 2016]); and in its suggestion that allelomorphic pairs represent the presence and absence of a factor (later known as the "presence and absence theory": see Swinburne [1962], S. Schwartz [2002]). At the end of Hurst's horse paper, the presence or absence of black pigment in the hairs is asserted to be "the critical feature which distinguishes chestnut from bay and brown"; Hurst (1906, 391–92, quotation on 391).

81. R. Hurst (1971, 867–69).
82. Weldon to Pearson, 8 and 10 November 1905 (quotation from latter), Pearson/11/1/22/40, PP.
83. Weldon to Galton, 19 November 1905, Galton/3/3/22/17, GP.
84. Weldon to Galton, 24 November 1905, Galton/3/3/22/17, GP. According to Punnett, even the Mendelians regarded Hurst as "over-apt to find the 3:1 ratio in everything he touched," to a degree that made Bateson mistrustful of him; Punnett (1950, 8).
85. Weldon to Galton, 10 December 1905, Galton/3/3/22/17, GP; R. Hurst (1971, 869–70).
86. Bateson to Hurst, 7 December 1905, typescript, D.21.a.85, BP, quoted in R. Hurst (1971, 870–71); Hurst (1906, 241–42). On Alice Hurst and her participation in this group, see R. Hurst (1971, 587–88, 872).
87. R. Harrison to Bateson, 13 December 1905, with an undated letter from Bateson in response, D.21.c.8–9, BP, quoted in R. Hurst (1971, 13–16). In the Bateson Papers, Hurst-generated correspondence and drafts relating to the 1905–6 push on horse coat colors can be found in G.5.c.1–7 and G.5.e.10 as well as in D.21.c.
88. Weldon to Galton, 10 December 1905, Galton/3/3/22/17, GP.
89. Weldon to Pearson, 23 December 1905, Pearson/11/1/22/40, PP; Weldon to Galton, 22 and 28 December 1905 (quotation in latter), Galton/3/3/22/17, GP. Weldon's horse-pedigree note cards can be found in Pearson/5/2/1/4, PP.
90. Weldon (1906b, quotations on 396, 398).
91. Bateson to Hurst, 19 and 23 January 1906, D.21.a.102 and 103, BP, quoted in R. Hurst (1971, 880); see also the very detailed account in Hurst to Bateson, 19 January 1906, G.5.c.7, BP.
92. Weldon to Pearson, 22 and 29 January 1906, Pearson/11/1/22/40, PP; Weldon to Galton, 4 February 1906, Galton/3/3/22/17, GP. Pearson told Galton that Bateson at the meeting had drawn "red herrings over the course as usual"; Pearson to Galton, 24 January 1906, Galton/3/3/16/9, GP. From Weldon's perspective, one of those red herrings concerned a horse named Ben Battle. At the 7 December meeting, Weldon had mentioned Ben Battle as an example of a chestnut sire whose offspring with chestnut mares were not themselves straightforwardly chestnutty. At the 18 January meeting, Bateson announced the discovery that, although recorded as chestnut in the stud book, Ben Battle had never run as chestnut, but always as bay or brown. Where Bateson and Hurst put that discrepancy down to a recording error, Weldon regarded it as but one of many examples of change of dominance as a horse grew older, in line with Galtonian expectations (but an impossibility if chestnut results only from the complete absence

of non-chestnut determinants). See, in addition to the letters cited above, Weldon to Pearson [February 1906], Pearson/11/1/22/40, PP; and Weldon to Galton, 15 February 1906, Galton/3/3/22/17, GP.

93. Galton to Weldon, 1 February 1906, Galton/3/3/7/28, GP.
94. Weldon to Galton, 4 February 1906, Galton/3/3/22/17, GP.
95. On Weldon's last weeks and days, see Pearson (1906, 48–49) (but ignore Pearson's mistakenly putting the Weldons' Roman holiday in February rather than December–January). On the visit to the records office, see Weldon to Galton, 15 February 1906, Galton/3/3/22/17, GP. On Weldon's rapid deterioration, see Bateson to Hurst, 23 April 1906, typescript, D.21.a.115, BP, quoted in R. Hurst 1971, 915.
96. For Weldon's reaction on seeing the new *Proceedings*, dated 12 April 1906 (the day before Weldon's death), see the extract from Pearson's letter to Hurst in Hurst to Bateson, 9 May 1906, D.21.c.13, BP, quoted in R. Hurst (1971, 916–17). The postscript in Hurst's paper is dated 31 January 1906, and seems to have been the joint work of Bateson and Hurst.
97. Mark Ridley (1986, 46) once claimed that Weldon "burned himself out, and went mad before he died," but I find no sign of anything along those lines in the surviving correspondence.

Chapter Ten

Epigraphs: Lawrence (1907, 70); Vilmorin (1907, 74).

1. Bateson (1907a, 91); Vilmorin (1907, 74); Bateson (1907b, 76).
2. Cambridge University Association (1906), quoted in Opitz (2011, 82).
3. Biffen (1907a, 35).
4. Biffen (1907b, 376–77).
5. For the most part, conference participants either dwelt constructively on Mendelism or ignored it. A notable exception was the Groningen naturalist C. L. W. Noorduijn (1907).
6. Although submitted to the Royal Society in spring 1907, Hurst's paper was not printed in the *Proceedings* until early 1908, months after Davenport's paper (written with his wife Gertrude) had been published in *Science*. See Davenport and Davenport (1907, quotation on 592); Hurst (1908).
7. Nominally in biology, Bateson's professorship was endowed for five years with the stipulation that its holder promote "enquiries into the physiology of heredity and variation" (i.e., genetics). When funds for a permanent endowment came in 1912, the professorship was renamed the Balfour Professorship in Genetics. By then, Bateson had left Cam-

bridge for the John Innes Horticultural Institution, and Punnett was elected to the chair.

8. Johannsen (1909, 124); the relevant passage in English translation appears in Carlson (1966, 20 and 22).
9. Bateson (1909). On the book's status as an authoritative guide for students, see Skopek (2008, 37).
10. On Bateson's visit to Brünn in 1910, see Cock and Forsdyke (2008, 561–64).
11. Bateson et al. (1910). The final entry in the committee's minute book is dated 6 July 1909: CMB/65/22, RS. According to Punnett, the committee formally ceased to exist only in December 1910, when the Royal Society declined to reappoint it. Punnett to B. Bateson, 25 October 1926, C.15.e.1, BP.
12. On the *Journal of Genetics*—the first issue of which came out in November 1910—as "a continuation of the *RS-Reports* in another form," see Cock and Forsdyke (2008, 337). Its American counterpart ran from 1910 to 1913, then was reborn as the *Journal of Heredity* when the American Breeders' Association rebranded itself the American Genetic Association; see Kimmelman (1983, 166).
13. A successful science, one might say, becomes independent of a founding research school, even one that spreads to other institutions via missionizing members. On how research schools become successful, the standard resource remains Geison and Holmes (1993; see esp. the list on 176), identifying fourteen common features: a charismatic leader, a pool of potential recruits, and so on. The Cambridge school under Bateson fits to a T.
14. Johannsen used "genotype" and "phenotype" in the same 1909 book in which he introduced "gene," promoting all three terms in a paper in English read to great acclaim at an American conference in December 1910 and published a few months later; see Johannsen (1911) and, for discussion, Roll-Hansen (2014, esp. 2432–33).
15. Bateson (1907a, 91).
16. I draw here on the analysis of intellectual ownership developed in a general way in MacLeod and Radick (2013) and in connection with Mendelism in Charnley and Radick (2013) and Radick (2013b).
17. Punnett wrote the 1911 "Mendelism" article; the zoologist Peter Chalmers Mitchell wrote the "Heredity" article. From the 1974 edition, I cite the short *Micropaedia* article (in the 1994 printing). On the need to treat the history of the twentieth-century science of heredity as more than the history of Mendelism, see esp. the introductions to Gausemeier, Müller-Wille, and Ramsden (2013) and Müller-Wille and Brandt (2016).

18. See, e.g., Punnett (1905, 1911).
19. Darbishire (1907, esp. 133–34). On D'Arcy Thompson's interest in this paper, see Esposito (2013b, 207n136).
20. A marvelous starting point for the investigations we still need into all that diversity is Mike Buttolph's database of 101 Mendelians, covering the years up to 1911; see Buttolph (2008).
21. Within historical and philosophical studies of the Mendelian gene concept, Jean Gayon's view that the "central supposition of genetics" is that "all inherited characters are determined by one or more pairs of factors that are not necessarily identical" comes closest to mine. He dated the emergence of this "heuristic proposition" to 1901–3, calling it "a kind of gamble upon which the Mendelian science of heredity was built." See Gayon (1998, 273–79, quotations on 276–77).
22. Attributed to the American embryologist—and heterodox Mendelian—Ernest Everett Just, in Harrison (1937, 372). I first learned about Just's remark from Maurizio Esposito's Leeds PhD thesis, published in revised form (2013b; see esp. 50, 134–43).
23. In my account of explanation I lean heavily on Geoffrey Hawthorn's adroit sifting and synthesizing (1991, 10–26, esp. 12; the leaning continues in this book's counterfactually engagé chapters 11 and 13). On contrastive explanation, see Garfinkel (1981, ch. 1) and Lipton (1990). Since Hawthorn wrote, the most influential philosophical treatment of explanation has been Woodward (2003), in which explanation, causation, and counterfactuals are similarly linked. But not all counterfactual-generating explanations are causal (see, on mathematical explanation, Saatsi and Pexton [2013]), and not all causal explanations generate counterfactuals that are as directly related to their explanations as in my examples (see Nolan [2013, 326]). My assimilating scientific and historical causal explanations is, in certain quarters, contentious; see, e.g., Arabatzis (2019).
24. For (1), see, e.g., Kim (1994, esp. 188) and Graham (1998, 17–31); for (2), see, e.g., Michie (1955, see esp. 27 for discussion of Bateson on "scientific Calvinism"); Lewontin (1993, 1–37); and Rose, Kamin, and Lewontin (1984, chs. 2–4); for (3), see, e.g., Thurtle (2007); Müller-Wille and Rheinberger (2012, with critical discussion in Radick 2014a); and Bonneuil (2016). I am grateful to Matthew Wright for the Michie reference and to Fernand Braudel for the marine metaphor (quoted in Hughes-Warrington [2000, 20]).
25. The classic list of nonexistent scientific entities is in Laudan (1981). A question related to mine, about whether the truth of a scientific theory can explain its acceptance, is the subject of a small philosophical literature responding to the sociology of scientific knowledge (dis-

cussed in the next section), in particular its "symmetry principle": that the explanation of why the winners in a scientific debate came to believe what they did should not be different in kind from the explanation of why the losers came to believe what they did (e.g., in MacKenzie and Barnes [1974, 14–15]). An excellent critical survey can be found in Bouterse (2016, ch. 5, 6–8).

26. For a general analysis of these problems, see Rosen (1996, esp. 52, 184–200). For an earlier application to the history of science, see Radick (2009, 159–61).
27. On Picasso's jettisoning of single-point perspective as belonging to the same cultural moment and tendency as Einstein's jettisoning of an absolute reference frame, see J. Jones (2004). Although the two strains I pick out were undoubtedly contemporary, the purification strain became culturally prominent only after the Great War; see, e.g., Loach (2018).
28. McDonald (2011, 34–36, quotations on 34, 35); Holmes and Loomis (1909, quotations on 51); Boas (1918).
29. In his introduction, McDonald (2011, 1) puts in bold print, "It is an embarrassment to the field of biology education that textbooks and lab manuals continue to perpetuate these myths" (see also his "summary for worried parents" checklist on 4). Of the characters he examines, only wet versus dry earwax survives scrutiny intact—but even here, he allows that, in rare cases, two dry-earwax parents might have a wet-earwax child.
30. Wilshire (1909, 5–6).
31. *Julius Caesar* (4.3.119–21); *The Merchant of Venice* (3.5.1–10). I have concentrated here on difficulties for the start of (3)'s story. What of the end? Especially impressive to (3)'s adherents, in my experience, is the seeming fit of the Mendelian gene concept with the needs of the scientistic modern state. But there is an obvious problem here, in the form of the mid-twentieth-century Soviet Union. No modern state embraced science more vigorously in guiding social reform; none was less enthusiastic about Mendelism. For further discussion, see ch. 11.
32. MacKenzie and Barnes (1974). Although widely circulated among historians of science, the paper appeared in full only in German translation. The versions published in English are MacKenzie and Barnes (1979) and MacKenzie (1981a,b). Valuable commentaries include Hesse (1978, 61); Roll-Hansen (1980, with a response from Barnes [1980]); Roll-Hansen (1983); and Kim (1994). Many thanks to Donald MacKenzie for sending me a copy of the original paper.
33. MacKenzie and Barnes (1974, quotations on 1, 22).

34. MacKenzie and Barnes (1974, esp. 22–29).
35. W. Coleman (1970); MacKenzie and Barnes (1974, 25–28). For more extensive discussion of Coleman's paper, the use it made of Mannheim, and the uses made of it by MacKenzie and Barnes, see Radick (2013b, esp. 287–90).
36. On Bateson's decades-long campaign on the utility of Mendelism, see Radick (2013b, esp. 281–85).
37. From the beginning, MacKenzie and Barnes acknowledged that Pearson objected not to Mendelism per se, but to Mendelian overreach, and that, given Weldon's silence on his political commitments, he was not Pearson's twin; see MacKenzie and Barnes (1974, 32–33 and 29, respectively). In subsequent years, however, MacKenzie came to appreciate that the differences went much further; see MacKenzie (1981a, 260n8; 1981b, 248n17, crediting Norton [1979]). For a more recent overview of the attacks by Pearson's London group on Mendelism applied to human racial, mental, and moral traits, see Porter (2018, 277–79).
38. It is striking how far even critics of MacKenzie and Barnes took for granted the idea that what was really at stake in the debate was continuity versus discontinuity. See, in addition to the works listed in note 32 above, Olby (1989a).
39. A variation on the theme is to suggest that, since American geneticists never had difficulty being Mendelian *and* biometric, the "debate over Mendel" was a storm in a British teacup. See, e.g., Wright (1966a, 248–49; 1978 [1986 reprint], 5); Kevles (1981); and Vicedo (1995). Again, the important point is that such easy meshing and selective appropriation was done on Mendelian terms, with Mendelism serving as the absorbing science, in an honorific way when not in a substantial one, and notwithstanding large flexibilities in what counted as "Mendelian" and "biometric."
40. Yule (1907). On the paper's foundational status for quantitative genetics, see—from its centennial year—Roff (2007, 1017). On Yule's achievement in remaking Galton's law as a multiple regression equation under polygenic Mendelism, see Wright (1978 [1986 reprint], 5) and Bulmer (2003, 268–70, esp. 270; 2004b, 17).
41. *Julius Caesar* 4.3.216–19.
42. Cf. Kim (1994, ch. 4, esp. 83), where Yule and Darbishire are characterized as card-carrying biometricians "converted" to Mendelism: an overstatement of the strength of their commitment in both directions.
43. On the absence of full Kuhnian incommensurability from the debate between the Mendelians and the biometricians, see Radick (2005a,

43–44); see also MacKenzie and Barnes (1979, 198–201). Commentary on Kuhn's book is legion, but the best expositor of his position is Kuhn himself, especially in the second edition (1970).

44. Here I take up Lorraine Daston's invitation to historians of science to see in reasoning from exemplars—the most distinctive as well as the most durable of Kuhn's emphases, and the binding element in my three-part explanation—an opportunity to reclaim something of Kuhn's interest in making historicized science generally instructive. See Daston (2016), discussed in Radick (2017).
45. Russell (2006, 33). In 2022, when this book was in press, the Leeds course was revamped along Weldonian lines, and *iGenetics* accordingly demoted from primary textbook to secondary further reading.
46. Kuhn (1970, esp. 137–43, but also 46–47, 187–89).
47. For an appreciation of how the process of communicating to outsiders about a science helps to constitute it, by transforming provisional research findings and divergent theorizing into simplified facts embedded in coherent theory, the classic book is one on which Kuhn drew, the Polish microbiologist Ludwik Fleck's *Genesis and Development of a Scientific Fact*; see Fleck (1979, esp. 111–25), updated influentially in Secord (2004). I take the word "teachable" from Bruce Hunt (2007, 145), who credited his teacher, Owen Hannaway. On the role of textbooks in consolidating the principles of quantum mechanics, see French (2020, ch. 8).
48. Hurst (1907).
49. Punnett (1907, 47–50, quotation on 50). In other words, what looked initially like a one-character cross concerning coat color (gray or black) was revealed to be a two-character cross, with the second, along-for-the-ride pair of allelomorphs determining pigmentation (*D*) and its absence, i.e., albinism (*R*). On what *Mendelism* in general, and Punnett's square in particular, did for Mendelism, see, respectively, Müller-Wille and Parolini (2020) and Wimsatt (2012).
50. Sturtevant (1910); J. Schwartz (2008, 190–91). Sturtevant also recalled reading Lock (Allen [1978, 308]), and in the paper cited Bateson's 1909 *Mendel's Principles of Heredity* (which discussed Hurst on horses). By 1910, most English-language surveys on heredity endorsed Mendelian segregation; see Kim (1994, 112–13).
51. Kuhn (1970, 47). "Normal science" was Kuhn's famous name for research in this exemplar-extending mode, whose fundamentally conservative and even dogmatic nature—"a strenuous and devoted attempt to force nature into the conceptual boxes supplied by professional education"—he saw as indispensable for scientific progress. See Kuhn (1970, esp. 5 [quotation], 23–42; 1963).

52. In his opening address at the conference, Bateson (1907a, 92) stressed how vital it was, at this early stage, not only that as many kinds of organisms as possible be studied but that everyone keep tabs on what everyone else was learning, since no one could know in advance whose investigation was going to furnish the next generally relevant breakthrough. For discussion of Mendelian problem solving as Kuhnian exemplar extension, see Shan (2020b).
53. Morgan (1910), though, as stressed in Allen (1978, 152), Morgan took care here to state merely that eye-color factors and the "sex factor" were associated. He symbolized the latter by "X," but without ever referring to chromosomes. For Morgan's pre-1910 skepticism about Mendelian factors, see, e.g., Morgan (1909). Extensive citations to Morgan's experimental studies of regeneration can be found in Weldon, "Theory of Inheritance," MS chs. 3–5, Pearson/5/2/10/4, PP. On Morgan's regeneration work, see Sunderland (2010). On his embryological work generally, including an 1897 book on frog-egg development reviewed in *Nature* by Weldon (1897b), see Maienschein (1991, ch. 8). On continuities between Morgan's embryological and genetic work, see Kingsland (2007) and Frezza and Capocci (2018). On Morgan and the fly room, see J. Schwartz (2008, chs. 9 and 10); Kohler (1994, chs. 2 and 3); and Allen (1978, chs. 1–5).
54. Russell (2006, 57).
55. Punnett (1909, 52–55); J. Schwartz (2008, 191–94).
56. Lock (1906, esp. 209–15, 232–63, quotation on 248–50); J. Schwartz (2008, 189–90, 200–204, 325–26n15). On Lock's book, which Muller read in preparation for the introductory course given by Morgan's Columbia colleague E. B. Wilson (a pioneer in research on sex determination by the chromosome), see Edwards (2012b). On the development of multifactorial, selection-friendly Mendelism between Yule and the Morgan group, especially in the work of the Swedish botanist Herman Nilsson-Ehle and the American maize biologist Edward East, see, e.g., Kim (1994, ch. 7); Allen (1978, 182–85).
57. Morgan et al. (1915, esp. chs. 2, 8, and 9, quotations on 45–46). On eye color as multifactorial, see 172–73, 208–9. On selection as factor recombination, sometimes facilitated by mutation, see 197–207. For the book's reception, see Brush (2002).
58. On this linking of understanding and utility, knowledge and power, as distinctive of what we call "science" since the early seventeenth century, when Francis Bacon articulated it as part of his case for the new, post-Aristotelian, experiment-avid natural philosophy, see Dear (2005).
59. Lock (1906, 215–21, with a first-hand testimonial to "the enthusiastic

interest with which practical men greeted [Biffen's] communication" at the 1906 London conference on 216; and 287–90, quotation on 288). The phrase "the science of genetics applied to human affairs" is from Lock's table of contents, xiii.

60. Punnett (1907, 58–59).
61. Punnett (1909, 78–82, quotation on 81–82).
62. On my reading, Kyung-Man Kim's *Explaining Scientific Consensus: The Case of Mendelian Genetics* (1994), although pioneering in its attention to breeders and medics in the making of Mendelism's success (see esp. ch. 5), in effect argues that Mendelian problem solving proved extendable in the wider world because Mendelians got nature right.
63. Kuhn (1970, x). Whatever Kuhn's rationale, others have seen in his omission a Cold War gesture, fitting in with American science policy and its emphasis on insulating science from external demands; see esp. Fuller (2000).
64. Surveys of historical studies of Mendelism among the breeders can be found in Charnley and Radick (2013, 223) and, more comprehensively, Berry (2021, 503–6). On Mendelism's modest value to commercial breeders at least up to the Great War, see Harwood (2015) and Paul and Kimmelman (1988, 295–96). On Vilmorin's slowness in incorporating Mendelian methods, see Kingsbury (2009, 178). For an analysis of Mendelism's treatment in successive editions of John Percival's textbook *Agricultural Botany*, widely used in Britain throughout the first half of the twentieth century, see Charnley and Radick (2013, 223–24, 229).
65. On technological futurism as potentially self-fulfilling, see Radick (2014b). Like "paradigm shift," "self-fulfilling prophecy" entered the English language from a piece of sociologically astute history and philosophy of science; see Merton (1948).
66. Bruce (1918, quotation on 12). I am grateful to Tina Barsby for sending me a copy of the pamphlet in which Bruce's encomium appeared, along with several other memoranda arguing the case for the establishment of the National Institute of Agricultural Botany.
67. H. C. (1926), interviewing "Professor Sir Boudleigh Bluffin." (Biffen was knighted in 1925.) Many thanks to Matt Holmes for locating this article, and to Bill Clark for helpful discussion.
68. See Morgan et al. (1915, 46–47, 226). On the Morgan group's construction of Mendelizing flies, see Kohler (1994, 28–29), with discussion in Radick (2003, 197–98).
69. Charnley and Radick (2013, 230–31). On Biffen and the emergence of a "Mendelian system" in Britain, see Charnley (2011). On the early history of NIAB, see Berry (2014a).

70. Hacking (1992, esp. 58–60, quotation on 59); see also Dear (2005, 403). What I have called the "traditional" and "alternative" accounts of experimental knowledge are helpfully contrasted in Schaffer (2009, 31–32). For a suggestive paralleling of Mendelism's agricultural extension with microbiology's medical extension as analyzed by Bruno Latour (also admired by Hacking), see Berry (2014b).
71. Punnett (1909, 86–89, quotations on 86, 88–89). The work on brachydactyly was due to William Farabee and, latterly, Harry Drinkwater; on cataracts to Edward Nettleship; and on alkaptonuria, as we have seen, to Archibald Garrod, who in 1909 included his studies of it in his classic book *Inborn Errors of Metabolism*. See also Bateson (1909, ch. 12) and, for historical overviews, Rushton (2017) and Kim (1994, 99–109). So keen were Bateson and Hurst on promoting applied Mendelism among the medics that in 1908 they helped found a new periodical, *The Mendel Journal*, in collaboration with George Mudge, a zoologist on the faculty of a London hospital; see Rushton (2017, 65) and Kim (1994, 81–82). Under the auspices of a new "Mendel Society," the journal ran until 1912; see Mazumdar (1992, 70).
72. Bateson (1909, 303–6, quotations on 305–6). MacKenzie and Barnes (1974, 28) noted Bateson's doubts about what would soon be called "positive eugenics," i.e., the breeding of superior humans, but were silent on his enthusiasm for the negative version. On Bateson's attitudes to eugenics and its proponents, I learned much from Jennifer Sutton's (2017) undergraduate dissertation at Leeds.

Chapter Eleven

Epigraphs: Bateson (1902a, 37–39); Weldon (1906a, 92–93). The "residual bubble" that Weldon mentioned was, as he explained, what was left when Henry Cavendish eliminated all the known gases from a sample of air, as reported in an account published in 1785.

1. Kuhn (1970, quotation on 4). On the strong, and controversial, contingentism in Kuhn's *Structure*, see Hacking (1999, 96–99).
2. For the background to the science wars, see Hacking (1999, esp. chs. 1–3). On the debt to Hacking's book among recent students of contingency and inevitability in the scientific past, see Soler (2015a, 4–5) and Radick (2016a, 160–61).
3. On "why historians (and everyone else) should care about counterfactuals," see Nolan (2013). On why historians of science in particular should care, see Radick (2005b, 2008a, and 2016a). Work by some historians and philosophers of science who do care can be sampled

in Radick (2008b), Soler and Sankey (2008) and Soler, Trizio, and Pickering (2015).

4. Florence Weldon, chronology of W. F. R. Weldon's life, Pearson/7/29/3, PP, 8.
5. My interpretation of the ties binding explanation, counterfactuals, and understanding owes a great deal to the account in Hawthorn (1991), esp. ch. 1.
6. D. Lewis (1973, 1). "Counterfactual" in the now-familiar sense was not prominent prior to Nelson Goodman's 1947 paper "The Problem of Counterfactual Conditionals." In a 1946 paper on the topic by Roderick Chisholm, titled "The Contrary-to-Fact Conditional," "counterfactual" occurs in the first paragraph and fitfully thereafter.
7. On Weldon's counterfactual, see Arabatzis and Gavroglu (2016)—a splendidly Kuhnian study of the argon discovery story, stressing the role of far-from-sudden rethinkings on a number of topics, including the very nature of an element.
8. In blending the fictive and the factual in aid of understanding, counterfactuals thus belong to a methodological genre that also includes idealizations, model making, simulations, thought experiments, and some physical experiments. See Radick (2008a, 551).
9. Kroeber (1917, quotation on 199).
10. Macaulay (1828, 36), discussed in Merton (1961, 353)—the classic article on "multiples" in the history of science—as well as in Radick (2016a, 161).
11. The best introduction to the debate remains Conway Morris and Gould (1998–99). See also Gould (1989, esp. ch. 5); and Conway Morris (1998, esp. ch. 8; 2003). For an even-handed updating from within evolutionary biology, see Losos (2017). For some helpful distinction-making within and across the categories "contingency" and "inevitability," see, for the history of life, Sterelny (2005), and for the history of science, Martin (2013) and Kinzel (2015).
12. On Mendel's "rediscovery" and what to make of it, see Brannigan (1981, ch. 6); see also Olby (1989b; 1985, ch. 6). The question of whether de Vries was in effect shamed into discussing Mendel has never been satisfactorily resolved. For details, see Brannigan (1981, 94).
13. Morgan et al. (1915, 226 [on de Vries], 137–38 [on Correns], 190 [on Tschermak]). See also, on de Vries, Gould (2002, 418–19); on Correns, Sapp (1987, 72–73) and Harwood (1993, 34, 62–74); and on Tschermak, Harwood (2000, 1064–65) and Šimůnek et al. (2012).
14. On whether Darwin and Wallace converged independently on natural selection, see Radick (2009, 153–54).

15. On convergence and the eye, see Gould (1999, 331–32); cf. Conway Morris (2003, 151–73, 193). On the large scope for disagreement about claims for convergence in biology, see Woolfson (2004).
16. Simpson (1953, quotations on 110). On the Baldwinian triple, see Richards (1987, 480–95). (For Osborn in 1904, see this book's fig. 9.1.)
17. For Lysenko in his own words, and in English, the indispensable volume is Lysenko (1954). The quotation is from Lysenko (1936–37, 186). Excellent overviews of Lysenkoism include Roll-Hansen (2005); deJong-Lambert (2012); and, for its international aspect, deJong-Lambert and Krementsov (2017). The phrase "Mendelism-Morganism" remains redolent of Lysenkoism, but another put-down label of Lysenko's, "classical genetics," was embraced by Western genetics; see Skopek (2008, ch. 4). Another terminological legacy is "Baldwin effect," since, as Simpson (1953, 110–11) made plain, his article on it was a neo-Darwinian's response to then-recent developments in the Soviet Union.
18. The quotation is from Lysenko (1948, 552). The full set of cartoons, along with an annotated translation of the article they illustrated, are reproduced in Studitski (1949). For discussion, see Krementsov (2018, 346). On the flourishing of Mendelian genetics in the early Soviet Union with Muller's help, see J. Schwartz (2008, 232–36).
19. Graham (1998, 17–31, quotation on 31). On Khrushchev's visit to Roswell Garst's farm, see Bershidsky (2016). On hybrid corn in Khrushchev's mid-1950s attempt to oust Lysenko, and the political fallout in the Soviet Union, see deJong-Lambert (2012, 148–53).
20. Graham (1998, quotation on 25). For Graham's reflections on the complex cultural politics surrounding the recent revival in Russia of the reputation of Lysenkoism, now seen as a forerunner of epigenetics, see Graham (2016).
21. Zirkle (1949, 25–26, quotation on 26); Huxley (1949, 181–82). On the 1950 "Golden Jubilee" for genetics (as measured from the 1900 triple rediscovery) in the context of the controversy, see Wolfe (2010, esp. 69–74; 2012).
22. See Levins and Lewontin (1976, esp. 171–74, 188–91) and Berlan and Lewontin (1986).
23. The methodology of counterfactual history is work in progress; but one of the ideals that enjoys robust support is what has been called the "minimal-rewrite rule." See, e.g., Levy (2015, 378, 382, with discussion on 389–94); Ferguson (1997, 83–90); Tetlock and Belkin (1996, esp. 7, 18, 23–25 [source of the "minimal rewrite" terminology]); and Hawthorn (1991). James Cushing's minimal rewrite of the history of quantum mechanics (1994, see esp. 174–93) pivots on the causal quantum

theory that, he argued, was a near miss in 1925–27. Closer to the subject of this book is Peter Bowler's (2013) minimal rewrite of the history of evolutionary theorizing, on what would have happened had Darwin drowned at sea while on the *Beagle* (1). I am making a case partly like Cushing's, in aiming to show that an alternative science of heredity was a near miss circa 1906, and partly like Bowler's, in that my counterfactual scenario involves a life-or-death moment going differently for an individual. But the debt to Bowler is greater, especially his refusal of the notion that, when conjecturing counterfactually, there are only two possibilities: either everything would have been more or less the same, or everything would have been more or less completely different (see Bowler [2008]). For further comparative discussion of Bowler's project and mine, along with critical reflections on rewrite minimalism in counterfactual history of science, see Hesketh (2016); Dagg (2019); and Tambolo (2020a,b).

24. On the defects of "dominance" talk, see Allchin (2000, 2002, 2005); see also Falk (2001a). On the defects of "gene for" talk, see Kitcher (1996, ch. 11); Burian and Kampourakis (2013); and Kampourakis (2017).
25. On the ideal "informational context" required to make well-informed decisions about genetic test results and "gene for" discovery claims, see Kitcher (1996, 70–72, quotation on 72, 246–54); see also Kampourakis (2017, ch. 12 [with discussion of *BRCA1* on 232–39]). For further reflections on the terms "genetic determinism" and "interaction," see postscript 1 in this book.
26. For a more detailed account of the project and its findings, see Jamieson and Radick (2013 and esp. 2017). Unbeknown to us, Michael Dougherty, then director of education at the American Society of Human Genetics, had proposed updating the genetics curriculum along similar lines—what he called "inverting the genetics curriculum"; see Dougherty (2009, 2010). Multiples again! Of course, as a matter of principle, no one should have needed Weldon's example in order to want to construct a curriculum along variability-and-context-emphasizing lines. Any devoted reader of Richard Lewontin (1982, 1991); Rose, Kamin, and Lewontin (1984); or Evelyn Fox Keller (2000, 2010, 2014) would be well-placed and well-motivated to come up with more or less the same thing.
27. Her teaching materials are freely available online at the project website: https://geneticspedagogies.leeds.ac.uk/.
28. For all that Dougherty's inverted curriculum takes a similar form, and to similar ends, developmental biology does not figure in a big way as a means to those ends. Yet our experience—and that of others now conducting their own Weldonian classroom experiments—is

that providing students with a good grounding in understanding how development works is crucial.

29. Although we did not know about it at the time, an earlier study, by Bates et al. (2003), corroborated our hunch that this example did not invite deterministic interpretations and so meshed well with our goals.

30. For all the shortcomings (see Donovan [2022, 6]), our curriculum study was the first to use science teaching as a means of putting a plausible speculation about a might-have-been successful science to empirical test, and so, in a modest way, of taking that science out of the counterfactual into the actual. A case for contingency in the scientific past need not go to such lengths to be convincing, however; see Soler (2015b); Kidd (2016); and Aylward (2019).

31. Warren (1900, 199). Although discussed in Vernon's *Variation in Animals and Plants* (1902, 179, 308–9), Warren's paper went on—as a little Googling reveals—to have a reputation as foundational not for heredity but for toxicology, ecotoxicology in particular.

32. Weldon, "[MS Contribution to *Lectures in Methods of Science*]," Pearson/5/2/10/7, PP, quotations on 2–4. In illustration of the point, Weldon next discussed his own chick-embryo experiments.

33. Woltereck (1909, esp. 138–40, with "*Reaktionsnorm*" first used on 135, emphasis in original). Although thoroughly Weldonian both in its means (Daphnian) and in its end (defending Darwinian selection against the saltationism of Johannsen and de Vries by stressing heredity-environment interactions, experimentally studied and quantitatively analyzed), the paper cited nothing by Weldon or Warren. So far as I know, their precedent has not been noticed before. On Woltereck's life and career, see Harwood (1996). On the 1909 paper, see Harwood (1996, 349–51); Sarkar (1999, 235–38); Falk (2001b, 119–24); and Gilbert (2011, 123) (instructively commented upon in Bonneuil [2016, 231]).

34. Woltereck (1911), cited in Morgan et al. (1915, 97).

35. For Altenburg's and Muller's recollections of these spurned offers from Morgan, see J. Schwartz (2008, 204–5, 330nn54–55).

36. Krafka (1920, quotation on 419). For discussion, see Sarkar (1999, 238–39).

37. Hogben (1933, ch. 5, esp. 96–98, 115–16, quotation on 121). For discussion, see Tabery (2014, 24–41, esp. 29–34); see also Sarkar (1999, 238–40). Ronald Fisher, the immediate target of Hogben's argumentation, regarded "the interdependence of nature and nurture" (in Hogben's phrase) as something that, if it happened at all, could be neglected, since the complications it introduced would cancel out at the population scale.

38. See, e.g., Lewontin (1982, 21–28; 2000, 20–38). My informal sampling of introductory university textbooks of genetics suggests that coverage of the norm-of-reaction concept is variable.
39. For a plant-breeding example (comparing maize hybrids), see Lewontin (2000, 25–27). For an animal-breeding example (comparing breeds of dairy cattle), see Bryant et al. (2006).
40. Weldon, "[MS Contribution to *Lectures in Methods of Science*]," 21–28, quotations on 24–25 (emphasis added), Pearson/5/2/10/7, PP.
41. An ardent Weldonian could well have noted, too, the match between agricultural Weldonism and the working methods of the most famous plant breeder in the world at that moment, the American Luther Burbank. See Burbank (1907) and, for a superb discussion, Gould (1996). On Burbank's non-enthusiasm for Mendelism, see Palladino (1994, 411–19, esp. 414–16).
42. Thomson (1908, with Weldon's UCL lectures cited contra Mendelism on 349, 382). In Thomson's bibliography, Weldon's publications from 1902 to 1906 are represented more or less in full.
43. Punnett (1909, 90–99). Punnett took particular delight in setting Thomson straight on how a 9:3:4 ratio in the F_2—a result from Tschermak that Thomson picked up via Weldon—had become an echt Mendelian result.
44. Thomson (1908, 242–49, quotations on 244–45, 247). At the time of the 1909 Daphnian paper, the Darwinian Woltereck, too, was open-minded about the possibility that modifications arising in response to environmental change could be transmitted to offspring, acquiring a germinal basis only later. For discussion, see Gilbert (2011, 123); Harwood (1996, 350–51).
45. Thomson (1908, 529–32, quotation on 531). Although the parallel should not be pushed too hard, it is striking that Burbank, with his similar outlook, promoted a comparably cultivational form of eugenics; see Gould (1996).
46. Galton's impression was that Weldon "needed some one very strong scientific end in view to compel him to concentrate his remarkable powers more steadily." Galton to Pearson, 7 May 1906, Galton/3/3/7/30, GP, transcribed in Pearson 1915–30, 3A:282.
47. Quotation from Weldon to Pearson, 22 July 1902, Pearson/11/1/22/40, PP. On Goodrich as Weldon's demonstrator, see Weldon to Pearson, 11 June 1905, Pearson/11/1/22/40, PP. Both letters touch on Weldon's attempts to help Goodrich secure a professorship, in Glasgow and in London, respectively. Goodrich got the Linacre Professorship only in 1921. On Goodrich's life and work, see De Beer (1947). On Oxford biology during Goodrich's time—when Darwinian research flourished as

perhaps nowhere else during the supposed "eclipse of Darwinism"—see Morrell (1997).

48. Goodrich (1913, 32–38, quotation on 37).
49. Another counterfactually promising ally is J. W. Jenkinson, who is sitting to Weldon's left in fig. 6.2. Jenkinson went on to be a star of Oxford zoology, publishing the first textbook in English on experimental embryology, and impressing the young Julian Huxley as among his best teachers (along with Goodrich). See Jenkinson (1909, esp. 71–74, for favorable coverage of Weldon on development and natural selection) and Dronamraju and Needham (1993, 8).

Chapter Twelve

Epigraphs: Bateson (1926, 869); DeJarnette (1921); Rostand (1965, 19).

1. J. Rostand, in Bynum and Porter (2005, 525); V. Vaverka, in Sosna (1966, xxi–xxii, quotation on xxi); Brink (1967).
2. Recent-ish reemphases include a timeline of progress in genetics in the shape of an unspooling DNA strand, from Mendel to the Human Genome Project (Leja [2003]); a *Time* special issue titled "Great Scientists" with a profile of Mendel that begins, "It's a delightful quirk of modern medicine that the rules of genetics that make up the foundation for how we understand and treat disease today were laid out not by a scientist but by a monk" (Park [2014, 93]); and Siddhartha Mukherjee's blockbuster *The Gene: An Intimate History* (2016a).
3. Stern (1966, 199); Dunn (1965, 84); Sturtevant (1965, 131–32). On Rostand's Mendelian enthusiasms—unusual in France before the 1940s—see Burian, Gayon, and Zallen (1988, 367). On his eugenic enthusiasms, see Fogarty and Osborne (2010, 341–42). On the decades of debate about whether the population genetics of recessive conditions did or did not make eugenic targeting of them futile, see Paul and Spencer (2001).
4. Beadle (1967, with "euphenics" on 349); Beadle and Tatum (1941). "Euphenics"—the improvement of human development (as distinct from its genetic underpinnings)—was the coinage of the bacterial geneticist Joshua Lederberg, a Stanford colleague and former graduate student of Tatum's; see Lederberg (1963).
5. On closed theories in general, and Newtonian mechanics as an example, see Heisenberg (1948), with discussion in Hacking (1992, 39–41) and Bokulich 2006; the latter includes excerpts from a fascinating 1963 exchange between Heisenberg and Kuhn.
6. For a concise overview of the "molecular revolution," see Olby (1990). On the emergence of sequencing and other technologies, including

recombinant DNA technologies, see Judson (1992, esp. 53–78). For a stimulating attempt to represent the practices and cultures of *mapping* as a thread running through the Mendelian, molecular, and genomic eras, see Rheinberger and Gaudillière (2004) and Gaudillière and Rheinberger (2004).

7. Morgan (1909, 365), quoted in Beadle (1967, 339), who mistakenly gave the date as 1908.
8. On Beadle up to the mid-1930s, and his collaboration with another postdoctoral fellow, Boris Ephrussi, that prompted Beadle to travel to Paris—and away from the corn and fruit-fly crossing research he was trained in—see Sapp (2003, 163–64).
9. Beadle (1967, 343). On the new microbial genetics, led by Joshua Lederberg, see Sapp (2003, 166–68). The molecular account of crossing-over on which Beadle drew is in Watson (1965, 281–85).
10. Beadle and Tatum (1941). For general discussion of their project, along with the subsequent resurrection of Archibald Garrod's reputation as neglected prophet of the "one gene, one enzyme" hypothesis, see Sapp (2003, 157–64). To show that disrupted synthetic ability was inherited, indeed, that it was "transmitted as it should be if it were differentiated from normal by a single gene" (1941, 505), Beadle and Tatum drew from the standard Mendelian tool kit, crossing their mutant strains with normal ones. Many thanks to Adrienne Jessop for a helpful exchange on this point.
11. Beadle (1967, 342); Watson and Crick (1953a,b); Olby (1974). Although Watson had trained in biology, it is hard to see that he brought anything distinctively biological, let alone genetic, to the partnership with the ex-physicist Crick.
12. The anecdote is told in Weiner (1999, 83).
13. Within the philosophy of science, the classic statement of the position adopted here—that the molecularizing of Mendelian characters typically brings with it such extensive reclassification of those characters that they are no longer the same characters—can be found in Hull (1974, ch. 1, esp. 39–44). For an overview of the ensuing debate about whether molecular genetics should nevertheless count as "reducing" Mendelian genetics, see Darden (2005, 351–58). Two excellent for-and-against papers, Kitcher (1984) and Waters (1990), are anthologized in Sober (1994, section 9). Whatever one's view of the reduction issue, the historical fact of molecularization-induced reclassification is not in doubt. For a fascinating study of that process at work over successive editions of a famous catalogue of human genetic disorders, *Mendelian Inheritance in Man*, to the point where it "ceased to be truly Mendelian," see McGovern 2021, esp. 227–28, quotation on 228.

14. From the Princeton Review MCAT study guide for 1994, 243.
15. See, e.g., McDonald (2011), 34–36; Bustamente (2010); Sturm and Frudakis (2004).
16. Bhattacharyya et al. (1990). For helpful summaries, see Guilfoile (1997, 92–93) and Clarke (2015). For chromosome 3 as the probable home of the gene encoding SBEI, see Ellis et al. (2011, 591). ("SBE" stands for Starch-Branching Enzyme.) On a visit to the John Innes Centre in July 2019, I learned that for Alison Smith, who discovered that wrinkling is due to lack of SBEI, a key paper was "The Molecular Basis of Dominance," by the Edinburgh geneticists Henrik Kacser and James Burns (1981). Kacser and Burns argued that, in general, heterozygous phenotypes are indistinguishable from homozygous-dominant phenotypes; that the facts of enzyme kinematics, which have nothing directly to do with evolution, explain why a 50% drop in enzyme production will be undetectable phenotypically; that dominance is nothing more than the effect of having a functional allele for enzyme production; and that recessiveness is the effect of loss of functionality in a mutant allele at that locus. Historically situated, this theory represents a fascinating molecular revival of Bateson's presence-and-absence theory of dominance, comprehensively criticized in Morgan et al. (1915, 216–22). (The fly-room folk also, unsurprisingly, found that the influence of the recessive factor on the hybrid phenotype is often eminently detectable by anyone who takes the trouble to look for it carefully; see Morgan et al. [1915, 31–32, 221].)
17. Fincham (1990). At the end, Fincham noted that the case for the SBEI gene being responsible for roundness and wrinkledness in the peas that Mendel had studied was hardly conclusive, but "it seems unlikely that teachers of genetics will allow that to spoil a good story" (209).
18. I follow convention in contrasting characters that are "qualitative" (i.e., binary, Mendelian) and "quantitative" (i.e., spectrum-spanning, Weldonian). But see Serpico (2020) for a superb discussion of why the "quantitative" side of this unsatisfactory dichotomy is as problematic as the qualitative side since there are several dimensions along which a character can fail to be qualitative.
19. Wang and Hedley (1991, quotation on 8). When, on my visit to the John Innes Centre, I showed Trevor Wang the 1902 image, he joked that he would surely have been accused of plagiarism had anyone known about the image when he published.
20. To see how, it helps to appreciate that the starch synthesis pathway in the garden pea has multiple steps, each one dependent on the functioning of a different enzyme. Geneticists represent the allele encod-

ing functional SBEI with *R* and the nonfunctional counterpart allele with *r*. An *rr* pea will be wrinkled. But an *RR* pea can be wrinkled, too, if elsewhere in its genome, at a different locus, the functional form of another starch-synthesis enzyme—say, starch synthase, due to a gene represented by *Rug5* (functional allele) or *rug5* (nonfunctional allele)—is not encoded. Suppose a breeder crosses one wrinkled variety with the genotype *RRrug5rug5* with another wrinkled variety with the genotype *rrRug5Rug5*. The offspring will all be heterozygous at both loci—*RrRug5rug5*—and so produce sufficient quantities of functional forms of both enzymes to look round. For this scenario, as well as the patient tuition I needed to grasp it, I am grateful to Claire Domoney and Trevor Wang.

21. Keller (2000). For a recent investigation of rogue peas in the John Innes collection, in which, it is suspected, the out-of-type characters are due not to DNA but to overlying, epigenetically inherited methylation patterns, see Santo, Pereira, and Leitão (2017) and, for discussion, Radick (2022a). Many thanks to Mike Ambrose, then at the John Innes Centre, for alerting me to this paper.
22. Keller (2014, 2427–28).
23. Mazumdar (2002). The historical literature on eugenics is large and complex, but valuable recent guides include Levine (2017) and, more extensively, Bashford and Levine (2010).
24. Mazumdar (2002). A similar picture emerges from part 3 of Porter (2018), which tracks human heredity as a "data science" rooted in the record keeping of asylums and other institutions in Britain, the United States, and Germany (for discussion, see Radick [2019b]). On Mendelism in German genetics, see Harwood (1993); Porter (2018, chs. 12 and 13); and Teicher (2020a, chs. 1 and 2). On Davenport—whom Theodore Porter calls "the driving force for universal Mendelism," who "never asked *if* Mendelian ratios applied, always *how*" (2018, 269, emphases in original)—and the ERO, see Porter (2018, ch. 11) and Pauly (2000, 222–26). For the Pearsonians' attack, see esp. Spencer and Paul (1998). For Goddard's recantation, see Gould (1997, 202–4). For a centennial lament using Goddard on feeble-mindedness to illustrate how many of Mendel's "intellectual descendants . . . reverse[d] the argument," proceeding not by identifying either/or traits and then postulating a single-gene explanation, but by presuming a single-gene explanation and then classifying heterogeneously caused, continuously varying traits into two categories, see Stern (1966, 210–11, quotation on 210).
25. Mazumdar (2002, 49). The most detailed survey of geneticists' atti-

tudes to the theory that feeble-mindedness was recessive found that, even into the late 1930s, outright rejection was relatively rare; see Barker (1989, esp. 362).

26. *Great Falls Tribune*, 18 March 1921, 6; *Chickasha Daily Express*, 6 April 1921, 1—both found via the Chronicling America website maintained by the Library of Congress.
27. Lorimer (1921), discussed in Okrent (2019, 269) and Churchwell (2019, 53). Although Grant did not say much about Mendelian heredity, he presented it as furnishing the physical basis for his views on race, in particular his idea that race characters are unit characters that combine harmoniously in pure races but disharmoniously whenever races become mixed; see Grant (1936, 13–14). On the wider uptake of that idea, see Teicher (2020a, 98–101).
28. On the Emergency Quota Act of 1921 as "the most important turning-point in American immigration policy," see Higham (1988, 311).
29. In all, there were three international congresses of eugenics, with the first held in 1912 in London (with Bateson, Punnett, Davenport, and Winston Churchill participating) and the third in Ithaca, NY, in 1932. For an excellent overview, see Carlson (2001, 265–73). On Morgan and eugenics, see Allen (1978, 227–34). On Muller and eugenics, see J. Schwartz (2008, 249–51, 257, 267, 289); and Teicher (2018).
30. For a concise overview of *Buck v. Bell*, see Kevles (1995, 110–12). For a more extensive one, see Cohen (2016).
31. Excerpts from the testimonies of DeJarnette and Estabrook are recorded in Laughlin (1930, 23–25; for Laughlin's deposition, see 16–19). On Laughlin's involvements, see Kevles (1995, 102–4, 110); Carlson (2001, 256–61). See also Cohen (2016, chs. 5 and 6).
32. Holmes's opinion is printed in Laughlin (1930, 50–52, quotation on 52), with the paraphrase I quoted above from DeJarnette's testimony quoted (in slightly amended form) on 51.
33. Later scrutiny of the evidence led Stephen Jay Gould to conclude that "there were no imbeciles, not a one, among the three generations of Bucks," and that Carrie had been committed to the colony not because of manifest mental defect but to cover up the shame of her having a child out of wedlock, after she had been raped by a relative of her foster parents. See Gould (2007, quotation on 572–73). On the extent of eugenic sterilizations in the United States after *Buck v. Bell*, under laws enacted in thirty states (though half of the sterilizations were carried out in California), see Paul (1998, 83).
34. On German eugenics before 1933, see, e.g., Paul (1998, 84–85).
35. Teicher (2020a, 125–26, quotation on 125). On Mendelism and human

genetics in Germany, see, alongside Teicher's important study, Porter (2018, 281–341); Weiss (2010); and Schmuhl (2008).

36. Teicher (2020a, 125–26, quotation on 126). Many thanks to Maren Meinhardt for advice on translation.
37. Teicher (2020a, 145–46, quotation on 146; 189–90).
38. Teicher (2020a, 21, 107–8, 113–16, 196–204, quotation on 201). Hitler was an enormous fan of Madison Grant's *Passing of the Great Race*, calling it "my Bible" in a fan letter to Grant: Spiro (2009, 357). On the notion of the seemingly healthy carrier as common to both germ-based public health campaigns and gene-based eugenic campaigns in the first half of the twentieth century, see also Teicher (2020b).
39. Evans (1997, 138). For an overview of social Darwinism, see Radick (2019a).
40. Teicher (2020a, 149–55).
41. Teicher (2020a, 130–43 and, more extensively, 2019). "The Germans are beating us at our own game" was DeJarnette's comment in 1934; quoted in Kevles (1995, 116). Acknowledging that the Nazi law was objectionable on a number of grounds, Rostand, in his 1939 book *Heredité et Racisme*, nevertheless judged that in introducing it, Hitler "was serving the genetic interests of the nation"; quoted, in French with English translation, in Lyle (2012).
42. Teicher (2020a, 155–64, quotation on 158). Teicher poses a fascinating counterfactual question: "What would the operation of Hereditary Courts look like if it was not Mendel's laws, but Galton's 'Law of Ancestral Heredity' that was the basis for launching a sterilization campaign?" The main difference, Teicher suggests, is that in the Galtonian past that might have been, a sterilization candidate could have adduced his or her own family history in disputing a verdict; whereas in the actual Mendelian past, that possibility was removed from the start. Between that removal, and the distinctive emphasis on dangerous recessives, Mendelism's effect in Germany, Teicher concludes, was "to radicalize eugenic thinking" (quotations on 163, 164).
43. Teicher (2020a, 90–102, 116–24, 165–77; Houtermans is quoted on vi).
44. Teicher (2020a, 224–31).
45. Lewontin (1993, vii).
46. I have stressed Morgan's background in accounting for this sophistication, but it was second nature to Muller from early days, as shown in a manuscript of his from 1911–12, titled "Erroneous Assumptions Regarding Genes," now held in the Muller Papers, II. Writings, LMC 1899, Lilly Library, Indiana University. Morgan's colleague and Muller's teacher E. B. Wilson was similarly minded; see Kingsland (2007, esp. 470).

47. Morgan et al. (1915, 38–39).
48. Morgan et al. (1915, 42).
49. Morgan et al. (1915, 44–45).
50. Morgan et al. (1915, 46).
51. Morgan et al. (1915, 95–96).
52. Morgan et al. (1915, 106–7).
53. Morgan et al. (1915, 208–9).
54. Morgan et al. (1915, 210).
55. Morgan et al. (1915, 211–12).
56. Morgan et al. (1915, 227).
57. For Richard Dawkins's defense, in the wake of controversy over *The Selfish Gene*, of his use of "gene for X" as professionally conventional shorthand for "a genetic contribution to variation in X," and his disavowal of the determinism attributed to him by critics, see Dawkins (1982, ch. 2, esp. 21–23). According to Dawkins, anyone prepared to speak "of a gene for tallness in Mendel's peas" should also be prepared to speak of a gene for reading in humans, "because the logic of the terminology is identical in the two cases" (23). Previous tracings of the difference-maker understanding to the Morgan group can be found in Kampourakis (2017, 31–34) and Waters (1990).
58. On "gene for" determinism in the media, in education, and in the promotional campaigns of "direct-to-consumer" genetic test companies, see Kampourakis (2017, ch. 5). The classic study of the gene as "cultural icon" is Nelkin and Lindee (1995).
59. Sinnott and Dunn (1925, pedigree problems on 394–97); Sinnott and Dunn (1932, quotation on 57, pedigree problems [reduced from twelve to six] on 387–88). On this textbook, including its innovative use of problem sets, its high standing in the field, and its wide use internationally, see Skopek (2008, 87–93, 113–15).
60. There is also *iGenetics: A Molecular Approach*—but it contains the same chapters, just reordered to put the DNA-to-genomics chapters before the Mendelism-and-its-extensions chapters.
61. Russell (2006, 14, emphases in original).
62. Russell (2006, 38).
63. On "genetic basis" and its troublesome assimilation to "gene for" talk and thinking, see Kitcher (1996, ch. 11, esp. 250–54).
64. To get this textbook-approved solution, first draw the three-generation pedigree, then work through the pedigree twice, supposing first that unreasonableness is a dominant trait and second that it is recessive. I am grateful to Andy Cuming for helpful discussion of this problem (and to the *iGenetics Study Guide* on which he drew).

65. Sarkar (1999, 241–42). For a detailed history of the origins of the concepts, see Laubichler and Sarkar (2002).
66. On the role of problem choice in helping to ensure that in practice geneticists gave little attention to the complexities of real-world interaction, see, e.g., Lamb (2011, 117) and Lewontin (2000, 30–31).

Chapter Thirteen

Epigraphs: Provine (1971, 64); Lyon and O'Rawe (2015, 291).

1. E. P. Thompson (1968, 13).
2. Lyon, pers. comm., 13 September 2012.
3. Lyon, pers. comm., 18 November 2012.
4. Lyon and O'Rawe (2015, 289), quoting Weldon (1902a, 252).
5. See Rope et al. (2011) and two press reports published in its wake: Maher (2011) and Riffkin (2011). The quotation is from the latter.
6. See, in addition to the sources in note 5 above, Lyon's own account (2011) of the discovery and its implications.
7. Quoted in Riffkin (2011).
8. Lyon (2011, esp. 1522–23); author interview with Lyon, 5 September 2019.
9. Maher (2011), quoting Eric Topol. I have tidied up some wayward punctuation in the quotation.
10. I follow Lyon's own presentation of his past at a talk he gave in Leeds on 1 July 2014, titled "Genetic Complexity and Neuropsychiatric Disorders." Lyon also said that naming the syndrome after the town of Ogden was intended to flag the potential significance of local environment to the syndrome's etiology. For Lyon's talk, see https://geneticspedagogies.leeds.ac.uk/events/.
11. Lyon and Wang (2012, 7). In an August 2016 issue of *Nature* that published a first report from the Exome Aggregation Consortium (ExAC), the editors noted, "The authors review evidence for 192 variants reported earlier to cause rare Mendelian disorders and [yet] found at a high frequency by ExAC, and uncover support for pathogenicity for only 9." "Rare Rewards," *Nature* (August 2016).
12. Lyon and Wang (2012, 12).
13. Lyon and O'Rawe (2015, 291). I have omitted (abundant) references from the passage.
14. Wu and Lyon (2018); Cheng et al. (2019). On the nomenclatural change, see Wu and Lyon (2018, 7). For the California family mutation as having an "indistinguishable phenotype," see Rope et al. (2011, 9; for the commonality-stressing table of symptom features, see 7), and cf. Wu

and Lyon (2018, 5); Cheng et al. (2019, 8). On canalization, see Waddington (1959); Lyon (2014); Lyon and O'Rawe (2015, 299). In 2018 Lyon moved from Cold Spring Harbor to the Institute for Basic Research in Developmental Disabilities, on nearby Staten Island—a state-funded facility where, under one roof, he can see patients and their families as well as conduct laboratory research on the genetics of their conditions. Before I visited him there in 2019, he wrote to me, "On some level, going back and reading the work of Weldon really helped me to understand just how much variability there is in outcrossed populations, and how much of phenotype is context-dependent." Lyon, pers. comm., 8 January 2019.

15. The indispensable guide for study of the controversy is Franklin et al. (2008). The cultural history I offer here (and more extensively in Radick [2022b]) is my interpretation of the chronology and bibliography set out in this remarkable volume, notably in ringmaster Allan Franklin's lengthy introduction.
16. See esp. Fisher (1925, 1930, 1935).
17. Fisher went beyond Weldon in claiming to show that Mendel's data are improbably good even when, by Fisher's lights, Mendel was in error about what his theory ought to predict. In other words, whether or not Mendel identified the right prediction, his data fell into line. See Fisher (1936, 125–26, 128–29, 132). On Fisher's knowing Weldon's results, see Radick (2022b, 41).
18. Fisher (1936, 122–24, quotation on 124).
19. Fisher (1936, 132).
20. Fisher to Ford, 2 January 1936, in Bennett (1983, 199–200), with further discussion in a letter of 15 January (200–201).
21. Gigerenzer et al. (1989, 149–52). For an earlier attempt, explicitly indebted to Weldon, to promote such tests among Mendelians, see Harris (1912), which aimed at improving the Weldon-inspired discussion in Johannsen (1909, 402–10).
22. Huxley (1949, 108).
23. On the 1950 conference—which included an extraordinary "New World Honors Mendel" ceremony, saluting the role of Mendelian breeding in American and, increasingly (via hybrid corn), Latin American agriculture—see Wolfe (2012). Whether the recording was broadcast is not known, but the proceedings were published in Dunn (1951).
24. Zirkle (1954).
25. Zirkle (1959). On Zirkle as a Cold Warrior, see DeJong-Lambert (2012, 153–58).
26. On these aspects of the culture of Cold War science in the West, see esp. Cohen-Cole (2014) and Rasmussen (2014, ch. 1). On that cul-

ture generally, see Wolfe (2019). An exception that proves the rule is Kuhn, whose picture of scientific progress as dependent on closed minds seriously alarmed some Cold Warriors in his circles; see Riesch (2016).

27. Zirkle (1964). For Zirkle, if not for the excellence of Mendel's results, "he might never have discovered Mendelism" (66).
28. Beadle (1967, 336–39); Sturtevant (1965, 12–16); Dunn (1965, 12–13); Wright (1966b); Dobzhansky (1967); De Beer (1964, 199–203); Hardy (1965, 89). For discussion of most of these commentators, and others, see Franklin (2008, 29–39).
29. See esp. Dobzhansky (1967, 1588).
30. Khrushchev's criticisms of Lysenko in the USSR began in 1957. A Czechoslovak geneticist, Jaroslav Kříženecký, right away published his own critique, which, he believed (perhaps mistakenly), led to his arrest and imprisonment. See Orel (1992, esp. 491).
31. In June 1959, Lysenko was elected to the Czechoslovak Academy of Sciences, whose general assembly in Prague he addressed the following April: Šimůnek and Hossfeld (2013, esp. 87). On the removal of Mendel's statue as ordered in 1959 or 1960, see, respectively, Orel (1996, 315) and Paleček (2014, 2). (The statue seems to have been moved twice: first, in 1950, after the monasteries were closed, from its prominent position in Mendel Square to the basilica next to the abbey; then, around 1960, from the basilica to the yard inside the abbey. Many thanks to Pavel Paleček and Ondřej Dostál for discussion.)
32. On Mendel's awkwardness for "an atheistic totalitarian regime that was fighting the bourgeoisie and clericalism," see Paleček (2016, 4–5, quotation on 5); see also Orel (1992, 492). On the institute planned for Brno, see Paleček (2014, 1–2, quotation on 1), and more extensively, Paleček (2004).
33. Orel (1966). The role of symposium organizer fell to Orel only after the death of Kříženecký, who was asked to take it on after the death of the initiator, the Prague geneticist K. Hrubý. See Orel (1992, 492). On Orel's life and work, see Paleček (2016).
34. Zirkle (1966).
35. Orel (1968). On Orel's career-long concern to protect Mendel's reputation, whatever his own research turned up, see Paleček (2016, 9).
36. Koestler (1971, 47–48). On the popular success of the book, see Buklijas and Taschwer (2019).
37. On this period in the cultural life of science, see, e.g., Agar (2012, ch. 17). On molecular biology as riding the up-and-down fortunes of Cold War science funding, see Rasmussen (2014), discussed in Radick (2016b).

38. Gardner (1977), though I quote from Gardner (1981, 123–24), as the 1977 version was rather heavily edited. In the 1981 version, Gardner cites Koestler (1971) and shows every sign of having learned about Fisher's 1936 paper initially from Koestler. As for the reported "ten-thousand-to-one shot," up from Weldon's initial 16-to-1 odds, it refers to Fisher's chi-square calculation for those cases where Mendel's results conformed too closely to what Fisher judged to be the wrong predicted value. See note 17 above and Franklin (2008, 24).
39. Broad and Wade (1985, 227).
40. Quotation from Montgomerie and Birkhead (2005, 17). I first took an interest in the history reconstructed here because, after every public talk I gave on Weldon's critique of Mendelism, someone asked about the data problem (though I had not mentioned it).
41. See, e.g., Judson (2004, 52–58, quotations on 56). Although Franklin et al. (2008) announced itself as "ending" the controversy, what it offered was not a resolution so much as a plea for, in Koestler's phrase, shrugging off the difficulties, as both irresolvable and unfairly hurting Mendel's reputation; see Stigler (2008). For the most recent attempt, at the time of this writing, at dispelling the odor of scandal, see Ellis et al. (2019).
42. Fisher (1911, 160).
43. Overlooked is not, however, the same as omitted. There have, to my knowledge, been two independent rediscoveries of Weldon's diagnosis of Mendel's data problem: Root-Bernstein (1983) and Weeden (2016).
44. Russell (2006, 17–19, quotation on 18).
45. Provine (2001, 71–72).
46. Goldschmidt (1952, 19–21). On Goldschmidt as championing "a physiological approach that emphasized how a single gene could produce many different phenotypes depending on differences in developmental and environmental interactions" via effects on the amount, rate, and location of enzyme production, see Dietrich (2016, 31).
47. Goldschmidt (1952, v–vi); Wolfe (2010, 73).
48. Provine (1971, 161–64, quotation on 164).
49. Medawar (1977, 171–72; the Mendel data problem is discussed on 181). ("There is," added Medawar, ". . . a profoundly important difference between the cases of Mendel and of Burt: Mendel was right.") For Haldane—from 1937 to 1957, the first Weldon Professor of Biometry at UCL (a position created courtesy of a bequest in Florence Weldon's will)—on what he called "the interaction of nature and nurture," see Haldane (1946–47) and, for an earlier statement, Haldane (1936); for discussion, see Radick (2007, 250, 450n26). The shrimp

study—undertaken at Oxford's Department of Zoology and Comparative Anatomy, thanks to the support of Weldon's successor, E. S. Goodrich, using shrimp from the Plymouth laboratory, but with Goldschmidt credited as inspiration—was by E. B. Ford and Julian Huxley (1927); for discussion, see Gould (1977, 205).

50. My remarks here on eugenics are much inspired by Porter (2018), discussed in Radick (2019b). For a superb study of how some animal breeders have come to bypass genes more or less entirely, concentrating instead on the predictive power of DNA marker profiles in improving particular breeds via "genomic selection," see Lowe and Bruce (2019).
51. Osborne (2002).
52. Xu and Lewis (1990); Hinman and Lewis (1992); author interview with Randy Lewis, 19 March 2014.
53. Hinman, Jones, and Lewis (2000, esp. 378); Osborne (2002); author interview with Randy Lewis, 19 March 2014.
54. Author interview with Randy Lewis, 19 March 2014.
55. Darwin (1859, 31); author interview with Randy Lewis, 19 March 2014. When I visited in 2014, I learned that silk-protein genes had also been made to work in silkworms and—more unexpectedly—alfalfa. On synthetic-biological gadgets and gene therapies as, like spider goats, technologies whose success beyond the laboratory depends on often invisible backstage work, see, respectively, Radick (2013c and 2016c, 4).
56. Kuhn (1963, 358–59).
57. Provine (2001, 197–98).
58. Provine (2001, 203–4). I did not know Provine (who died in 2015), but he attended a conference talk that I gave on Weldon in 2008, and was gracious and encouraging when I told him about my project. I salute his example, and cherish my signed copy of *The Origins of Theoretical Population Genetics*.
59. But is it not timelessly good pedagogy to start simply—which, in genetics, means starting with Mendel's peas? The question makes two dubious presumptions: first, that misinformation absorbed early on can be corrected later with no lingering cognitive consequences (for the case against that presumption, see Lewandowsky et al. 2012); and second, that complexity in nature must be presented in a complex way (for the case against, see my overview in ch. 11 of Annie Jamieson's opening lecture in our Weldonian course).
60. Quotations from the "Refresh Your Genetics Curriculum" online workshop, 1 July 2020, and subsequent feedback. At the time of this writing, there have been two more workshops, generating similar responses.

61. Pearl (1911; 1941, 86–89). Pearl's fifteen observers were not students but trained biological professionals, including experts in corn breeding and judging. For discussion of Pearl and his change of allegiance on Mendelism in these years, see Kim (1994, 123–42).
62. Root-Bernstein (1983, 282–85); see also Stansfield (2016).
63. Freudenberg-Hua et al. (2014). Consider, too, the Resilience Project, which aimed to sequence the genomes of healthy people to identify "genetic superheroes" unaffected by mutations hitherto classed as harmful, in order to work out how, in these individuals, nature and/or nurture had neutralized the harm, then to develop new preventive or therapeutic drugs drawing on the conclusions reached; see Friend (2014); Friend and Schadt (2014); Swartz (2014). At pilot scale, the project was a success, identifying thirteen normal adults whose childhoods should have been blighted by severe Mendelian diseases: Chen et al. (2016). The basic research strategy is now part of the business plan for the Connecticut-based health care company Sema4. Author interview with Eric Schadt, 6 September 2019.
64. For polygenic risk scores, and the big-data genome-wide association studies (GWAS) on which the scores are based, as a game changer in extracting a determinist message from DNA, see Plomin (2018). For deflationist commentary, see Kampourakis (2020, esp. 57–76, 109–10). On a role for Galton's quincunx in pressing home the GWAS-deflationist case, see Kaplan and Turkheimer (2021).
65. A Weldonian observation: between the Mendelian and Weldonian pedagogic extremes, we should expect to find intermediates—and we do. At Dalhousie University, for example, Joe Bielawski's genetics course starts with Mendelian genetics but then goes straight to quantitative genetics "to try and build a more balanced (natural) picture of the relationship between phenotype and genotype by the time we hit the ⅓ point in the course." Bielawski, pers. comm., 20 October 2020.
66. On the case for beginning a Weldonized introduction to genetics with the debate over Mendel, by way of putting students on notice from the first about the "culturally laden nature" of what has counted as genetic knowledge (and the concern of their inclusivity-minded teachers to ensure that the injustices of the past are not repeated), see Sparks, Baldwin, and Darner (2020).
67. Under the banner of "ecological developmental biology," Scott F. Gilbert has synthesized the state of the art in this area for both researchers (Gilbert [2012]) and students (Gilbert and Epel [2015]).
68. For a masterful introduction to human genetics along exactly these lines, see Lewontin (1982). For a concise update for the era of post-genomic medicine, see Riordan and Nadeau (2017).

69. Here I echo Evelyn Fox Keller's call (2010, ch. 4) for greater emphasis on phenotypic plasticity and its multiple sources as an excellent strategy for dissolving what she called "the mirage of a space between nature and nurture."
70. For a concise survey of Darwinian evolution without genetic determinism—what is now often called the "extended evolutionary synthesis"—see Lewontin (1983); for a fuller exposition, see Lewontin (2000).
71. See Porter (2018) for a history of human heredity in which the era of Mendelian enthusiasm appears as a kind of blip within the otherwise continuous history of predictive statistics and its ever-evolving technologies, documentary and computational.
72. For both of my examples, the pedagogic aim is the same: to reduce the scope for students to become so mesmerized by "gene for" talk that they fail to notice how important context is for genes having the effects they do—and relatedly, to increase their reflectiveness about the larger ramifications of different ways of matching genes and contexts.
73. On the sickle cell therapy, developed by Stuart Orkin, see McKie (2016). In September 2019, on a visit to Harvard where I had a chance to speak with Dr. Orkin, I learned from Chirag Patel, a biomedical informatician developing "genome-environment-wide association studies" (Patel and Ioannidis [2014, 3]), that one of the major findings so far is that air pollution is a common factor in lots of diseases. For a thoughtful discussion of the claims on our attention (and funding) of genetic versus other ameliorating interventions, see Kitcher (1996, ch. 14).
74. Saini (2019, 259), citing Garcia (2004, 1395).
75. Donovan (2016, 2017).
76. Donovan et al. (2019). The image in fig. 13.5 draws on Rosenberg (2011).
77. For a superb survey of research on the essentialist thinking and other psychological biases that make humans so prone to determinism about genes, see Heine (2017).
78. *Honoring the Complexity of Genetics: Exploring How Undergraduate Learning of Multifactorial Genetics Affects Belief in Genetic Determinism*, NSF Award ID 1914843. Also working with us on the project are Andy Brubaker, Jean Flanagan, Dennis Lee, Kelly Schmid, Awais Syed, and Monica Weindling.

Conclusion

1. For an example, see Radick (2016a, 157–60).
2. Nozick (1981, 12, emphasis in original), discussed in Hawthorn (1991, 10, 17–18).

3. Porter (2018, 146).
4. S. Sparks (2019).
5. For evidence that people with a more deterministic understanding of genes “are more likely to view some groups as being inferior to others, support government eugenic policies to control who reproduces, and have fears about GMO foods,” see Cheung, Schmalor, and Heine (2021, quotation on 8; on that evidence as supporting curriculum reform along Weldonian lines, see 9).

Postscript One

Epigraphs: Dawkins (2003b, 231); Oyama (1985, 26–27, quoted as an epigraph in Kitcher [2001, 396]); J. Lewis (2011, 175).

1. Detailed studies of genetic determinism in school biology textbooks include Castéra, Bruguière, and Clément (2008); Castéra et al. (2008); Carvalho dos Santos, Joaquim, and El-Hani (2012); and Aivelo and Uitto (2015). For Dawkins’s classic discussion of why genetic determinism is no part of his “selfish gene” perspective on selection, see Dawkins (1982, ch. 1, cited in a different connection in ch. 12 of this book, note 57). On Dawkins as indeed “adopt[ing] a fairly standard ‘interactionist’ approach” to development, see Gray (2001, 192). To the likes of Gray and Oyama, that approach nevertheless falls well short of the maximally interactionist “developmental systems theory” perspective they defend, in which genes are of no more interest than other developmentally relevant causes.
2. Kampourakis (2017, 6). Kampourakis distinguishes genetic determinism from two related terms: “genetic essentialism” (“genes are fixed entities, which are transferred unchanged across generations and which are the essence of what we are by specifying characters from which their existence can be inferred”) and “genetic reductionism” (“genes alone provide the ultimate explanation for characters, and the best approach to explain these is by studying phenomena at the level of genes”).
3. On genetic determinism as “mythical,” because utterly lacking in adherents, see Dennett (2004). For predictably inconclusive attempts to meet that argument head-on, see the lengthy exchanges initiated by Anthony Gordon at ResearchGate, “Examples, please, of biological or genetic determinism?” 26 August 2013, https://www.researchgate.net/post/Examples-please-of-biological-or-genetic-determinism and by Angelique Richardson, “Elbowed off the Pavement,” *LRB* (blog), *London Review of Books*, 20 Aug 2020, https://www.lrb.co.uk/blog/2020/august/elbowed-off-the-pavement.

4. On the dispute over genetic determinism as turning—in common with other disputes in biology—not on matters of principle but on the kinds of models used in practice as defaults, see Kitcher (2001, esp. 407, 413n7).
5. See Carver et al. (2017) for an illuminating history of this genre of questionnaire as well as, at the time of this writing, the most impressive contribution to it.
6. Mukherjee (2016a, 298–300, 441–46, quotation on 446); Radick (2016c). In a subsequent *New Yorker* article about the genetics of schizophrenia, Mukherjee (2016b) chronicled the history of the investigation leading to the *Nature* paper whose claims excited him: "A magnificently simple theory began to convulse out of the results." For the paper itself, see Sekar et al. (2016). For a more measured commentary, noting that "accelerated synaptic pruning may be only one of many mechanisms underlying what we call schizophrenia, may not be unique to this illness, and may not be central to this collection of disease entities," see Keshavan, Lizano, and Prasad (2020, quotation on 111).
7. See, e.g., Pinker (2002, esp. 112–13, 122–24); Dennett (2004). Until the late 1940s—the start of the Lysenkoist controversy—"genetic determinism" was barely in use in English-language publications. It became notably more prominent in the period circa 1965 to circa 1980, coinciding with controversies over the new human ethology and then sociobiology, and more prominent still in the period circa 1990 to circa 2000, heyday of the Human Genome Project. When Pinker and Dennett complained about the term, it had already entered into a decline, which, as of 2019, showed little sign of abating.
8. For an outstanding discussion of genetic determinism, generic determinism, and free will, see Lipton (2004).
9. Punnett (1907, 47–53, quotation on 50).
10. Bateson (1909, 60).
11. Rose, Kamin, and Lewontin (1984, 268–87).
12. See esp. Pinker (2004; on the "everything-affects-everything diagram," see 17). On "determinism" as a "scare word" that should not "get in the way of understanding our genetic roots," see Pinker (2010, 140).
13. Pinker (2011).
14. Pinker (2019). I take it the 1990s are picked out as when "gene for" human genetics expired because of the introduction in the mid-1990s of the genome-wide association study (GWAS). Certainly advocates of GWAS, such as the behavioral geneticist Robert Plomin, have been keen to emphasize the cleanness of the break with the "gene for" past; see, e.g., Plomin (2018, 221–24). On the case for skepticism, see Comfort (2018).

15. On G×E interaction in historical and philosophical perspective, from Hogben to the present, see Tabery (2014).
16. A recent study of genetics teaching materials at an open-access website, CourseSource, found that a minority of assessment questions touch on the wider environment, and none at all on genotype-by-environment interactions; see Schmid et al., in press.
17. The difficulties involved in making it a thing of the past should not be underestimated. Not even the Lysenkoist Soviet Union succeeded in fully de-Mendelizing its teaching materials: Peacock (2015). For a field report from an interactionist educator among determinist audiences, see Moore (2008).
18. Chang (2004).
19. For future development of the notion of cognitive centrality versus marginality in science, a promising resource is Ronald Giere's (2006) perspectival philosophy of science.
20. Heine (2017, 264).

Postscript Two

What follows is my own reconstruction of Bernard Norton's reconstruction of Karl Pearson's reconstruction of Weldon's reasoning. See Norton (1979, 190–93) and Pearson (1908, esp. 85 [case 2]). See also Bulmer (2003, 260–61).

1. There are affinities here with what Sharon Kingsland has called the "quantitative hypothesis" of chromosomal sex determination in the work of E. B. Wilson at exactly this time. On this hypothesis, in her summary, she writes, "The difference in the *quantity of chromosomal material* determined whether a male or a female would be produced by affecting the physiological processes of a cell: the chromosomes themselves were not qualitatively different." Kingsland (2007, 476, emphasis in original).
2. Pearson (1908, 84) attributed something like this idea to de Vries, writing of his "view of an interchange of chromomeres bearing the determinants [the term Weldon uses for "factors"] occurring at random between paternal and maternal chromosomes at the reducing division."
3. Many thanks to Amir Teicher for help in simplifying my presentation of the reasoning here.
4. In 1905, Hurst passed the materials from his garden-pea crossing experiment to Darbishire, who, with help from Weldon's Oxford assistant, Frank Sherlock, kept it going until Darbishire's death at the front during the Great War in 1915. Wrote Darbishire in 1915: "I consider

the Mendelian principles to be still *sub judice*, and they are so attractive by reason of their simplicity that they need to be under a very stern judge." According to Yule, who published Darbishire's results in 1923, the experiments had been conducted in a spirit of "pure scientific scepticism," and the records he had collected were exemplary in their "care and accuracy." So it was, in Yule's view, the more telling that on analyzing those records, he found divergences from Mendelian expectations far too large to be explained away in the standard Mendelian ways. "The mechanism at work," Yule concluded, "appears to be more complex than is commonly postulated." Yule (1923, quotations on 257, 262).

Postscript Three

1. Cowan 1976, 252.

References

Archives

American Eugenics Society. Papers. American Philosophical Society, Philadelphia.

Archive, Stazione Zoologica Anton Dohrn, Naples (SZN).

Archive and Library, Royal Society of London (RS). Many items can be viewed online via the Royal Society website.

Balfour Family Papers, National Records of Scotland, Edinburgh.

Bateson, William. Papers. Cambridge University Library (BP). Photocopied sets of these papers, along with other Batesoniana, are available in Special Collections at the John Innes Centre in Norwich and in the library of Queen's University, Kingston, Canada. A partial set is available on microfilm from the American Philosophical Society Library, Philadelphia.

Darwin, Charles. Papers. Cambridge University Library. The entire collection, apart from the correspondence, can be viewed online at the Darwin Online website. Much of Darwin's correspondence can be viewed online at the Darwin Correspondence Project website; the associated published edition is nearly complete.

Davenport, Charles Benedict. Papers. American Philosophical Society Library, Philadelphia.

Galton, Francis. Papers. University College London (GP). Much of this collection is now available online through the Wellcome Library website.

Hurst, Charles Chamberlain. Papers. Cambridge University Library. A partial photocopied set is available at the American Philosophical Society Library, Philadelphia.

Larmor, Joseph. Collection. Special Collections, Library, St. John's College, University of Cambridge.

Muller, Hermann. Papers. Lilly Library, Indiana University, Bloomington.

National Marine Biological Library. Marine Biological Association, Plymouth (NMBL).

Newton, Alfred. Papers. Cambridge University Library.

Pearson, Karl. Papers. University College London (PP). Summaries of many items can be viewed online via the UCL Archives website.

Sedgwick, Adam. Correspondence. Department of Manuscripts and University Archives, Cambridge University Library.

Thompson, D'Arcy Wentworth. Papers. University of St. Andrews (TP).

Tschermak[-Seysenegg], Erich von. Papers. Archive of the Austrian Academy of Sciences, Vienna.

Weldon, Walter Frank Raphael. Papers. University College London. These are incorporated into the Pearson Papers. Summaries of many items can be viewed online via the UCL Archives website.

Interviews

Domoney, Claire. Norwich. 22 July 2019.

Lewis, Randy. Logan, UT. 19 March 2014.

Lyon, Gholson. Staten Island, NY. 5 September 2019.

Orkin, Stuart. Cambridge, MA. 12 September 2019.

Patel, Chirag. Cambridge, MA. 12 September 2019.

Schadt, Eric. Stamford, CT. 6 September 2019.

Smith, Alison. Norwich. 22 July 2019.

Wang, Trevor. Norwich. 19 September 2019.

Books and Articles

Agar, Jon. 2012. *Science in the Twentieth Century and Beyond*. London: Polity.

Agassiz, Alexander. 1881. "Paleontological and Embryological Development." *Proceedings of the American Association for the Advancement of Science, 29th Meeting, August 1880*, 389–414.

Aivelo, Tuomas, and Anna Uitto. 2015. "Genetic Determinism in the Finnish Upper Secondary School Biology Textbooks." *Nordic Studies in Science Education* 11:139–52.

Allchin, Douglas. 2000. "Mending Mendelism." *American Biology Teacher* 62:633–39.

Allchin, Douglas. 2002. "Dissolving Dominance." In *Mutating Concepts, Evolving Disciplines: Genetics, Medicine, and Society*, edited by Lisa Parker and Rachel Ankeny, 43–61. Dordrecht: Kluwer.

Allchin, Douglas. 2005. "The Dilemma of Dominance." *Biology and Philosophy* 20:427–51.

Allen, Garland E. 1978. *Thomas Hunt Morgan: The Man and His Science*. Princeton, NJ: Princeton University Press.

Ankeny, Rachel A. 2000. "Marvelling at the Marvel: The Supposed Conversion of A. D. Darbishire to Mendelism." *Journal of the History of Biology* 33:315–47.

Arabatzis, Theodore. 2019. "Explaining Science Historically." *Isis* 110:354–59.

Arabatzis, Theodore, and Kostas Gavroglu. 2016. "From Discrepancy to Discovery: How Argon Became an Element." In *The Philosophy of Historical Case Studies*, edited by Tilman Sauer and Raphael Scholl, 203–22. Cham: Springer International.

Ariew, André, Yasha Rohwer, and Collin Rice. 2017. "Galton, Reversion and the Quincunx: The Rise of Statistical Explanation." *Studies in History and Philosophy of Biological and Biomedical Sciences* 66:63–72.

Arnold, Nick, and Tony De Saulles. 2009. *Horrible Science Annual 2010*. London: Scholastic.

Aylward, Alex. 2019. "Against Defaultism and Towards Localism in the Contingency/Inevitability Conversation: Or, Why We Should Shut Up about Putting-Up." *Studies in History and Philosophy of Science* 74:30–41.

Babbage, Charles. 1830. *Reflections on the Decline of Science in England and on Some of Its Causes*. London: B. Fellowes.

Bailey, Liberty Hyde. 1903. "Some Recent Ideas on the Evolution of Plants." *Science* 17:441–54.

Bailey, Liberty Hyde. 1904. "A Medley of Pumpkins." In *Proceedings: International Conference on Plant Breeding and Hybridization*, 117–24. Memoirs of the Horticultural Society of New York, vol. 1.

Balfour, Francis M. 1875. "A Comparison of the Early Stages in the Development of Vertebrates." Reprinted in Balfour 1885, vol. 1, 112–34.

Balfour, Francis M. 1879. "On the Early Development of the Lacertilia, Together with Some Observations on the Nature and Relations of the Primitive Streak." Reprinted in Balfour 1885, vol. 1, 644–45.

Balfour, Francis M. 1880. "[Address to the Department of Anatomy and Physiology.]" In *Report of the Fiftieth Meeting of the British Association for the Advancement of Science, Held at Swansea in August and September 1880*, 636–44. London: John Murray. Also in *Nature* 22 (2 September): 417–20, and Balfour 1885, vol. 1, 698–713.

Balfour, Francis M. 1885. *The Works of Francis Maitland Balfour*. Edited by Michael Foster and Adam Sedgwick. 4 vols. London: Macmillan.

Barker, David. 1989. "The Biology of Stupidity: Genetics, Eugenics and Mental Deficiency in the Inter-War Years." *British Journal for the History of Science* 22:347–75.

Barnes, Barry. 1980. "On the Causal Explanation of Scientific Judgment." *Social Science Information* 19:685–95.

Barnes, Barry. 1996. Review of Kim 1994. *Isis* 87:198–99.

Barrett, Paul H., Peter J. Gautrey, Sandra Herbert, David Kohn, and Sydney Smith, eds. 1987. *Charles Darwin's Notebooks 1836–1844*. Cambridge: Cambridge University Press.

Bartenstein, E., F. Teindl, J. Hirsch, and C. Lauer. 1837. "Protokol über die Verhandlungen bei der Schafzüchter-Versammlung in Brünn am 1. und 2. Mai 1837." *Mittheilungen der k.k. Mährisch-Schlesischen Gesellschaft zur Beförderung des Ackerbaues, der Natur- und Landeskunde in Brünn* 29:225–32.

Bartley, Mary M. 1992. "Darwin and Domestication: Studies on Inheritance." *Journal of the History of Biology* 25:307–33.

Barton, Ruth. 2018. *The X Club: Power and Authority in Victorian Science*. Chicago: University of Chicago Press.

Bashford, Alison, and Philippa Levine, eds. 2010. *The Oxford Handbook of the History of Eugenics*. Oxford: Oxford University Press.

Bates, Benjamin R., Alan Templeton, Paul J. Achter, Tina M. Harris, and Celeste M. Condit. 2003. "What Does 'A Gene for Heart Disease' Mean? A Focus Group Study of Public Understandings of Genetic Risk Factors." *American Journal of Medical Genetics* 119A:156–61.

Bateson, Beatrice, ed. 1928a. *William Bateson, F.R.S., Naturalist: His Essays and Addresses, Together with a Short Account of His Life*. Cambridge: Cambridge University Press. Facsimile reprint, New York: Garland, 1984.

Bateson, Beatrice, ed. 1928b. *Letters from the Steppe[,] Written in the Years 1886–1887 By William Bateson*. London: Methuen.

Bateson, Gregory. 1972. "A Re-examination of Bateson's Rule." In his *Steps to an Ecology of Mind*, 379–99. Chicago: University of Chicago Press.

Bateson, William. 1883. "Abstract of Observations on the Development of Balanoglossus." *Johns Hopkins University Circulars* 3 (27): 4.

Bateson, William. 1884a. "On the Early Stages in the Development of Balanoglossus Aurantiacus." *Proceedings of the Cambridge Philosophical Society* 5 (11 February): 107.

Bateson, William. 1884b. "The Early Stages in the Development of Balanoglossus (sp. Incert.)." *Quarterly Journal of Microscopical Science* 24 (April): 208–36.

Bateson, William. 1884c. "Note on the Later Stages in the Development of *Balanoglossus Kowalevskii* (Agassiz), and on the Affinities of the Enteropneusta." *Proceedings of the Royal Society of London* 38 (18 December): 23–30.

Bateson, William. 1885a. "On the Types of Excretory System Found in the Enteropneusta." *Proceedings of the Cambridge Philosophical Society* 5 (16 February): 225.

Bateson, William. 1885b. "The Later Stages in the Development of Balanoglossus Kowalevskii, with a Suggestion as to the Affinities of the

Enteropneusta." *Quarterly Journal of Microscopical Science* 25 (April supplement): 81–122.

Bateson, William. 1885c. "Suggestions with Regard to the Nervous System of the Chordata." *Proceedings of the Cambridge Philosophical Society* 5 (9 November): 321.

Bateson, William. 1886a. "Continued Account of the Later Stages in the Development of Balanoglossus Kowalevskii, and of the Morphology of the Enteropneusta." *Quarterly Journal of the Microscopical Society* 26:511–33.

Bateson, William. 1886b. "The Ancestry of the Chordata." Reprinted in Bateson 1971, vol. 1, 1–31.

Bateson, William. 1888. "On Variations of *Cardium edule* from the Aral Sea." *Proceedings of the Cambridge Philosophical Society* 6 (13 February): 181–82.

Bateson, William. 1889. "On Some Variations of *Cardium edule* Apparently Correlated to the Conditions of Life." Reprinted in Bateson 1971, vol. 1, 34–71.

Bateson, William. 1890a. "Notes and Memoranda [from the *Journal of the Marine Biological Association*]." Reprinted in Bateson 1971, vol. 1., 71–78.

Bateson, William. 1890b. "The Sense-Organs and Perceptions of Fishes; with Remarks on the Supply of Bait." Reprinted in Bateson 1971, vol. 1, 79–112.

Bateson, William. 1890c. "On Some Cases of Abnormal Repetition of Parts in Animals." Reprinted in Bateson 1971, vol. 1, 113–23.

Bateson, William. 1891. "On Variations in the Colour of Cocoons (*Saturnia carpini* and *Eriogaster lanestris*), with Reference to Recent Theories of Protective Coloration." *Proceedings of the Cambridge Philosophical Society* 7:251.

Bateson, William. 1892a. "On Variation in the Colour of Cocoons of *Eriogaster lanestris* and *Saturnia carpini*." Reprinted in Bateson 1971, vol. 1, 162–68.

Bateson, William. 1892b. "Variation in the Colour of Cocoons, Pupae, and Larvae: Further Experiments." Reprinted in Bateson 1971, vol. 1, 169–77.

Bateson, William. 1892c. "The Alleged 'Aggressive Mimicry' of *Volucellae*." Reprinted in Bateson 1971, vol. 1, 202–5.

Bateson, William. 1892d. "The Alleged 'Aggressive Mimicry' in *Volucellae* [Part 2]." Reprinted in Bateson 1971, vol. 1, 206–7.

Bateson, William. 1892e. "Numerical Variation in Teeth." Reprinted in Bateson 1971, vol. 1, 178–92.

Bateson, William. 1894. *Materials for the Study of Variation, Treated with Especial Regard to Discontinuity in the Origin of Species*. London: Macmillan. Reprint, Baltimore: Johns Hopkins University Press, 1992.

Bateson, William. 1895a. "The Origin of the Cultivated *Cineraria*." *Nature* 51 (25 April): 605–7.

Bateson, William. 1895b. "On the Colour-Variations of a Beetle of the Family *Chrysomelidae*, Statistically Examined." Reprinted in Bateson 1971, vol. 1, 331–43.

Bateson, William. 1897. "Progress in the Study of Variation, 1." In Bateson 1971, vol. 1, 344–56.

Bateson, William. 1898a. "Progress in the Study of Variation, 2." In Bateson 1971, vol. 1, 357–70.

Bateson, William. 1898b. "Experiments in the Crossing of Local Races of Lepidoptera." In Bateson 1971, vol. 1, 371–73.

Bateson, William. 1900. "Hybridisation and Cross-Breeding as a Method of Scientific Investigation." Hybrid Conference Report 1900, *Journal of the Royal Horticultural Society* 24:59–66. Reprinted in B. Bateson 1928a, 161–70.

Bateson, William. 1901a. "Problems of Heredity as a Subject for Horticultural Investigation." *Journal of the Royal Horticultural Society* 25:54–61. Reprinted in B. Bateson 1928a, 171–80.

Bateson, William. 1901b. "Heredity, Differentiation, and Other Conceptions of Biology: A Consideration of Professor Karl Pearson's Paper 'On the Principle of Homotyposis.'" Reprinted in Bateson 1971, vol. 1, 404–18.

Bateson, William. 1901c. [Introduction to Mendel 1901.] *Journal of the Royal Horticultural Society* 26:1–3. Reprinted in Bateson 1971, vol. 2, 1–3.

Bateson, William. 1902a. *Mendel's Principles of Heredity: A Defence*. Cambridge: Cambridge University Press.

Bateson, William. 1902b. "Note on the Resolution of Compound Characters by Cross-Breeding." *Proceedings of the Cambridge Philosophical Society* 12:50–54. Reprinted in Bateson 1971, vol. 2, 69–73.

Bateson, William. 1903a. "Mendel's Principles of Heredity in Mice." *Nature* 67 (19 March): 462–63. Reprinted in Bateson 1971, vol. 2, 109–11.

Bateson, William. 1903b. "The Present State of Knowledge of Colour-Heredity in Mice and Rats." *Proceedings of the Zoological Society of London* 2:71–99. Reprinted in Bateson 1971, vol. 2, 76–108.

Bateson, William. 1904a. "Practical Aspects of the New Discoveries in Heredity." In *Proceedings: International Conference on Plant Breeding and Hybridization*, 1–9. Memoirs of the Horticultural Society of New York, vol. 1.

Bateson, William. 1904b. "Section D. Zoology. Opening Address by William Bateson, M.A., F.R.S., President of the Section." *Report of the Seventy-Fourth Meeting of the British Association for the Advancement of Science, Held at Cambridge in August 1904*, 574–79. London: John Murray. Also in *Nature* 70:406–13; B. Bateson 1928a, 233–59.

Bateson, William. 1905. [Comment on Galton 1905]. *Sociological Papers* 1:64–65.

Bateson, William. 1907a. “The Progress of Genetic Research.” In Wilks 1907, 90–97. Also in Bateson 1971, vol. 2, 142–51.

Bateson, William. 1907b. “[Toast to the Board of Agriculture, Horticulture, and Fisheries.]” In Wilks 1907, 75–77.

Bateson, William. 1908. “The Methods and Scope of Genetics.” Reprinted in B. Bateson 1928a, 317–33.

Bateson, William. 1909. *Mendel's Principles of Heredity*. Cambridge: Cambridge University Press. [I consulted the 1913 printing, which includes additional, updating appendices.]

Bateson, William. 1910. “1883–84.” In “William Keith Brooks: A Sketch of His Life by Some of His Former Pupils and Associates.” *Journal of Experimental Zoology* 9 (1): 5–8.

Bateson, William. 1913. *Problems of Genetics*. New Haven: Yale University Press.

Bateson, William. 1917. “Gamete and Zygote: A Lay Discourse.” In B. Bateson 1928a, 201–14.

Bateson, William. 1922. “Evolutionary Faith and Modern Doubt.” In B. Bateson 1928a, 389–98.

Bateson, William. 1924. “Progress in Biology.” In B. Bateson 1928a, 399–408.

Bateson, William. 1926. “Mendelism.” In *Encyclopaedia Britannica*, 13th ed., supplementary vol. 2, 867–70.

Bateson, William. 1971. *Scientific Papers of William Bateson*. 2 vols. Edited by R. Punnett. Reprint, Sources of Science, 136. New York: Johnson Reprint. First published in 1928 by Cambridge University Press.

Bateson, William, and Anna Bateson. 1891. “On Variations in the Floral Symmetry of Certain Plants Having Irregular Corollas.” Reprinted in Bateson 1971, vol. 1, 126–61. [I have here corrected a small misprint in the title.]

Bateson, William, and H[arold] H[ulme] Brindley. 1892. “Some Cases of Variation in Secondary Sexual Characters Statistically Examined.” Reprinted in Bateson 1971, vol. 1, 193–201.

Bateson, William, and Edith R. Saunders. 1902. *Reports to the Evolution Committee of the Royal Society. Report 1. Experiments undertaken by W. Bateson, F.R.S., and Miss E. R. Saunders*. London: Harrison. Also in Bateson et al. 1910.

Bateson, William, E. R. Saunders, and R. C. Punnett. 1906. “Further Experiments on Inheritance in Sweet Peas and Stocks: Preliminary Account.” *Proceedings of the Royal Society of London* 77 (26 February): 236–38. Also in Bateson 1971, vol. 2, 139–41.

Bateson, William, E. R. Saunders, Reginald C. Punnett, and C. C. Hurst.

1905. *Reports to the Evolution Committee of the Royal Society. Report 2. Experimental Studies in the Physiology of Heredity; Experiments with Poultry*. London: Harrison & Sons. Also in Bateson et al. 1910.

Bateson, William, Edith R. Saunders, Reginald C. Punnett et al. 1910. *Reports to the Evolution Committee of the Royal Society: Reports 1–5. 1902–09*. London: Royal Society.

Baxter, Alice, and John Farley. 1979. "Mendel and Meiosis." *Journal of the History of Biology* 12:137–73.

Beach, S. A. 1904. "Correlation between Different Parts of the Plant in Form, Color, Size and Other Characteristics." In *Proceedings: International Conference on Plant Breeding and Hybridization*, 63–67. Memoirs of the Horticultural Society of New York, vol. 1.

Beadle, George W. 1967. "Mendelism, 1965." In *Heritage from Mendel*, edited by R. Alexander Brink, 335–50. Madison: University of Wisconsin Press.

Beadle, George W., and Edward L. Tatum. 1941. "Genetic Control of Biochemical Reactions in *Neurospora*." *Proceedings of the National Academy of Sciences of the United States of America* 27:499–506.

Bedford, Thomas. 1951. "Obituary: H. M. Vernon." *Occupational Medicine* 1:47–48.

Bennett, J. A., ed. 1983. *Natural Selection, Heredity, and Eugenics: Including Selected Correspondence of R. A. Fisher with Leonard Darwin and Others*. Oxford: Clarendon Press.

Benson, Keith R. 1985. "American Morphology in the Late Nineteenth Century: The Biology Department at Johns Hopkins University." *Journal of the History of Biology* 18:163–205.

Benson, Keith R. 2010. "William Keith Brooks (1848–1908) and the Defense of Late Nineteenth Century Darwinian Evolutionary Theory." In *The Hereditary Hourglass: Genetics and Epigenetics, 1868–2000*, edited by Ana Barahona, Edna Suarez-Díaz, and Hans-Jörg Rheinberger, 23–33. MPI Preprint 392. Berlin: Max Planck Institute for the History of Science.

Berlan, Jean-Pierre, and Richard C. Lewontin. 1986. "The Political Economy of Hybrid Corn." *Monthly Review* 38:35–47.

Berry, Dominic J. 2014a. "Genetics, Statistics, and Regulation at the National Institute of Agricultural Botany, 1919–1969." PhD dissertation, University of Leeds.

Berry, Dominic J. 2014b. "Bruno to Brünn; or the *Pasteurization* of Mendelian Genetics." *Studies in History and Philosophy of Biological and Biomedical Sciences* 48B:280–86.

Berry, Dominic J. 2021. "Historiography of Plant Breeding and Agriculture." In *Handbook of the Historiography of Biology*, edited by Michael

Dietrich, Mark E. Borrello, and Oren Harman, 499–525. Dordrecht: Springer.

Bershidsky, Leonid. 2016. "What Happened When Khrushchev Came to Iowa." *Atlanta Journal-Constitution*, 30 January. https://www.ajc.com/news/national-govt-politics/what-happened-when-khrushchev-came-iowa/Znb59NDsu6AMOdkKcXAIpK/.

Bhattacharyya, Madan K., Alison M. Smith, T. H. Noel Ellis, Cliff Hedley, and Cathie Martin. 1990. "The Wrinkled-Seed Character of Pea Described by Mendel Is Caused by a Transposon-Like Insertion in a Gene Encoding Starch-Branching Enzyme." *Cell* 60:115–22.

Biffen, Rowland H. 1907a. "Exhibit of Hybrid Wheats and Barleys." In Wilks 1907, 33–37. London: Royal Horticultural Society.

Biffen, Rowland H. 1907b. "Experiments on the Breeding of Wheats for English Conditions." In Wilks 1907, 90–97. London: Royal Horticultural Society.

Blackman, Helen. 2004. "A Spiritual Leader? Cambridge Zoology, Mountaineering and the Death of F. M. Balfour." *Studies in History and Philosophy of Biological and Biomedical Sciences* 35:93–117.

Blackman, Helen. 2007a. "Lampreys, Lungfish and Elasmobranchs: Cambridge Zoology and the Politics of Animal Selection." *British Journal for the History of Science* 40:413–37.

Blackman, Helen. 2007b. "The Natural Sciences and the Development of Animal Morphology in Late-Victorian Cambridge." *Journal of the History of Biology* 40:71–108.

Blackman, V. H. 1902. "Some Recent Work on Hybrids in Plants [1 & 2]." *New Phytologist* 1:73–80; 97–106.

Boas, Helene M. 1918. "Inheritance of Eye-Color in Man." *American Journal of Physical Anthropology* 2:15–20.

Bokulich, Alisa. 2006. "Heisenberg Meets Kuhn: Closed Theories and Paradigms." *Philosophy of Science* 73:90–107.

Bonneuil, Christophe. 2016. "Pure Lines as Industrial Simulacra: A Cultural History of Genetics from Darwin to Johannsen." In *Heredity Explored: Between Public Domain and Experimental Science, 1850–1930*, edited by Staffan Müller-Wille and Christina Brandt, 213–42. Cambridge, MA: MIT Press.

Bos, L. 1999. "Beijerinck's Work on Tobacco Mosaic Virus: Historical Context and Legacy." *Philosophical Transactions: Biological Sciences* 354: 675–85.

Bourne, Gilbert C. 1906. [Obituary notice for W. F. R. Weldon.] *Proceedings of the Linnean Society of London* 118:109–14.

Bouterse, Jeroen. 2016. "Nature and History: Towards a Hermeneutic Philosophy of Historiography of Science." PhD dissertation, University of Leiden.

Boveri, Theodor. 1902. "Über mehrpolige Mitosen als Mittel zur Analzyse des Zellkerns." *Verhandlungen der Physikalisch-Medizinischen Gesellschaft zu Würzburg* 35:67–90. Published in English as "On Multipolar Mitosis as a Means of Analysis of the Cell Nucleus." In *Foundations of Experimental Embryology*, 2nd ed., edited by Benjamin H. Willier and Jane M. Oppenheimer, 74–97. Englewood Cliffs, NJ: Prentice Hall, 1964.

Bowler, Peter J. 1983. *The Eclipse of Darwinism: Anti-Darwinian Evolution Theories in the Decades around 1900*. Baltimore: Johns Hopkins University Press. [The 1992 paperback edition has a valuable updating preface.]

Bowler, Peter J. 1989a. *The Mendelian Revolution: The Emergence of Hereditarian Concepts in Modern Science and Society*. Baltimore: Johns Hopkins University Press.

Bowler, Peter J. 1989b. "Development and Adaptation: Evolutionary Concepts in British Morphology, 1870–1914." *British Journal for the History of Science* 22:283–97.

Bowler, Peter J. 1992. "Foreword." In William Bateson, *Materials for the Study of Variation, Treated with Especial Regard to Discontinuity in the Origin of Species*, xvii–xxvii. Reprint, Baltimore: Johns Hopkins University Press. First published 1894, London: Macmillan.

Bowler, Peter J. 1996. *Life's Splendid Drama: Evolutionary Biology and the Reconstruction of Life's Ancestry, 1860–1940*. Chicago: University of Chicago Press.

Bowler, Peter J. 2004. "Lankester, Edwin Ray." In *Dictionary of Nineteenth-Century British Scientists*, edited by Bernard Lightman, vol. 3, 1175–80. Bristol: Thoemmes Continuum.

Bowler, Peter J. 2008. "What Darwin Disturbed: The Biology That Might Have Been." *Isis* 99:560–67.

Bowler, Peter J. 2009. *Science for All: The Popularization of Science in Early Twentieth-Century Britain*. Chicago: University of Chicago Press.

Bowler, Peter J. 2013. *Darwin Deleted: Imagining a World without Darwin*. Chicago: University of Chicago Press.

Bowler, Peter J. 2014. "Francis Galton's Saltationism and the Ambiguities of Selection." *Studies in History and Philosophy of Biological and Biomedical Sciences* 48B:272–79.

Brannigan, Augustine. 1979. "The Reification of Mendel." *Social Studies of Science* 9:423–54.

Brannigan, Augustine. 1981. *The Social Basis of Scientific Discoveries*. Cambridge: Cambridge University Press.

Brautigam, Jeffrey C. 1993. "Inventing Biometry, Inventing 'Man': *Bio-*

metrika and the Transformation of the Human Sciences." PhD dissertation, University of Florida.

Brink, R. Alexander, ed. 1967. *Heritage from Mendel*. Madison: University of Wisconsin Press.

Broad, William, and Nicholas Wade. 1985. *Betrayers of the Truth*. Oxford: Oxford University Press.

Brooks, William Keith. 1882–83. "Speculative Zoology." *Popular Science Monthly* 22:195–204, 364–80.

Brooks, William Keith. 1883. *The Law of Heredity. A Study of the Cause of Variation, and the Origin of Living Organisms*. Baltimore: John Murray. [A bibliographic curiosity: The copies that I have seen indicate that they are the revised second edition; but all signs are that (a) the book came out in autumn 1883, not leaving much time for a revised second edition that same year, and (b) it sold poorly, so that a second edition would not have been called for in any case.]

Bruce, A. B. 1918. "The Economic Results of Plant Breeding." In *Memoranda on the Establishment of a National Institute of Agricultural Botany*, 12–15. [Pamphlet printed and circulated in November 1918 by "the Organiser of the Institute," Lawrence Weaver. A copy is held in the Library at NIAB in Cambridge.]

Brush, Stephen. 2002. "How Theories Become Knowledge: Morgan's Chromosome Theory of Heredity in America and Britain." *Journal of the History of Biology* 35:471–535.

Bryant, J. R., N. López-Villalobos, J. E. Pryce, C. W. Holmes, and D. L. Johnson. 2006. "Reaction Norms Used to Quantify the Responses of New Zealand Dairy Cattle of Mixed Breeds to Nutritional Environment." *New Zealand Journal of Agricultural Research* 49:371–81.

Buklijas, Tatjana, and Klaus Taschwer. 2019. "A Feeling for Lamarckism: The Making, Reception and Impact of Arthur Koestler's *The Case of the Midwife Toad* (1971)" [Unpublished MS, available online at academia .edu]

Bulmer, Michael. 2003. *Francis Galton: Pioneer of Heredity and Biometry*. Baltimore: Johns Hopkins University Press.

Bulmer, Michael. 2004a. "Did Jenkin's Swamping Argument Invalidate Darwin's Theory of Natural Selection?" *British Journal for the History of Science* 37:281–97.

Bulmer, Michael. 2004b. "Galton's Theory of Ancestral Inheritance." In *A Century of Mendelism in Human Genetics*, edited by Milo Keynes, A. F. W. Edwards, and Robert Peel, 13–18. London: CRC Press.

Burbank, Luther. 1907. *The Training of the Human Plant*. New York: Century.

Burbridge, David. 2001. "Francis Galton on Twins, Heredity and Social Class." *British Journal for the History of Science* 34:323–40.

Burian, Richard M., Jean Gayon, and Doris Zallen. 1988. "The Singular Fate of Genetics in the History of French Biology, 1900–1940." *Journal of the History of Biology* 21:357–402.

Burian, Richard M., and Kostas Kampourakis. 2013. "Against 'Genes For': Could an Inclusive Concept of Genetic Material Effectively Replace Gene Concepts?" In *The Philosophy of Biology: A Companion for Educators*, edited by Kostas Kampourakis, 597–628. Dordrecht: Springer.

Burkhardt, Frederick, et al., eds. 1985–. *The Correspondence of Charles Darwin*. 30 vols. Cambridge: Cambridge University Press.

Burkhardt, Richard W. Jr. 1979. "Closing the Door on Lord Morton's Mare: The Rise and Fall of Telegony." *Studies in History of Biology* 3:1–21.

Bustamente, Erika. 2010. "Eye Color." *Stanford at the Tech: Understanding Genetics*, 23 September. https://genetics.thetech.org/ask/ask377. Accessed 29 July 2020.

Buttolph, Mike. 2008. "One Hundred and One Mendelians." MSc dissertation, London Centre for the History of Science, Medicine and Technology. Downloadable, along with his database, at http://made-in-sts-ucl.uk/dissertations-research-reports-msc/buttloph-2008/. Accessed 3 March 2020.

Buttolph, Mike. 2015. "Retrodictions after the Rediscovery: Twentieth-Century Theory Sustained by Nineteenth-Century Data." Unpublished MS.

Button, Clare. 2018. "James Cossar Ewart and the Origins of the Animal Breeding Research Department in Edinburgh, 1895–1920." *Journal of the History of Biology* 51:445–77.

Bynum, W. F. 1993. "The Historical Galton." In *Sir Francis Galton, FRS: The Legacy of His Ideas*, edited by Milo Keynes, 33–44. London: Macmillan/Galton Institute.

Bynum, W. F., and Roy Porter, eds. 2005. *Oxford Dictionary of Scientific Quotations*. Oxford: Oxford University Press.

Cambridge University Association. 1906. "A Plea for Cambridge." *Quarterly Review* 204:521–22.

Campbell, Neil A. 1993. *Biology*. 3rd ed. New York: Benjamin/Cummings.

Candolle, Alphonse de. 1873. *Histoire des Sciences et des Savants depuis Deux Siècles*. Geneva: Georg.

Cannon, William Austin. 1902. "A Cytological Basis for the Mendelian Laws." *Bulletin of the Torrey Botanical Club* 29:657–61.

Cannon, William Austin. 1904. "Some Cytological Aspects of Hybrids." In *Proceedings: International Conference on Plant Breeding and Hybridization*, 89–92. Memoirs of the Horticultural Society of New York, vol. 1.

Carlson, Elof Axel. 1966. *The Gene: A Critical History*. Philadelphia: W. B. Saunders.

Carlson, Elof Axel. 2001. *The Unfit: A History of a Bad Idea*. Cold Spring Harbor, NY: Cold Spring Harbor Laboratory Press.

Carter, Bill, and Merri Sue Carter. 2002. *Latitude: How American Astronomers Solved the Mystery of Variation*. Annapolis, MD: Naval Institute Press.

Carvalho dos Santos, Vanessa, Leyla M. Joaquim, and Charbel N. El-Hani. 2012. "Hybrid Deterministic Views about Genes in Biology Textbooks: A Key Problem in Genetics Teaching." *Science and Education* 21: 543–78.

Carver, Rebecca Bruu, Jérémy Castéra, Niklas Gericke, Neima Alice Menezes Evangelista, and Charbel N. El-Hani. 2017. "Young Adults' Belief in Genetic Determinism, and Knowledge and Attitudes towards Modern Genetics and Genomics: The PUGGS Questionnaire." *PLoS ONE* 12 (1): 1–24, with supplements.

Castéra, Jérémy, Mondher Abrougui, Olympia Nisiforou et al. 2008. "Genetic Determinism in School Textbooks: A Comparative Study Conducted Among Sixteen Countries." *Science Education International* 19:163–84.

Castéra, Jérémy, Catherine Bruguière, and Pierre Clément. 2008. "Genetic Diseases and Genetic Determinism Models in French Secondary School Biology Textbooks." *Journal of Biological Education* 42:53–55.

Castle, William. 1903a. "Note on Mr Farabee's Observations." *Science* 17 (9 January): 75–76.

Castle, William. 1903b. "The Laws of Heredity of Galton and Mendel, and Some Laws Governing Race Improvement." *Proceedings of the American Academy of Arts and Sciences* 39:223–42.

Castle, William, and Glover M. Allen. 1903. "The Heredity of Albinism." *Proceedings of the American Academy of Arts and Sciences* 38:603–22.

Chang, Hasok. 2004. *Inventing Temperature: Measurement and Scientific Progress*. Oxford: Oxford University Press.

Charnley, Berris. 2011. "Agricultural Science, Plant Breeding and the Emergence of a Mendelian System in Britain, 1880–1930." PhD dissertation, University of Leeds.

Charnley, Berris, and Gregory Radick. 2013. "Intellectual Property, Plant Breeding, and the Making of Mendelian Genetics." *Studies in History and Philosophy of Science* 44:222–33.

Chen, Rong, et al. 2016. "Analysis of 589,306 Genomes Identifies Individuals Resilient to Severe Mendelian Childhood Diseases." *Nature Biotechnology* 34 (May): 531–38.

Cheng, Hanyin, et al. 2019. "Phenotypic and Biochemical Analysis of an

International Cohort of Individuals with Variants of *NAA10* and *NAA15*." *Human Molecular Genetics* 28 (17): 2900–19.

Cheung, Benjamin Y., Anita Schmalor, and Steven J. Heine. 2021. "The Role of Genetic Essentialism and Genetics Knowledge in Support for Eugenics and Genetically Modified Foods." *PLoS ONE* 16 (9): e0257954.

Chisholm, Roderick. 1946. "The Contrary-to-Fact Conditional." *Mind* 55:289–307.

Churchill, Frederick B. 1969. "From Machine-Theory to Entelechy: Two Studies in Developmental Teleology." *Journal of the History of Biology* 2:165–85.

Churchill, Frederick B. 1987. "From Heredity Theory to *Vererbung*: The Transmission Problem, 1850–1915." *Isis* 78:337–64.

Churchwell, Sarah. 2019. "American Immigration: A Century of Racism." *New York Review of Books*, 26 September, 53–55.

Clarke, Belinda. 2015. "Round or Wrinkled; Mendel's Peas Explained." Website of Steven M. Carr, Memorial University of Newfoundland. http://www.mun.ca/biology/scarr/Round_&_Wrinkled.htm. Accessed 31 July 2020.

Cobb, Matthew. 2006. "Heredity before Genetics: A History." *Nature Reviews: Genetics* 7:953–58.

Cock, Alan G. 1983. "William Bateson's Rejection and Eventual Acceptance of Chromosome Theory." *Annals of Science* 40:19–59.

Cock, Alan G., and Donald R. Forsdyke. 2008. *Treasure Your Exceptions: The Science and Life of William Bateson*. New York: Springer.

Cohen, Adam. 2016. *Imbeciles: The Supreme Court, American Eugenics, and the Sterilization of Carrie Buck*. New York: Penguin.

Cohen-Cole, Jamie. 2014. *The Open Mind: Cold War Politics and the Sciences of Human Nature*. Chicago: University of Chicago Press.

Coleman, Clare. 2021. "Plant Hybridity before Mendelism: Diversity and Debate in British Botany, 1837–1899." PhD thesis, University of Leeds.

Coleman, William. 1965. "Cell, Nucleus, and Inheritance: An Historical Study." *Proceedings of the American Philosophical Society* 109:124–58.

Coleman, William. 1970. "Bateson and Chromosomes: Conservative Thought in Science." *Centaurus* 15:228–314.

Comfort, Nathaniel. 2018. "Genetic Determinism Redux" [review of Plomin 2018]. *Nature* 561 (27 September): 461–63.

Conklin, Edwin Grant. 1913. "William Keith Brooks[,] 1848–1908." *Biographical Memoirs of the National Academy of Sciences* 7:23–88.

Conway Morris, Simon. 1998. *The Crucible of Creation: The Burgess Shale and the Rise of Animals*. Oxford: Oxford University Press.

Conway Morris, Simon. 2003. *Life's Solution: Inevitable Humans in a Lonely Universe*. Cambridge: Cambridge University Press.

Conway Morris, Simon, and Stephen Jay Gould. 1998–99. "Showdown on the Burgess Shale." *Natural History* 107 (December–January): 48–55.

Corcos, Alain F., and Floyd V. Monaghan. 1993. *Gregor Mendel's* Experiments on Plant Hybrids: *A Guided Study*. New Brunswick, NJ: Rutgers University Press.

Correns, Carl. 1900a. "G. Mendel's Regel über das Verhalten der Nachkommenschaft der Rassenbastarde." *Berichte der Deutschen Botanischen Gesellschaft* 18:158–68. Published in English as "G. Mendel's Law Concerning the Behavior of Progeny of Varietal Hybrids," translated by Leonie K. Pitternick. In *The Origin of Genetics: A Mendel Source Book*, edited by Curt Stern and Eva R. Sherwood, 119–32. London: W. H. Freeman, 1966.

Correns, Carl. 1900b. "Gregor Mendel's 'Versuche über Pflanzen-Hybriden' und die Bestätigung ihrer Ergebnisse durch die neuesten Untersuchungen." *Botanische Zeitung* 58, cols. 229–35.

Correns, Carl. 1900c. "Ueber Levkojenbastarde." *Botanisches Centralblatt* 84:97–113.

Cowan, Ruth Schwartz. 1969. "Sir Francis Galton and the Study of Heredity in the Nineteenth Century." PhD dissertation, Johns Hopkins University. Reprint, New York: Garland Publishing, 1985.

Cowan, Ruth Schwartz. 1976. "Weldon, Walter Frank Raphael." In *Dictionary of Scientific Biography*, edited by C. C. Gillispie, vol. 14, 251–52. New York: Charles Scribner's Sons.

Cowan, Ruth Schwartz. 1977. "Nature and Nurture: The Interplay of Biology and Politics in the Work of Francis Galton." *Studies in History of Biology* 1:133–208. Baltimore: Johns Hopkins University Press.

Cuénot, Lucien. 1902. "La loi de Mendel et l'heredité de la pigmentation chez les souris." *Archives de zoologie expérimentale et génerale*, 3rd series, 10:27–30. Reprinted in *Fundamenta Genetica*, edited by Jaroslav Křížnecký, 158–60. Brno: Moravian Museum, 1965.

Cushing, James T. 1994. *Quantum Mechanics: Historical Interpretation and the Copenhagen Hegemony*. Chicago: University of Chicago Press.

Dagg, Joachim L. 2019. "Motives and Merits of Counterfactual Histories of Science." *Studies in History and Philosophy of Biological and Biomedical Sciences* 73:19–26.

Darbishire, Arthur D. 1902. "Note on the Results of Crossing Japanese Waltzing Mice with European Albino Races." *Biometrika* 2:101–4.

Darbishire, Arthur D. 1903a. "Second Report on the Result of Crossing Japanese Waltzing Mice with European Albino Races." *Biometrika* 2 (February): 165–73.

Darbishire, Arthur D. 1903b. "Third Report on Hybrids between Waltzing Mice and Albino Races[:] On the Result of Crossing Japanese Waltzing Mice with 'Extracted' Recessive Albinos." *Biometrika* 2:282–85.

Darbishire, Arthur D. 1904. "On the Result of Crossing Japanese Waltzing with Albino Mice." *Biometrika* 3:1–51.

Darbishire, Arthur D. 1907. "Recent Advances in Animal Breeding and Their Bearing on Our Knowledge of Heredity." In Wilks 1907, 130–37.

Darden, Lindley. 1985. "Hugo de Vries's Lecture Plates and the Discovery of Segregation." *Annals of Science* 42:233–42.

Darden, Lindley. 2005. "Relations Among Fields: Mendelian, Cytological and Molecular Mechanisms." *Studies in History and Philosophy of Biological and Biomedical Sciences* 36:349–71.

Darwin, Charles. 1859. *On the Origin of Species by Means of Natural Selection*. London: John Murray.

Darwin, Charles. 1863. *Über die Entstehung der Arten im Thier- und Pflanzen-Reich durch natürliche Züchtung*. Trans. H. G. Bronn. Stuttgart: E. Schweizerbart'sche Verlagshandlung und Druckerei.

Darwin, Charles. 1868. *The Variation of Animals and Plants under Domestication*. 2 vols. London: John Murray.

Darwin, Charles. 1871a. *The Descent of Man, and Selection in Relation to Sex*. London: John Murray.

Darwin, Charles. 1871b. "Pangenesis." *Nature* 3 (27 April): 502–3.

Darwin, Charles. 1875. *The Variation of Animals and Plants under Domestication*. 2nd ed. 2 vols. London: John Murray.

Darwin, Charles. 1881. "Inheritance." *Nature*, 21 July, 257.

Daston, Lorraine. 2016. "History of Science without *Structure*." In *Kuhn's* Structure of Scientific Revolutions *at Fifty: Reflections on a Science Classic*, edited by Robert J. Richards and Lorraine Daston, 115–32. Chicago: University of Chicago Press.

Davenport, Charles B. 1899. *Statistical Methods with Special Reference to Biological Variation*. New York: John Wiley & Sons.

Davenport, Charles B. 1900a. "A History of the Development of the Quantitative Study of Variations." *Science* 12:864–70.

Davenport, Charles B. 1900b. "Review of von Guaita's Experiments in Breeding Mice." *Biological Bulletin* 2:121–28.

Davenport, Charles B. 1901. "Mendel's Law of Dichotomy in Hybrids." *Biological Bulletin* 2:307–10.

Davenport, Charles B. 1904a. "Color Inheritance in Mice." *Science* 19 (15 January): 110–14.

Davenport, Charles B. 1904b. "Wonder Horses and Mendelism." *Science* 19 (22 January): 151–53.

Davenport, Gertrude, and Charles B. Davenport. 1907. "Heredity of Eye-Color in Man." *Science* 26 (1 November): 589–92.

Dawkins, Richard. 1982. *The Extended Phenotype: The Long Reach of the Gene*. Oxford: Oxford University Press.

Dawkins, Richard. 2003a. "An Early Flowering of Genetics." *Guardian*, 8 February, 34–36. A modified version is available under the title "Light Will Be Thrown" in *A Devil's Chaplain: Selected Essays by Richard Dawkins*, edited by Latha Menon, 73–90. London: Phoenix.

Dawkins, Richard. 2003b. "The Art of the Developable." In *A Devil's Chaplain: Selected Essays by Richard Dawkins*, edited by Latha Menon, 227–37. London: Phoenix.

Dawkins, Richard. 2010. "Foreword." In *Darwin . . . Off the Record*, by Peter J. Bowler, 6–7. London: Watkins.

Dawkins, Richard. 2017. "Science and Sensibility." In *Science in the Soul: Selected Writing of a Passionate Rationalist*, 76–96. London: Bantam Press.

Dear, Peter. 2005. "What Is the History of Science the History *Of*? Early Modern Roots of the Ideology of Modern Science." *Isis* 96:390–406.

De Beer, Gavin. 1947. "Edwin Stephen Goodrich 1868–1946." *Biographical Memoirs of Fellows of the Royal Society* 15:477–90.

De Beer, Gavin. 1964. "Mendel, Darwin and Fisher (1865–1965)." *Notes and Records of the Royal Society of London* 19:192–226.

De Bont, Raf. 2014. *Stations in the Field: A History of Place-Based Animal Research, 1870–1930*. Chicago: University of Chicago Press.

DeJarnette, Joseph. 1921. "Mendel's Law: A Plea for a Better Race of Men." Typescript. American Philosophical Society Library. Scan available at www.eugenicsarchive.org/html/eugenics/static/images/1235.html. Accessed 21 July 2020.

DeJong-Lambert, William. 2012. *The Cold War Politics of Genetic Research: An Introduction to the Lysenko Affair*. Dordrecht: Springer.

DeJong-Lambert, William, and Nikolai Krementsov, eds. 2017. *The Lysenko Controversy as a Global Phenomenon: Genetics and Agriculture in the Soviet Union and Beyond*. 2 vols. London: Palgrave Macmillan.

De Marrais, Robert. 1974. "The Double-Edged Effect of Sir Francis Galton: A Search for the Motives in the Biometrician-Mendelian Debate." *Journal of the History of Biology* 7:141–74.

Dennett, Daniel C. 2004. "The Mythical Threat of Genetic Determinism." In *The Best American Science and Nature Writing 2004*, edited by Steven Pinker, 45–50. New York: Houghton Mifflin. First published in *Chronicle of Higher Education*, 31 January 2003.

Depew, David J., and Bruce H. Weber. 1995. *Darwinism Evolving: Systems Dynamics and the Genealogy of Natural Selection*. Cambridge, MA: Bradford Books/MIT Press.

Desmond, Adrian, and James Moore. 2009. *Darwin's Sacred Cause: Race, Slavery and the Quest for Human Origins*. London: Allen Lane.

De Vries, Hugo. 1889. *Intracellulare Pangenesis*. Jena: Fischer. Published in English as *Intracellular Pangenesis, Including a Paper on Fertilization and Hybridization*, translated by C. S. Gager. Chicago: Open Court, 1910.

De Vries, Hugo. 1900a. "Hybridising of Monstrosities." Hybrid Conference Report 1900, *Journal of the Royal Horticultural Society* 24:69–75.

De Vries, Hugo. 1900b. "Sur la loi de disjonction des hybrides." *Comptes Rendus de l'Académie des Sciences Paris* 130:845–47. Published in English as "Concerning the Law of Segregation of Hybrids," translated by Aloha Hannah. *Genetics* 35 (1950): 30–32. Reprinted in Olby 1987, 418–20.

De Vries, Hugo. 1900c. "Das Spaltungsgesetz der Bastarde," *Berichte der Deutschen Botanischen Gesellschaft* 18:83–90. Published in English as "The Law of Segregation of Hybrids," translated by Evelyn Stern. In *The Origin of Genetics: A Mendel Source Book*, edited by Curt Stern and Eva R. Sherwood, 107–17. London: W. H. Freeman, 1966.

De Vries, Hugo. 1900d. "Sur les unités des caractères spécifiques et leur application à l'étude des hybrides." *Revue Géneral de Botanique* 12:257–71.

De Vries, Hugo. 1901. *Die Mutationstheorie, Versuche und Beobachtungen über die Enstehung der Arten im Pflanzenreich*. Vol. 1. Leipzig. Published in English as *The Mutation Theory: Experiments and Observations on the Origin of Species in the Vegetable Kingdom*, vol. 1, translated by J. B. Farmer and A. D. Darbishire. London: Kegan Paul; Trench, Trübner, 1910.

De Vries, Hugo. 1904. "On Artificial Atavism." In *Proceedings: International Conference on Plant Breeding and Hybridization*, 17–24. Memoirs of the Horticultural Society of New York, vol. 1.

Di Cesnola, A. P. 1904. "Preliminary Note on the Protective Value of Colour in *Mantis religiosa*." *Biometrika* 3:58–59.

Di Cesnola, A. P. 1907. "A First Study of Natural Selection in *'Helix arbustorum' (helicogena)*." *Biometrika* 5:387–99.

Dietrich, Michael R. 2016. "Experimenting with Sex: Four Approaches to the Genetics of Sex Reversal before 1950." *History and Philosophy of the Life Sciences* 38 (1): 23–41.

Di Gregorio, Mario A. 1990. *Charles Darwin's Marginalia*. With N. W. Gill. Vol. 1. London: Garland.

D[ixey], F. A. 1902. "Mendel's Theory of Heredity" [review of Bateson and Saunders 1902 and Bateson 1902a]. *Nature* 66 (9 October): 573.

Dobzhansky, Theodosius. 1967. "Looking Back at Mendel's Discovery." *Science* 156 (23 June): 1588–89.

Doncaster, Leonard. 1904. "Experiments in Hybridization, with Special Reference to the Effects of Conditions on Dominance." *Philosophical Transactions of the Royal Society of London B* 196:119–73.

Donovan, Brian M. 2016. "Framing the Genetics Curriculum for Social Justice: An Experimental Exploration of How the Biology Curriculum Influences Beliefs about Racial Difference." *Science Education* 100:586–616.

Donovan, Brian M. 2017. "Learned Inequality: Racial Labels in the Biology Curriculum Can Affect the Development of Racial Prejudice." *Journal of Research in Science Teaching* 54:379–411.

Donovan, Brian M. 2022. "Ending Genetic Essentialism Through Genetics Education." *HGG Advances* 3:1–13.

Donovan, Brian M., et al. 2019. "Toward a More Humane Genetics Education: Learning About the Social and Quantitative Complexities of Human Genetic Variation Research Could Reduce Racial Bias in Adolescent and Adult Populations." *Science Education* 103 (3): 529–60.

Dougherty, Michael J. 2009. "Closing the Gap: Inverting the Genetics Curriculum to Ensure an Informed Public." *American Journal of Human Genetics* 85:6–12.

Dougherty, Michael J. 2010. "It's Time to Overhaul Our Outdated Genetics Curriculum." *American Biology Teacher* 72:218.

Dronamraju, Krishna R., and Joseph Needham, eds. 1993. *If I Am To Be Remembered: Correspondence of Julian Huxley*. London: World Scientific.

Dröscher, Ariane. 2015a. "Gregor Mendel, Franz Unger, Carl Nägeli and the Magic of Numbers." *History of Science* 53:492–508.

Dröscher, Ariane, ed. 2015b. "Crossroads between Nature and Culture: Papers in Honour of Christiane Groeben." Special issue, *History and Philosophy of the Life Sciences* 36, no. 3 (January).

Dunn, L. C., ed. 1951. *Genetics in the 20th Century: Essays on the Progress of Genetics During Its First 50 Years*. New York: Macmillan.

Dunn, L. C. 1965. *A Short History of Genetics: The Development of Some of the Main Lines of Thought: 1864–1939*. New York: McGraw-Hill.

Edwards, A. W. F. 2008. "G. H. Hardy (1908) and Hardy-Weinberg Equilibrium." *Genetics* 179:1143–50.

Edwards, A. W. F. 2012a. "Punnett's Square." *Studies in History and Philosophy of Biological and Biomedical Sciences* 43:219–24.

Edwards, A. W. F. 2012b. "Reginald Crundall Punnett: First Arthur Balfour Professor of Genetics, Cambridge, 1912." *Genetics* 192:3–13.

Edwards, A. W. F. 2013. "Robert Heath Lock and His Textbook of Genetics, 1906." *Genetics* 194:529–37.

Edwards, A. W. F. 2016. "Punnett's Square: A Postscript." *Studies in History and Philosophy of Biological and Biomedical Sciences* 57:69–70.

Eimer, G. H. Theodor. 1890. *Organic Evolution as the Result of the Inheritance of Acquired Characters According to the Laws of Organic Growth*. Translated by Joseph T. Cunningham. London: Macmillan.

Ellis, T. H. Noel, Julie M. I. Hofer, Martin T. Swain, and Peter J. Van Dijk. 2019. "Mendel's Pea Crosses: Varieties, Traits and Statistics." *Hereditas* 156 (33): 1–11.

Ellis, T. H. Noel, Julie M. I. Hofer, Gail M. Timmerman-Vaughan, Clarice J.

Coyne, and Roger P. Hellen. 2011. "Mendel, 150 Years On." *Trends in Plant Science* 16:590–96.

Endersby, Jim. 2007. *A Guinea Pig's History of Biology: The Plants and Animals Who Taught Us the Facts of Life*. London: William Heinemann.

Esposito, Maurizio. 2013a. "Weismann versus Morgan Revisited: Clashing Interpretations on Animal Regeneration." *Journal of the History of Biology* 46:511–41.

Esposito, Maurizio. 2013b. *Romantic Biology, 1890–1945*. London: Pickering & Chatto.

Evans, Richard J. 1997. "In Search of German Social Darwinism." In *Rereading German Social History: From Unification to Reunification, 1800–1996*, 119–44. London: Routledge.

Ewart, James Cossar. 1899. "Experimental Contributions to the Theory of Heredity. A. Telegony." *Proceedings of the Royal Society of London* 65: 243–51.

Ewart, James Cossar. 1901. "Variation: Germinal and Environmental." *Scientific Transactions of the Royal Dublin Society* 7:353–78.

F. W. R. 1889. "Walter Weldon, F.R.S." *Proceedings of the Royal Society of London* 46:xix–xxiv.

Fairbanks, Daniel J. 2020. "Mendel and Darwin: Untangling a Persistent Enigma." *Heredity* 124:263–73.

Fairbanks, Daniel J., and Scott Abbott. 2016. "Darwin's Influence on Mendel: Evidence from a New Translation of Mendel's Paper." *Genetics* 204:401–5.

Fairbanks, Daniel J., and Bruce Rytting. 2001a. "Mendel's Marginalia in Darwin's *Origin of Species*." Online "supplementary data" appendix to Fairbanks and Rytting 2001b. Available at the *American Journal of Botany* website, http://ajbsupp.botany.org/v88/fairbanks.html.

Fairbanks, Daniel J., and Bruce Rytting. 2001b. "Mendelian Controversies: A Botanical and Historical Review." *American Journal of Botany* 88:737–52. Reprinted with an update in Franklin et al. 2008, 264–311.

Falk, Raphael. 2001a. "The Rise and Fall of Dominance." *Biology and Philosophy* 16:285–323.

Falk, Raphael. 2001b. "Can the Norm of Reaction Save the Gene Concept?" In *Thinking about Evolution: Historical, Philosophical, and Political Perspectives*, edited by Rama S. Singh, Costas B. Krimbas, Diane B. Paul, and John Beatty, 119–40. Cambridge: Cambridge University Press.

Fancher, Raymond E. 1979. "A Note on the Origin of the Term 'Nature and Nurture.'" *Journal of the History of the Behavioral Sciences* 15:321–22.

Fancher, Raymond E. 1983a. "Francis Galton's African Ethnography and Its Role in the Development of His Psychology." *British Journal for the History of Science* 16:67–79.

Fancher, Raymond E. 1983b. "Alphonse de Candolle, Francis Galton, and the Early History of the Nature-Nurture Controversy." *Journal of the History of the Behavioral Sciences* 19:341–52.

Fancher, Raymond E. 1998. "Biography and Psychodynamic Theory: Some Lessons from the Life of Francis Galton." *History of Psychology* 1:99–115.

Farmer, J[ohn] B[retland]. 1901. "The Present Aspect of Some Cytological Problems" [review of Wilson 1900]. *Nature* 63 (7 March): 437–38.

Farrall, Lyndsay Andrew. 1969. "The Origins and Growth of the English Eugenics Movements 1865–1925." PhD dissertation, Indiana University. Reprint, New York: Garland Publishing, 1985.

Farrall, Lyndsay Andrew. 1975. "Controversy and Conflict in Science: A Case Study—The English Biometric School and Mendel's Laws." *Social Studies of Science* 5:269–301.

Ferguson, Niall. 1997. "Virtual History: Towards a 'Chaotic' Theory of the Past." In *Virtual History: Alternatives and Counterfactuals*, edited by N. Ferguson, 1–90. London: Basic Books.

Fincham, J. R. S. 1990. "Mendel—Now Down to the Molecular Level." *Nature* 343 (18 January): 208–9.

Fisher, Ronald A. 1911. "Heredity." For the version recorded in the minute book of the Cambridge University Eugenics Society, see Bernard Norton and E. S. Pearson, "A Note on the Background to, and Refereeing of, R. A. Fisher's 1918 Paper 'On the Correlation between Relatives on the Supposition of Mendelian Inheritance,'" *Notes and Records of the Royal Society of London* 31 (1976): 151–62. For the slightly different version in Fisher's own copy, see Bennett 1983, 51–63.

Fisher, Ronald A. 1925. *Statistical Methods for Research Workers*. Edinburgh: Oliver and Boyd.

Fisher, Ronald A. 1930. *The Genetical Theory of Natural Selection*. Oxford: Clarendon Press. 2nd ed. 1958. New York: Dover.

Fisher, Ronald A. 1935. *The Design of Experiments*. Edinburgh: Oliver and Boyd.

Fisher, Ronald A. 1936. "Has Mendel's Work Been Rediscovered?" *Annals of Science* 1:115–37. Reprinted in Stern and Sherwood 1966, 139–72, and in Franklin et al. 2008, 117–40. Also available from the R. A. Fisher Digital Archive.

Fleck, Ludwik. 1979. *Genesis and Development of a Scientific Fact*. Edited by Thaddeus J. Trenn and Robert K. Merton. Translated by Fred Bradley and Thaddeus J. Trenn. Chicago: University of Chicago Press. First published 1935.

Fletcher, H. R. 1969. *The Story of the Royal Horticultural Society 1804–1968*. Oxford: Oxford University Press.

Fogarty, Richard S., and Michael A. Osborne. 2010. "Eugenics in France and the Colonies." In *The Oxford Handbook of the History of Eugenics*, edited by Alison Bashford and Philippa Levine, 332–46. Oxford: Oxford University Press.

Ford, E. B., and Julian S. Huxley. 1927. "Mendelian Genes and Rates of Development in *Gammarus chevreuxi*." *British Journal of Experimental Biology* 5:112–34.

Forrest, D. W. 1974. *Francis Galton: The Life and Work of a Victorian Genius*. London: Paul Elek.

Franklin, Allan. 2008. "The Mendel-Fisher Controversy: An Overview." In Franklin et al. 2008, 1–77.

Franklin, Allan, A. W. F. Edwards, Daniel J. Fairbanks, Daniel L. Hartl, and Teddy Seidenfeld. 2008. *Ending the Mendel-Fisher Controversy*. Pittsburgh: University of Pittsburgh Press.

French, Steven. 2020. *There Are No Such Things as Theories*. Oxford: Oxford University Press.

Freudenberg-Hua, Yun, et al. 2014. "Disease Variants in Genomes of 44 Centenarians." *Molecular Genetics & Genomic Medicine* 2:438–50.

Frezza, Giulia, and Mauro Capocci. 2018. "Thomas Hunt Morgan and the Invisible Gene: The Right Tool for the Job." *History and Philosophy of the Life Sciences* 40 (31): 1–18.

Friend, Stephen H. 2014. "The Hunt for 'Unexpected Genetic Heroes.'" TED talk, March 2014. https://www.ted.com/talks/stephen_friend_the_hunt_for_unexpected_genetic_heroes#t-622778.

Friend, Stephen H., and Eric Schadt. 2014. "Clues from the Resilient." *Science* 344 (30 May): 970–72.

Froggatt, P., and N. C. Nevin. 1971a. "The 'Law of Ancestral Heredity' and the Mendelian-Ancestrian Controversy in England, 1889–1906." *Journal of Medical Genetics* 8:1–36.

Froggatt, P., and N. C. Nevin. 1971b. "Galton's 'Law of Ancestral Heredity': Its Influence on the Early Development of Human Genetics." *History of Science* 10:1–27.

Fuller, Steve. 2000. *Thomas Kuhn: A Philosophical History for Our Times*. Chicago: University of Chicago Press.

Galton, Francis. 1857. "Negroes and the Slave Trade." *Times* (London), 26 December, 10d.

Galton, Francis. 1865. "Hereditary Talent and Character." *Macmillan's Magazine* 12:157–66; 318–27.

Galton, Francis. 1869. *Hereditary Genius: An Inquiry into Its Laws and Consequences*. London: Macmillan.

Galton, Francis. 1871a. "Experiments in Pangenesis, by Breeding from Rabbits of a Pure Variety, into Whose Circulation Blood Taken from

Other Varieties Had Previously Been Largely Transfused." *Proceedings of the Royal Society of London* 19:393–410.

Galton, Francis. 1871b. "Pangenesis." *Nature* 4 (4 May): 5–6.

Galton, Francis. 1872a. "On Blood-Relationship." *Proceedings of the Royal Society of London* 20:392–402.

Galton, Francis. 1872b. *The Art of Travel; or, Shifts and Contrivances Available in Wild Countries*. 5th ed. London: John Murray.

Galton, Francis. 1873a. "Hereditary Improvement." *Fraser's Magazine* 7:116–30.

Galton, Francis. 1873b. "On the Causes Which Operate to Create Scientific Men." *Fortnightly Review* 13:345–51.

Galton, Francis. 1874a. *English Men of Science: Their Nature and Nurture*. London: Macmillan.

[Galton, Francis.] 1874b. "Men of Science: Their Nature and Nurture." *Nature* 9 (5 March): 344–45.

Galton, Francis. 1875a. "A Theory of Heredity." *Contemporary Review* 27:80–95.

Galton, Francis. 1875b. "The History of Twins, as a Criterion of the Relative Powers of Nature and Nurture." *Fraser's Magazine* 12:566–76. [Published with additional material in 1876 in *Journal of the Anthropological Institute* 5:391–406.]

Galton, Francis. 1876. "A Theory of Heredity." *Journal of the Anthropological Institute* 5:329–48.

Galton, Francis. 1877. "Typical Laws of Heredity." *Nature* (5, 12 and 19 April): 492–95; 512–14; 532–33.

Galton, Francis. 1883. *Inquiries into Human Faculty and Its Development*. London: Macmillan. 2nd ed. 1908, London: J. M. Dent.

Galton, Francis. 1884. "Free-will—Observations and Inferences." *Mind* 9:406–13.

Galton, Francis. 1885. [Presidential Address to Section H.] In *Report of the Fifty-Fifth Meeting of the British Association for the Advancement of Science Held at Aberdeen in September 1885*, 1206–14. London: John Murray.

Galton, Francis. 1887. "Address Delivered at the Anniversary Meeting of the Anthropological Institute of Great Britain and Ireland, January 25th 1887." *Journal of the Anthropological Institute of Great Britain and Ireland* 16:387–402.

Galton, Francis. 1888. "Co-relations and Their Measurements, Chiefly from Anthropometric Data." *Proceedings of the Royal Society of London* 45:135–45.

Galton, Francis. 1889. *Natural Inheritance*. London: Macmillan.

Galton, Francis. 1894. "Discontinuity in Evolution." *Mind* 3:362–72.

Galton, Francis. 1895. "A New Step in Statistical Science." *Nature*, 31 January, 319.

Galton, Francis. 1897a. "The Average Contribution of Each Several Ancestor to the Total Heritage of the Offspring." *Proceedings of the Royal Society of London* 61:401–13.

Galton, Francis. 1897b. "Hereditary Colour in Horses." *Nature* 56 (21 October): 598–99.

Galton, Francis. 1898. "A Diagram of Heredity." *Nature* 57:293.

Galton, Francis. 1901. "The Possible Improvement of the Human Breed under the Existing Conditions of Law and Sentiment." *Nature* 64 (31 October): 659–65.

Galton, Francis. 1903. "Our National Physique: Prospects of the British Race: Are We Degenerating?" *Daily Chronicle*, 29 July 1903.

Galton, Francis. 1904. "Eugenics: Its Definition, Scope and Aims." *American Journal of Sociology* 10:1–6.

Galton, Francis. 1905. "Eugenics: Its Definition, Scope and Aims." *Sociological Papers* 1:45–51 [with discussion, a response, and press commentary on 52–84].

Galton, Francis. 1906. "Studies in National Eugenics." *Sociological Papers* 2:14–17. First published as "National Eugenics," *British Medical Journal* 1 (25 February 1905): 440–41. [It was printed immediately before the first overview of Weldon's 1904–5 lecture on the hereditary process.]

Galton, Francis. 1908. *Memories of My Life*. London: Methuen.

Garcia, Richard S. 2004. "The Misuse of Race in Medical Diagnosis." *Pediatrics* 111 (5): 1394–95.

Gardner, Martin. 1977. "Great Fakes of Science." *Esquire*, October 1977, 88–91.

Gardner, Martin. 1981. "Great Fakes of Science." In his *Science: Good, Bad and Bogus*, 123–30. Buffalo: Prometheus. [The pre-edit version of Gardner 1977, with references and a postscript.]

Garfinkel, Alan. 1981. *Forms of Explanation: Rethinking the Questions in Social Theory*. New Haven: Yale University Press.

Garland, Martha McMackin. 1980. *Cambridge before Darwin: The Ideal of a Liberal Education, 1800–1860*. Cambridge: Cambridge University Press.

Garrod, Archibald E. 1902. "The Incidence of Alkaptonuria: A Study of Chemical Individuality." *Lancet* 2:1616–20.

Garrod, Archibald E. 1903. "Ueber chemische Individualität und chemische Missbildungen." *Pflügers Archiv für die gesammte Physiologie des Menschen und der Tiere* 97:410–18.

Gaudillière, Jean-Paul, and Hans-Jörg Rheinberger, eds. 2004. *From Mo-*

lecular Genetics to Genomics: The Mapping Cultures of Twentieth-Century Genetics. London: Routledge.

Gausemeier, Bernd, Staffan Müller-Wille, and Edmund Ramsden, eds. 2013. *Human Heredity in the Twentieth Century*. London: Pickering & Chatto.

Gayon, Jean. 1998. *Darwinism's Struggle for Survival: Heredity and the Hypothesis of Natural Selection*. Translated by Matthew Cobb. Cambridge: Cambridge University Press.

Gayon, Jean. 2007. "Karl Pearson ou: les enjeux du phénoménalisme dans les sciences biologique vers 1900." In *Conceptions de la Science: Hier, Aujourd'hui, Demain: Hommage à Marjorie Grene*, 305–24. Brussels: Ousia. [I consulted an unpublished English-language version supplied by the author.]

Gayon, Jean, and Richard M. Burian. 2000. "France in the Era of Mendelism (1900–1930)." *Comptes Rendus de l'Académie des Sciences Paris, Sciences de la Vie* 323:1097–1106.

Geddes, Patrick. 1889. "Mr Francis Galton on Natural Inheritance." *Scottish Leader* 14 March, 2.

Geison, Gerald L. 1969. "Darwin and Heredity: The Evolution of His Hypothesis of Pangenesis." *Bulletin of the History of Medicine* 24:375–411.

Geison, Gerald L. 1976. "Thiselton-Dyer, William Turner." In *Dictionary of Scientific Biography*, edited by C. C. Gillispie, vol. 13, 341–44. New York: Charles Scribner's Sons.

Geison, Gerald L. 1978. *Michael Foster and the Cambridge School of Physiology: The Scientific Enterprise in Late Victorian Society*. Princeton, NJ: Princeton University Press.

Geison, Gerald L., and Frederic L Holmes, eds. 1993. "Research Schools: Historical Reappraisals." *Osiris* 8.

Giere, Ronald N. 2006. *Scientific Perspectivism*. Chicago: University of Chicago Press.

Gigerenzer, Gerd, Zeno Swijtink, Theodore Porter, Lorraine Daston, John Beatty, and Lorenz Krüger. 1989. *The Empire of Chance: How Probability Changed Science and Everyday Life*. Cambridge: Cambridge University Press.

Gilbert, Scott. 2011. "The Decline of Soft Inheritance." In *Transformations of Lamarckism: From Subtle Fluids to Molecular Biology*, edited by Snait B. Gissis and Eva Jablonka, 121–25. Cambridge, MA: MIT Press.

Gilbert, Scott. 2012. "Ecological Developmental Biology: Environmental Signals for Normal Animal Development." *Evolution & Development* 14 (1): 20–28.

Gilbert, Scott F., and David Epel. 2015. *Ecological Developmental Biology:*

The Environmental Regulation of Development, Health, and Evolution. 2nd ed. Sunderland, MA: Sinauer Associates.

Gillham, Nicholas Wright. 2001. *A Life of Sir Francis Galton: From African Exploration to the Birth of Eugenics*. Oxford: Oxford University Press.

Gitschier, Jane. 2014. "In Pursuit of the Gene: An Interview with James Schwartz." *PLoS Genetics* 10:1–5.

Gliboff, Sander. 1999. "Gregor Mendel and the Laws of Evolution." *History of Science* 37:217–35.

Gliboff, Sander. 2015. "Breeding Better Peas, Pumpkins, and Peasants: The Practical Mendelism of Erich Tschermak." In *New Perspectives on the History of the Life Sciences and Agriculture*, edited by Denise Phillips and Sharon E. Kingsland, 419–39. Cham: Springer.

Gökyiğit, Emel Aileen. 1994. "The Reception of Francis Galton's *Hereditary Genius* in the Victorian Periodical Press." *Journal of the History of Biology* 27:215–40.

Goldschmidt, Richard B. 1952. *Understanding Heredity: An Introduction to Genetics*. London: Chapman & Hall.

Goodman, Nelson. 1947. "The Problem of Counterfactual Conditionals." *Journal of Philosophy* 44:113–28.

Goodrich, Edwin S. 1913. *The Evolution of Living Organisms*. London: T. C. & E. C. Jack.

Gould, Stephen Jay. 1977. *Ontogeny and Phylogeny*. Cambridge, MA: Belknap Press.

Gould, Stephen Jay. 1989. *Wonderful Life: The Burgess Shale and the Nature of History*. London: Hutchinson Radius.

Gould, Stephen Jay. 1991. "Fleeming Jenkin Revisited." In his *Bully For Brontosaurus: Reflections in Natural History*, 340–53. New York: W. W. Norton.

Gould, Stephen Jay. 1996. "Does the Stoneless Plum Instruct the Thinking Reed?" In his *Dinosaur in a Haystack: Reflections in Natural History*, 285–95. London: Jonathan Cape.

Gould, Stephen Jay. 1997. *The Mismeasure of Man*. Revised and expanded edition. London: Penguin. First edition published in 1981.

Gould, Stephen Jay. 1999. "Brotherhood by Inversion (or, As the Worm Turns)." In his *Leonardo's Mountain of Clams and the Diet of Worms: Essays on Natural History*, 319–35. London: Vintage.

Gould, Stephen Jay. 2002. *The Structure of Evolutionary Theory*. Cambridge, MA: Belknap Press/Harvard University Press.

Gould, Stephen Jay. 2007. "Carrie Buck's Daughter." In his *The Richness of Life: The Essential Stephen Jay Gould*, edited by Steven Rose, 564–73. New York: Norton.

Graham, Loren R. 1998. *What Have We Learned about Science and Technology from the Russian Experience?* Stanford: Stanford University Press.

Graham, Loren R. 2016. *Lysenko's Ghost: Epigenetics and Russia*. Cambridge, MA: Harvard University Press.

Grant, Madison. 1936. *The Passing of the Great Race, or the Racial Basis of European History*. 4th rev. ed. With prefaces by Henry Fairfield Osborn. New York: Charles Scribner's Sons.

Gray, Russell D. 2001. "Selfish Genes or Developmental Systems?" In *Thinking about Evolution: Historical, Philosophical, and Political Perspectives*, edited by Rama S. Singh, Costas B. Krimbas, Diane B. Paul, and John Beatty, 184–207. Cambridge: Cambridge University Press.

Guilfoile, Patrick. 1997. "Wrinkled Peas and White-Eyed Fruit Flies: The Molecular Basis of Two Classical Genetic Traits." *American Biology Teacher* 59:92–95.

H. C. 1926. "Yeoman XXXIII. Successful Result of Research in Plant Breeding. Professor Sir Boudleigh Bluffin Interviewed." *Cambridge University Agricultural Society Magazine*, 23–25.

Hacking, Ian. 1990. *The Taming of Chance*. Cambridge: Cambridge University Press.

Hacking, Ian. 1992. "The Self-Vindication of the Laboratory Sciences." In *Science as Practice and Culture*, edited by Andrew Pickering, 29–64. Chicago: University of Chicago Press.

Hacking, Ian. 1999. *The Social Construction of What?* Cambridge, MA: Harvard University Press.

Haldane, J. B. S. 1936. "Some Principles of Causal Analysis in Genetics." *Erkenntnis* 6:346–57.

Haldane, J. B. S. 1946–47. "The Interaction of Nature and Nurture." *Annals of Eugenics* 13:197–205.

Haldane, J. B. S. 1957. "Karl Pearson, 1857–1957." *Biometrika* 44:303–13.

Hale, Piers J. 2014. *Political Descent: Malthus, Mutualism, and the Politics of Evolution in Victorian England*. Chicago: University of Chicago Press.

Hall, Brian K. 2004. "Balfour, Francis Maitland." In *Dictionary of Nineteenth-Century British Scientists*, edited by Bernard Lightman, vol. 1, 97–100. Bristol: Thoemmes Continuum.

Hall, Brian K. 2005. "Betrayed by *Balanoglossus*: William Bateson's Rejection of Evolutionary Embryology as the Basis for Understanding Evolution." *Journal of Experimental Zoology* 304B:1–17.

Hall, Kersten, and Staffan Müller-Wille. 2013. "Legumes and Linguistics." *Viewpoint: Magazine of the British Society for the History of Science*, no. 100 (February): 6.

Hall, Marie Boas. 1984. *All Scientists Now: The Royal Society in the Nineteenth Century*. Cambridge: Cambridge University Press.

Hardy, Alister. 1965. *The Living Stream: A Restatement of Evolution Theory and Its Relation to the Spirit of Man*. London: Collins.

Harris, J. Arthur. 1912. "A Simple Test of the Goodness of Fit of Mendelian Ratios." *American Naturalist* 46:741–45.

Harrison, R. G. 1937. "Embryology and Its Relations." *Science* 85:369–74.

Harwood, Jonathan. 1993. *Styles of Scientific Thought: The German Genetics Community 1900–1933*. Chicago: University of Chicago Press.

Harwood, Jonathan. 1996. "Weimar Culture and Biological Theory: A Study of Richard Woltereck (1877–1944)." *History of Science* 34:347–77.

Harwood, Jonathan. 2000. "The Rediscovery of Mendelism in Agricultural Context: Erich von Tschermak as Plant-Breeder." *Comptes Rendus de l'Académie des Sciences Paris, Sciences de la Vie* 323:1061–67.

Harwood, Jonathan. 2015. "Did Mendelism Transform Plant Breeding? Genetic Theory and Breeding Practice, 1900–1945." In *New Perspectives on the History of Life Sciences and Agriculture*, edited by Denise Phillips and Sharon Kingsland, 345–70. Dordrecht: Springer.

Hawthorn, Geoffrey. 1991. *Plausible Worlds: Possibility and Understanding in History and the Social Sciences*. Cambridge: Cambridge University Press.

Hays, Willet M. 1904. "Breeding for Intrinsic Qualities." In *Proceedings: International Conference on Plant Breeding and Hybridization*, 55–62. Memoirs of the Horticultural Society of New York, vol. 1.

Heffer, Simon. 2013. *High Minds: The Victorians and the Birth of Modern Britain*. London: Random House.

Heggie, Vanessa. 2008. "Lies, Damn Lies, and Manchester's Recruiting Statistics: Degeneration as an 'Urban Legend' in Victorian and Edwardian Britain." *Journal of the History of Medicine and Allied Sciences* 63:178–216.

Heimans, J. 1978. "Hugo de Vries and the Gene Theory." In *Human Implications of Scientific Advance*, edited by E. G. Forbes, 469–80. Edinburgh: Edinburgh University Press.

Heine, Steven J. 2017. *DNA Is Not Destiny: The Remarkable, Completely Misunderstood Relationship between You and Your Genes*. New York: W. W. Norton.

Heisenberg, Werner. 1948. "The Notion of a 'Closed Theory' in Modern Science." In *Across the Frontiers*, 39–46. New York: Harper & Row. First published as "Der Begriff 'Abgeschlossene Theorie' in der Modernen Naturwissenschaft." *Dialectica* 2 (1948): 331–36.

Henig, Robin Marantz. 2000. *The Monk in the Garden: The Lost and Found Genius of Gregor Mendel, the Father of Genetics*. New York: Houghton Mifflin.

Herbst, Curt. 1901. *Formative Reize in der tierischen Ontogenese. Ein Beitrag zum Verständnis der tierischen Embryonalentwicklung*. Leipzig: Georgi.

Hesketh, Ian. 2016. "Counterfactuals and History: Contingency and Convergence in Histories of Science and Life." *Studies in History and Philosophy of Biological and Biomedical Sciences* 58:41–48.

Higgitt, Rebekah, and Charles W. J. Withers. 2008. "Science and Sociability: Women as Audience at the British Association for the Advancement of Science, 1831–1901." *Isis* 99:1–27.

Higham, John. 1988. *Strangers in the Land: Patterns of American Nativism, 1860–1925*. 2nd ed. New Brunswick, NJ: Rutgers University Press. First published in 1955.

Hill, E. G. 1904. "Breeding of Florists' Flowers." In *Proceedings: International Conference on Plant Breeding and Hybridization*, 111–16. Memoirs of the Horticultural Society of New York, vol. 1.

Hilts, Victor L. 1975. "A Guide to Francis Galton's *English Men of Science*." *Transactions of the American Philosophical Society* 65:1–85.

Hinman, Michael B., Justin A. Jones, and Randolph V. Lewis. 2000. "Synthetic Spider Silk: A Modular Fiber." *Trends in Biotechnology* 18:374–79.

Hinman, Michael B., and Randolph V. Lewis. 1992. "Isolation of a Clone Encoding a Second Dragline Silk Fibroin." *Journal of Biological Chemistry* 267:19320–24.

Hobhouse, Leonard T. 1901. *Mind in Evolution*. London: Macmillan.

Hobhouse, Leonard T. 1913. *Development and Purpose: An Essay Towards a Philosophy of Evolution*. London: Macmillan.

Hodge, [M.] Jonathan [S.]. 1977. "The Structure and Strategy of Darwin's 'Long Argument.'" *British Journal for the History of Science* 10:237–46. Reprinted in his *Before and After Darwin: Origins, Species, Cosmogonies, and Ontologies*, ch. 7. Aldershot: Ashgate, 2008.

Hodge, [M.] Jonathan [S.]. 1985. "Darwin as a Lifelong Generation Theorist." In *The Darwinian Heritage*, edited by David Kohn, 207–43. Princeton, NJ: Princeton University Press. Reprinted in his *Darwin Studies: A Theorist and His Theories in Their Context*, ch. 6. Aldershot: Ashgate, 2009.

Hodge, [M.] Jonathan [S.]. 1989. "Generation and the Origin of Species (1837–1937): A Historiographical Suggestion." *British Journal for the History of Science* 22:267–81. Reprinted in his *Before and After Darwin: Origins, Species, Cosmogonies, and Ontologies*, ch. 12. Aldershot: Ashgate, 2008.

Hodge, [M.] Jonathan [S.]. 2009. "The Notebook Programmes and Projects of Darwin's London Years." In *The Cambridge Companion to Darwin*, 2nd ed., edited by Jonathan Hodge and Gregory Radick, 44–72. Cambridge: Cambridge University Press.

Hodge, [M.] Jonathan [S.]. 2010. "The Darwin of Pangenesis." *Comptes Rendus Biologies* 33:129–33.

Hogben, Lancelot. 1933. *Nature and Nurture.* London: George Allen & Unwin.

Holladay, April. 2004. "Blue-Eyed Parents Can Have Brown-Eyed Kids and Other Eye-Oddities." *USA Today*, 15 October 2004. http://usatoday30.usatoday.com/tech/columnist/aprilholladay/2004-10-14-wonderquest_x.htm.

Holmes, S. J., and H. M. Loomis. 1909. "The Heredity of Eye Color and Hair Color in Man." *Biological Bulletin* 18:50–65.

Holt, Jim. 2005. "Measure for Measure: The Strange Science of Francis Galton." *New Yorker* (24 and 31 January): 84–90.

Holterhoff, Kate. 2014. "The History and Reception of Charles Darwin's Hypothesis of Pangenesis." *Journal of the History of Biology* 47:661–95.

Hopwood, Nick. 2009. "Embryology." In *The Cambridge History of Science*, vol. 6: *The Modern Biological and Earth Sciences*, edited by Peter J. Bowler and John V. Pickstone, 285–315. Cambridge: Cambridge University Press.

Hoquet, Thierry. 2018. *Revisiting the* Origin of Species: *The Other Darwins*. London: Routledge.

Huggins, William. 1906. *The Royal Society, or, Science in the State and in the Schools*. London: Methuen.

Hughes-Warrington, Marnie. 2000. "Fernand Braudel." In her *Fifty Key Thinkers on History*, 17–24. London: Routledge.

Hull, David L. 1973. *Darwin and His Critics: The Reception of Darwin's Theory of Evolution by the Scientific Community*. Cambridge, MA: Harvard University Press.

Hull, David L. 1974. *Philosophy of Biological Science*. Englewood Cliffs, NJ: Prentice Hall.

Hull, David L. 2009. "Darwin's Science and Victorian Philosophy of Science." In *The Cambridge Companion to Darwin*, 2nd ed., edited by Jonathan Hodge and Gregory Radick, 173–96. Cambridge: Cambridge University Press.

Hunt, Bruce J. 2007. [Review of David Kaiser, ed., *Pedagogy and the Practice of Science*.] *British Journal for the History of Science* 40:144–45.

Hurst, C. Chamberlain. 1900. "Notes on Some Experiments in Hybridisation and Cross-Breeding." Hybrid Conference Report 1900, *Journal of the Royal Horticultural Society* 24:90–126.

Hurst, C. Chamberlain. 1902. "Mendel's 'Law' Applied to Orchid Hybrids." *Journal of the Royal Horticultural Society* 26:688–95.

Hurst, C. Chamberlain. 1904. "Notes on Mendel's Methods of Cross-Breeding." In *Proceedings: International Conference on Plant Breeding and Hybridization*, 11–16. Memoirs of the Horticultural Society of New York, vol. 1.

Hurst, C. Chamberlain. 1906. “On the Inheritance of Coat Colour in Horses.” *Proceedings of the Royal Society of London B* (12 April): 388–94. Also in Hurst 1925, 239–45.

Hurst, C. Chamberlain. 1907. “Eye-Colours of Parents and Offspring in a Leicestershire Village.” In Hurst 1925, 284–85. [The title is given in the volume’s table of contents.]

Hurst, C. Chamberlain. 1908. “On the Inheritance of Eye-Colour in Man.” *Proceedings of the Royal Society of London B* 80:85–96. Reprinted in Hurst 1925, 272–83.

Hurst, C. Chamberlain. 1925. *Experiments in Genetics*. Cambridge: Cambridge University Press.

Hurst, Rona. 1971. *The Evolution of Genetics*. Unpublished MS. Copies are available in Cambridge University Library and the American Philosophical Society Library.

Hutchinson, John M. C. 1990. “Stabilising Selection in Weldon’s Snails: A Reappraisal.” *Heredity* 64:113–20.

Huxley, Julian. 1949. *Soviet Genetics and World Science: Lysenko and the Meaning of Heredity*. London: Chatto & Windus.

Huxley, Julian. 1960. “The Emergence of Darwinism.” Reprinted in his *Essays of a Humanist*, 13–38. Harmondsworth: Penguin, 1964.

Hynes, Samuel. 1968. *The Edwardian Turn of Mind*. London: Pimlico.

Iltis, Hugo. 1932. *Life of Mendel*. Translated from the 1924 German original by Eden and Cedar Paul. Reprint 1966. London: George Allen & Unwin.

Inter-Departmental Committee on Physical Deterioration. 1904. *Report*. 3 vols. London: HMSO.

Jamieson, Annie, and Gregory Radick. 2013. “Putting Mendel in His Place: How Curriculum Reform in Genetics and Counterfactual History of Science Can Work Together.” In *The Philosophy of Biology: A Companion for Educators*, edited by Kostas Kampourakis, 577–95. Dordrecht: Springer.

Jamieson, Annie, and Gregory Radick. 2017. “Genetic Determinism in the Genetics Curriculum: An Exploratory Study of the Effects of Mendelian and Weldonian Emphases.” *Science and Education* 26:1261–90.

Jenkin, Fleeming. 1867. “The Origin of Species.” *North British Review* 46:277–318.

Jenkinson, J. W. 1909. *Experimental Embryology*. Oxford: Clarendon Press.

Johannsen, Wilhelm. 1903a. “Concerning Heredity in Populations and in Pure Lines: A Contribution to the Elucidation of Current Problems of Selection.” Translated from the German by Harold Gall and Elga Putschar. In *Selected Readings in Biology for Natural Sciences*, vol. 3, 172–215. Chicago: University of Chicago Press, 1955. Originally published as *Ueber Erblichkeit in Populationen und in reinen Linien. Ein Beitrag zur Beleuchtung schwebender Selektionsfragen* (Jena: Gustav Fischer, 1903).

Johannsen, Wilhelm. 1903b. "About Darwinism, Seen from the Point of View of the Science of Heredity." Translated from the Danish by Nils Roll-Hansen. *BSHS Translations* 2018. https://www.bshs.org.uk/bshs-translations/johannsen#translation. Originally published as "Om Darwinismen, set fra Arvelighedslærens Standpunkt," *Tilskueren* 1903: 525–41.

Johannsen, Wilhelm. 1909. *Elemente der exakten Erblichkeitslehre*. Jena: G. Fischer.

Johannsen, Wilhelm. 1911. "The Genotype Conception of Heredity." *American Naturalist* 45:129–59.

Jones, H. F. 1919. *Samuel Butler, Author of* Erewhon *(1835–1902): A Memoir*. 2 vols. London: Macmillan.

Jones, Jonathan. 2004. "Fragments of the Universe." *Guardian* 22 May 2004 (supplement): 16–17.

Judson, Horace. 1992. "A History of the Science and Technology Behind Gene Mapping and Sequencing." In *The Code of Codes: Scientific and Social Issues in the Human Genome Project*, edited by Daniel J. Kevles and Leroy Hood, 37–80. Cambridge, MA: Harvard University Press.

Judson, Horace. 2004. *The Great Betrayal: Fraud in Science*. New York: Harcourt.

Kacser, Henrik, and James A. Burns. 1981. "The Molecular Basis of Dominance." *Genetics* 97:639–66.

Kampourakis, Kostas. 2017. *Making Sense of Genes*. Cambridge: Cambridge University Press.

Kampourakis, Kostas. 2020. *Understanding Genes*. Cambridge: Cambridge University Press.

Kaplan, Jonathan M., and E. Turkheimer. 2021. "Galton's Quincunx: Probabilistic Causation in Developmental Behavior Genetics." *Studies in History and Philosophy of Science* 88:60–69.

Keller, Evelyn Fox. 2000. *The Century of the Gene*. Cambridge, MA: Harvard University Press.

Keller, Evelyn Fox. 2010. *The Mirage of a Space between Nature and Nurture*. Durham, NC: Duke University Press.

Keller, Evelyn Fox. 2014. "From Gene Action to Reactive Genomes." *Journal of Physiology* 592:2423–29.

Keshavan, Matcheri, Paulo Lizano, and Konasale Prasad. 2020. "The Synaptic Pruning Hypothesis of Schizophrenia: Promises and Challenges." *World Psychiatry* 19:110–11.

Kevles, Daniel J. 1981. "Genetics in the United States and Great Britain 1890–1930: A Review with Speculations." In *Biology, Medicine and Soci-*

ety 1840–1940, edited by Charles Webster, 193–215. Cambridge: Cambridge University Press.

Kevles, Daniel J. 1995. *In the Name of Eugenics: Genetics and the Uses of Human Heredity*. With a new preface. Cambridge, MA: Harvard University Press. First published in 1985.

Kidd, Ian James. 2016. "Inevitability, Contingency, and Epistemic Humility." *Studies in History and Philosophy of Science* 55:12–19.

Kim, Kyung-Man. 1994. *Explaining Scientific Consensus: The Case of Mendelian Genetics*. With commentaries by Nils Roll-Hansen and Robert Olby. London: Guildford Press.

Kimmelman, Barbara A. 1983. "The American Breeders' Association: Genetics and Eugenics in an Agricultural Context, 1903–13." *Social Studies of Science* 13:163–204.

Kimmelman, Barbara A. 1987. "A Progressive Era Discipline: Genetics at American Agricultural Colleges and Experiment Stations, 1900–1920." PhD dissertation, University of Pennsylvania.

Kingsbury, Noel. 2009. *Hybrid: The History and Science of Plant Breeding*. Chicago: University of Chicago Press.

Kingsland, Sharon E. 2007. "Maintaining Continuity through a Scientific Revolution: A Rereading of E. B. Wilson and T. H. Morgan on Sex Determination and Mendelism." *Isis* 98:468–88.

Kinzel, Katherina. 2015. "State of the Field: Are the Results of Science Contingent or Inevitable?" *Studies in History and Philosophy of Science* 52:55–66.

Kitcher, Philip. 1984. "1953 and All That: A Tale of Two Sciences." *Philosophical Review* 93:335–73. Reprinted in his *In Mendel's Mirror: Philosophical Reflections on Biology*, 3–30. Oxford: Oxford University Press, 2003.

Kitcher, Philip. 1996. *The Lives to Come: The Genetic Revolution and Human Possibilities*. London: Allen Lane/Penguin Press.

Kitcher, Philip. 2001. "Battling the Undead: How (and How Not) to Resist Genetic Determinism." In *Thinking about Evolution: Historical, Philosophical, and Political Perspectives*, edited by Rama S. Singh, Costas B. Krimbas, Diane B. Paul, and John Beatty, 396–414. Cambridge: Cambridge University Press. Reprinted in his *In Mendel's Mirror: Philosophical Reflections on Biology*, 283–300, Oxford: Oxford University Press, 2003.

Klette, Rebecka. 2015. "Depicting Decay: The Reception and Representation of Degeneration Theory in *Punch*, 1869–1910." The Victorianist: British Association for Victorian Studies Postgraduates Pages, https://victorianist.wordpress.com.

Koestler, Arthur. 1971. *The Case of the Midwife Toad*. London: Hutchinson.

Kohler, Robert E. 1994. *Lords of the Fly:* Drosophila *Genetics and the Experimental Life*. Chicago: University of Chicago Press.

Kottler, Malcolm J. 1979. "Hugo de Vries and the Rediscovery of Mendel's Laws." *Annals of Science* 36:517–38.

Krafka, Joseph. 1920. "The Effect of Temperature upon Facet Number in the Bar-Eyed Mutant of Drosophila [1–3]." *Journal of General Physiology* 2:409–32; 433–44; 445–64.

Kragh, Helge. 2002. "The Vortex Atom: A Victorian Theory of Everything." *Centaurus* 44:32–114.

Krementsov, Nikolai. 2018. *With and Without Galton: Vasilii Florinskii and the Fate of Eugenics in Russia*. Cambridge: Open Book Publishers.

Kroeber, Alfred. 1917. "The Superorganic." *American Anthropologist*, n.s., 19:163–213.

Kuhn, Thomas S. 1963. "The Function of Dogma in Scientific Research." In *Scientific Change*, edited by A. C. Crombie, 347–69. London: Heineman.

Kuhn, Thomas S. 1970. *The Structure of Scientific Revolutions*. 2nd ed. Chicago: University of Chicago Press.

Labby, Zacariah. 2009. "Weldon's Dice, Automated." *Chance* 22 (4): 6–13.

Lamb, Marion J. 2011. "Attitudes to Soft Inheritance in Great Britain, 1930s–1970s." In *Transformations of Lamarckism: From Subtle Fluids to Molecular Biology*, edited by Snait B. Gissis and Eva Jablonka, 109–20. Cambridge, MA: MIT Press.

Laubichler, Manfred D., and Sahotra Sarkar. 2002. "Flies, Genes, and Brains: Oskar Vogt, Nikolai Timoféef-Ressovsky, and the Origin of the Concepts of Penetrance and Expressivity." In *Mutating Concepts, Evolving Disciplines: Genetics, Medicine and Society*, edited by L. S. Parker and R. A. Ankeny, 63–85. Dordrecht: Kluwer.

Laudan, Larry. 1981. "A Confutation of Convergent Realism." *Philosophy of Science* 48:19–49.

Laughlin, Harry H. 1930. *The Legal Status of Eugenical Sterilization*. Chicago: Municipal Court of Chicago.

Lawrence, Sir Trevor. 1907. "Presentation of Medals." In Wilks 1907, 69–71.

Laxton, Thomas. 1866a. "Observations on the Variations Effected by Crossing in the Colour and Character of the Seed of Peas." In *The International Horticultural Exhibition, and Botanical Congress, Held in London from May 22nd to May 31st, 1866: Report of Proceedings*, 156. London: Truscott, Son, & Simmons.

Laxton, Thomas. 1866b. "Crossing Papilionaceous Flowers." *Gardeners' Chronicle and Agricultural Gazette*, 22 September, 900–901.

Lederberg, Joshua. 1963. "Molecular Biology, Eugenics and Euphenics." *Nature* 198 (4 May): 428–29.

Leja, Darryl. 2003. "Human Genome Project Timeline." Image available at https://commons. wikimedia.org/wiki/File:Human_Genome_Project _Timeline_ (26964377742).jpg. Accessed 21 July 2020.

Lenay, Charles. 2000. "Hugo de Vries: From the Theory of Intracellular Pangenesis to the Rediscovery of Mendel." *Comptes Rendus de l'Académie des Sciences Paris, Sciences de la Vie* 323:1053–60.

Lester, Joe. 1995. *E. Ray Lankester and the Making of Modern British Biology*. Edited by Peter J. Bowler. BSHS Monographs 9. Oxford: British Society for the History of Science.

Levine, Philippa. 2017. *Eugenics: A Very Short Introduction*. Oxford: Oxford University Press.

Levins, Richard, and Richard Lewontin. 1976. "The Problem of Lysenkoism." In *The Radicalisation of Science*, edited by H. Rose and S. Rose, 32–64. Critical Social Studies. London: Palgrave. Reprinted in their *The Dialectical Biologist*, ch. 7. Cambridge, MA: Harvard University Press, 1985.

Levy, Jack. S. 2015. "Counterfactuals, Causal Inference, and Historical Analysis." *Security Studies* 24:378–402.

Lewandowsky, Stephen, Ullrich K. H. Ecker, Colleen M. Seifert, Norbert Schwartz, and John Cook. 2012. "Misinformation and Its Correction: Continued Influence and Successful Debiasing." *Psychological Science in the Public Interest* 13 (3): 106–31.

Lewens, Tim. 2010. "Natural Selection Then and Now." *Biological Reviews* 85:829–35.

Lewes, George Henry. 1859–60. *The Physiology of Common Life*. 2 vols. Edinburgh: William Blackwood and Sons.

Lewis, David. 1973. *Counterfactuals*. Oxford: Basil Blackwell.

Lewis, Jenny. 2011. "Genetics and Genomics." In *Teaching Secondary Biology*, edited by Michael Reiss, 173–214. London: Hodder Education.

Lewontin, Richard C. 1982. *Human Diversity*. New York: Scientific American Library.

Lewontin, Richard C. 1983. "Gene, Organism and Environment." In *Evolution from Molecules to Men*, edited by D. S. Bendall, 273–85. Cambridge: Cambridge University Press.

Lewontin, Richard C. 1993. *The Doctrine of DNA: Biology as Ideology*. London: Penguin.

Lewontin, Richard C. 1996. "In the Blood." *New York Review of Books*, 23 May, 31–32.

Lewontin, Richard C. 2000. *The Triple Helix: Gene, Organism, and Environment*. Cambridge, MA: Harvard University Press.

Lidwell-Durnin, John. 2020. "William Benjamin Carpenter and the Emerging Science of Heredity." *Journal of the History of Biology* 53:81–103.

Lindberg, David R. 1998. "William Healey Dall: A Neo-Lamarckian View of Molluscan Evolution." *Veliger* 41 (3): 227–38.

Lints, F. A., and Jean Delcour. 1968. "Galton and the Mendelian Ratios." *Heredity* 23:153–55.

Lipton, Peter. 1990. "Contrastive Explanation." Supplement, *Royal Institute of Philosophy* 27:247–66.

Lipton, Peter. 2004. "Genetic and Generic Determinism: A New Threat to Free Will?" In *The New Brain Sciences: Perils and Prospects*, edited by Dai Rees and Steven Rose, 88–100. Cambridge: Cambridge University Press.

Liu, Yongsheng, and Xiuju Li. 2014. "Has Darwin's Pangenesis Been Rediscovered?" *Bioscience* 64:1037–41.

Livio, Mario. 2013. *Brilliant Blunders: From Darwin to Einstein—Colossal Mistakes by Great Scientists That Changed Our Understanding of Life and the Universe*. New York: Simon and Schuster.

Loach, Judi. 2018. "Architecture, Science and Purity." In *Being Modern: The Cultural Impact of Science in the Early Twentieth Century*, edited by Robert Bud, Paul Greenhalgh, Frank James, and Morag Shiach, 207–44. London: UCL Press.

Lock, R[obert] H[eath]. 1904. "Studies in Plant Breeding in the Tropics. I.—Introductory: The Work of Mendel and an Account of Recent Progress on the Same Lines, with Some New Illustrations." *Annals of the Royal Botanic Gardens, Peradeniya* 2 (2): 299–356.

Lock, R[obert] H[eath]. 1905. "Studies in Plant Breeding in the Tropics. II.—Experiments with Peas." *Annals of the Royal Botanic Gardens, Peradeniya* 2 (3): 357–414.

Lock, Robert Heath. 1906. *Recent Progress in the Study of Variation, Heredity, and Evolution*. London: John Murray.

Lock, R[obert] H[eath]. 1908. "The Present State of Knowledge in *Pisum*." *Annals of the Royal Botanic Gardens, Peradeniya* 4:93–111.

López-Beltrán, Carlos. 1994. "Forging Heredity: From Metaphor to Cause, a Reification Story." *Studies in History and Philosophy of Science* 25: 211–35.

López-Beltrán, Carlos. 2004. "In the Cradle of Heredity: French Physicians and *L'Hérédité Naturelle* in the Early 19th Century." *Journal of the History of Biology* 37:39–72.

Lorenzano, Pablo. 2011. "What Would Have Happened if Darwin Had Known Mendel (or Mendel's Work)?" *History and Philosophy of the Life Sciences* 33:3–48.

Lorimer, George Horace. 1921. "The Great American Myth." *Saturday Evening Post*, 7 May, 20.

Losos, Jonathan B. 2017. *Improbable Destinies: Fate, Chance, and the Future of Evolution*. New York: Riverhead Books.

Lowe, James W. E., and Anne Bruce. 2019. "Genetics without Genes? The

Centrality of Genetic Markers in Livestock Genetics and Genomics." *History and Philosophy of the Life Sciences* 41 (50): 1–29.

Lyle, Louise. 2012. "*Les grandes terres et l'intérêt national:* French Literary Responses to the Nazi Eugenic Legislation of July 1933." Abstract of a talk given at the workshop "Puériculture, Biotypology and 'Latin' Eugenics in Comparative Context," Oxford, 20 April. Available online at www.brookes.ac.uk. Accessed 6 August 2020.

Lynch, R. I. 1904. "Classification of Hybrids." In *Proceedings: International Conference on Plant Breeding and Hybridization*, 29–33. Memoirs of the Horticultural Society of New York, vol. 1.

Lyon, Gholson J. 2011. "Personal Account of the Discovery of a New Disease Using Next-Generation Sequencing." Interview with Natalie Harrison. *Pharmacogenomics* 12:1519–23.

Lyon, Gholson J. 2014. "Genetic Complexity and Neuropsychiatric Disorders." Talk given on 1 July at the conference "Nurturing Genetics: Reflections on a Century of Scientific and Social Change," University of Leeds. Video available at https://geneticspedagogies.leeds.ac.uk/events/.

Lyon, Gholson J., and Jason O'Rawe. 2015. "Human Genetics and Clinical Aspects of Neurodevelopmental Disorders." In *The Genetics of Neurodevelopmental Disorders*, edited by Kevin J. Mitchell, 289–317. New York: John Wiley & Sons.

Lyon, Gholson J., and Kai Wang. 2012. "Identifying Disease Mutations in Genomic Medicine Settings: Current Challenges and How to Accelerate Progress." *Genomic Medicine* 4 (58): 1–16.

Lysenko, Trofim D. 1936–37. "Two Trends in Genetics." In Lysenko 1954, 160–94.

Lysenko, Trofim D. 1948. "The Situation in Biological Science." In Lysenko 1954, 515–54.

Lysenko, Trofim D. 1954. *Agrobiology: Essays on Problems of Genetics, Plant Breeding and Seed Growing*. Moscow: Foreign Languages Publishing House.

Macaulay, Thomas Babington. 1828. "Dryden." In his *Essays, Critical and Miscellaneous*, 35–50. Philadelphia: Hart, Carey & Hart, 1850.

MacKenzie, Donald. 1981a. *Statistics in Britain 1865–1930: The Social Construction of Scientific Knowledge*. Edinburgh: Edinburgh University Press.

MacKenzie, Donald. 1981b. "Sociobiologies in Competition: The Biometrician-Mendelian Debate." In *Biology, Medicine and Society 1840–1940*, edited by Charles Webster, 243–88. Cambridge: Cambridge University Press.

MacKenzie, Donald, and Barry Barnes. 1974. "Biometrician Versus Mendelian: A Controversy and Its Explanation." Science Studies Unit, University of Edinburgh. Unpublished MS. Published in German in 1975 as "Biometriker versus Mendelianer. Eine Kontroverse und ihre Erklarung," *Kölner Zeitsschrift für Soziologie und Sozialpsychologie* 13:165–96.

MacKenzie, Donald, and Barry Barnes. 1979. "Scientific Judgment: The Biometry-Mendelism Controversy." In *Natural Order: Historical Studies of Scientific Culture*, edited by Barry Barnes and Steven Shapin, 191–210. Beverly Hills: Sage.

MacLeod, Christine, and Gregory Radick. 2013. "Claiming Ownership in the Technosciences: Patents, Priority and Productivity." *Studies in History and Philosophy of Science* 44:188–201.

MacLeod, Roy M. 1971. "Of Medals and Men: A Reward System in Victorian Science, 1826–1914." *Notes and Records of the Royal Society of London* 26:81–105.

MacLeod, Roy M. 1994. "Embryology and Empire: The Balfour Students and the Quest for Intermediate Forms in the Laboratory of the Pacific." In *Darwin's Laboratory: Evolutionary Theory and Natural History in the Pacific*, edited by Roy MacLeod and Philip E. Rehbock, 140–65. Honolulu: University of Hawai'i Press.

Magnello, Eileen. 1993. "Karl Pearson: Evolutionary Biology and the Emergence of a Modern Theory of Statistics (1884–1936)." DPhil thesis, University of Oxford.

Magnello, Eileen. 1996. "Karl Pearson's Gresham Lectures: W. F. R. Weldon, Speciation and the Origins of Pearsonian Statistics." *British Journal for the History of Science* 29:43–63.

Magnello, Eileen. 1998. "Karl Pearson's Mathematization of Inheritance: From Ancestral Heredity to Mendelian Genetics (1895–1909)." *Annals of Science* 55:35–94.

Magnello, Eileen. 2004. "The Reception of Mendelism by the Biometricians and the Early Mendelians (1899–1909)." In *A Century of Mendelism in Human Genetics*, edited by Milo Keynes, A. W. F. Edwards and Robert Peel, 19–32. London: Galton Institute/CRC Press.

Maher, Brendan. 2011. "Software Pinpoints Cause of Mystery Genetic Disorder." *Nature*, 23 June. https://doi.org/10.1038/news.2011.382.

Maienschein, Jane. 1991. *Transforming Traditions in American Biology, 1880–1915*. Baltimore: Johns Hopkins University Press.

Marks, Jonathan. 2008. "The Construction of Mendel's Laws." *Evolutionary Anthropology* 17:250–53.

Martin, Joseph D. 2013. "Is the Contingentist/Inevitabilist Debate a Matter of Degree?" *Philosophy of Science* 80:919–30.

Martins, Lilian Al-Chueyr Pereira. 1999. "Did Sutton and Boveri Propose the So-Called Sutton-Boveri Chromosome Hypothesis?" *Genetics and Molecular Biology* 22:261–71.

Martins, Lilian Al-Chueyr Pereira. 2020. "Weldon's Unpublished Manuscript: An Attempt at Reconciliation between Mendelism and Biometry?" In *Life and Evolution*, edited by L. Baravalle and L. Zaterka, 11–28. Dordrecht: Springer.

Mawer, Simon. 2006. *Gregor Mendel: Planting the Seeds of Genetics*. New York: Abrams.

Maxwell, James Clerk. 1873. "Essay for the Eranus Club on Science and Free Will." In *The Scientific Letters and Papers of James Clerk Maxwell*, edited by P. M. Harman, vol. 2, 814–23. Cambridge: Cambridge University Press.

Maxwell, James Clerk. 1875. "Atom." Reprinted in facsimile in *Maxwell on Molecules and Gases*, edited by E. Garber, S. G. Brush, and C. W. F. Everitt, 176–215. Cambridge, MA: MIT Press, 1986.

Mazumdar, Pauline M. H. 1992. *Eugenics, Human Genetics and Human Failings: The Eugenics Society, Its Sources and Its Critics in Britain*. London: Routledge.

Mazumdar, Pauline M. H. 2002. "'Reform' Eugenics and the Decline of Mendelism." *Trends in Genetics* 18:48–52.

McClung, Clarence E. 1902. "The Accessory Chromosome—Sex Determination?" *Biological Bulletin* 3:43–84.

McCullough, Dennis M. 1969. "W. K. Brooks's Role in the History of American Biology." *Journal of the History of Biology* 2:411–38.

McDonald, John H. 2011. *Myths of Human Genetics*. Baltimore: Sparky House.

McGovern, Michael F. 2021. "Genes Go Digital: *Mendelian Inheritance in Man* and the Genealogy of Electronic Publishing in Biomedicine." *British Journal for the History of Science* 54:213–31.

McKie, Robin. 2016. "Gene Therapy Hope for Sickle Cell Breakthrough." *Observer*, 2 October, 20.

Medawar, Peter. 1977. "Unnatural Science." In his *Pluto's Republic*, 167–83. Oxford: Oxford University Press, 1992. First published in *New York Review of Books*, 3 February 1977.

Mehler, Barry. 1996. "Heredity and Hereditarianism." In *Philosophy of Education: An Encyclopedia*, edited by J. J. Chambliss, 260–63. London: Routledge.

Meijer, Onno G. 1983. "The Essence of Mendel's Discovery." In *Gregor Mendel and the Foundation of Genetics*, edited by Vítězslav Orel and Anna Matalová, 123–78. Brno: Mendelianum.

Meijer, Onno G. 1985. "Hugo de Vries no Mendelian?" *Annals of Science* 42:189–232.

Mendel, Gregor. 1866. "Versuche über Pflanzen-Hybriden." *Verhandlungen des naturforschenden Vereines in Brünn* 4, second part (*Abhandlungen*): 3–47.

Mendel, Gregor. 1870. "Über einige aus künstlicher Befruchtung gewonnenen Hieracium-Bastarde." *Verhandlungen des naturforschenden Vereines in Brünn* 8, second part (*Abhandlungen*): 26–31. Available in English translation in Stern and Sherwood 1966, 49–55.

Mendel, Gregor. 1901. "Experiments in Plant Hybridisation." Translated by Charles Druery [uncredited], with introduction and notes by William Bateson. *Journal of the Royal Horticultural Society* 26 (August): 1–32.

Mendel, Gregor. 2016. *Experiments on Plant Hybrids (1866)*. Translation and commentary by Staffan Müller-Wille and Kersten Hall. British Society for the History of Science Translation Series. Available at http://www.bshs.org.uk/bshs-translations/mendel. [A print edition was published in 2020 by Masaryk University Press, Brno.]

Merton, Robert. 1948. "The Self-Fulfilling Prophecy." *Antioch Review* 8: 193–210.

Merton, Robert. 1961. "Singletons and Multiples in Science." Reprinted in his *The Sociology of Science: Theoretical and Empirical Investigations*, edited by N. W. Storer. Chicago: University of Chicago Press, 343–70.

Merz, J. T. 1965. *A History of European Scientific Thought in the Nineteenth Century.* 2 vols. New York: Dover. First published in 1903.

Meunier, Robert. 2016. "The Many Lives of Experiments: Wilhelm Johannsen, Selection, Hybridization, and the Complex Relations of Genes and Characters." *History and Philosophy of the Life Sciences* 38:42–64.

Michie, Donald. 1955. "Professor Darlington and the Mighty Gene." *Marxist Quarterly* 2:27–36.

Michie, Donald. 1958. "The Third Stage in Genetics." In *A Century of Darwin*, edited by S. A. Barnett, 56–84. London: Heinemann.

Mitchell, Peter Chalmers. 1937. *My Fill of Days.* London: Faber and Faber.

Montgomerie, Bob, and Tim Birkhead. 2005. "A Beginner's Guide to Scientific Misconduct." *ISBE Newsletter* 17:16–24.

Moore, David S. 2008. "Espousing Interactions and Fielding Reactions: Addressing Laypeople's Beliefs about Genetic Determinism." *Philosophical Psychology* 21:331–48.

Morant, G. M. 1939. *A Bibliography of the Statistical and Other Writings of Karl Pearson*. With the assistance of B. L. Welch. Cambridge: Cambridge University Press; University College London.

Morgan, T. H. 1909. "What Are 'Factors' in Mendelian Explanations?" *American Breeders' Association Report* 5:365–68.

Morgan, T. H. 1910. "Sex Limited Inheritance in Drosophila." *Science* 32 (22 July): 120–22.

Morgan, T. H., A. H. Sturtevant, H. J. Muller, and C. B. Bridges. 1915. *The Mechanism of Mendelian Heredity.* New York: Henry Holt. Reprinted in 1972 with an introduction by Garland Allen. New York: Johnson Reprint Corporation.

Morrell, Jack. 1997. *Science at Oxford, 1914–1939: Transforming an Arts University*. Oxford: Clarendon Press.

Morris, Susan W. 1994. "Fleeming Jenkin and *The Origin of Species*: A Reassessment." *British Journal for the History of Science* 1994:313–43.

Mukherjee, Siddhartha. 2016a. *The Gene: An Intimate History*. London: Bodley Head.

Mukherjee, Siddhartha. 2016b. "Runs in the Family: New Findings about Schizophrenia Rekindle Old Questions about Genes and Identity." *New Yorker*, 28 March. https://www.newyorker.com/magazine/2016/03/28/the-genetics-of-schizophrenia.

Müller-Wille, Staffan. 2021. "Gregor Mendel and the History of Heredity." In *Handbook of the Historiography of Biology*, edited by Michael R. Dietrich, Mark Borrello, and Oren Harman, 105–26. Dordrecht: Springer.

Müller-Wille, Staffan, and Christina Brandt, eds. 2016. *Heredity Explored: Between Public Domain and Experimental Science, 1850–1930*. Cambridge, MA: MIT Press.

Müller-Wille, Staffan, and Giuditta Parolini. 2020. "Punnett Squares and Hybrid Crosses: How Mendelians Learned Their Trade by the Book." *BJHS Themes* 5:149–65.

Müller-Wille, Staffan, and Hans-Jörg Rheinberger, eds. 2007. *Heredity Produced: At the Crossroads of Biology, Politics, and Culture, 1500–1870.* Cambridge, MA: MIT Press.

Müller-Wille, Staffan, and Hans-Jörg Rheinberger. 2012. *A Cultural History of Heredity*. Chicago: University of Chicago Press.

Müller-Wille, Staffan, and Marsha L. Richmond. 2016. "Revisiting the Origins of Genetics." In *Heredity Explored: Between Public Domain and Experimental Science, 1850–1930*, edited by Staffan Müller-Wille and Christina Brandt, 367–94. Cambridge, MA: MIT Press.

Nelkin, Dorothy, and M. Susan Lindee. 1995. *The DNA Mystique: The Gene as a Cultural Icon.* New York: W. H. Freeman.

Newman, Stuart A. 2007. "William Bateson's Physicalist Ideas." In *From Embryology to Evo-Devo: A History of Developmental Evolution*, edited by M. D. Laubichler and J. Maienschein, 83–107. Cambridge, MA: MIT Press.

Noguera-Solano, Ricardo, and Rosaura Ruiz-Gutiérrez. 2009. "Darwin and

Inheritance: The Influence of Prosper Lucas." *Journal of the History of Biology* 42:685–714.

Nolan, Daniel. 2013. "Why Historians (and Everyone Else) Should Care about Counterfactuals." *Philosophical Studies* 163:317–35.

Noorduijn, C. L. W. 1907. "The Hereditary Transmission of Colour in Cross-Breeding." In Wilks 1907, 210–12.

Nordau, Max. 1895. *Degeneration*. Translated from the German second edition. London: William Heineman.

Norton, Bernard J. 1973. "The Biometric Defense of Darwinism." *Journal of the History of Biology* 6:283–316.

Norton, Bernard J. 1975. "Biology and Philosophy: The Methodological Foundations of Biometry." *Journal of the History of Biology* 8:85–93.

Norton, Bernard J. 1979. "Karl Pearson and the Galtonian Tradition: Studies in the Rise of Quantitative Social Biology." PhD thesis, University College London.

Nozick, Robert. 1981. *Philosophical Explanations*. Oxford: Clarendon Press.

"Obituary: Walter Weldon, F.R.S." 1885. *Journal of the Society of Chemical Industry* 4 (20 October): 577–81.

Okrent, Daniel. 2019. *The Guarded Gate: Bigotry, Eugenics, and the Law That Kept Two Generations of Jews, Italians, and Other European Immigrants Out of America.* New York: Scribner.

Olby, Robert C. 1963. "Charles Darwin's Manuscript of *Pangenesis*." *British Journal for the History of Science* 1:251–63.

Olby, Robert C. 1974. *The Path to the Double Helix: The Discovery of DNA*. London: Macmillan. Reprint, Mineola, NY: Dover Press, 1994.

Olby, Robert C. 1979. "Mendel no Mendelian?" *History of Science* 17:53–72. Reprinted in Olby 1985, 234–68.

Olby, Robert C. 1985. *Origins of Mendelism*. 2nd ed. Chicago: University of Chicago Press.

Olby, Robert C. 1987. "William Bateson's Introduction of Mendelism to England: A Reassessment." *British Journal for the History of Science* 20:399–420.

Olby, Robert C. 1989a. "The Dimensions of Scientific Controversy: The Biometric-Mendelian Debate." *British Journal for the History of Science* 22:299–320.

Olby, Robert C. 1989b. "Rediscovery as an Historical Concept." In *New Trends in the History of Science*, edited by R. P. W. Visser, H. J. M. Bos, L. C. Palm, and H. A. M. Snelders, 197–208. Amsterdam: Rodopi.

Olby, Robert C. 1990. "The Molecular Revolution in Biology." In *Companion to the History of Modern Science*, edited by R. C. Olby, G. N. Cantor, J. R. R. Christie, and M. J. S. Hodge, 503–20. London: Routledge.

Olby, Robert C. 1993. "Constitutional and Hereditary Disorders." In *Com-*

panion Encyclopedia of the History of Medicine, edited by William F. Bynum and Roy Porter, vol. 1, 412–37. London: Routledge; 1993.

Olby, Robert C. 2000a. "Mendelism: From Hybrids and Trade to a Science." *Comptes Rendus de l'Académie des Sciences Paris, Sciences de la Vie* 323:1043–51.

Olby, Robert C. 2000b. "Horticulture: The Font for the Baptism of Genetics." *Nature Reviews: Genetics* 1:65–70.

Olby, Robert C. 2009. "Variation and Inheritance." In *The Cambridge Companion to the "Origin of Species,"* edited by Michael Ruse and Robert J. Richards, 30–46. Cambridge: Cambridge University Press.

Olby, Robert C. 2013. "Darwin and Heredity." In *The Cambridge Encyclopedia of Darwin and Evolutionary Thought*, edited by Michael Ruse, 116–23. Cambridge: Cambridge University Press.

Opitz, Donald L. 2011. "Cultivating Genetics in the Country: Whittinghame Lodge, Cambridge." In *Geographies of Nineteenth-Century Science*, edited by David N. Livingstone and Charles W. J. Withers, 73–98. Chicago: University of Chicago Press.

Oppenheimer, Jane M. 1970. "Some Diverse Backgrounds for Curt Herbst's Ideas about Embryonic Induction." *Bulletin of the History of Medicine* 44:241–50.

Orel, Vítězslav. 1966. "Opening of the Mendel Memorial attached in the Moravian Museum: Opening Address, Gregor Mendel Memorial in Brno." In Sosna (1966, 41–44).

Orel, Vítězslav. 1968. "Will the Story on 'Too Good' Results of Mendel's Data Continue?" *BioScience* 18:776–78.

Orel, Vítězslav. 1984. *Mendel*. Translated by Stephen Finn. Oxford: Oxford University Press.

Orel, Vítězslav. 1992. "Jaroslav Kříženecký (1896–1964), Tragic Victim of Lysenkoism in Czechoslovakia." *Quarterly Review of Biology* 67:487–94.

Orel, Vítězslav. 1996. *Gregor Mendel: The First Geneticist*. Translated by Stephen Finn. Oxford: Oxford University Press.

Orel, Vítězslav, and Margaret H. Peaslee. 2015. "Mendel's Research Legacy in the Broader Historical Network." *Science and Education* 24:9–27.

Orel, Vítězslav, and Roger J. Wood. 2000. "Scientific Animal Breeding in Moravia Before and After the Rediscovery of Mendel's Theory." *Quarterly Review of Biology* 75:149–57.

Osborne, Lawrence. 2002. "Got Silk." *New York Times Magazine*, 16 June. Available online at https://www.nytimes.com/2002/06/16/magazine/got-silk.html. Reprinted in *The Best American Science Writing 2003*, edited by Oliver Sacks, 186–93. New York: HarperCollins.

Oyama, Susan. 1985. *The Ontogeny of Information: Developmental Systems and Evolution*. Cambridge: Cambridge University Press.

Paleček, Pavel. 2004. "Project of Gregor Mendel-Forschungsinstitut at Brno planned during WWII." *Verhandlungen zur Geschichte und Theorie der Biologie* 10:159–62.

Paleček, Pavel. 2014. "One Hundred Years of Efforts to Establish an International Mendel Institute in Brno." Unpublished MS.

Paleček, Pavel. 2016. "Vítězslav Orel (1926–2015): Gregor Mendel's Biographer and the Rehabilitation of Genetics in the Communistic Bloc." *History and Philosophy of the Life Sciences* 38 (4): 1–12.

Palladino, Paolo. 1994. "Wizards and Devotees: On the Mendelian Theory of Inheritance and the Professionalization of Agricultural Science in Great Britain and the United States, 1880–1930." *History of Science* 32:409–44.

Papillon, Fernand. 1873. "The Phenomena of Heredity" and "Heredity and Race-Improvement." Translated by J. Fitzgerald. *Popular Science Monthly* 4 (November and Dec): 55–64 and 170–81.

Park, Alice. 2014. "Gregor Mendel." In *Great Scientists: The Geniuses, Eccentrics and Visionaries Who Transformed Our World.* New York: Time Books.

Patel, Chirag J., and John P. A. Ioannidis. 2014. "Studying the Elusive Environment in Large Scale." *Journal of the American Medical Association* 311 (21): 1–4.

Paul, Diane B. 1998. *Controlling Human Heredity: 1865 to the Present*. Amherst, NY: Humanity Books.

Paul, Diane B., and Barbara A. Kimmelman. 1988. "Mendel in America: Theory and Practice, 1900–1919." In *The American Development of Biology*, edited by Ronald Rainger, Keith R. Benson, and Jane Maienschein, 281–310. Philadelphia: University of Pennsylvania Press.

Paul, Diane, and Hamish G. Spencer. 2001. "Did Eugenics Rest on an Elementary Mistake?" In *Thinking about Evolution: Historical, Philosophical, and Political Perspectives*, edited by Rama S. Singh, Costas B. Krimbas, Diane B. Paul, and John Beatty, 103–18. Cambridge: Cambridge University Press.

Pauly, Philip J. 2000. *Biologists and the Promise of American Life: From Meriwether Lewis to Alfred Kinsey*. Princeton, NJ: Princeton University Press.

Peacock, Margaret. 2015. "Mendel Lives: The Survival of Mendelian Genetics in the Lysenkoist Classroom, 1937–1964." *Science and Education* 24:101–14.

Pearl, Raymond. 1911. "The Personal Equation in Breeding Experiments Involving Certain Characters in Maize." *Biological Bulletin* 21:339–66.

Pearl, Raymond. 1941. *Introduction to Medical Biometry and Statistics*. 3rd ed. Philadelphia: W. B. Saunders.

Pearson, Egon S. 1938. *Karl Pearson: An Appreciation of Some Aspects of His Life and Work*. Cambridge: Cambridge University Press.

Pearson, Egon S. 1965. "Studies in the History of Probability and Statistics.14. Some Incidents in the Early History of Biometry and Statistics, 1890–94." *Biometrika* 52:3–18.

Pearson, Karl. 1888. *The Ethic of Freethought: A Selection of Essays and Lectures*. London: T. Fisher Unwin.

Pearson, Karl. 1892. *The Grammar of Science*. London: Walter Scott.

Pearson, Karl. 1893. "Asymmetrical Frequency Curves." *Nature* 48:615–16.

Pearson, Karl. 1894a. "Contributions to the Mathematical Theory of Evolution." *Philosophical Transactions of the Royal Society of London A* 185:71–110. Abstract published in *Proceedings of the Royal Society of London* 54 (1893): 329–33. Reprinted in Pearson 1948, 1–40.

Pearson, Karl. 1894b. "Science and Monte Carlo." *Fortnightly Review* 55:183–93. Reprinted with slight modifications in Pearson 1897a, vol. 1, 42–62.

Pearson, Karl. 1894c. "Socialism and Natural Selection." *Fortnightly Review* 56:1–21. Reprinted with slight modifications in Pearson 1897a, vol. 1, 103–39.

Pearson, Karl. 1895a. "Contributions to the Mathematical Theory of Evolution. II: Skew Variation in Homogeneous Material." *Philosophical Transactions of the Royal Society of London A* 186:343–414. Abstract published in *Proceedings of the Royal Society of London* 57:257–60. Reprinted in Pearson 1948, 41–112.

Pearson, Karl. 1895b. "Note on Regression and Inheritance in the Case of Two Parents." *Proceedings of the Royal Society of London* 58:240–42.

Pearson, Karl. 1896a. "Mathematical Contributions to the Theory of Evolution. 3. Regression, Heredity, and Panmixia." *Philosophical Transactions of the Royal Society of London A* 187:253–318. Abstract published in *Proceedings of the Royal Society of London* 59:69–71. Reprinted in Pearson 1948, 113–78.

Pearson, Karl. 1896b. "Contribution to the Mathematical Theory of Evolution. Note on Reproductive Selection." *Proceedings of the Royal Society of London* 59:301–5.

Pearson, Karl. 1897a. *The Chances of Death and Other Essays in Evolution*. 2 vols. London: Edward Arnold.

Pearson, Karl. 1897b. "Mathematical Contributions to the Theory of Evolution. On Telegony in Man, &c." *Proceedings of the Royal Society of London* 60:273–83.

Pearson, Karl. 1898. "Mathematical Contributions to the Theory of Evolution: On the Law of Ancestral Heredity." *Proceedings of the Royal Society of London* 62:386–412.

Pearson, Karl. 1900a. *The Grammar of Science*. 2nd ed. London: Adam and Charles Black.

Pearson, Karl. 1900b. "Mathematical Contributions to the Theory of Evolution: On the Law of Reversion." *Proceedings of the Royal Society of London* 66:140–64.

Pearson, Karl. 1900c. "Data for the Problem of Evolution in Man. 4. Note on the Effect of Fertility Depending on Homogamy." *Proceedings of the Royal Society of London* 66:316–23.

Pearson, Karl. 1900d. "Mathematical Contributions to the Theory of Evolution. 8. On the Correlation of Characters Not Quantitatively Measurable." *Proceedings of the Royal Society of London* 66:241–44.

Pearson, Karl. 1900e. "On the Criterion That a Given System of Deviations from the Probable in the Case of a Correlated System of Variables Is Such That It Can Be Reasonably Supposed to Have Arisen from Random Sampling." *The London, Edinburgh and Dublin Philosophical Magazine and Journal of Science* 50:157–75. Reprinted in Pearson 1948, 339–57.

Pearson, Karl. 1901a. "Mathematical Contributions to the Theory of Evolution. 9. On the Principle of Homotyposis and Its Relation to Heredity, to the Variability of the Individual, and to That of the Race. Part 1.—Homotyposis in the Vegetable Kingdom." With the assistance of Alice Lee, Ernest Warren, and others. *Philosophical Transactions of the Royal Society of London* 197:285–379.

Pearson, Karl. 1901b. *National Life from the Standpoint of Science*. London: Adam & Charles Black.

Pearson, Karl. 1902. "On the Fundamental Conceptions of Biology." *Biometrika* 1:320–44.

Pearson, Karl. 1903a. "The Law of Ancestral Heredity." *Biometrika* 2:211–36.

Pearson, Karl. 1903b. "Prof. Johannsen on Heredity." *Nature* 69 (17 December): 149–50.

Pearson, Karl. 1903c. "Inheritance of Psychical and Physical Characters in Man." *Nature* 68 (22 October): 607–8.

Pearson, Karl. 1904a. "Mathematical Contributions to the Theory of Evolution. 12. On a Generalised Theory of Alternative Inheritance, with Special Reference to Mendel's Laws." *Philosophical Transactions of the Royal Society A* 203:53–86. First published in *Journal of the Anthropological Institute of Great Britain and Ireland* 33:179–237.

Pearson, Karl. 1904b. "On the Laws of Inheritance in Man. II. On the Inheritance of the Mental and Moral Characters in Man, and Its Comparison with the Inheritance of the Physical Characters." *Biometrika* 3:131–90.

Pearson, Karl. 1904c. "On Differentiation and Homotyposis in the Leaves of *Fagus sylvatica*." *Biometrika* 3:104–7.

Pearson, Karl. 1904d. "A Mendelian's View of the Law of Ancestral Inheritance." *Biometrika* 3:109–12.

Pearson, Karl. 1906. "Walter Frank Raphael Weldon. 1860–1906." *Biometrika* 5:1–52.

Pearson, Karl. 1908. "On a Mathematical Theory of Determinantal Inheritance, From Suggestions and Notes of the Late W. F. R. Weldon." *Biometrika* 6:80–93.

Pearson, Karl. 1914–30. *The Life, Letters, and Labours of Francis Galton*. 4 vols. Cambridge: Cambridge University Press.

Pearson, Karl. 1948. *Karl Pearson's Early Statistical Papers*. Cambridge: Cambridge University Press.

Pearson, Karl, and Alice Lee. 1900a. "Mathematical Contributions to the Theory of Evolution. 7. On the Application of Certain Formulae in the Theory of Correlation to the Inheritance of Characters Not Capable of Quantitative Measurement." *Proceedings of the Royal Society of London* 66:324–27.

Pearson, Karl, and Alice Lee. 1900b. "Mathematical Contributions to the Theory of Evolution. 8. On the Inheritance of Characters Not Capable of Exact Quantitative Measurement." *Philosophical Transactions of the Royal Society of London A* 195:79–150.

Pence, Charles H. 2011. "'Describing Our Whole Experience': The Statistical Philosophies of W. F. R. Weldon and Karl Pearson." *Studies in History and Philosophy of Biological and Biomedical Sciences* 42:475–85.

Pence, Charles H. 2022a. *The Rise of Chance in Evolutionary Theory: A Pompous Parade of Arithmetic*. London: Academic Press.

Pence, Charles H. 2022b. "Of Stirps and Chromosomes: Abstraction Through Detail." *Studies in History and Philosophy of Science*.

Peterson, Erik L. 2008. "William Bateson from *Balanoglossus* to *Materials for the Study of Variation*: The Transatlantic Roots of Discontinuity and the (Un)naturalness of Selection." *Journal of the History of Biology* 41:267–305.

Pick, Daniel. 1989. *Faces of Degeneration: A European Disorder, c. 1848–1918*. Cambridge: Cambridge University Press.

Pinker, Steven. 2002. *The Blank Slate: The Modern Denial of Human Nature*. London: BCA/Penguin.

Pinker, Steven. 2004. "Why Nature and Nurture Won't Go Away." *Daedalus* 133:5–24. Reprinted in his *Language, Cognition, and Human Nature: Selected Articles*, 214–27. Oxford: Oxford University Press, 2014.

Pinker, Steven. 2010. "My Genome, My Self." In *The Best American Essays 2010*, edited by Christopher Hitchens, 136–55. New York: Mariner/Houghton Mifflin. First published in *New York Times Magazine*, 11 January 2009.

Pinker, Steven. 2011. *The Better Angels of Our Nature: A History of Humanity and Violence*. London: Penguin.

Pinker, Steven. 2019. "What Can We Expect from the 2020s? Look Beyond the Gloom of the Daily Headlines and the Case for Progress Is Still Strong." *Financial Times*, 27 December. https://www.ft.com/content/e448f4ae-224e-11ea-92da-f0c92e957a96.

Plomin, Robert. 2018. *Blueprint: How DNA Makes Us Who We Are*. London: Allen Lane.

Poczai, Péter, Neil Bell, and Jaakko Hyvönen. 2014. "Imre Festetics and the Sheep: Breeders' Society of Moravia: Mendel's Forgotten 'Research Network.'" *PLoS Biology* 12:1–5.

Porter, Theodore M. 2004. *Karl Pearson: The Scientific Life in a Statistical Age*. Princeton, NJ: Princeton University Press.

Porter, Theodore M. 2014. "The Curious Case of Blending Inheritance." *Studies in History and Philosophy of Biological and Biomedical Sciences* 46:125–32.

Porter, Theodore M. 2018. *Genetics in the Madhouse: The Unknown History of Human Heredity*. Princeton, NJ: Princeton University Press.

Poulton, Edward B. 1890. *The Colours of Animals*. London: Kegan, Paul et al.

Poulton, Edward B. 1937. "The History of Evolutionary Theory as Recorded in Meetings of the British Association." *Nature* 140:395–407.

Provine, William B. 1971. *The Origins of Theoretical Population Genetics*. Chicago: University of Chicago Press.

Provine, William B. 2001. *The Origins of Theoretical Population Genetics*. 2nd ed., with a new afterword. Chicago: University of Chicago Press.

Punnett, Reginald C. 1905. *Mendelism*. Cambridge: Macmillan and Bowes.

Punnett, Reginald C. 1907. *Mendelism*. 2nd ed. Cambridge: Bowes and Bowes.

Punnett, Reginald C. 1909. *Mendelism*. American edition (reprint of the British 2nd ed., with additions). New York: Wilshire.

Punnett, Reginald C. 1910. *Mendelism*. 2nd ed. Cambridge: Bowes and Bowes.

Punnett, Reginald C. 1911. "Mendelism." In *Encyclopaedia Britannica*, 11th ed., 115–21.

Punnett, Reginald C. 1950. "Early Days of Genetics." *Heredity* 4:1–10.

Radick, Gregory. 2003. "Cultures of Evolutionary Biology." *Studies in History and Philosophy of Biological and Biomedical Sciences* 34:187–200.

Radick, Gregory. 2005a. "Other Histories, Other Biologies." In *Philosophy, Biology and Life*, edited by Anthony O'Hear, 21–47. Supplement to *Philosophy*. Royal Institute of Philosophy Supplement 56. Cambridge: Cambridge University Press.

Radick, Gregory. 2005b. "The Case for Virtual History." *New Scientist* 187, no. 2513 (20 August 2005): 34–35.

Radick, Gregory. 2007. *The Simian Tongue: The Long Debate about Animal Language*. Chicago: University of Chicago Press.

Radick, Gregory. 2008a. "Why What If?" *Isis* 99:547–51.

Radick, Gregory, ed. 2008b. Focus: "Counterfactuals and the Historian of Science." *Isis* 99:547–84.

Radick, Gregory. 2009. "Is the Theory of Natural Selection Independent of Its History?" In *The Cambridge Companion to Darwin*, 2nd ed., edited by Jonathan Hodge and Gregory Radick, 147–72. Cambridge: Cambridge University Press.

Radick, Gregory. 2011. "Physics in the Galtonian Sciences of Heredity." *Studies in History and Philosophy of Biological and Biomedical Sciences* 42:129–38.

Radick, Gregory. 2013a. "Should 'Heredity' and 'Inheritance' Be Biological Terms? William Bateson's Change of Mind as a Historical and Philosophical Problem." *Philosophy of Science* 79:714–24.

Radick, Gregory. 2013b. "The Professor and the Pea: Lives and Afterlives of William Bateson's Campaign for the Utility of Mendelism." *Studies in History and Philosophy of Science* 44:280–91.

Radick, Gregory. 2013c. "Biomachine Dreams." *Studies in History and Philosophy of Biological and Biomedical Sciences* 44:790–92.

Radick, Gregory. 2014a. [Review of Müller-Wille and Rheinberger 2012 etc.] *British Journal for the History of Science* 47:747–48.

Radick, Gregory. 2014b. "Consciously Digital." *Times Literary Supplement*, 20 June, 32.

Radick, Gregory. 2016a. "Presidential Address: Experimenting with the Scientific Past." *British Journal for the History of Science* 49:153–72.

Radick, Gregory. 2016b. [Review of Rasmussen 2014]. *Medical History* 60: 115–17.

Radick, Gregory. 2016c. "The Enemy Within: Simplifying the Vital Story of How Genes Determine Our Characteristics and Chances of Mortality" [review of Mukherjee 2016a]. *Times Literary Supplement*, 25 November, 3–4.

Radick, Gregory. 2017. [Review of Richards and Daston 2016]. *British Journal for the History of Science* 50 (3): 562–3.

Radick, Gregory. 2018. "How and Why Darwin Got Emotional about Race." In *Historicizing Humans: Deep Time, Evolution and Race in Nineteenth-Century British Sciences*, edited by Efram Sera-Shriar, 139–71. Pittsburgh: University of Pittsburgh Press.

Radick, Gregory. 2019a. "Darwinism and Social Darwinism." In *The Cambridge History of Modern European Thought*, edited by Warren Breckman and Peter E. Gordon, vol. 1, *The Nineteenth Century*, 279–300. Cambridge: Cambridge University Press.

Radick, Gregory. 2019b. "Genes and Genocide: The Questionable Use of Scientific Endeavour," review of Porter 2018. *Times Literary Supplement*, 10 May, 29.

Radick, Gregory. 2020a. "Breeding Back to Former Glory: The Role of Eugenics in Nazi Germany," review of Teicher, *Social Mendelism. Times Literary Supplement*, 14 August, 23–24.

Radick, Gregory. 2020b. "Making Sense of Mendelian Genes." In "Making Sense of Metaphor: Evelyn Fox Keller and Commentators on Language and Science," edited by Marga Vicedo and Denis Walsh. Special issue, *Interdisciplinary Science Reviews* 45:299–314.

Radick, Gregory. 2022a. "Theory-Ladenness as a Problem for Plant Data Linkage." In *Towards Responsible Plant Data Linkage: Global Data Challenges for Agricultural Research and Development*, edited by Hugh Williamson and Sabina Leonelli. Dordrecht: Springer.

Radick, Gregory. 2022b. "Mendel the Fraud? A Social History of Truth in Genetics." In "New Directions in the Historiography of Genetics," edited by Yafeng Shan, Ehud Lamm, and Oren Harman. Special issue, *Studies in History and Philosophy of Science* 93:39–46.

Ramsay, William. 1905. *The Gases of the Atmosphere*. 3rd ed. London: Macmillan.

"Rare Rewards." 2016. *Nature* 536 (18 August): 249.

Rasmussen, Nicolas. 2014. *Gene Jockeys: Life Science and the Rise of Biotech Enterprise*. Baltimore: Johns Hopkins University Press.

Reisch, George A. 2016. "Aristotle in the Cold War: On the Origins of Thomas Kuhn's *The Structure of Scientific Revolutions*." In *Kuhn's* Structure of Scientific Revolutions *at Fifty: Reflections on a Science Classic*, edited by Robert J. Richards and Lorraine Daston, 12–30. Chicago: University of Chicago Press.

Renouvier, Charles. 1876. *Uchronie (L'Utopie dans L'Histoire)*. 2nd ed. Paris: Bureau de la Critique Philosophique. Reprint, Librairie Arthème Fayard, 1988.

Renwick, Chris. 2012. *British Sociology's Lost Biological Roots: A History of Futures Past*. London: Palgrave Macmillan.

Renwick, Chris. 2018. "Lee, Alice Elizabeth." *Oxford Dictionary of National Biography*.

"Report of the Anthropometric Committee." 1880. In *Report of the Fiftieth Meeting of the British Association for the Advancement of Science, Held at Swansea in August and September 1880*, 120–59. London: John Murray.

Rheinberger, Hans-Jörg. 2003. "Carl Correns' Experiments with *Pisum*, 1896–1899." In *Reworking the Bench: Research Notebooks in the History of Science*, edited by Frederic L. Holmes, Jürgen Renn, and Hans-Jörg Rheinberger, 221–52. Dordrecht: Kluwer.

Rheinberger, Hans-Jörg, and Jean-Paul Gaudillière, eds. 2004. *Classical Genetic Research and Its Legacy: The Mapping Cultures of Twentieth-Century Genetics*. London: Routledge.

Richards, Robert J. 1987. *Darwin and the Emergence of Evolutionary Theories of Mind and Behavior*. Chicago: University of Chicago Press.

Richards, Robert J. 2008. *The Tragic Sense of Life: Ernst Haeckel and the Struggle over Evolutionary Thought*. Chicago: University of Chicago Press.

Richmond, Marsha. 1997. "'A Lab of One's Own': The Balfour Biological Laboratory for Women at Cambridge University, 1884–1914." *Isis* 88:422–55.

Richmond, Marsha. 2001. "Women in the Early History of Genetics: William Bateson and the Newnham College Mendelians, 1900–1910." *Isis* 92:55–90.

Richmond, Marsha. 2006. "The 'Domestication' of Heredity: The Familial Organization of Geneticists at Cambridge University, 1895–1910." *Journal of the History of Biology* 39:565–605.

Richmond, Marsha. 2008. "William Bateson's Pre- and Post-Mendelian Research Program in 'Heredity and Development.'" In *A Cultural History of Heredity IV: Heredity in the Century of the Gene*, 213–41. Preprint 343. Berlin: Max Planck Institute for the History of Science.

Riddle, Oscar. 1947. "Biographical Memoir of Charles Benedict Davenport 1866–1944." *Biographical Memoirs of the National Academy of Sciences* 25:75–110.

Ridley, Mark. 1986. "Embryology and Classical Zoology in Great Britain." In *A History of Embryology*, edited by T. J. Horder, J. A. Witkowsky, and C. C. Wylie, 35–67. Cambridge: Cambridge University Press.

Riffkin, Rebecca. 2011. "Rare Genetic Mutation Causes Infant Deaths in Small Town." American Association for the Advancement of Science, "In-Depth" (blog), 5 August. https://www.aaas.org/rare-genetic-mutation-causes-infant-deaths-small-town.

Riordan, Jesse D., and Joseph H. Nadeau. 2017. "From Peas to Disease: Modifier Genes, Network Resilience, and the Genetics of Health." *American Journal of Human Genetics* 101:177–91.

Roarty, Dan, and Edward Bryan. 2014. "Using Genomics to Find Olympic and Investment Gold." *Context: The AB Blog on Investing*. 19 February 2014. https://www.alliancebernstein.com/post/en/2014/02/using-genomics-to-find-olympic-and-investment-gold.

Roberts, H. F. 1929. *Plant Hybridization before Mendel*. Princeton, NJ:

Princeton University Press. Facsimile reprint, New York: Hafner, 1965.

Robinson, Gloria. 1979. *A Prelude to Genetics: Theories of Material Substance of Heredity; Darwin to Weismann.* Lawrence, KS: Coronado Press.

Roff, Derek A. 2007. "A Centennial Celebration for Quantitative Genetics." *Evolution* 61:1017–32.

Rolfe, R. Allen. 1900. "Hybridisation Viewed from the Standpoint of Systematic Botany." Hybrid Conference Report 1900, *Journal of the Royal Horticultural Society* 24:181–202.

Roll-Hansen, Nils. 1980. "The Controversy between Biometricians and Mendelians: A Test Case for the Sociology of Scientific Knowledge." *Social Science Information* 19:501–17.

Roll-Hansen, Nils. 1983. "The Death of Spontaneous Generation and the Birth of the Gene: Two Case Studies of Relativism." *Social Studies of Science* 13:481–519.

Roll-Hansen, Nils. 1989. "The Crucial Experiment of Wilhelm Johannsen." *Biology and Philosophy* 4:303–29.

Roll-Hansen, Nils. 2005. *The Lysenko Effect: The Politics of Science.* New York: Humanity Books.

Roll-Hansen, Nils. 2014. "The Holist Tradition in Twentieth Century Genetics[:] Wilhelm Johannsen's Genotype Concept." *Journal of Physiology* 592:2431–38.

Root-Bernstein, Robert Scott. 1983. "Mendel and Methodology." *History of Science* 21:275–95.

Rope, Alan F., et al. 2011. "Using VAAST to Identify an X-Linked Disorder Resulting in Lethality in Male Infants Due to N-Terminal Acetyltransferase Deficiency." *American Journal of Human Genetics* 89 (15 July): 28–43.

Rose, Steven, Leon J. Kamin, and R. C. Lewontin. 1984. *Not in Our Genes: Biology, Ideology and Human Nature*. Harmondsworth: Penguin.

Rosen, Michael. 1996. *On Voluntary Servitude: False Consciousness and the Theory of Ideology*. Cambridge: Cambridge University Press.

Rosenberg, Noah A. 2011. "A Population-Genetic Perspective on the Similarities and Differences among Worldwide Human Populations." *Human Biology* 83:659–84.

Rostand, Jean. 1965. "Johann Gregor Mendel." *UNESCO Courier* 18 (4): 16–19.

Rushton, Alan R. 2014. "William Bateson and the Chromosome Theory of Heredity: A Reappraisal." *British Journal for the History of Science* 47:147–71.

Rushton, Alan R. 2017. "Bateson and the Doctors: The Introduction of Mendelian Genetics to the British Medical Community 1900–1910." In

History of Human Genetics: Aspects of Its Development and Global Perspectives, edited by Heike I. Petermann, Peter S. Harper, and Susanne Doetz, 59–71. Dordrecht: Springer.

Russell, Nicholas. 1986. *Like Engend'ring Like: Heredity and Animal Breeding in Early Modern England*. Cambridge: Cambridge University Press.

Russell, Peter J. 2006. *iGenetics: A Mendelian Approach*. San Francisco: Pearson/Benjamin Cummings.

Saatsi, Juha, and Mark Pexton. 2013. "Reassessing Woodward's Account of Explanation: Regularities, Counterfactuals, and Noncausal Explanations." *Philosophy of Science* 80:613–24.

Saini, Angela. 2019. *Superior: The Return of Race Science*. London: 4th Estate.

Santo, Tatiana E., Ricardo J. Pereira, and José M. Leitão. 2017. "The Pea (*Pisum sativum* L.) Rogue Paramutation Is Accompanied by Alterations in the Methylation Pattern of Specific Genomic Sequences." *Epigenomes* 1 (6): 1–11.

Sapp, Jan. 1987. *Beyond the Gene: Cytoplasmic Inheritance and the Struggle for Authority in Genetics.* Oxford: Oxford University Press.

Sapp, Jan. 2003. *Genesis: The Evolution of Biology*. Oxford: Oxford University Press.

Sarkar, Sahotra. 1999. "From the *Reaktionsnorm* to the Adaptive Norm: The Norm of Reaction, 1909–1960." *Biology and Philosophy* 14:235–52.

Saunders, Rebecca. 1897. "On a Discontinuous Variation Occurring in *Biscutella lavigata*." *Proceedings of the Royal Society of London* 62: 11–26.

Schaffer, Simon J. 2000. "Who's Minding the Garden?" In *No1SE* [exhibit catalogue], edited by Adam Lowe and Simon J. Schaffer. Cambridge: Whipple Museum; London: Wellcome Institute.

Schaffer, Simon J. 2009. "Knowledge Is a Social Institution." In *Ideas on the Nature of Science*, edited by David Cayley, 17–33. Fredericton, New Brunswick: Goose Lane Editions.

Schmid, Kelly M., et al. In press. "Mendelian or Multifactorial? Current Undergraduate Genetics Assessments Focus on Genes and Rarely Include the Environment." *Journal of Microbiology & Biology Education.*

Schmuhl, Hans-Walter. 2008. *The Kaiser Wilhelm Institute for Anthropology, Human Heredity, and Eugenics, 1927–1945*. Dordrecht: Springer.

Schuster, E. H. J. 1905. "Results of Crossing Grey (House) Mice with Albinos." *Biometrika* 4:1–12.

Schwartz, James. 2008. *In Pursuit of the Gene: From Darwin to DNA*. Cambridge, MA: Harvard University Press.

Schwartz, Sara. 2002. "Characters as Units and the Case of the Presence and Absence Hypothesis." *Biology and Philosophy* 17:369–88.

Secord, James A. 2004. "Knowledge in Transit." *Isis* 95:654–72.

Sekar, Aswin, et al. 2016. "Schizophrenia Risk from Complex Variation of Complement Component 4." *Nature* 530:177–83.

Serpico, Davide. 2020. "Beyond Quantitative and Qualitative Traits: Three Telling Cases in the Life Sciences." *Biology & Philosophy* 35 (34): 1–26.

Shakespeare, William. 1993. *The Oxford Shakespeare: The Merchant of Venice*. Edited by Jay L. Halio. Oxford: Oxford University Press.

Shakespeare, William. 1998. *The Arden Shakespeare: Julius Caesar*. Edited by David Daniell. London: Arden Shakespeare/Thomson Learning.

Shan, Yafeng. 2020a. *Doing Integrated History and Philosophy of Science: A Case Study of the Origin of Genetics*. Dordrecht: Springer.

Shan, Yafeng. 2020b. "Kuhn's 'Wrong Turn' and Its Legacy." *Synthese* 197:381–406.

Shan, Yafeng. 2021. "Beyond Mendelism and Biometry." *Studies in History and Philosophy of Science* 89:155–63.

Simpson, George Gaylord. 1953. "The Baldwin Effect." *Evolution* 7:110–17.

Šimůnek, Michal, and Uwe Hossfeld. 2013. "Trofim D. Lysenko in Prague 1960: A Historical Note." *Studies in History of Biology* 5:84–87.

Šimůnek, Michal, Uwe Hossfeld, and Olaf Breidbach. 2012. "'Further Development' of Mendel's Legacy? Erich von Tschermak-Seysenegg in the Context of Mendelian-Biometry Controversy, 1901–1906." *Theory in Biosciences* 131:243–52.

Sinnott, Edmund W., and L. C. Dunn. 1925. *Principles of Genetics: An Elementary Text, with Problems*. New York: McGraw-Hill.

Sinnott, Edmund W., and L. C. Dunn. 1932. *Principles of Genetics: A Textbook, with Problems.* 2nd ed. With an appendix by D. R. Charles. New York: McGraw-Hill.

Skopek, Jeffrey M. 2008. "Shaping Science with the Past: Textbooks, History, and the Disciplining of Genetics." PhD thesis, University of Cambridge.

Sloan, Phillip R. 1985. "Darwin's Invertebrate Program, 1826–1836: Preconditions for Transformism." In *The Darwinian Heritage*, edited by David Kohn, 71–120. Princeton, NJ: Princeton University Press.

Sloan, Phillip R. 2000. "Mach's Phenomenalism and the British Reception of Mendelism." *Comptes Rendus de l'Académie des Sciences Paris, Sciences de la Vie* 323:1069–79.

Sober, Elliott, ed. 1994. *Conceptual Issues in Evolutionary Biology*. 2nd ed. Cambridge, MA: MIT Press.

Soler, Léna. 2015a. "Introduction." In Soler, Trizio, and Pickering 2015, 1–42.

Soler, Léna. 2015b. "Why Contingentists Should Not Care about the Inevitabilist Demand to 'Put-Up-or-Shut-Up': A Dialogic Reconstruction of the Argumentative Network." In Soler, Trizio, and Pickering 2015, 45–98.

Soler, Léna, and Howard Sankey, eds. 2008. "The Contingentism versus Inevitabilism Issue." Special section, *Studies in History and Philosophy of Science* 39:220–46.

Soler, Léna, Emiliano Trizio, and Andrew Pickering, eds. 2015. *Science As It Could Have Been: Discussing the Contingency/Inevitability Problem*. Pittsburgh: University of Pittsburgh Press.

Sollas, William J. 1900. "Opening Address as President of the Geological Section." *Nature* 62:481–89. Also in *Report of the Seventieth Meeting of the British Association for the Advancement of Science, Held at Bradford in September 1900*, 711–30. London: John Murray. Also in *Nature* 62:481–89.

Soloway, Richard A. 1995. *Demography and Degeneration: Eugenics and the Declining Birthrate in Twentieth-Century Britain*. With a new preface. Chapel Hill: University of North Carolina Press. First published in 1990.

Sosna, Milan, ed. 1966. *G. Mendel Memorial Symposium 1865–1965: Proceedings of a Symposium held in Brno in August 4–7, 1965*. Prague: Academia.

Sparks, Rachel A., Kara Esther Baldwin, and Rebekka Darner. 2020. "Using Culturally Relevant Pedagogy to Reconsider the Genetics Canon." *Journal of Microbiology & Biology Education* 21:1–6, plus a four-page appendix outlining a proposed genetics module.

Sparks, Sarah D. 2019. "Genetics Lessons Can Spark Racism in Students. This Change Can Prevent It." *Education Week* 27 March. Available at https://www.edweek.org/teaching-learning/genetics-lessons-can-spark-racism-in-students-this-change-can-prevent-it/2019/03.

Spencer, Hamish G., and Diane B. Paul. 1998. "The Failure of a Scientific Critique: David Heron, Karl Pearson and Mendelian Eugenics." *British Journal for the History of Science* 31:441–52.

Spencer, Herbert. 1864. *The Principles of Biology*. 2 vols. London: Williams & Norgate.

Spillman, William J. 1902a. "Quantitative Studies on the Transmission of Parental Characters to Hybrid Offspring." *Proceedings of the 15th Annual Convention of the Association of American Agricultural Colleges and Experiment Stations, November 12–14, 1901*. *OES Bulletin* 115 (Washington, DC: GPO): 88–97.

Spillman, William J. 1902b. "Exceptions to Mendel's Law." *Science* 16 (31 October): 709–10.

Spillman, William J. 1902c. "Exceptions to Mendel's Law." *Science* 16 (14 November): 794–96.

Spillman, William J. 1903. "Mendel's Law." *Popular Science Monthly* 62: 269–80.

Spiro, Jonathan. 2009. *Defending the Master Race: Conservation, Eugenics, and the Legacy of Madison Grant.* Chicago: University of Chicago Press.

Stamhuis, Ida H. 2015. "Why the Rediscoverer Ended up on the Sidelines: Hugo de Vries's Theory of Inheritance and the Mendelian Laws." *Science and Education* 24:29–49.

Stamhuis, Ida H., Onno G. Meijer, and Erik J. A. Zevenhuizen. 1999. "Hugo de Vries on Heredity, 1889–1903." *Isis* 90:238–67.

Standfuss, Max. 1896. *Handbuch der Paläarktischen Gross-Schmetterlinge für Forscher und Sammler*. Jena: Fischer.

Stanford, Kyle P. 2006. *Exceeding Our Grasp: Science, History, and the Problem of Unconceived Alternatives*. Oxford: Oxford University Press.

Stansfield, William D. 2016. "Letter to the Editor." *American Biology Teacher* 78 (March): 185.

Sterelny, Kim. 2005. "Another View of Life." *Studies in History and Philosophy of Biological and Biomedical Sciences* 36:585–93.

Stern, Curt. 1966. "Mendel and Human Genetics." In Sosna 1966, 199–218.

Stern, Curt, and Eva R. Sherwood, eds. 1966. *The Origin of Genetics: A Mendel Source Book*. London: W. H. Freeman.

Stigler, Stephen M. 1986. *The History of Statistics: The Measurement of Uncertainty before 1900*. Cambridge, MA: Belknap Press/Harvard University Press.

Stigler, Stephen M. 2008. "CSI: Mendel." *American Scientist* 96 (September–October): 425.

Stigler, Stephen M. 2010. "Darwin, Galton and the Statistical Enlightenment." *Journal of the Royal Statistical Society A* (173): 469–82.

Stoltzfus, Arlin, and Kele Cable. 2014. "Mendelism-Mutationism: The Forgotten Evolutionary Synthesis." *Journal of the History of Biology* 47: 501–46.

Stomps, T. J. 1954. "On the Rediscovery of Mendel's Work by Hugo de Vries." *Journal of Heredity* 45:293–94.

Studitski, A. N. 1949. "Fly-Lovers and Man-Haters." Translated by Rudolf Emanuel. Introduced and annotated by Robert C. Cook. *Journal of Heredity* 40:307–14.

Sturm, Richard A., and Tony N. Frudakis. 2004. "Eye Colour: Portals into Pigmentation Genes and Ancestry." *Trends in Genetics* 20:327–32.

Sturtevant, Alfred H. 1910. "On the Inheritance of Color in the American Harness Horse." *Biological Bulletin* 19:2014–16.

Sturtevant, Alfred H. 1965. *A History of Genetics*. New York: Harper & Row.

Sunderland, Mary Evelyn. 2010. "*Regeneration:* Thomas Hunt Morgan's Window into Development." *Journal of the History of Biology* 43: 325–61.

Sutton, Jennifer. 2017. "William Bateson and the British Eugenics Movement." Undergraduate dissertation, University of Leeds.

Sutton, Walter S. 1902. "On the Morphology of the Chromosome Group in *Brachystola magna*." *Biological Bulletin* 4:24–39.

Sutton, Walter S. 1903. "The Chromosomes in Heredity." *Biological Bulletin* 4:231–51.

Swallow, Dallas. 2020–21. "My Life in Genetics: An Interview with Professor Dallas Swallow." By Robert Johnston. *Galton Review*, no. 14, 4–9.

Swartz, Aimee. 2014. "The Search for Genes That Prevent Disease." *Atlantic*, 29 May. https://www.theatlantic.com/health/archive/2014/05/searching-for-the-genes-that-prevent-disease/371256/

Swinburne, R. G. 1962. "The Presence-and-Absence Theory." *Annals of Science* 18:131–45.

Swinburne, R. G. 1965. "Galton's Law: Formulation and Development." *Annals of Science* 21:15–31.

Tabery, James G. 2004. "The 'Evolutionary Synthesis' of George Udny Yule." *Journal of the History of Biology* 37:73–101.

Tabery, James G. 2014. *Beyond Versus: The Struggle to Understand the Interaction of Nature and Nurture*. Cambridge, MA: MIT Press.

Tambolo, Luca. 2020a. "So Close No Matter How Far: Counterfactuals in History of Science and the Inevitability/Contingency Controversy." *Synthese* 197:2111–41.

Tambolo, Luca. 2020b. "An Unappreciated Merit of Counterfactual Histories of Science." *Studies in History and Philosophy of Biological and Biomedical Sciences* 81 (101183): 1–6.

Teicher, Amir. 2014. "Mendel's Use of Mathematical Modelling: Ratios, Predictions and the Appeal to Tradition." *History and Philosophy of the Life Sciences* 36:187–208.

Teicher, Amir. 2018. "Caution, Overload: The Troubled Past of Genetic Load." *Genetics* 210:747–55.

Teicher, Amir. 2019. "Why Did the Nazis Sterilize the Blind? Genetics and the Shaping of the Sterilization Law of 1933." *Central European History* 52:289–309.

Teicher, Amir. 2020a. *Social Mendelism: Genetics and the Politics of Race in Germany, 1900–1948.* Cambridge: Cambridge University Press.

Teicher, Amir. 2020b. "Medical Bacteriology and Medical Genetics, 1880–1914: A Call for Synthesis." *Medical History* 64:325–54.

Tetlock, Philip, and Aaron Belkin. 1996. "Counterfactual Thought Experiments in World Politics: Logical, Methodological, and Psychological Perspectives." In *Counterfactual Thought Experiments in World Politics*, edited by P. Tetlock and A. Belkin, 1–38. Princeton, NJ: Princeton University Press.

Theunissen, Bert. 1994. "Knowledge Is Power: Hugo de Vries on Science, Heredity and Social Progress." *British Journal for the History of Science* 27:291–311.

Thiselton-Dyer, William T. 1895. "Variation and Specific Stability." *Nature* 51 (14 March): 459–61.

Thompson, D'Arcy Wentworth. 1917. *On Growth and Form*. Cambridge: Cambridge University Press.

Thompson, E. P. 1968. *The Making of the English Working Class*. Revised version of the 1963 first edition. London: Pelican.

Thompson, Herbert. 1896. "On Certain Changes Observed in the Dimensions of Parts of the Carapace of *Carcinus maenas*." *Proceedings of the Royal Society of London* 60:195–98.

Thomson, J. Arthur. 1899. *The Science of Life: An Outline of the History of Biology and Its Recent Advances*. London: Blackie & Son.

Thomson, J. Arthur. 1900a. "Facts of Inheritance." *Nature* 62 (2 August): 331–34. [An abridgment of Thomson 1900b.]

Thomson, J. Arthur. 1900b. "Facts of Inheritance." *Veterinary Journal* 50:253–64.

Thomson, J. Arthur. 1908. *Heredity*. London: John Murray.

Thurtle, Phillip. 2007. *The Emergence of Genetic Rationality: Space, Time, & Information in American Biological Science, 1870–1920*. Seattle: University of Washington Press.

Tschermak, Erich. 1900a. "Ueber künstliche Kreuzung bei *Pisum sativum*." *Berichte der Deutschen Botanischen Gesellschaft* 18:232–39. Published in English in 1950 as "Concerning Artificial Crossing in *Pisum sativum*," translated by Aloha Hannah. *Genetics* 35:42–47.

Tschermak, Erich. 1900b. "Ueber künstliche Kreuzung bei *Pisum sativum*." *Zeitschrift für das Landwirtschaftliche Versuchswesen in Österreich* 3: 465–555.

Tucker, Jennifer. 1997. "Photography as Witness, Detective, and Impostor: Visual Representation in Victorian Science." In *Victorian Science in Context*, edited by Bernard Lightman, 378–408. Chicago: University of Chicago Press.

Turner, Herbert Hall. 1904. *Astronomical Discovery*. London: Edward Arnold.

Van Dijk, Peter J., and T. H. Noel Ellis. 2022. "Mendel's Reaction to Darwin's Provisional Hypothesis of Pangenesis and the Experiment That Could Not Wait." *Heredity*. https://doi.org/10.1038/s41437-022-00546-w.

Van Dijk, Peter J., Franz J. Weissing, and T. H. Noel Ellis. 2018. "How Mendel's Interest in Inheritance Grew out of Plant Improvement." *Genetics* 210:347–55.

Vernon, H[orace] M. 1895. "The Effect of Environment on the Development of Echinoderm Larvae: An Experimental Inquiry into the Causes of Variation." *Proceedings of the Royal Society of London* 57:382–85.

Vernon, H[orace] M. 1900. "Certain Laws of Variation. 1. The Reaction of Developing Organisms to Environment." *Proceedings of the Royal Society of London* 67:85–101.

Vernon, H[orace] M. 1902. *Variation in Animals and Plants*. New York: Henry Holt.

Vicedo, Marga. 1995. "What Is That Thing Called Mendelian Genetics?" [review of Kim 1994]. *Social Studies of Science* 25:370–82.

Vilmorin, Philippe de. 1907. [Response to the Toast of the Foreign and British Members of the Conference.] In Wilks 1907, 73–74.

Von Guaita, Georg. 1898. "Versuche mit Kreuzungen von verschiedenen Rassen der Hausmaus." *Berichte der Naturforschenden Gesellschaft zu Freiburg* 10:317–32.

Von Guaita, Georg. 1900. "Zweite Mittheilung über Versuche mit Kreuzungen von verschiedenen Hausmausrassen." *Berichte der Naturforschenden Gesellschaft zu Freiburg* 11:131–38.

Vorzimmer, Peter J. 1963. "Charles Darwin and Blending Inheritance." *Isis* 54:371–90.

Vorzimmer, Peter J. 1970. *Charles Darwin: The Years of Controversy—The* Origin of Species *and Its Critics 1859–1882*. Philadelphia: Temple University Press.

Waddington, C. H. 1959. "Canalization of Development and Genetic Assimilation of Acquired Characters." *Nature* 183:1654–55.

Wallace, Alfred Russel. 1889. *Darwinism: An Exposition of the Theory of Natural Selection and Some of Its Applications*. Macmillan: London.

Wallace, Alfred Russel. 1895. "The Method of Organic Evolution, Parts 1 and 2." *Fortnightly Review* 63:211–24, 435–45.

Waller, John C. 2001a. "Ideas of Heredity, Reproduction and Eugenics in Britain, 1800–1875." *Studies in History and Philosophy of Biological and Biomedical Sciences* 32:457–89.

Waller, John C. 2001b. "Gentlemanly Men of Science: Sir Francis Galton and the Professionalization of the British Life-Sciences." *Journal of the History of Biology* 34:83–114.

Waller, John C. 2002. "Putting Method First: Re-appraising the Extreme Determinism and Hard Hereditarianism of Sir Francis Galton." *History of Science* 40:35–62.

Waller, John C. 2004. "Becoming a Darwinian: The Micro-politics of Sir Francis Galton's Career 1859–65." *Annals of Science* 61:141–63.

Wang, T. L., and C. L. Hedley. 1991. "Seed Development in Peas: Knowing Your Three 'R's' (or Four, or Five)." *Seed Science Research* 1:3–14.

Ward, C. W. 1904. "The Improvement of Carnations." In *Proceedings: Inter-*

national Conference on Plant Breeding and Hybridization, 151–55. Memoirs of the Horticultural Society of New York, vol. 1.

Warren, Ernest. 1899. "An Observation on Inheritance in Parthenogenesis." *Proceedings of the Royal Society of London* 65:154–58.

Warren, Ernest. 1900. "On the Reaction of *Daphnia magna (Straus)* to Certain Changes in Environment." *Quarterly Journal of Microscopical Science* 43:199–224.

Waters, C. Kenneth. 1990. "Why the Antireductionist Consensus Won't Survive the Case of Classical Mendelian Genetics." *PSA 1990* 1:125–39.

Watson, J. D. 1965. *Molecular Biology of the Gene*. New York: W. A. Benjamin.

Watson, J. D., and F. H. C. Crick. 1953a. "Molecular Structure of Nucleic Acids: A Structure of Deoxyribose Nucleic Acid." *Nature* 171 (25 April): 737–38.

Watson, J. D., and F. H. C. Crick. 1953b. "Genetical Implications of the Structure of Deoxyribonucleic Acid." *Nature* 171 (30 May): 964–67.

Weeden, Norman F. 2016. "Are Mendel's Data Reliable? The Perspective of a Pea Geneticist." *Journal of Heredity* 107:635–46.

Weiner, Jonathan. 1999. *Time, Love, Memory: A Great Biologist and His Quest for the Origins of Behavior*. New York: Knopf.

Weismann, August. 1886. "Retrogressive Development in Nature." In Weismann 1889–92, vol. 2, 1–30.

Weismann, August. 1888. "The Supposed Transmission of Mutilations." In Weismann 1889–92, vol. 1, 419–48.

Weismann, August. 1889–92. *Essays upon Heredity and Kindred Biological Problems*. Edited by Edward B. Poulton, Selmar Schönland, and Arthur E. Shipley. 2 vols. Oxford: Clarendon Press.

Weiss, Sheila Faith. 2010. *The Nazi Symbiosis: Human Genetics and Politics in the Third Reich*. Chicago: University of Chicago Press.

Weldon, W. F. R. 1883. "Note on the Early Development of Lacerta Muralis." *Quarterly Journal of Microscopical Science* 23:134–44.

Weldon, W. F. R. 1884a. "Note on the Origin of the Suprarenal Bodies of Vertebrates." *Proceedings of the Royal Society of London* 37:422–25.

Weldon, W. F. R. 1884b. "On the Head Kidney of Bdellostoma, with a Suggestion as to the Origin of the Suprarenal Bodies." *Quarterly Journal of Microscopical Science* 24:171–82.

Weldon, W. F. R. 1885. "On the Suprarenal Bodies of Vertebrata." *Quarterly Journal of Microscopical Science* 25:137–150.

Weldon, W. F. R. 1887a. "Preliminary Note on a Balanoglossus Larva from the Bahamas." *Proceedings of the Royal Society of London* 42:146–50.

Weldon, W. F. R. 1887b. "Note on Communication Entitled 'Preliminary

Note on Balanoglossus Larva from the Bahamas' ('Roy. Soc. Proc.,' vol. 42, p. 146)." *Proceedings of the Royal Society of London* 42:473.

Weldon, W. F. R. 1890a. "The Variations Occurring in Certain Decapod Crustacea.—1. *Crangon vulgaris*." *Proceedings of the Royal Society of London* 47:445–53.

Weldon, W. F. R. 1890b. "*Palaemonetes varians* in Plymouth." *Journal of the Marine Biological Association* 1 (November): 459–61.

Weldon, W. F. R. 1892. "Certain Correlated Variations in *Crangon vulgaris*." *Proceedings of the Royal Society of London* 51:1–21.

Weldon, W. F. R. 1893. "On Certain Correlated Variations in *Carcinus maenas*." *Proceedings of the Royal Society of London* 54:318–29.

Weldon, W. F. R. 1894a. "The Study of Animal Variation" [review of Bateson 1894]. *Nature* 50 (10 May): 25–26.

Weldon, W. F. R. 1894b. "Panmixia." *Nature* 50 (3 May): 5.

Weldon, W. F. R. 1895a. "Report of the Committee [. . .] for Conducting Statistical Inquiries into the Measurable Characteristics of Plants and Animals." Part 1. "An Attempt to Measure the Death-Rate due to the Selective Destruction of *Carcinus maenas* with Respect to a Particular Dimension." *Proceedings of the Royal Society of London* 57:360–79, with corrections to the labeling of the key selection diagram published in 58:lxv.

Weldon, W. F. R. 1895b. "Remarks on Variation in Animals and Plants. To Accompany the First Report of the Committee for Conducting Statistical Inquiries into the Measurable Characteristics of Plants and Animals." *Proceedings of the Royal Society of London* 57:379–82. Also published as "Variations in Animals and Plants," *Nature* 51 (7 March 1895): 449–50.

Weldon, W. F. R. 1897a. "Karl Pearson on Evolution." *Natural Science* 11 (July): 50–54.

Weldon, W. F. R. 1897b. "Experimental Embryology." *Nature* 56 (23 September): 489–90.

Weldon, W. F. R. 1898. Presidential Address to BAAS Section D, delivered Thursday 8 September. In *Report of the Sixty-Eighth Meeting of the British Association for the Advancement of Science, Held at Bristol in September 1898*, 887–902. London: John Murray. Also in *Nature* 58 (22 September): 499–506.

Weldon, W. F. R. 1901a. "Editorial: The Scope of *Biometrika*." *Biometrika* 1:1–2. [Unsigned, but listed among Weldon's publications in Pearson 1906.]

Weldon, W. F. R. 1901b. "A First Study of Natural Selection in *Clausilia laminata* (Montagu)." *Biometrika* 1:109–24.

Weldon, W. F. R. 1901c. "Change in Organic Correlation of *Ficaria ranunculoides* during the Flowering Season." *Biometrika* 1:125–28.

Weldon, W. F. R. 1902a. "Mendel's Laws of Alternative Inheritance in Peas." *Biometrika* 1:228–54.

Weldon, W. F. R. 1902b. "Professor de Vries on the Origin of Species." *Biometrika* 1:365–74.

Weldon, W. F. R. 1902c. "Variation and Selection." *Encyclopaedia Britannica*, 10th ed., vol. 33, 633–37.

Weldon, W. F. R. 1902d. "On the Ambiguity of Mendel's Categories." *Biometrika* 2:44–55.

Weldon, W. F. R. 1903. "Mr Bateson's Revisions of Mendel's Theory of Heredity." *Biometrika* 2:286–98.

Weldon, W. F. R. 1904a. "Albinism in Sicily and Mendel's Laws." *Biometrika* 3:107–9.

Weldon, W. F. R. 1904b. "Note on a Race of *Clausilia itala (von Martens)*." *Biometrika* 3:299–307.

Weldon, W. F. R. 1904–5. "Theory of Inheritance." Unpublished MS. Pearson/5/2/10/4, PP.

Weldon, W. F. R. 1905. "[Comment on Galton 1905.]" *Sociological Papers* 1:56–58.

Weldon, W. F. R. 1906a. "Inheritance in Animals and Plants." In *Lectures on the Method of Science*, edited by T. B. Strong, 81–109. Oxford: Clarendon Press.

Weldon, W. F. R. 1906b. "Note on the Offspring of Thoroughbred Chestnut Mares." *Proceedings of the Royal Society of London B* 77:394–98.

Weldon, W. F. R. 1907. "On Heredity in Mice from the Records of the Late W. F. R. Weldon. On the Inheritance of Sex-Ratio and of the Size of Litter." *Biometrika* 5:436–49.

[Weldon, W. F. R.] 1915. "Appendix to *Biometrika*: W. F. R. Weldon's Mice Breeding Experiments. Records of Matings." *Biometrika* 11:1–60.

Weldon, W. F. R., and G. H. Fowler. 1890. "The Rearing of Lobster Larvae." *Journal of the Marine Biological Association* 1 (November): 367–70.

Weldon, W. F. R., and Karl Pearson. 1903. "Inheritance in *Phaseolus vulgaris*." *Biometrika* 2:499–503.

Wilks, W., ed. 1907. *Report of the Third International Conference 1906 on Genetics; Hybridisation (the Cross-Breeding of Genera or Species), the Cross-Breeding of Varieties, and General Plant-Breeding.* London: Royal Horticultural Society/Spottiswoode.

"William Keith Brooks: A Sketch of His Life By Some of His Former Pupils and Associates." 1910. *Journal of Experimental Zoology* 9 (1): 1–52.

Williams, Patricia J. 2021. "They, the People: Social Darwinism in the United States." *Times Literary Supplement*, 6 August: 3–5.

Wilshire, Gaylord. 1909. "Preface." In Punnett 1909, 3–6.

Wilson, Edmund B. 1900. *The Cell in Development and Inheritance*. 2nd ed. New York: Macmillan.

Wilson, Edmund B. 1902. "Mendel's Principles of Heredity and the Maturation of the Germ-Cells." *Science* (19 December), 991–93.

Wimsatt, W. C. 2012. "The Analytic Geometry of Genetics, Part 1: The Structure, Function, and Early Evolution of Punnett Squares." *Archives of the History of the Exact Sciences* 66:359–96.

Winther, Rasmus G. 2001. "August Weismann on Germ-Plasm Variation." *Journal of the History of Biology* 34:517–55.

Wolfe, Audra J. 2010. "What Does It Mean to Go Public? The American Response to Lysenkoism, Reconsidered." *Historical Studies in the Natural Sciences* 40:48–78.

Wolfe, Audra J. 2012. "The Cold War Context of the Golden Jubilee, Or, Why We Think of Mendel as the Father of Genetics." *Journal of the History of Biology* 45:389–414.

Wolfe, Audra J. 2019. *Freedom's Laboratory: The Cold War Struggle for the Soul of Science*. Baltimore: Johns Hopkins University Press.

Woltereck, Richard. 1909. "Weitere experimentelle Untersuchungen über Artveränderung, speziell über das Wesen quantitativer Artunterschiede bei Daphniden." *Verhandlungen der Deutschen Zoologischen Gesellschaft* 19:110–73.

Woltereck, Richard. 1911. "Über Veränderung der Sexualität bei Daphniden." *International Review of Hydrobiology* 4:91–128.

Wood, Roger J. 2007. "The Sheep Breeders' View of Heredity Before and After 1800." In *Heredity Produced: At the Crossroads of Biology, Politics and Culture, 1500–1870*, edited by Staffan Müller-Wille and Hans-Jörg Rheinberger, 229–50. Cambridge, MA: MIT Press.

Wood, Roger J. 2015. "Darbishire Expands His Vision of Heredity from Mendelian Genetics to Inherited Memory." *Studies in History and Philosophy of Biological and Biomedical Sciences* 53:16–39.

Wood, Roger J., and Vítězslav Orel. 1982. "The Sheep Breeders' Legacy to Gregor Mendel." In *Gregor Mendel and the Foundation of Genetics*, edited by Vítězslav Orel and Anna Matalová, 57–69. Brno: Mendelianum.

Wood, Roger J., and Vítězslav Orel. 2001. *Genetic Prehistory in Selective Breeding: A Prelude to Mendel*. Oxford: Oxford University Press.

Wood, Roger J., and Vítězslav Orel. 2005. "Scientific Breeding in Central Europe during the Early Nineteenth Century: Background to Mendel's Later Work." *Journal of the History of Biology* 38:239–72.

Woodward, James. 2003. *Making Things Happen: A Theory of Causal Explanation*. Oxford: Oxford University Press.

Woolfson, Adrian. 2004. "How to Make a Mermaid." *London Review of Books*, 5 February, 25–26.

Worboys, Michael, Julie-Marie Strange, and Neil Pemberton. 2018. *The Invention of the Modern Dog: Breed and Blood in Victorian Britain*. Baltimore: Johns Hopkins University Press.

Wright, Sewall. 1966a. "The Foundations of Population Genetics." In *Heritage from Mendel*, edited by R. Alexander Brink, 245–63. Madison: University of Wisconsin Press.

Wright, Sewall. 1966b. "Mendel's Ratios." In *The Origin of Genetics: A Mendel Source Book*, edited by Curt Stern and Eva R. Sherwood. London: W. H. Freeman.

Wright, Sewall. 1978. "The Relation of Livestock Breeding to Theories of Evolution." *Journal of Animal Science* 46:1192–1200. Reprinted in his *Evolution: Selected Papers*, edited by William B. Provine, 1–11. Chicago: University of Chicago Press, 1986.

Wu, Yiyang, and Gholson J. Lyon. 2018. "*NAA*-10-Related Syndrome." *Experimental & Molecular Medicine* 50 (85): 1–10.

Xu, Ming, and Randolph V. Lewis. 1990. "Structure of a Protein Superfiber: Spider Dragline Silk." *Proceedings of the National Academy of Sciences of the United States of America* 87:7120–24.

Yule, G. Udny. 1902. "Mendel's Laws and Their Probable Relations to Intra-racial Heredity." *New Phytologist* 1 (9): 193–207; 222–38.

Yule, G. Udny. 1903. "Professor Johannsen's Experiments in Heredity: A Review." *New Phytologist* 2:235–42.

Yule, G. Udny. 1907. "On the Theory of the Inheritance of Quantitative Compound Characters on the Basis of Mendelian Laws: A Preliminary Note." In Wilks 1907, 140–42.

Yule, G. Udny. 1923. "The Progeny, in Generation F12 to F17[,] of a Cross between a Yellow-Wrinkled and a Green-Round Seeded Pea; A Report on Data afforded by Experiments initiated by the late A. D. Darbishire, M.A., in 1905, and Conducted by Him until His Death in 1915." *Journal of Genetics* 13:255–331.

Yule, G. Udny, and L. N. G. Filon. 1936. "Karl Pearson 1857–1936." *Obituary Notices of Fellows of the Royal Society* 2:72–110.

Zeven, A. C. 1970. "Martinus Willem Beijerinck [:] A Hybridizer of *Triticum* and *Hordeum* Species at the End of the 19th Century and His Investigations into the Origin of Wheat." *Euphytica* 19:263–75.

Zevenhuizen, Erik. 1998. "The Hereditary Statistics of Hugo de Vries." *Acta Botanica Neerlandica* 47:427–63.

Zirkle, Conway. 1949. *Death of a Science in Russia: The Fate of Genetics as*

Described in Pravda *and Elsewhere*. Philadelphia: University of Pennsylvania Press.

Zirkle, Conway. 1954. "Citation of Fraudulent Data." *Science* 120 (30 July): 189–90.

Zirkle, Conway. 1959. *Evolution, Marxian Biology, and the Social Scene.* Philadelphia: University of Pennsylvania Press.

Zirkle, Conway. 1964. "Some Oddities in the Delayed Discovery of Mendelism." *Journal of Heredity* 55:65–72.

Zirkle, Conway. 1966. "Some Anomalies in the History of Mendelism." In Sosna 1966, 31–37.

Index

Page numbers in *italics* refer to figures. References to a page number include text in figure captions on that page.